P9-EFG-021

A History of the International Chemical Industry

The Chemical Sciences in Society Series

ALREADY PUBLISHED

Peter J. T. Morris, *The American Synthetic Rubber Research Program.* 1989

A History of the International Chemical Industry

Fred Aftalion

Translated by Otto Theodor Benfey

Preface by Patrick P. McCurdy

UNIVERSITY OF PENNSYLVANIA PRESS Philadelphia

Printed in the United States of America

Library of Congress Cataloging-in-Publication-Data

Aftalion, Fred.
[Histoire de la chimie. English]
A history of the international chemical industry / Fred Aftalion; translated by Otto Theodor Benfey; preface by Patrick P. McCurdy.
p. cm.—(The Chemical sciences in society series)
Translation of: Histoire de la chimie.
Includes indexes.
ISBN 0-8122-8207-8 (cloth).—ISBN 0-8122-1297-5 (paper)
1. Chemical industry—History. 2. Chemistry—History. I. Title. II. Series.
HD9650.5.A3813 1991
338.4'766'009—dc20 91-8179
CIP

Second casebound printing 1991

To Béatrice

Contents

Illustrations

Foreword

The Chemical Sciences in Society: what may one expect in a series with such a title? The question is legitimate. The answer is important.

Science and technology have transformed our world and our ideas. Our physical environment, our material culture, our conceptual systems, and our manner of living have been changed, are being changed, and will be changed by science. That much is familiar. Less noticed is the way in which, of all the scientific and technological domains, those associated with the chemical sciences have most often been ignored or misunderstood—subject to ignorance or fear. The very word "chemical" has become synonymous with "undesirable"—as in "chemical dependency." Chemistry is seen as somehow the cause of acid rain, air pollution, carcinogens, drugs, environmental problems, ozone depletion, and smog, to offer only an abbreviated list. At the same time, the chemical sciences are often viewed as intellectually unimportant—the mere detail that fills in the lofty work of physics, the fine print that obscures the grand designs of biology.

The aim of *The Chemical Sciences in Society* is to redress this balance, by bringing into prominence the growing body of scholarship which testifies to the importance of the chemical sciences in society. Chemistry has always been the most earthy and the most central of the sciences. Its earthiness—its connection to the colors, smells, and sounds of substance and of change—stretches backward into the mists of alchemy. But its earthiness also extends into the huge, complex business enterprises of our day, in petrochemicals, in pharmaceuticals, in the cornucopia of polymers and plastics, in composites and "advanced materials," in agriculture and biotechnology, and in cryogenics and electronics. Likewise, the centrality of the chemical sciences may be seen in their role in areas as varied as molecular biology, materials science, and the clinical sciences.

Also worthy of note is the way in which science and technology are increasingly to be understood as political. Science today is an arena in which actors compete for scarce resources—of prestige, of intellectual turf, of government support, of student interest, of corporate involvement, and of public acclaim. The chemical sciences offer a particularly rich area of study when seen in this way—thanks to their long history, their ubiquity, their commercial significance, and their involvement with government and medicine (no monarch could ignore someone able to transmute base metals into gold, or to offer the elixir of life—either in the Renaissance or in the National Institutes of Health).

The chemical sciences in society thus offer a rewarding field for serious study. The aim of the present series is to capture the best of the scholarship developing in this field, whether it deals with recent events or with more traditionally "historical" periods. Our desire is not to be partisan, but to be serious and balanced. The Beckman Center for the History of Chemistry is committed to helping bring into being a deeper, more reflective analysis of the chemical sciences. We trust *The Chemical Sciences in Society* will assist toward this end.

Arnold Thackray

Preface: The Chemical Revolution(s)

Although it may be comforting to view our lives as if they were the result of some long sweep of gradual, evolutionary change, starting from a mythical "good old days," it has been revolution rather than evolution that has set the pace. When I was born (1928), neither the transistor nor the electronic computer existed; nor had the genetic code been broken; and space odysseys were still in the realm of science fiction. When my father was born (1893), neither the radio nor the airplane existed; the automobile was a curiosity at best; and the inner space of matter—protons, electrons, photons, and X-rays—was just beginning to be understood. When my grandfather was born (1856), neither the telephone nor the electric light existed; and the periodic order of the elements had not yet been satisfactorily worked out. And when my great grandfather was born (1829), the telegraph was not yet in place and the steam locomotive and steamship were in their infancy; nor had chemists yet embraced Dalton's concept of atoms.

That just about gets us back to the eighteenth century. No more revolutionary period has existed, especially its second half. Before this period, people lived and worked in essentially the way they had since Roman times; indeed, farming was little different from that practiced in ancient Egypt and Babylonia.

What happened? There were, of course, those two momentous events—the American and the French Revolutions. Libraries are stacked with books on the subject. And then there was the so-called Industrial Revolution, commonly said to have started in England about 1760. The steam engine was its driving force, figuratively and literally. Increased productivity was its hallmark. A changing way of life has been its lasting legacy.

But at about the same time, another—and in many ways, more far-reaching—revolution was taking place, though it is not commonly referred to as such. It is now clear that a Chemical Revolution was beginning in the mid-eighteenth century, was gathering strength in the nineteenth, and was running full force in the twentieth. Not a single convulsion, this

Chemical Revolution has been a continuing series of interconnected developments, going off like a string of Chinese firecrackers, which show no signs of letting up. If steam leveraged the Industrial Revolution, human intellect throughout the civilized world has been driving this Chemical Revolution.

At the beginning of this period, chemistry, if it could be called that, was not all that far removed from alchemy. And there was no such thing as a "chemical industry." People did things with things that seemed to work: They made soap from animal fat and plant ashes; they derived substances with apparent therapeutic value from bark, roots, or plants; they made glass from sand, soda ash, and other materials; they made do with mineral colorants (iron oxide, for example) much as did their Stone Age ancestors; they used natural preservatives and fertilizers. But they did not understand why or, for the most part, how these substances worked. And they were restricted to mixing the basic materials at hand, in the basic form in which they existed at hand. That all started to change a little over two hundred years ago, when humans began to unravel the basic mysteries of matter—how it is subdivided, how it is held together, how it is transformed, how it can be reformed into "new" substances.

Unfortunately, much of the written history of chemistry, chemical technology, and the chemical industry has been fragmented and piecemeal. This is a pity, inasmuch as chemistry, more than any other physical science, has touched people directly and beneficially in almost every aspect of their daily lives through the vast array of products and systems produced by the chemical industry.

In his *History of the International Chemical Industry* (originally published in French as *Histoire de la chimie*), Fred Aftalion, whom I have known as a good friend and esteemed colleague for almost thirty years, has encapsulated a huge amount of material in a relatively few, readable pages.

Chemistry has succeeded almost too well. Some of its products have become so commonplace and necessary that they are "invisible" to the public or even viewed with contempt by some. On the other hand, accidents and poor waste management, together with sensationalized media coverage, have made the industry highly visible and have sparked public fear. Much misinformation about chemicals and the chemical industry has been issued in recent years. Many people have been understandably unsettled.

This book will be best understood and welcomed by participants in the revolution in chemistry described by Fred Aftalion—chemists, chemical engineers, biochemists, biotechnologists—all who have grasped the centrality of chemistry as a science and as a prime mover in modern life. But in truth, all citizens have been participating in the revolution, as often as not without knowing it.

Those wanting perspective on an industry critical to their daily lives and well-being—mostly for the better—will find an informative, reliable guide in Fred Aftalion's book.

PATRICK P. MCCURDY
Washington, D.C.

Acknowledgments

In preparing this book for the press I have been helped by many individuals. A preliminary anonymous translation served as the basis for my work, which then was further developed by Arnold Thackray. The manuscript was typed by Judy Weglarz, while chemical structures and equations were prepared by Thomas Feist and Rose Cantave. At the University of Pennsylvania Press I was much helped by the expertise and thoughtful advice of Patricia Smith and her associates Rita Colanzi, Eileen Mullen, Alison Anderson, Carl Gross, and Kathleen Moore.

My part in this book I dedicate to the French part of my family, Elisabeth and Philip.

OTTO THEODOR BENFEY
Beckman Center for the History of Chemistry

Introduction

Chemistry is a science that is at the same time fascinating and mysterious.

It is fascinating for those who have chosen it as a profession, since the chemist is a true architect of matter. Chemists not only reproduce by synthesis a great number of natural substances; they also create new molecules capable of curing diseases, protecting crops, and, more generally, improving our living conditions.

Chemistry, nonetheless, remains mysterious for the great majority of citizens who, approaching it from the outside, are put off by the complex structures of molecules, by the strange representations chemists give them, and by the complicated names of chemical substances. To add to the mystery, most direct products of chemical industry are ''invisible'' to ordinary mortals. These products serve, instead, as intermediates between the basic raw materials nature has provided and the final consumer goods of our daily life. Unless one should deliberately want to, one may spend a whole lifetime without ever seeing the actual products of the chemical industry. Even in the factories where they are made, chemicals are invisible. They are stored out of sight in barrels or tanks, circulated through pipes in distillation columns or closed reactors, and passed discreetly into tankers to be processed elsewhere into consumer goods; only in this final form do they appear in our drugstores and on the shelves of our shops.

Today, it is only after an accident that has caused gas to leak into the atmosphere or a few kilograms of synthetic substances to spill into a river that public opinion is informed, this time with all the inevitable hue and cry, of the existence of the products of the chemical industry.

More than any other science, chemistry cannot be dissociated from the chemical industry which is its twin, for chemical manufacturers cannot do without the scientific knowhow provided by research chemists, while the latter can no longer afford to disregard the practical consequences of their work.

This book is intended for three kinds of readers: researchers who are increasingly required to specialize in a particular discipline but, nevertheless, want an overall view of the various branches of the chemical industry; industry leaders whose involvement with chemicals has often been due more to circumstances than to actual vocation but who should not be indifferent to the industry's progress since its beginnings; and, finally, the noninitiated who view the industry from the outside and generally know only of its more forbidding aspects.

In writing this book, I wanted to point to the sequence of events that has shaped the landscape of chemical science and industry into its present state. Even when the origin of certain major discoveries was accidental, it was to the judicious interpretation of chance phenomena by keen and well-prepared minds that we owe the benefits such discoveries bring us today. This book is, therefore, essentially the history of human intelligence applied to a particular field of knowledge. It has been written to the glory of scholars and industrial leaders who by their daring, their shrewdness, and their tenacity have provided us with the means for a better life. May those who have taken over from them today be inspired by their stirring example.

Prelude: From Prehistory to the Dawn of Quantitative Chemistry

The human race has always attempted to unlock the secrets of matter, whether it be to transform this matter for utilitarian purposes or to penetrate into the philosophical mysteries of existence.

Prelude to Chemistry

The control of fire was one of the most important factors that distinguished humans from animals. At Chou Kou Tien in China, there is clear evidence that early humans, *Sinanthropus pekinensis*, made use of fire nearly half a million years ago. Large accumulations of charcoal, ashes, and burnt bones have been found in the caves they occupied.

Fire enabled humans, from the very beginning of Paleolithic times, to have some light, to cook food, and to provide warmth. It also enabled people to transform some materials, such as changing yellow ochre into red ochre by heating it. Later, Neolithic people knew how to bake pottery and to extract such metals as *copper* from malachite by reduction using charcoal.

Bronze, which appeared around 3000 B.C., is the first "synthetic" metal, since it is an alloy (90 percent copper and 10 percent tin) that does not exist in nature. To obtain bronze required a reaction between cassiterite, a tin oxide, and copper ore. By 2500 B.C., the reduction by charcoal of minerals containing iron oxide had begun in the Middle East. The Hittites still had a monopoly on producing *iron* in 1200 B.C. This period also saw the appearance of *brass*, an alloy of copper and zinc obtained by reducing calamine (zinc silicate) along with molten copper. As for *gold, silver*, and *mercury*, they were discovered long before copper because they occur as native metals without having to be extracted from mineral ores. Their scarcity and their low chemical activity help explain the fact that they were highly valued but not much used.

By mastering fire and through it the reduction of minerals, humanity moved successively from the Stone Age into the Copper, Bronze, and Iron Ages. Further empirical achievements, linked with a better control over matter, made possible the improvement of everyday life. The tanning of hides, using such vegetable substances as bark from oak trees, dates back to prehistory. We owe to the Egyptian civilization several discoveries between 6000 and 2000 B.C.: *plaster* produced by heating gypsum, writing *paper* made from papyrus stiffened with starch, *glass* produced with silica and carbonate of soda, various *medicines* extracted from plants or derived from minerals, *incense* made from crushed spices mixed with resin, and *potash* (derived from the ashes of cooking fires), which, when heated with fats, gives soft soap.

Some *colors*, such as carbon black and clays containing metallic oxides, were used thousands of years ago for the rock paintings found in the Spanish Altamira caves or the French Lascaux caves. The Egyptians added many hues to the palette and created the first known varnishes around 1000 B.C., using beeswax or styrax as a binder. (From the biblical story we are familiar with Joseph's coat of many colors and the envy it produced among his brothers.)

Chemical processes became more refined with the Greeks and Romans but with few radical innovations. *Minium* (lead oxide) was used by the Greeks in their paint and to protect the woodwork of their ships. *Alum* (aluminum potassium sulfate) was familiar as a "mordant" for dyes, helping to attach them to the objects being dyed. The Romans used lime as mortar in construction and prepared white lead from lead and vinegar (as described by Pliny). To decorate the togas of their dignitaries, they extracted the purple dye of the Phoenicians from the mollusc *Murex branderis* as well as the scarlet dye *kermes* from female scale insects found on oak trees. The word *perfume* comes from the Latin term *perfumare*, which means to fill with smoke, that is, to burn incense as the Egyptians did in their temples. The Romans were the first to use *ointments*, which they made from spices and flowers macerated with oils or fats to retain the fragrance. The Romans also used *beauty products* such as cinnabar (mercury sulfide) for lipstick, and white lead (basic lead carbonate) for makeup.

The *domestic economy* had greatly improved by the end of ancient times. Butter and cheese were prepared from milk; beer was brewed from cereals; wine was obtained by the alcoholic fermentation of the sugars in grape juice; and vinegar in turn by the further fermentation of wine. *Salt* acquired both a practical and symbolic value. States imposed special taxes on salt and made sure they kept the monopoly of its distribution. The use of salt increased tremendously when sedentary ways of life replaced nomadic ones. Since people were no longer eating raw foods, they had to salt meat and fish to preserve them, and to add salt to cooked food to compensate for the loss of this mineral in the cooking juices. The Roman army distributed salt rations called *salarius* to their soldiers. These rations were later replaced by coins, thereby giving the origin of our word "salary."

Mastering new techniques to improve well-being was not enough for the people of the time; they sought to formulate an intellectually satisfying reason for the existence of things. Greek philosophers considered nature to be composed of four fundamental elements, which are described in the works of Plato and Aristotle: *fire, air, water*, and *earth*. Four qualities, hot and cold, wet and dry, were associated with these elements. Aristotle also discussed the transmutation of one element into another, an idea developed by the alchemists. According to another philosophical concept, conceived by Empedocles and Democritus and further elaborated by Epicurus and Lucretius, all matter was formed of particles that could not be further reduced, the *atoms* (from the Greek word ατομος meaning indivisible), which move in a void.

From Alchemy to Chemistry

The concept of transmutation of base metals into silver and gold is found in ancient China and Egypt and comes to us from the Greeks living in Alexandria in the second century B.C. The *early alchemists* included a certain mystical and allegorical quality in this concept. Some of their works were translated into Arabic and were later discovered by Renaissance scholars.

For over a thousand years, two main trends characterized alchemy. On the one hand, there was a philosophical pursuit of the transmutation of metals, of the philosopher's stone, and of the elixir of life. This pursuit was shrouded in esoteric terms associated with secret rituals. On the other hand, practical operations were being carried out with constantly improving instruments. Different heating methods were developed, such as sand and water baths. The *alembic*, for instance, was not a very efficient condenser, but its introduction signaled the first rudimentary beginnings of the art of distillation.

Research methods were imprecise. Sensitive scales to weigh things accurately did not exist, and the substances used were impure. Furthermore, the materials being treated were often radically transformed by the alchemists' operations. The use of fire usually was sufficient to destroy all organic substances. The result was that alchemy was largely a form of inorganic or mineral chemistry, using the "dry" processes of calcination (heating) and melting. Nevertheless, by the twelfth century A.D. certain adepts knew how to distill rose petals to produce *essence of roses*, and to prepare alcohol concentrates called *aqua vitae*. The spread of knowledge of distillation in medieval Europe may be traced through the appearance of ordinances against drunkenness.

Mineral acids were discovered in the thirteenth century: sulfuric acid, called "oil of vitriol," was produced by burning sulfur in a humid atmosphere, or by calcining certain sulfates; *aqua fortis* and "spirit of salt," that is, nitric acid and hydrochloric acid, were obtained by burning native nitrates and chlorides with copper or iron sulfates. *Mercury*, which with *sulfur* was one of the two basic principles of alchemy, had been used for a long time because of its easy production from its ores and its ability to create amalgams for extracting metals from other ores.

In the long run, repeated failures to achieve transmutation, coupled with the occult nature of many alchemical texts, caused an unfavorable impression to settle in the minds of the less credulous regarding the validity of alchemical theories. The lasting contributions of alchemy came through the development of manipulatory techniques and later, during the Renaissance in the sixteenth century, through a focus on the medical field.

The dominance of medicine based solely on herbal remedies came to an end with *Paracelsus*, who practiced in Basel and elsewhere. He introduced the radical idea (for Europe) of treatments using chemical substances such as mercury, antimony, iron, copper and their salts. Paracelsus' teaching was highly controversial. Nevertheless he acquired quite a reputation because of spectacular cures using laudanum. He believed in the four basic elements, which he related to three principles, sulfur, mercury, and salt. Paracelsus also emphasized the notion that one should always seek the essence of matter, its "quintessence," by isolating the raw material from its matrix. It is partly due to his influence that chemical practitioners concentrated more carefully on the purification of substances which were to be used as medicines.

Iatrochemistry, or medical chemistry, thus came into being, exerting a considerable influence on the future of chemistry in general. From then on, alchemists, physicians, and apothecaries worked hand in hand to promote the notion of *pure substances* that could be chemically identified. Until the nineteenth century, many prominent chemists began their careers by studying medicine or pharmacy. Chemistry was taught for a very long time as one of the four or five essential subjects in medical schools. The first separate chemistry department was created in 1609 at the University of Marburg in Prussia. In 1648, William Davison gave the first chemistry lecture at the royal botanical gardens in Paris (the Jardin du Roi, now known as the Jardin des Plantes). Chemistry treatises that gave valuable information for the preparation of medicines became very popular. In fact the *Cours de chymie* written by Nicolas Lémery in 1675 was a best-seller in its day.

Empirical chemistry made significant progress during the seventeenth century. The Belgian physician Johann Baptista van Helmont recognized the identity of a *gas sylvestre*, known as carbon dioxide today, by burning combustible substances like charcoal or alcohol. Johann Glauber, a younger German medical contemporary of van Helmont, while working on mineral salts, described for the first time double decomposition reactions. He improved the method of preparing hydrochloric acid by the action of sulfuric acid on salt. The *sodium sulfate* produced during the reaction is still called Glauber's salt. Glauber used this (essentially harmless) substance as his *sal mirabile*, which formed the basis of many medicines.

The Englishman Robert Boyle, the author of the law governing the pressure-volume relationship in gases, was the first to distinguish acids from alkalis using indicators like syrup of violets. By chemical analysis he was able to determine three types of substances: *salts, acids*, and *bases*. The foundations of analytical chemistry were derived directly from these experiments using aqueous reactions, whereas the alchemists had operated almost exclusively with dry substances.

Sulfuric acid was the only mineral acid of commercial importance until the nineteenth century. It was obtained at Nordhausen in Saxony by the dry distillation of green or blue vitriol (iron or copper sulfate). Sulfuric acid was used in the preparation of dyes and for the purification of gold and silver.

Niter, the potassium salt of nitric acid, was vital in the production of gunpowder, which was invented in China during the ninth century but took three centuries to reach Europe. The composition of gunpowder, a mixture of sulfur, charcoal, and potassium nitrate, varied very little over time. A very pure quality of sulfur could be obtained in Sicily. Charcoal, which had been used for thousands of years as fuel, was readily available. On the other hand, niter, also called *saltpeter*, was not easily found. One well-recognized source was its presence along the walls of stables and pigsties as a result of the interaction of bacteria with manure. Saltpeter was collected by the arduous process of scraping the walls, and purifying the scrapings by crystallization in water.

The Chemical Study of Gases and the Phlogiston Theory

The study of gases, initiated by van Helmont and Boyle, had its real start after the invention of the pneumatic trough in 1719, which enabled Stephen Hales to measure quantitatively the gases collected from fermentation and putrefaction. Around 1754 Joseph Black chemically identified for the first time *carbon dioxide*, which he produced by calcining carbonate of magnesium or of calcium, or by treating them with acids. Before Black's work it was believed that all gases were the element air more or less filled with various impurities.

Hydrogen, previously known as "inflammable air," was discovered in 1766 by Henry Cavendish, who generated it from the action of an acid on a metal such as iron. Hydrogen was later used in balloons and dirigibles.

An English clergyman, Joseph Priestley, a remarkable but impetuous self-taught scientist, discovered "dephlogisticated air." On August 1, 1774, Priestley obtained a gas by heating red calx of mercury (now recognized as mercuric oxide) in a closed vessel by means of a lens focusing sunlight. He found that a candle burned brilliantly in this gas. A little earlier the Swedish apothecary Carl Wilhelm Scheele had isolated such a gas, but his work was not published until 1777. During this period, the French scientist Antoine Laurent Lavoisier was also experimenting with combustion and respiration. He

Joseph Priestley (1733–1804). Courtesy Edgar Fahs Smith Memorial Collection.

met Priestley in Paris in October 1774 and called Priestley's new gas "vital air." Later it was called *oxygen*.

Lavoisier gave a totally new explanation of the chemical action of oxygen. Before Lavoisier, combustion was explained by the phlogiston theory enunciated by Georg Ernst Stahl, a German physician and chemist of the latter part of the seventeenth century, who was heavily influenced by ancient allegories and by alchemical thought. For Stahl, phlogiston (from the Greek word *phlogistos*, blazing) was a kind of fire mingled in matter itself which escaped during combustion. Thus when calcined a metal would lose its phlogiston and be transformed into calx (today we would say oxide). If on the other hand phlogiston were put into calx by means of charcoal, the calx would again become a metal. Carl Wilhelm Scheele believed in phlogiston and referred in his works to a "fire-air" (*Feuerluft*). Priestley was also a follower of Stahl's theory; for him the air that was especially fit for respiration was "dephlogisticated." He believed, like most of his contemporaries, that when air was saturated with phlogiston combustion was prevented.

Lavoisier, on the other hand, observed that the phlogiston theory could not explain why a metal became heavier when it was changed into calx. He proved in 1776 that the oxygen found in the air, which combined with the metals when they were calcined, accounted for the increase in weight. From that time phlogiston, which could not be weighed, gradually became discredited as an entity. Air was no longer an irreducible element; combustion could be explained by the presence of oxygen, which was a measurable gas. The "antiphlogiston" theory of oxidation came into being. Many familiar concepts gave way at this stage to the newly rigorous approach of quantitative chemistry.

The Life and Work of Lavoisier

Antoine Laurent Lavoisier came from a wealthy family, and thus never had to face material problems. He at first followed the family tradition of studying law, and passed the bar examination in 1764. But soon he felt the attraction of his true vocation, the sciences. He began studying botany with Bernard de Jussieu at the Jardin du Roi and later became a student of Guillaume François Rouelle, whose chemistry courses were famous throughout Europe.

Lavoisier's first research essays, concerning the problems of lighting a large city such as Paris and of distributing drinkable water, brought him to the attention of the French Académie des Sciences, which admitted him into their ranks in 1769 when he was just twenty-five years old.

In this period also, the economist and financier Anne Robert Jacques Turgot instituted La Régie des Poudres. Lavoisier was one of the first persons appointed as *régisseur*. He rapidly improved the quality and production of saltpeter in France by creating artificial saltpeter-beds.

Lavoisier later became a Fermier Général. This appointment was to have important consequences, providing the financial resources that helped him pursue his scientific experiments, but also rendering him politically vulnerable in due course. La Compagnie des Fermiers Généraux was a group, independent of the government, created by Jean-Baptiste Colbert, comptroller general of finances under Louis XIV and founder of the Académie; the Compagnie had the privilege of collecting all of the taxes and duties to the king. Needless to say, this function was not a popular one.

Lavoisier was married on December 4, 1771 to the daughter of the Fermier Général Jacques Paulze, and set up his laboratory in the Arsenal. It is there that he carried out his experiments on combustion and respiration as well as his studies of the composition of water, thereby laying the foundation of modern chemistry. By 1775, Lavoisier had become convinced that air combined with metals and combustible substances during their calcination. He proceeded to demonstrate that air was a mixture of oxygen and another substance, which he called *nitrogen*. According to Lavoisier, the phlogiston theory could be totally discarded, since combustion and calcination were simply the combination of an active substance with oxygen.

In a memoir to the Académie des Sciences dated 1783, he argued at length against the phlogiston theory and was supported by the French mathematician and astronomer Pierre Simon de Laplace, as well as by the chemists Claude Louis Berthollet, Louis Bernard Guyton de Morveau, and Antoine François Fourcroy. These three chemists joined Lavoisier in establishing a new method of chemical nomenclature that was published in 1787, soon to be widely accepted. The compounds of oxygen with metals were called *oxides*, while those between oxygen and non-metals such as sulfur were denominated *acids*, such as sulfuric acid. The salts of sulfuric acid were *sulfates*. Sulfurous acid was sulfur combined with a lesser quantity of oxygen, and its salts were called *sulfites*. The names of combinations without oxygen would end with ''ide'' (lead sulfide, copper phosphide).

In order to make known his ideas, Lavoisier published in 1789 a brilliant *Traité Élémentaire de Chimie* which was soon translated into several languages. This treatise had a profound impact throughout Europe. In the *Traité*, he listed thirty-three substances that he proposed as elemental, the majority of which still appear in our present table of elements. Nicolas Lémery fifty years earlier had already defined an element as an irreducible substance, that is, anything that chemical analysis could not break down into simpler entities. Lavoisier, however, was insistent on enunciating clearly the concept of *conservation of mass*. He stated in the *Traité*: ''Nothing is created, whether it be in the fields of art or in those of nature, and one can set as a principle that in every process there is an equal amount of matter in the beginning and at the end of the process.''

Quantitative chemistry had come to the forefront. However, Lavoisier, thrown into the turbulence of the French Revolution, found himself with preoccupations remote from chemistry. He was part of the committee in charge of setting up the metric system; he contributed to the reform of the monetary system; and he continued to serve as a Fermier Général and member of the Académie des Sciences. Many of his colleagues in the Académie were jealous of his financial means and of his prerogatives as a Fermier Général. The revolutionary Jean Paul Marat hated Lavoisier, who had questioned his scientific competence. Little by little, Lavoisier found himself abandoned by the entire scientific community in spite of the fact that he was one of its leading members. Instead of supporting him, colleagues such as Guyton de Morveau and Fourcroy were swept by a revolutionary zeal and heaped abuse upon him.

In 1791, Lavoisier was removed from his post as Régisseur des Poudres. In the same year he witnessed the suppression of the Compagnie des Fermiers Généraux and in 1793 that of the Académie des Sciences. Soon Lavoisier was arrested, along with the thirty other Fermiers Généraux. He was tried and guillotined on May 8, 1794, at the Place de la Révolution (known today as the Place de la Concorde).

Jean Baptiste Coffinhal, the presiding judge, is said to have rebuked the few courageous friends who came to defend Lavoisier with the remark, ''The Republic has no need of scientists, let justice take its course.'' Thus perished a man who literally revolutionized the field of chemistry while rendering immense services to his country.

Chapter 1
The First Chemical Manufacturers

The chemical industry was born in the middle of the eighteenth century out of the demands created by other industries.

Sulfuric Acid

Sulfuric acid was needed for the preparation of dyes as well as for the production of hydrochloric and nitric acids used in the treatment of metals. The small amounts of acid available from Nordhausen in Saxony were very expensive, however. In 1736 in Twickenham an Englishman, Joshua Ward, first succeeded in reducing the manufacturing cost of sulfuric acid by burning a mixture of sulfur and saltpeter[1] above a thin layer of water in large, wide-necked glass jars. The dilute acid produced was concentrated through distillation. A decisive breakthrough occurred in 1746 when one of his fellow countrymen, John Roebuck of Birmingham, replaced the fragile glass containers by vast *lead-coated* chambers set in an imposing installation. At the time, lead was the only cheap acid-proof metal.

Roebuck thought to protect himself by keeping his process secret instead of taking out a patent. But he soon had imitators. The first sulfuric acid works set up in Rouen in France by John Holker in 1766 were fitted a few years later with lead chambers. The number of these "lead cathedrals," which were to give a distinctive look to chemical plants for over a century, grew rapidly, the more so as demand for sulfuric acid, a product hard to transport over great distances, was spurred by new uses.

The lead chamber process was gradually improved, mainly through the work of the Frenchmen A. de la Follie, Nicolas Clément, Charles-Bernard Désormes, and Jean Antoine Chaptal. By the end of the eighteenth century, the cost of sulfuric acid had dropped to £30 per ton instead of the £280 per ton it had fetched in Roebuck's day.

[1] The process was still empirical. It was not understood at the time that, by producing oxides of nitrogen, saltpeter was playing a purely catalytic role in sulfur oxidation.

The Demand for Alkalis

The second half of the eighteenth century saw an ever-growing demand for alkalis for glassmaking and soapmaking, for dyes for textiles, and for bleach for raw cloth. Demand was particularly strong in France, which already had a firmly established glassmaking industry. Indeed, in France as early as 1665 Colbert had granted a twenty-year license to Manufactures des Glaces et Miroirs in an effort to get around Venice's near monopoly. When on April 26, 1692, Louis XIV inaugurated a factory set up on the ruins of the Saint-Gobain chateau, he could himself watch the Venetian workers enticed away from the Murano glass and mirror factories blow a mirror, polish it, and coat it with quicksilver. Glassmaking increased the need for *potash* (potassium carbonate) and then for *soda* (sodium carbonate).

France, like England, introduced both mineral and plant-based alkalis to satisfy these requirements. As living standards rose in the towns, not only glass but also soap were in increased demand. A natural sodium sesquicarbonate called *natron* (or trona) was imported from Egypt by soapmakers and mixed with lime to saponify olive oil. But natron was expensive. *Potash* from wood ash came from areas richly endowed with forests such as Russia, Scandinavia, and North America. But it was not cheap, and the rebellion of the settlers in America hindered supplies. *Barilla*, imported from Spain and produced from the ashes of plants like saltwort (*Salsola*, hence the name "sal soda") was a cheaper source of alkalis, but its sodium carbonate content varied, seldom exceeding 15 percent. Rougher kinds of soda with a strong salt content were also obtained through calcination of seaweed (kelp) collected along the coasts of Brittany, Ireland, and later Scotland.

These different sources of sodium carbonate barely added up to 10,000 tons per year, and France, more than other countries, was feeling the need for supplies steady in quantity, quality, and price. In 1775 the Académie des Sciences offered a prize of 2,400 livres for "anyone who should find the most simple and economical method to decompose in bulk salts from the sea, extract the alkali which forms their base in its pure state, free from any acid or neutral combination, in such a manner that the value of this mineral alkali shall not exceed the price of the product extracted from the best foreign sodas." It fell to Nicolas Leblanc, a Frenchman, to take up the Académie challenge in 1789.

The Tragic Destiny of Nicolas Leblanc

Like Lavoisier but in different circumstances, Nicolas Leblanc was a victim of the Revolution. Born in a poor family, he was orphaned at the age of nine. He was able to study to be a surgeon thanks to the protection of a friend of his father's, Jean d'Arcet, his teacher at the Collège de France, who awoke his interest in chemistry and set him to study the possibility of extracting soda from common salt.

While still a young man, Leblanc was appointed surgeon to the House of the Duc d'Orléans, a position that was to help him find the funds he would need later to set up a soda plant. His research brought him into contact with the work of Duhamel Dumonceau, the first to demonstrate (in 1736) that soda and common salt (sodium chloride) had a common base; he even prepared a compound similar to barilla from Glauber's salt (sodium sulfate).

Some time later, the German chemist Andreas Sigismund Marggraf was to improve the preparation of soda from common salt, thus starting in Prussia a soda industry that was to last into the nineteenth century. On the other hand, the double decomposition of salt or sodium sulfate with potash was already being performed in England to boost soda production. Finally, Abbé Malherbe had obtained in 1774 a soda solution prepared through the action of sodium sulfate on coal in the presence of iron.

The originality of Leblanc's process was in the use of chalk (calcium carbonate) which, in certain proportions, can be reacted with coal and sodium sulfate to produce sodium carbonate. Leblanc had also provided for recovery of the hydrochloric acid fumes emitted in the reaction of salt on sulfuric acid during the preliminary production of sodium sulfate from common salt.[2]

On September 25, 1791, Leblanc was granted a patent for his process on behalf of Louis XVI, who had become the constitutional monarch. That same year, with 200,000 livres obtained from the Duc d'Orléans, he built his first soda plant at Saint-Denis, which produced up to 320 tons a year. Two years later, however, the Duc d'Orléans was sent to the gallows, even though he had opted for the Republic and taken the name of Philippe Égalité. Considered the Duc's property, the Saint-Denis plant was seized in 1794 by the Comité de Salut Public. Production was halted, and Leblanc's patent became public property.

This was a windfall for others but not for the inventor: soda plants using the Leblanc process were built in Lille and in Marseilles to supply the soap-making industry, and also, beyond the Channel, where for over a century they were to make the fortunes of England's chemical industry. As for Nicolas Leblanc, not only did he never receive the prize money, which the new Académie had raised to 12,000 livres in 1789, but he also lost his plant and his legitimate patent rights. A Consulat decree in April 1801 ordered the Saint-Denis plant to be returned to him, but without the compensation he had been claiming and therefore without the means to make it work. On January 16,

[2] The following equations represent the reactions:

a) $2NaCl + H_2SO_4 \longrightarrow Na_2SO_4 + 2HCl$

b) $2Na_2SO_4 + 2CaCO_3 + 4C \longrightarrow 2Na_2CO_3 + 2CaS + 4CO_2$

1806, after twenty years of hopeless struggle against his country's administration, Leblanc in desperation shot himself.

Textile Bleaching

By the beginning of the eighteenth century, Holland had acquired a monopoly of sorts in linen bleaching, which was carried out near Haarlem and occupied large expanses of land. Following alkali treatment, cloth was repeatedly exposed to sunlight and dipped in buttermilk until it was suitably bleached. The whole process used up much milk, time, and space, and depending on weather conditions, required five to six months for successful completion.

It was a great step forward in 1758 when, in England, at the very time when the use of cotton fabrics was developing, dilute sulfuric acid began to be substituted for buttermilk. The same bleaching effect could now be produced in twelve to twenty-four hours. The real breakthrough came from work carried out by the Frenchman Claude Louis Berthollet at the Gobelins tapestry works near Paris. In a communication read at the Académie des Sciences in 1785, he revealed the powerful and previously unsuspected bleaching effect of chlorine on linen. Gaseous chlorine was difficult to transport, however, as well as dangerous and awkward to handle. Louis Alban conceived the idea of dissolving chlorine in caustic potash, and thus began the manufacture in 1796 at Javel of the potassium hypochlorite that French households still use today as *eau de Javel*. For the textile industry, a *bleaching powder* based on calcium hypochlorite proved to be more economical and easier to handle. A Scottish bleacher, Charles Tennant, obtained the powder by passing chlorine into slaked lime. He took out a patent for his process in 1797.

These various improvements led to spectacular consequences: with the bleaching process for cotton reduced to a few hours, less capital was needed to build up stocks of raw cloth, thus production was boosted, along with the development of the textile industry. Fields and manpower were freed for other productive jobs. The fledgling chemical industry thrived on these changes, for they sparked off a great need for sulfuric acid to be used in bleaching and to make chlorine by reacting the acid with common salt. As for Tennant's powder, while his Saint-Rollox works near Glasgow turned out barely 57 tons in its first year, some 13,000 tons were being produced in Britain by 1852.

Further Consequences of the French Revolution

The French Revolution quickly led to war. The isolation that peaked under the "continental blockade" had an initially stimulating effect on the chemical industry in France. Substitute products had to be created as imports of sulfur from Sicily, barilla from Spain, potash from Scotland, saltpeter from India, cochineal from Peru, and sugarcane from the West Indies became ever scarcer.

Forming an elite corps of engineers, and calling upon the great chemists who had survived the Reign of Terror, became a necessity. Antoine François de Fourcroy had the Convention Nationale approve the establishment in 1794 of the École Polytechnique and, with Louis Nicolas Vauquelin, Louis Bernard Guyton de Morveau, Claude Louis Berthollet, and Jean Antoine Chaptal, he lectured there on chemistry. A Conservatoire des Arts et Métiers, focusing on the applied sciences, was set up at about the same time in Paris.

The chemical industry pioneers Leblanc, Antoine Baumé, N. de Carny, and Chaptal were roped into the quest for French sources of raw materials. Local pyrites began to replace sulfur for the manufacture of sulfuric acid; saltpeter plants helped supplement the supplies of organic residues collected for saltpeter production needed for munitions making; Leblanc soda came in as a timely substitute for plant-based alkalis; madder grown in the south of France was used as a dye; beet sugar, discovered by Andreas Sigismund Marggraf in 1747, was introduced into France by Benjamin Delessert following a decree by Napoleon I.

By the end of 1810, France's industry was capable of producing annually 20,000 tons of sulfuric acid, 10,000–15,000 tons of Leblanc soda, and 600 tons of hydrochloric acid. But with its technical and higher education system heavily concentrated in Paris and its economy geared to self-sufficiency regardless of production costs, France would be in no position to take up the British challenge. Britain emerged victorious from the Napoleonic wars, and also enjoyed the advantages of a very decentralized educational system and an industry wide open to international currents.

The Rise of England's Chemical Industry

Unlike France, where a centralizing administration had taken charge of the scientific education of citizens who wished to become researchers or engineers, Britain left that job to private institutions and learned societies scattered over the land. Such were Birmingham's Lunar Society, which sponsored Priestley's work; the Manchester Academy, where John Dalton lectured; the London Royal Institution, which benefited from the teachings of Humphry Davy and Frederick Accum; and Glasgow University, where chemistry was first taught by Joseph Black and then by his pupils, Thomas Charles Hope and Thomas Thomson.

France remained faithful to centrally regulated industry, from the ancient Manufactures Royales such as Saint-Gobain and Les Gobelins, which still enjoyed privileges, to the state-owned *Régies* (Salpêtre, Poudre), which functioned as monopolies. But the England of the Industrial Revolution promoted the private enterprise from which were to emerge the first daring captains of the chemical industry.

It had been the hope of the unfortunate Leblanc to sell his synthetic soda to England. But his process rather than his products crossed the Channel. William Losh was the first to benefit from the generosity of the Comité de Salut Public. Before setting up his soda plant in Tyneside in 1816, he had full leisure to read up on the details of the Leblanc process, published in 1797 in the *Annales de Chimie*. He even took advantage of the short-lived ''peace of Amiens'' to visit the Saint-Denis soda plant in 1802.

The reputation of French chemists was so high that when Charles Tennant, who was already making sulfuric acid and chlorine at his Saint-Rollox plant near Glasgow, which was to become the leader in its time, decided to manufacture Leblanc soda, Chaptal and Jean Pierre Joseph d'Arcet were the two men he came to consult in 1816. Another pioneer was the Irish industrialist James Muspratt, who set up his own soda plant in Liverpool in 1822 to supply Lancashire soapmakers. Spurred on by the cotton and soapmaking industrial boom and by the steady needs of glassmakers, the ''Leblanc system'' became the hard core of England's chemical production from 1830 on, despite a slow start because of the salt tax and the reluctance of English soapmakers to give up plant-based alkalis.

On the other hand, Leblanc soda turned up in Germany only in 1840. In America, it was never manufactured. And while in France production had soared to 45,000 tons by 1852, England had forged ahead with 140,000 tons produced that year. At the same time, the price of soda crystals fell from £35 per ton just before the Leblanc process breakthrough to £15 by mid-century.

The Leblanc system was instrumental in turning Britain's soda plants, scattered around the Glasgow area and along the Mersey and Tyne riverbanks, into a truly integrated industry. These plants produced their own sulfuric acid, selling part of it to the metals industry and to textile bleaching works. Sodium sulfate (salt-cake) was supplied to glassmakers as a partial substitute for soda. The Leblanc system exacted heavy costs from the environment. The hydrochloric acid produced as a gaseous by-product poisoned the air around soda plants for a long while, as a truly toxic form of ''acid rain.'' Only with the passage of the Alkali Act in 1863 were manufacturers required to absorb this acid in the towers first designed by William Gossage in 1836.

It was a long time, however, before the residue of the Leblanc process, a cake containing calcium sulfide, was put to use. Instead hundreds of thousands of tons piled up in neighboring fields. Leblanc's system turned out a single product, sodium carbonate, by shedding two in the process—calcium sulfide and hydrochloric acid (which continued to be discharged into rivers even after condensation in diluted form in Gossage's tower).

England built up its chemical industry on the Leblanc process. The required assets were there to develop it successfully. Salt was extracted in

Cheshire, coal came from the Tyne basin, and lime from the chalk quarries of Derbyshire; a network of rivers and canals made it an easy matter to transport base chemicals to the soda plants, and the finished products to the ports from which they were exported.

By 1840 most of England's sulfur was being extracted from iron and copper pyrites, initially from home sources and then from Norway and Spain. A group of French merchants had obtained a trade monopoly over Sicilian sulfur in 1830, whereupon its price had doubled. When the monopoly was scrapped ten years later, the Sicilian market had disappeared in favor of pyrites.

By the middle of the nineteenth century England's constantly developing downstream outlets and abundant upstream raw materials which could easily be transported, combined with the entrepreneurial spirit of its industry leaders, and thanks to the Leblanc system, made it the first country to employ over 10,000 people in the manufacture of chemicals.

The Birth of America's Chemical Industry

It was from their forests that American settlers (like the Scandinavians) drew their first chemical resources: potash, tanning products, and also ''naval stores,'' so called because the residues of coniferous wood distillation (pitch and tar) were used for building and repairing ships, while rosin and oil of turpentine served to make lacquers.

The first sulfuric acid unit with lead chambers was built in 1773 in Philadelphia by John Harrison after he had spent two years in England working with Joseph Priestley.

Only when it came to making gunpowder did the United States awake to the fact that it was in its interest to become chemically independent. Because of the importance of gunpowder, Alexander Hamilton had removed barriers to importing sulfur and saltpeter and set up a 10 percent customs duty on the finished product. In 1802, Eleuthère Irénée du Pont de Nemours, aged 31, brought on stream a black powder factory on the Brandywine Creek, near Wilmington, Delaware. His starting capital was $36,000 provided partly by his father, the physiocrat Pierre Samuel du Pont de Nemours who had fled the excesses of the French Revolution by emigrating to the United States in 1799. Eleuthère Irénée had worked at the Arsenal under Lavoisier, and both his factory plans and his equipment came directly from France. After a difficult start, he became America's leading munitions supplier in 1812, obtaining most of his saltpeter from Kentucky's natural nitrate deposits.

Antoine Laurent Lavoisier (1743–1794) and Irénée du Pont de Nemours (1771–1834).

Chapter 2

Advances in Chemistry During the First Half of the Nineteenth Century

As a science, chemistry had already gained from Lavoisier the concept of conservation of mass and the notion of the element. It had yet to formulate plausible assumptions regarding the structure of matter, to develop general laws to describe the phenomena observed through experience, and to acquire a system of universal notation for use by all chemists.

The Laws of Combining Weights

A Frenchman, Joseph Proust, enunciated the *law of definite proportions*, which distinguishes mixtures (solid or liquid), whose composition may vary, from pure compounds having a fixed composition. Proust had been Guillaume François Rouelle's pupil at the Jardin du Roi before establishing himself in Madrid, where he succeeded in demonstrating through a series of careful analyses that "the proportions in which two simple elements combine to form a chemical species is fixed." Proust was another of the Revolution's victims. Napoleon's troops destroyed his laboratory in 1808, putting an end to his scientific work and reducing him to poverty. He had to await the return of the Bourbons to be granted a pension and be made a member of the French Academy.

John Dalton, the Englishman who established the atomic theory, was certainly influenced by the work of Proust when he expressed his law of *multiple proportions*. Although a self-taught man of science, Dalton obtained a post as a teacher in Manchester in 1793 and remained in that city to the end of his life. He became interested in the vapor pressure of gases, and in the solubility of gases in water, formulating what was then a daring assumption, namely that "two elements may combine in different proportions but the weights of the second element which combine with a fixed weight of the first are simple multiples of each other.''

Dalton's Atomic Theory

Dalton considered that a compound cannot be fragmented beyond a certain limit. This quite naturally led him to attribute to matter a discontinuous structure and to conceive the idea of ultimate, indivisible particles for each kind of

John Dalton (1766–1844). Courtesy Edgar Fahs Smith Memorial Collection.

chemical, called *atoms*. They had already been conceived, albeit vaguely, by the Greek philosophers. Thus, all atoms in one element have the same weight and the same properties, but differ from the atoms of any other element. A chemical combination occurs when the atoms of one or several elements join to each other.

Since it is obviously impossible to weigh single atoms, any system of atomic weights can be established only on a comparative basis. Experimentally, only combining or "equivalent" weights are available.

Dalton had the idea of assigning 1 to the weight of hydrogen. Using Lavoisier's analysis of water, according to which it contains 85 percent oxygen and 15 percent hydrogen, he calculated an atomic weight 5.5 for oxygen. Likewise, he determined an atomic weight for nitrogen from the composition of oxides of nitrogen. Dalton was also an early scientist to advocate the use of arbitrary symbols for representing the atoms. But it fell to Jöns Jakob Berzelius to round out Dalton's ideas and to determine the chemical composition of a number of compounds, particularly in the cases where the compounds formed by two elements did not involve one atom only of each.

Berzelius and His Work

Jöns Jakob Berzelius (1779–1848) was a Swedish chemist who very quickly established his authority within the scientific circles of the time. He is considered one of the founders of modern chemical science. His early years were hard. He earned a living analyzing minerals, and his first teaching post was an unpaid one at Stockholm's medical faculty. This was to become the Karolinska Institute in 1810, the very year when he was himself appointed president of Sweden's Academy of Sciences.

Berzelius was painstakingly accurate in his experimental work, which involved determining the validity of the laws of chemical weights. He was aided in this by several great contemporary chemists, some of whom, like Pierre Dulong and Eilhardt Mitscherlich, had studied under him. Support came from *Joseph Louis Gay-Lussac's law* on combining volumes of gases (1808), the *rules of Pierre Louis Dulong and Alexis Thérèse Petit* on the specific heats of elements (1818), *Mitscherlich's law of isomorphism* on the properties of elements whose compounds have the same crystalline form (1822), and *the method of Jean Baptiste André Dumas* for the determination of vapor densities (1826).

The range of work carried out provided Berzelius with the means to crosscheck reason by analogy, helping him to determine the atomic weights of the forty elements known at the time. His table, published in 1828, provided values that are close to those still accepted today. His chemical symbolism, using letters to represent the elements, followed by numbers, is still in use, although it was not immediately accepted by other chemists. Berzelius

Jöns Jakob Berzelius (1779–1848). Courtesy Edgar Fahs Smith Memorial Collection.

also discovered new elements (cerium, selenium, silicon, thorium). He encompassed within the term *halogens* (meaning substances which generate salts), chlorine, bromine, iodine, which are all extracted from sea water and he introduced in 1831 the important concept of *isomerism* to indicate substances which have the same chemical composition but different physical properties. Finally he perceived the phenomenon of *catalysis*, using for the first time the expression "catalytic force" to explain that an element like platinum, even in small quantities and while remaining unchanged, can activate chemical reactions.

Despite the improvements brought about by Berzelius, Dalton's system of proportional numbers, connected with the laws of combining weights, was still unrelated to Gay-Lussac's law of combining volumes of gases. The Italian physicist Amedeo Avogadro has to be credited with making a synthesis of these two concepts by introducing the notion of the *molecule*.

Avogadro's Intuition

Amedeo Avogadro (1776–1856) was born in Turin, where he taught at the Academy and where he died at eighty years of age. Having studied physics and mathematics, he approached chemistry more on a theoretical than an experimental basis. This explains why his work was given little notice by the great chemists of his time.

To simplify, Dalton assumed that binary combinations (involving only two atoms) took place when two elements form a firm union. For the reaction leading to the formation of water, he wrote:

$$H + O = HO$$

Gay-Lussac came nearer to the truth. As the reaction occurs between two volumes of hydrogen and one volume of oxygen, he represented water synthesis with the equation:

$$2H + O = H_2O$$

In 1811, Avogadro made the hypothesis, known since as *Avogadro's Law*, that "equal volumes of simple or compound gases contain the same number of *molecules*."[1] Under this assumption and using the simplest hypothesis consistent with it, that the molecules of hydrogen and oxygen are made up of two atoms, the equation becomes:

$$2H_2 + O_2 = 2H_2O$$

which, with Avogadro's law, correctly states that two volumes of hydrogen unite with one volume of oxygen to produce two volumes of water vapor, the total volume contracting by one third in the process.

Whereas Gay-Lussac had considered only atoms, the introduction of the molecule concept (diatomic in the case of the common gases) in Avogadro's hypothesis made it possible not only to calculate with some confidence the

[1] Of course, under identical conditions of temperature and pressure.

Amedeo Avogadro (1776–1856). Courtesy Edgar Fahs Smith Memorial Collection.

percentage composition of chemical compounds from the formulas but also to know how many atoms of each element make up a given molecule.

Dalton refused to accept both Gay-Lussac's laws and Avogadro's hypothesis, which at the time received support only from the physicist André Marie Ampère. Moreover, Avogadro himself made no mention of the word "atom" in his papers. Accordingly, a certain amount of confusion continued among chemists. The atomic theory emerged from chaos only around 1850, when France's Auguste Laurent and Charles Frédéric Gerhardt, and then Italy's Stanislas Cannizzaro, reasserted the molecule concept as a fundamental

Joseph Louis Gay-Lussac (1778–1850). Courtesy Edgar Fahs Smith Memorial Collection.

unit in chemistry and brilliantly reinstated the Avogadro-Ampère law. Finally it became possible to establish correct chemical formulas.

Davy, Faraday, and the Laws of Electrochemistry

Alessandro Volta's invention of the electric pile in 1800 provided chemists with a method other than heat to decompose matter. Berzelius became interested in the new phenomenon of electricity very early on and in 1811 formulated a theory of the dualistic structure of chemical substances.

England's Humphry Davy (1778–1829) isolated for the first time, in 1807, two metal elements, sodium and potassium, through electrolysis, by causing the current of a voltaic pile to pass through molten soda and potash, respectively. Davy, it is believed, became interested in chemistry after reading Lavoisier's *Traité Elémentaire* in 1797. He first studied the therapeutic properties of gases such as nitrous oxide (N_2O)[2] and was appointed professor at London's Royal Institution in 1801.

Through his studies on the alkali metals he demonstrated that, as Lavoisier had stated, bases contain oxygen. But analyzing muriatic acid (HCl), he showed that, contrary to what Lavoisier had supposed, it was hydrogen and not oxygen that was common to all acids. Davy also recognized the elementary nature of chlorine and iodine. It has been said however that Davy's greatest discovery was Michael Faraday (1791–1867), who became his laboratory assistant at the Royal Institution in 1813. Davy took him with him on a tour of Europe, giving him the opportunity to become acquainted with Volta, Ampère, and Gay-Lussac. Faraday became famous both as a physicist for his work on electromagnetism, and as a chemist for original contributions ranging from the isolation of benzene to the liquefaction of carbon dioxide and to the use of platinum as a catalyst. From 1830 onwards, Faraday was interested in the electrical nature of matter. He proposed that in the electrolysis of an aqueous *electrolyte* solution, the positively charged entities or *cations* are drawn towards the negatively charged electrode or *cathode*, while the negative charges or *anions* are attracted to the positively charged electrode or *anode*.

This led him to state the *laws of electrochemical equivalents*, which established a relationship between electricity and chemical equivalents (1833): a) "The amount of substance decomposed in electrolysis is proportional to the quantity of electricity that passes through the substance;" b) "The same quantity of electricity passing through salt solutions, always decomposes equivalent quantities of salt." A further half century would pass before sufficiently powerful generators were available to develop the industrial applications of electrochemistry.

The Advent of Artificial Fertilizers

While a remarkable group of scientists was providing chemistry with its first chemical laws, significant progress was taking place in the application of chemistry to agriculture. Ever since humans have worked the soil, they have known that manure is beneficial in stimulating plant growth. Well before the Spanish conquest, the Incas had observed the fertilizing value of *guano* deposits. Made up of bird excrement, huge guano deposits existed in the islands close to Peru's coastline. Because of dry weather, these deposits were never

[2] Laughing gas.

washed away by rainwater. Humboldt brought a sample back to Europe in 1804, and it was discovered that, besides organic matter, guano contained nitrogen and phosphorus.

The use of guano to supplement manure soared considerably, especially in England, which began importing it by 1840, as well as in the United States. Trade in this natural fertilizer lasted thirty-five years, until sources diminished because the deposits were worked too intensively; it was replaced by *Chile saltpeter* ($NaNO_3$).

The role of *bones* and *wood ash* for soil improvement had also been known for a long time, but it was not until the first half of the nineteenth century that a relationship was established between plant growth and the nature of soil ingredients such as phosphorus, nitrogen, and potassium. This was done thanks to the work of France's Jean Baptiste Joseph Dieudonné Boussingault, Germany's Justus von Liebig, and England's John Bennett Lawes and Joseph Henry Gilbert.

Liebig's fame mainly stems from the school of chemistry he set up at the University of Giessen, where for twenty-seven years he trained a great number of talented pupils in organic chemistry. Although he failed clearly to understand the nitrogen fixation mechanism of plants or the need for soluble fertilizers for better soil absorption, he did nevertheless propose in 1840 the treatment of bones with sulfuric acid and identified in England the first mineral source of calcium phosphate (coprolite). It was also he who suggested extracting from the gas manufacturing plants the ammonia needed by plants, by combining it with sulfuric acid.

J. B. Lawes, assisted by J. H. Gilbert, who had worked at Giessen, was the first in 1841 to obtain a patent for *superphosphates* produced from bone meal treated with concentrated sulfuric acid. His Deptford factory, which came on stream in 1843, was soon capable of supplying superphosphates from mineral phosphates (coprolite, then apatite from Norway).

Because the acid produced in the lead chambers of the Leblanc plants was not sufficiently concentrated, superphosphate manufacturers in England integrated their manufacturing process by concentrating the acid themselves. They thereby developed practically permanent sulfuric acid overcapacities in the country. By 1854 there were six superphosphate plants with a 30,000-ton bone- or mineral-processing capacity, giving England a monopoly in this area until 1890.

The rise of *ammonium sulfate* production, suggested by Liebig, really began only after 1870 with the development of gas plants, although it was used by Lawes in his Rothamsted estate.

The use of *potash* in agriculture only began to spread after discovery of the Stassfurt salt deposits in Saxony in 1856. While Glauber had indeed demonstrated that saltpeter improved plant growth, for a long while this source of potassium was used entirely for the production of black gunpowder.

As artificial manures gradually replaced organic excrements, which necessarily were incapable of meeting growing needs, there was a distinct improvement in agricultural production. At Rothamsted Experimental Farm, the average wheat crop per hectare improved from 18.3 hectoliters (1793–1815) to 25.7 hectoliters (1855–1894).

Chevreul's Scientific Studies of Fats

Michel Eugène Chevreul, who died in Paris in 1889 at the age of 103, belonged to the great French tradition of his time as a scientist, a teacher, and a practicing chemist. He was a pupil of L. N. Vauquelin, who set him to work on a study of indigo, and he succeeded Vauquelin in 1830 at the Natural History Museum of the Jardin des Plantes. He also headed the dyes department of the Manufacture des Gobelins from 1824 onwards.

It was as Antoine François de Fourcroy's assistant that he became interested in fats in 1809. Saponification had been an entirely empirical process until Chevreul was able to demonstrate in 1816 that soap was not a mixture of fats and alkalis but true salts that can be produced by a base reacting with a natural fatty acid.

Together with Gay-Lussac, Chevreul discovered in 1825 the chemical process of fats splitting into glycerin and fatty acids by *hydrolysis*, for example:

glyceryl stearate + base (NaOH) = sodium stearate + glycerol
sodium stearate + acid (HCl) = stearic acid + salt (NaCl).

In this way he isolated a number of fatty acids from plant and animal substances, and patented a process to make candles from stearic acid, obtaining a cleaner and less smelly product than tallow candles. The first stearic acid factory was set up in 1831 by Adolphe de Milly at Barrière de l'Étoile and production spread throughout France, peaking at 30,000 tons in 1872.

Chevreul also introduced inert solvents, such as alcohol, for making separations; and he advocated the use of melting points as a criterion of chemical purity. He saw his work on fats vindicated by Marcellin Berthelot, who, in 1854, prepared glycerides from fatty acids and glycerol.

Rubber Becomes an Industrial Product

Ever since the French explorer Charles Marie de La Condamine had sent the Académie des Sciences in 1736, from Peru, a sample of an elastic rubber which the Indians drew from the sap of the Hevea tree and which they used to waterproof their moccasins and to make syringes (hence its popular name of syringe tree at the time), no practical use had been found for rubber. True,

Joseph Priestley had discovered in England, in 1770, that it had the peculiarity of rubbing out pencil marks on paper. It was, therefore, used as a "rubber." It was only in 1823 that the Scottish chemist Charles MacIntosh managed to dissolve rubber in a coal-based volatile liquid with which he coated cloth, obtaining the early raincoats after evaporation of the solvent. A breakthrough in the use of this still mysterious elastic material occurred in 1839 when Charles Goodyear, in the United States, found that he could obtain a strong, unalterable product that could be molded, by mixing sulfur with rubber. Improving upon Goodyear's makeshift process, Thomas Hancock, in England, heated the rubber with molten sulfur. His friend William Brockedon termed the reaction *vulcanization*, after the God of fire and metallurgy, Vulcan.

When Goodyear took out his patent in 1844, Hancock was already making vulcanized rubber in England. Two years later, by increasing the amount of sulfur and extending the period of reaction, he managed to produce ebonite, a kind of artificial tortoise shell as hard as horn. World rubber production, which did not exceed 300 tons in 1838, was to rise to 1,000 tons in 1850 and to 40,000 tons by the end of the century, when the use of rubber tires developed as one facet of "the bicycle craze."

The Dawn of Synthetic Organic Chemistry

Until early in the nineteenth century, it was believed that organic compounds, meaning substances that contained atoms of *carbon* together with atoms of hydrogen, oxygen, or nitrogen, could only come from living organisms, whether animals or plants, nature's "vital force" being required to produce them. Germany's Friedrich Wöhler can be credited with throwing doubt on this belief. In 1828 he succeeded in achieving the *synthesis* of urea, an organic compound, from two substances considered minerals, ammonium chloride and silver cyanate.[3] Wöhler had studied medicine at the University of Marburg before going to Stockholm, where he worked with Berzelius, who aroused his interest in chemistry. The work he carried out on silver fulminate and cyanate, at the same time as Liebig, from 1830 on, led both scientists to delve deeper into the phenomenon of *isomerism*. A joint study they performed on the oil of

[3] $AgOCN + NH_4Cl \longrightarrow NH_4OCN + AgCl$

$$NH_4OCN \xrightarrow{heat} O{=}C\begin{matrix} \diagup NH_2 \\ \diagdown NH_2 \end{matrix}$$

ammonium cyanate — urea

bitter almonds in 1832 suggested the existence of *radicals* in organic chemistry.[4]

Before vitalism's final invalidation by means of the total synthesis of a number of organic compounds carried out by Marcellin Berthelot by 1860, the concept declined further through the work of Germany's Adolf Kolbe. Kolbe was the eldest of fifteen children, and he had been Wöhler's pupil. He afterwards became Robert Wilhelm Bunsen's assistant at Marburg and took over from him as chemistry professor.

During Kolbe's research on chlorination carried out between 1843 and 1847 he obtained carbon tetrachloride, CCl_4, by chlorinating carbon disulfide, the latter having been previously prepared by heating iron pyrites with coal. He later, by heating, converted CCl_4 to tetrachloroethylene C_2Cl_4, and from the latter obtained trichloracetic acid by the action of chlorine and water in sunlight and further converted trichloracetic acid to acetic acid using a reaction already studied by Louis Henri Frédéric Melsens in 1842. Kolbe thus proved irrefutably that a product, which so far had been extracted only from vinegar, could be obtained through synthesis from mineral substances.

The Problem of Organic Formulas

Organic formulas were far more complex than those of mineral substances. It became essential to clarify the internal patterning of the formulas in order to achieve the synthesis of the substances represented. A first step had been made thanks to Liebig and Wöhler's concept of the radical group of atoms retaining its identity through a series of chemical reactions. Their fellow countryman, Robert Bunsen (inventor of the Bunsen burner) who was a professor at Marburg from 1839 onwards, developed it further. By the 1840s, chemists were talking as they would today of methyl, acetyl, benzoyl, or hydroxyl radicals. Further progress in determining the structure of organic molecules came from Gay-Lussac's demonstration that an atom of chlorine could be substituted for an atom of hydrogen in the preparation of cyanogen chloride.

While studying the chlorination of ethyl alcohol in 1834, J.B.A. Dumas devised his *substitution theory* to explain the replacement of hydrogen by a halogen (chlorine, bromine, iodine). His theory received the support of his assistant Laurent despite the objection of Berzelius, who was still tied up in his "dualistic" theory.

If one believed with Berzelius that chemical combinations resulted from

[4] The benzaldehyde contained in the oil was subjected to reactions, which demonstrated the persistence of the "benzoyl" radical as if it were a simple entity analogous to an inorganic atom.

atoms that carried opposite electrical charges, it was difficult to understand how two identical oxygen atoms could be capable of forming an oxygen molecule (O_2) according to Avogadro's law, nor how a positive hydrogen atom could be replaced by a negative chlorine atom as stated in the substitution theory. Laurent and Dumas's theory was validated, nonetheless, by the accumulation of experimental results well before Cannizzaro demonstrated the value of Avogadro's law.

A further step forward in the establishment of formulas for organic substances was achieved in 1839 by the Alsatian chemist Charles Frédéric Gerhardt, who also enjoyed Laurent's support. With his *theory of residues*, Gerhardt explained the principle of *condensation reactions*, in which hydrogen and oxygen atoms separated from organic molecules during a reaction, to produce water on the one hand as well as "residues," radicals of sorts which cannot be isolated but combine to form a new stable organic compound.[5]

Gerhardt also devised the concept of *homologous* compounds, which differ only in the number of CH_2 groups.[6] This concept made it possible to recognize the missing compounds in a *homologous series*, to predict their properties and to devise a method for their isolation.

Moreover, Gerhardt and Laurent were highly instrumental in developing the *type theory*. According to their theory, organic substances derive from simple molecules or *types*, which are either elements (H_2), or compound bodies made up of a few atoms like ammonia (NH_3). Accordingly, one only needed to replace one or several of the hydrogen atoms by radicals to obtain a series of compounds related to the base type. August Wilhelm von Hofmann and Charles Adolphe Wurtz' experiments brought out the usefulness of the theory by linking amines to the ammonia type.

August von Hofmann, who had studied law at the University of Giessen, became interested in chemistry through Liebig. He remained Liebig's assistant until leaving for the newly founded Royal College of Chemistry in London. As for Charles Wurtz, who was at school with Gerhardt in Strasbourg, he was also Liebig's assistant before working at the École de Médecin with Dumas, whom he succeeded in 1853.

Studying methyl- and ethylamines, Wurtz linked these compounds to ammonia in which hydrogen had been replaced by a methyl and an ethyl radical.

[5] In more modern terms, the reaction of benzene with nitric acid would be:

$$C_6H_5\text{-}H \quad + \quad HO\text{-}NO_2 \longrightarrow C_6H_5NO_2 \quad + \quad H_2O$$

benzene nitric acid nitrobenzene water

[6] e.g. CH_4 (methane), C_2H_6 (ethane), C_3H_8 (propane), ... C_nH_{2n+2}

Pursuing this line of thought, Hofmann recognized the *ammonia type* in various substitution products, all of which retained their character as bases.[7]

Another of Liebig's pupils, England's Alexander William Williamson, who studied ethers from 1850 to 1852, explained the *water type*, which he connected with ethyl alcohol and sulfuric (or ethyl) ether by substituting ethyl radicals for hydrogen in the water molecule.[8]

It was left to Gerhardt to systematize these different studies by establishing a four-type classification: hydrogen, hydrogen chloride, water, and ammonia. While such a classification still contained imperfections, it served to draw up an initial representation of organic molecules and to bring some order to the intricacies of this new chemistry.

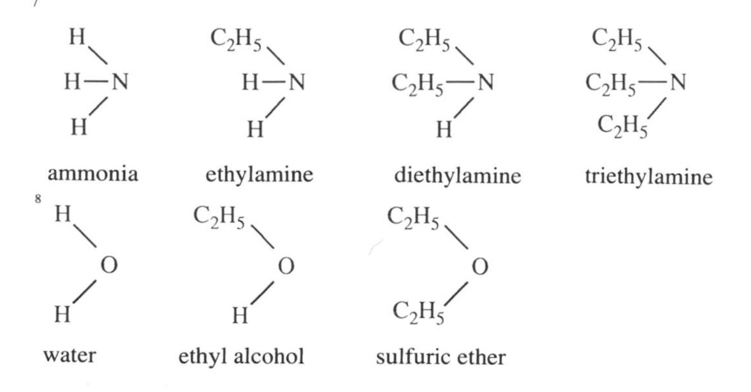

Chapter 3
The Great Scientific Breakthroughs from 1850 to 1914

The Chemical Industry in 1850

By 1850, the chemical industry was essentially an inorganic or mineral industry controlled by France and England. Both countries enjoyed the advantages of markets in which the Industrial Revolution had boosted activities like soapmaking, glassmaking, or textiles, which required increasing quantities of chemical products. There were also available the talents of exceptional men who, in France, consisted of scientists, professors, and industry leaders, while in England they were self-made entrepreneurs eager to develop their business by exporting, if need be, their products overseas.

In France, the distinguished chemist Nicolas Clément, improver of the sulfuric acid manufacturing process together with Charles-Bernard Désormes, took over management of the Saint-Gobain glassworks after a Leblanc soda plant was set up at Chauny in 1822. Gay-Lussac, who succeeded him from 1832 to 1850, improved the process even further by fitting out the Chauny works in 1835 with the first condensation towers he had himself designed. These towers served to recover nitrous substances that until then had escaped into the atmosphere.

In 1850, another great chemist, Théophile Jules Pelouze, set up a unit to manufacture sodium sulfate (salt-cake) for use as a base for glassmaking. By then, the glassworks were only using up to 15 percent of Chauny's chemical production and Saint-Gobain was firmly moving towards the marketing of sodium carbonate, chlorine, sulfuric acid, calcium chloride and even of potassium chlorate for fireworks.

The enterprising chemist *Frédéric Kuhlmann* had set up a small lead chamber unit in 1825 at Loos, near Lille, and by the following year was able to supply first sodium sulfate, then Leblanc soda, for the northern textile industries. Before setting up his own firm, Kuhlmann had studied in Vauquelin's laboratory, after which he had created an applied chemistry professorship at the Faculté de Lille. In 1834, he conceived the idea of treating bones with the hydrochloric acid recovered in his works, and soon became France's foremost supplier of superphosphates, which he sold as fertilizers to neighboring sugar beet producers. Business grew to such an extent that Kuhlmann bought a second plant in 1847 located at La Madeleine lès Lille.

France's oldest surviving chemical firm is the one *Philippe Charles Kestner* established in Thann, near Mulhouse, in 1808. By 1816, Kestner Frères

Frédéric Kuhlmann (1803–1881).

were employing thirty workers to supply the neighboring calico printing industry with sulfuric, nitric, and tartaric acids, as well as the tin salts required for dyeing cloth. Kestner was joined in 1824 by his son Charles, who had studied law and chemistry in Frankfurt and Göttingen. Having weathered the 1847–48 economic slump, Kestner was employing 240 people at the Thann plant by 1849, producing, besides inorganic and tartaric acids, sodium carbonate and acetic acid obtained through lime-processing of the acidic juice extracted from charcoal.

In the south of France, where Leblanc soda plants were supplying Marseilles' soapmaking industry with some 40,000 tons per year of carbonate, another manufacturer, J. Perret, who had developed and patented a pyrites roasting oven, set up a sulfuric acid plant at Saint-Fons, near Lyons in 1837. That plant became the major one in France. While such factories were being set up to satisfy needs that were almost exclusively domestic, a competitive chemical industry was developing in *England*. Taking advantage of improved transportation networks, English industrialists managed to expand their markets overseas. Between 1847 and 1851, England's alkali exports soared from 16,500 tons to 44,400 tons and kept on growing.

English industry leaders were men like *Charles Tennant*, who was joined by his sons at the Saint-Rollox plant; *James Muspratt* and *Josias Gamble*, who, as early as 1828, had introduced at Saint-Helens the Leblanc process; Gossage the soapmaker, who had invented the hydrochloric acid absorption tower; and the *Chance* brothers, glassmakers who started around 1830 to produce their own acid and soda in the Midlands. These pioneers were soon joined by *Henry Deacon* and *William Pilkington*, who settled at Widnes from 1850 onwards, as well as by *Edmond Sturge*, who was manufacturing potassium chlorate in Birmingham at about the same time. As stated earlier, a superphosphate industry had already been started in England by the middle of the nineteenth century; the country was preparing to consume for its own use as much fertilizer as all other European countries put together.

In *Germany*, a customs union (*Zollverein*) favored creation of an enlarged economic area. Lack of tariff protection, however, left the field wide open for alkali imports from England. Despite efforts made by the Italian manufacturers *Grasselli* in Mannheim and *Giulini* in Wohlgelegen, no significant Leblanc soda plant was established before 1851, when one was set up in Heilbronn. Similarly, in other areas of Continental Europe, the chemical industry was still in a fledgling state in the 1850s.

The chemical industry had not developed much in the United States, either. For one thing, the free trade policy of American administrations before the Civil War favored importation of alkalis and bleaching powders from England, and, for another, soapmakers and glassmakers were still using plant-based potash. Nevertheless, a chemical industry was gradually building up around Philadelphia. The pioneer was *John Harrison*, followed by *Charles Lennig*, who set up a sulfuric acid plant in 1829 at Bridesburg. Ten years later he was to use platinum in boilers that concentrated sulfuric acid to oleum on a continuous basis. A little further south, *Du Pont's* Brandywine black gunpowder factory had become the main supplier to the federal arsenals, while, in Baltimore, in 1832, *William Davison* was beginning production of superphosphates by treatment of bones with acids.

As early as 1806 in New York, *William Colgate* had started making soap and candles. Thirty years later, the Englishman *William Procter* and the Irishman *James Gamble*, a member of England's alkali industry pioneer family, established in Cincinnati, in 1837, the firm that still operates today under the name of *Procter and Gamble*. Also in Cincinnati, Eugene Ramiro Grasselli, following the example of what his father had done in the Rhineland, built a lead chamber unit in 1839, at the same time as England's *Thomas Emery* was establishing a tallow candle factory near the town's slaughterhouse. Despite the efforts of these pioneers, however, only about a thousand people were being employed in the 170 plants of America's chemical industry in 1850.

From Dye Extracts to Synthetic Dyestuffs

Since the seventeenth century the blue *pastel* from Provence, obtained from woad, and *indigo* extracted from the leaves of a vegetable from the coast of Coromandel, India, were used to make blue dyes. *Carmine red* was extracted from female cochineal found on cacti in Mexico and Peru and *vermillion red* from kermes insects. There were also extracts from tinted wood like *brazilwood*, which produced a glowing red and from which the country Brazil took its name. Trading firms, like the one Samuel Debar founded in 1801 in Lyon, supplied these exotic dyes to the silk manufacturers.

Madder was also cultivated in Europe as a dyestuff. First concentrated in Flanders and Zeeland, the growing of madder strongly developed in southeast France, where the red trousers of the French army provided a ready market for the dyestuff. Sales to England were also brisk, its imports of natural dyes rising to 75,000 tones in 1856.

Murexide, or Roman purple, was used as early as 1817 with an appropriate mordant for the dyeing of cotton or silk. It was manufactured in Manchester starting in 1853 by treating guano-based uric acid with nitric acid. *Picric acid* was the first synthetic dye (trinitrophenol) to be produced. It colored silk a bright though not very fast yellow. N. Guinon began to manufacture it industrially at Saint-Fons in 1847.

William Henry Perkin (1838–1907) succeeded in preparing the first true synthetic dye, later to be known as *mauveine*. Perkin became a student at the Royal College of Chemistry in London eight years after its founding in 1845. Two years later he became one of Hofmann's assistants. Hofmann suggested that he study the synthesis of quinine, an antimalarial drug essential for the colonial forces. Starting from quinine's rough formula ($C_{20}H_{24}N_{2}0_{2}$), Perkin reasoned that by oxydizing allyl toluidine ($C_{10}H_{13}N$), he would obtain the de-

William Henry Perkin (1836–1907), discoverer of mauve. Courtesy Edgar Fahs Smith Memorial Collection.

sired substance.[1] What he did obtain turned out to be a blackish red deposit. Replacing alllyl toluidine with aniline, Perkin then obtained a dark lump. Washing out the vessel with alcohol, he chanced upon a substance that could color silk purple. Perkin had succeeded by accident in synthesizing mauve. He took out a patent for it on August 26, 1856.

Running counter to Hofmann's advice, he resigned from the Royal College and at eighteen years of age set out with the help of his father and his brother to manufacture his own dyes at Greenford Green, near London. He had to overcome the reluctance of English dyers to use anything but natural dyes. They finally came around when mauve-dyed silks became all the rage at the court of Napoleon III. Perkin also had to prepare his own intermediates, such as nitrobenzene produced by reacting fuming nitric acid with benzene, which itself was extracted from coal tar distillate, and aniline through reduction of nitrobenzene according to a method which A. Béchamp developed in Lyon. Perkin then oxidized the aniline with potassium dichromate to obtain mauve. The exact composition of the dye was established only in 1888 by Otto Fischer, when it had already gone out of fashion.

Discovery of this first aniline dye was followed in 1858 with a red dye called *fuchsin* developed by France's E. Verguin and named after the Lyon firm *Renard Frères* for which he worked (Renard means fox, or *Fuchs* in German). Fuchsin soon became Magenta red to commemorate Napoleon III's victory over the Austrians in the Italian campaign.

Because it was red, fuchsin was more generally used than mauve. It was also easier to manufacture, and various preparative processes were suggested, particularly by the Société Lyonnaise *Monnet et Dury*, by the English firm of *Nicholson, Maule & Co.* as well as by France's *Girard et Georges de Laire* who, in 1860, also produced an aniline blue (*bleu de Lyon*). Fuchsin served as a base for the production of other dyes, as well, such as Hofmann violet, Perkin green, and Manchester brown.

This all led to a patent battle in England between Nicholson Maule and Read Holliday, and in France between Renard Frères and Monnet et Dury.

Renard Frères obtained confirmation of their sole rights to Verguin's patent, and on the strength of it they set up in 1864 the *Société la Fuchsine*, in which the Crédit Lyonnais bank took a share. But instead of trying to lower their prices through market extension and process improvement, they followed a narrow policy of high prices and spent valuable time chasing counterfeiters. The outcome was a return to natural dyes in France, while fuchsin manufacturers chose rather to set up their factories abroad, which were to rival Renard's. At the time, French legislation protected products and not

[1] According to the following equation:

$$2C_{10}H_{13}N + 3O \longrightarrow C_{20}H_{24}N_2O_2 + H_2O$$

processes, a fact that was hardly an inducement to improving methods of preparation.

In 1868 the Société la Fuchsine went bankrupt, and Verguin's patents and licenses were bought up by J. Poirrier, a Saint-Denis manufacturer who had started making mauve in 1861 and Paris violet in 1865. By the time the Franco-Prussian war broke out, he was employing 300 people in what was to become in 1881 the *S.A. des Matières Colorantes et Produits Chimiques de Saint Denis*. In Lyon, Monnet et Dury managed to survive by associating with Cartier and with Marc Gilliard, Debar's nephew, who had succeeded his uncle in 1856. Later Gilliard, Monnet, and Cartier were to start the *Société des Usines du Rhône*.

France, which had been at the center, first of natural and then of synthetic dyes, was soon to leave the field to its neighbors, the Swiss and the Germans. It was chilly protectionism that caused the decline, whether through patent legislation that granted unwarranted monopolies, or through a wish to protect the madder producers of southeast France, a wish that could lead only to disaster, for they were bound to disappear.

England, also, after a promising start, was forced, for other reasons, to admit its defeat in the race to be leader of the synthetic dye industry. To be sure, a number of manufacturers had followed in Perkin's footsteps and launched out into dyestuff production, successfully taking advantage of the knowhow of German chemists Hofmann attracted to the country. Moreover, England was beginning to produce benzene by-products from coal tar distillation. Some producers, such as *Read Holliday* or *Burt, Boulton, Haywood* used those by-products to manufacture synthetic dyes themselves.

It was also in England that the first *azo dyes* were developed. They were to supplant aniline dyes which were not light-fast. J. P. Griess, who had been Kolbe's pupil in Marburg before being recommended to Hofmann in London, had published in 1858 a study of diazotization by which nitrous acid reacts with aniline. Two other Germans, Heinrich Caro, who was a Mulheim cloth printer, and Carl Martius, the future founder of AGFA, completed the work started by Griess, who had failed to perceive the prospects that his discovery of diazonium salts would open up. Caro had settled in Manchester in 1859, and learned all about dyestuffs at Robert Dale & Co. He became a partner in the firm, which Martius joined in 1863.

By coupling diazonium salts with aromatic amines, both men produced azo dyes, such as Manchester brown, in 1865. One of their fellow countrymen, Ivan Levinstein, who was trained in Berlin, also settled in Manchester in 1864, building up a company that was quickly to become one of the foremost in the country.

From 1880, natural dyes were practically wiped out by synthetic products. By then, however, England's dye industry had already started to decline. It had been developed empirically by daring industry leaders, but there were no others to take over from them. Moreover, England's patent legislation protected inventions without making it compulsory to exploit them locally.

Finally, inasmuch as the dye itself only accounted for 1 percent of the price of a piece of cloth, English dyers saw no particular reason for favoring the national suppliers, the more so as with the slackening of research efforts, the dyes locally produced were not of consistent quality. Given the circumstances, German producers had no trouble preparing their onslaught.

The Rise of the German Chemical Industry

The progress accomplished by the German chemical industry during the second half of the nineteenth century was the result of long-standing efforts in the training of scientists. In France, while the great chemists had their own personal laboratories, there were no public experimental laboratories to back up the teaching of chemistry at the Jardin des Plantes, the Conservatoire des Arts et Métiers, the École Normale Supérieure, where Henri Sainte-Claire Deville succeeded Antoine-Jérôme Balard as chemistry professor in 1851, or at the École Polytechnique, which trained engineers for government service. Furthermore, appointments and promotions were made in Paris by a few ''official chemists'' such as J.B.A. Dumas, and Marcellin Berthelot, a procedure that relied as much on the candidate's worldly connections as on his talent.

In England, scientists were diehard individualists with no interest in creating a following. While London's Royal College of Chemistry, established in 1845, had in Hofmann its hour of glory, enrollment dwindled when he left in 1865. The private sponsors who supported it, in the absence of government subsidies, were interested only in scientific research that yielded immediate returns.

In Germany matters were quite different, and Liebig's role was a crucial one. He had studied in Paris with Vauquelin. On Alexander von Humboldt's recommendation he was able to work in Gay-Lussac's laboratory at the Arsenal, then to lecture on chemistry at Giessen, where he was appointed by the Grand Duke of Hesse. He remained there for twenty-seven years and created a model institution. He used the laboratories which he had built there to train his students in qualitative and quantitative analysis and then in organic synthesis. He followed their work closely, imparting his enthusiasm for chemistry, establishing a true esprit de corps within the school. A great number of talented chemists were trained at Giessen, including Wilhelm von Hofmann, Karl Remegius Fresenius, Emil Erlenmeyer, Friedrich August Kekulé, Charles Adolphe Wurtz, and Henri Victor Regnault. Friedrich Wöhler applied the Giessen pattern to Göttingen, as did Robert Bunsen to Heidelberg, and Kekulé to Bonn. Hofmann applied it to Berlin when he succeeded Eilhardt Mitscherlich there in 1865, after leaving England. Germany had not yet been unified, and a spirit of competition pitted states against one another to finance

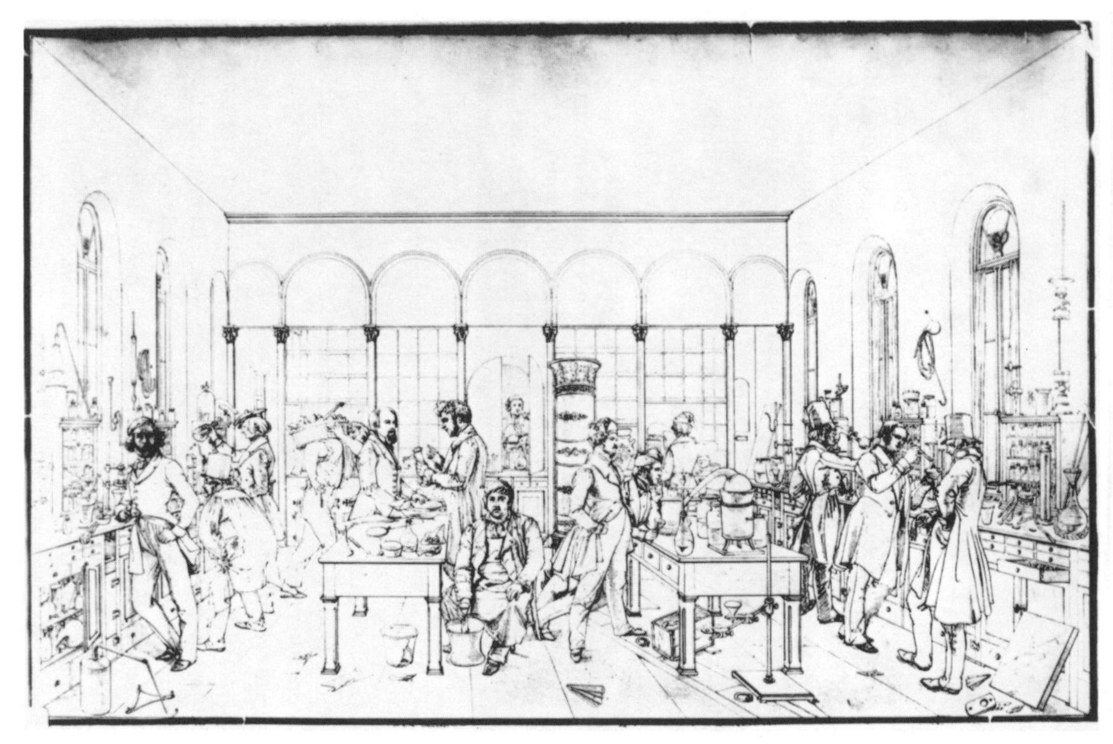

Justus Liebig's (1803–1873) laboratory in Giessen, 1842. Courtesy Edgar Fahs Smith Memorial Collection.

schools and laboratories, to enroll students and select professors. Such professors were well paid, and the most distinguished, such as Liebig, Hofmann, Kekulé, and Bunsen, were granted titles of nobility. There were also *Technische Hochschulen* in Germany, such as the one first set up in Karlsruhe. State-subsidized, they provided high-level technical training for students planning to work in industry. In fact, there was no airtight separation in Germany between university research and industrial research.

Also, several German chemists had the opportunity of learning about industrial applications of organic chemistry when they went to work in England's dye factories, which benefited in turn from their scientific expertise. Returning to Germany, those chemists used the experience they had acquired by contributing to the birth and development of the first dye plants built in their country, in the 1860s. The patent jungle prevailing within the thirty-nine States of the German Confederation made their task easier. It was only after unification in 1871, when Germany's technical leadership had been well established, that the Reich thought it advisable to protect its patents through stiff legislation. A patent decree was published in 1876.

In 1856, a dealer in dyewoods, K.G.R. Oehler, established a tar distillery in Offenbach and started making picric acid, fuchsin, and mauve. *Griesheim Elektron*, a company set up the same year to manufacture Leblanc soda, bought in 1905 what had been the first synthetic dye business. Another dealer who specialized in dyestuff extracts, Friedrich Bayer, also began producing fuchsin and its aniline dye derivatives in 1863, in a plant located in Elberfeld, Westphalia. In 1881, following the death of its founder, it was reorganized into a joint-stock company and became *Farben Fabrik vormals Friedrich Bayer*.

A similar process led, in 1880, to the *Farbwerke Hoechst*. Initially it had been set up in 1862 by Eugen Lucius in joint venture with two dealers, Wilhelm Meister and Adolf Brüning, to manufacture aniline dyes at Hoechst near Frankfurt.

The other German chemical giant, *Badische Anilin und Soda Fabrik* (BASF), developed from the association in 1861 of a businessman, Friedrich Engelhorn, who was a tar distiller in Mannheim, with the Clemm brothers, two chemists trained by Liebig in Giessen. Their first fuchsin and aniline plant was built in Mannheim. A soda and mineral acid factory was added to it in 1865 located on the other bank of the Main at Ludwigshafen where BASF established its head office. Three years later, Engelhorn had the chemist Heinrich Caro join BASF when he returned from England.

Also on the Main, at Biebrich, Paul Wilhelm Kalle established *Kalle & Co.* in 1864. It concentrated on the production of specialty dyes, while in the same area, a firm of dye dealers dating back to 1807, *Leopold Cassella & Cie.* also launched out into synthetic dyes in 1867. As for Carl Martius, also returned from England, he joined the same year with Paul Mendelssohn-Bartholdy, the son of the composer, to set up an aniline dye factory in

Rummelsberg near Berlin that became in 1873 the Aktien Gesellschaft für Anilin Fabrikation, better known as *AGFA*.

With the exception of AGFA, all these new companies had settled in the Rhineland-Westphalia region, close to the waterways and to supplies of mineral acids, alkalis, and aromatics obtained through coal-tar distillation. They also enjoyed the benefits of private bank financing, of a body of chemists well trained both in scientific and technical disciplines, of competent technical and marketing management teams, and of a privileged relationship with university professors who kept them abreast of the most recent developments in organic chemical research.

From Empirical Work to Systematic Research

Early synthetic dyes, as we have seen, were produced empirically. While their properties could be determined immediately, it took longer to find out their structures. The task was made far easier when new concepts were elaborated. Through Stanislas Cannizzaro, who pointed out the value of his fellow countryman Avogadro's hypothesis at the Karlsruhe Congress in 1860, atom and molecule were finally clearly distinguished one from the other, so that it became possible to calculate *atomic and molecular weights*.

Likewise, the *valence concept*, which Wurtz had defined as the ability of an atom to add to or substitute others, based on hydrogen's valence as 1, turned out to be decisive in organic chemistry thanks to Kekulé's insight. Kekulé had come to chemistry through Liebig's influence, having originally gone to Giessen to study architecture. In 1857, Kekulé was led to add the methane type (CH_4) to four types previously selected by Gerhardt, confirming his notion that the constant valence of carbon is 4.

He further made the hypothesis, as did Archibald Scott Couper at the same time, that carbon atoms can link with one another to form a chain. It is likely that their work, published in 1858, helped Josef Loschmidt, Emil Erlenmeyer, and Lothar Meyer between 1861 and 1864, to develop the concept of unsaturation, according to which two adjacent carbon atoms with a free valence each are linked by a *double or ethylenic bond*.[2] Other substances can be added due to this unsaturation to form a network of singly-bonded atoms.[3] The same

[2] Ex: ethylene

$$H_2C{=}CH_2$$

[3]

$$H_2C{=}CH_2 + HCl \longrightarrow H{-}CH_2{-}CHCl{-}H \quad \text{monochloroethane}$$

Friedrich August Kekulé (1829–1896). Courtesy Edgar Fahs Smith Memorial Collection.

applies to *triple bonds* as found in acetylene. But Kekulé's most imaginative vision was, undoubtedly, his *benzene ring*, which revolutionized aromatic chemistry. In order to explain that the six benzene carbons (C_6H_6) are linked to only six hydrogen atoms despite their tetravalence, Kekulé suggested in 1865 a closed chain structure of alternate double and single bonds in a ring according to the formula that is now a classic:

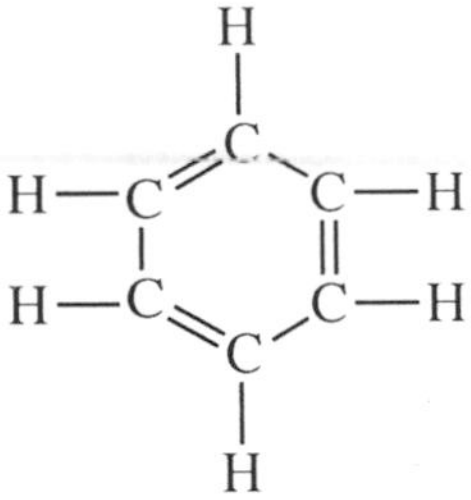

or more schematically:

With this formula he managed to give a clear representation of the *isomers* of polysubstituted derivatives in the benzene series:

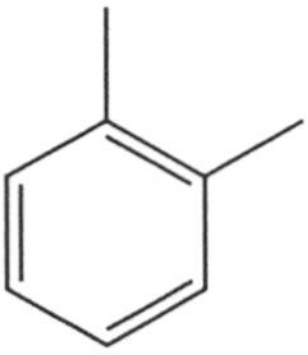

ortho 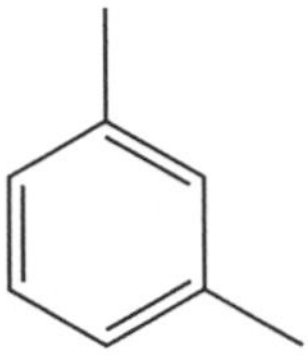 meta 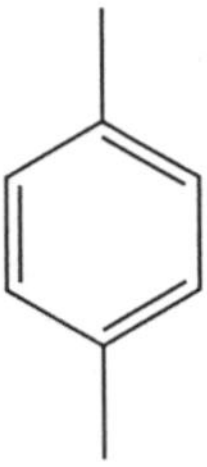 para

Such new concepts were to help organic chemistry develop in a rational manner and complex syntheses could be made. At the same time the structures of many compounds were being elucidated.

Erlenmeyer was thus able to propose a structure with two rings for naphthalene[4] in 1866, while Carl Graebe and Carl Theodor Liebermann proposed a three-ring structure for anthracene.[5]

In the same way, the chemistry of *heterocyclic compounds* was clarified

4

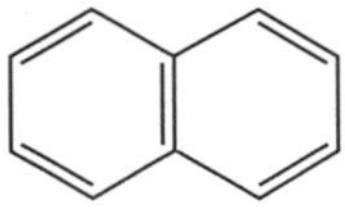

naphthalene

5

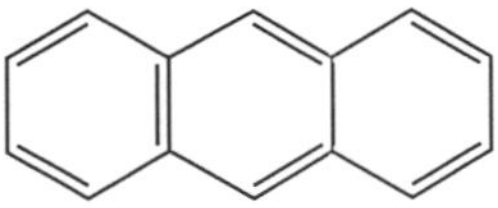

anthracene

when, working from analogy, William Körner and, independently, James Dewar proposed ring structures for pyridine C_5H_5N and quinoline C_9H_7N, having observed their similarity in properties to benzene C_6H_6 and naphthalene $C_{10}H_8$.[6]

The *synthesis of alizarin* is a particularly good illustration of the progress made in understanding the structure of organic molecules.[7]

Adolf von Baeyer succeeded Liebig as chemistry professor in Munich and later took the same post at the University of Berlin. Very early on in his career he had become interested in the composition of natural dyes and had, in particular, obtained isatin through oxidation of indigo. His two pupils in Berlin, Carl Graebe and Carl Liebermann, were engaged in trying to convert alizarin into anthracene in 1868 when Baeyer advised them to synthesize alizarin from anthracene, since the latter was a by-product of tar distillation.

A synthetic method was developed by using fuming sulfuric acid instead of bromine during the first stage, a process that was industrialized by Heinrich Caro. Perkin also had studied the synthesis of alizarin and he had taken out his patent on June 26, 1869, barely a day after Caro, Graebe, and Liebermann had taken out theirs. An agreement was reached between Perkin and BASF, which had sole rights over the patent of the German chemists. But what was to be Perkin's last successful venture turned out to be the start of the dominant position Germany would acquire in the dye business. The process underwent further improvements. In particular, the use of fuming sulfuric acid, which had to be brought at great expense from Nordhausen in carboys, was made unnecessary. By 1893, synthetic alizarin had totally supplanted the madder harvested in France's Languedoc province, forcing peasants there to cultivate vineyards instead.

The *synthesis of indigo*, which was even harder to achieve than the synthesis of alizarin, was a remarkable feat of Germany's fledgling chemical

6

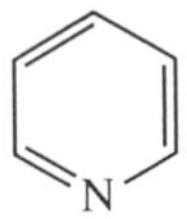

N
pyridine

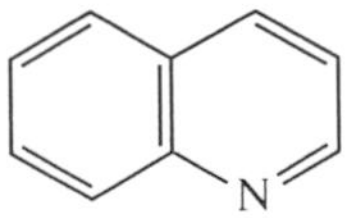

N
quinoline

7

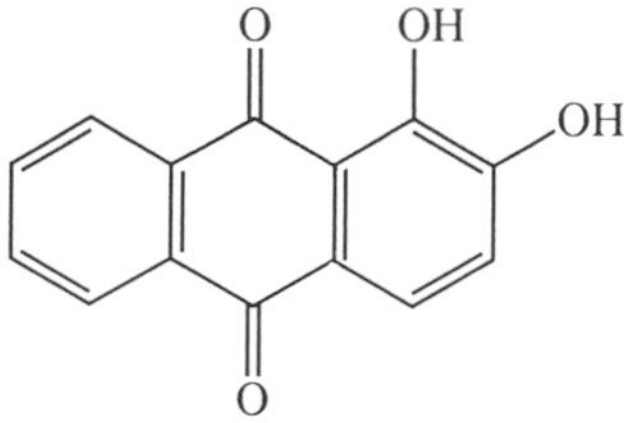

O OH
OH
O
alizarin

industry at the end of the nineteenth century. There again, resuming his early research, it was Baeyer who managed to produce indigo from isatin in 1870, then to synthesize isatin itself in 1878. In 1883, he was able to reveal the structure of indigo,[8] and by 1880 he had entered into an agreement with BASF and Hoechst under which both firms were to share in the fruits of this research. Only when Karl Heumann, who was working for BASF, developed in 1890 a process from naphthalene did it become commercially practical.

BASF finally started industrial production of indigo in 1897, but the cost of the research it had conducted up to that time had eaten badly into its capital. Hoechst, which started its own production a year later, brought distinct improvements to the process in 1901. By 1910, natural indigo had disappeared from the market. German supremacy, based on the close relationship between the university and chemical industry, received its consecration: in recognition of his work on organic dyes, Adolf von Baeyer was awarded the Nobel Prize for Chemistry in 1905.

Even when *organic synthesis* failed to produce successful industrial applications, its development was spectacular during the second half of the nineteenth century because of the new concepts and the consensus chemists gradually achieved about them. One of the most prolific researchers in this area was Marcellin Berthelot, who was Balard's assistant at the Collège de France and was influenced by him before going on to work with Théophile Jules Pelouze, Jean Baptiste André Dumas and Henri Victor Regnault. Berthelot's preparation of fatty acid esters by synthesis confirmed Michel Eugène Chevreul's ideas on the subject, and his work on glycerol and polyalcohols was epoch-making. He achieved the synthesis of ethyl alochol based on the reaction of sulfuric acid with ethylene followed by hydrolysis and conversely prepared olefins from alcohols. He also produced acetylene in 1862 by passing hydrogen gas through an electric carbon arc and in 1866 obtained benzene by passing acetylene through a red-hot tube. It was he who introduced the word "synthesis," and his accomplishments discredited the last champions of "vitalism."

Between 1850 and 1900 other distinguished chemists broadened the range of organic compounds that could be obtained from easily available start-

[8]

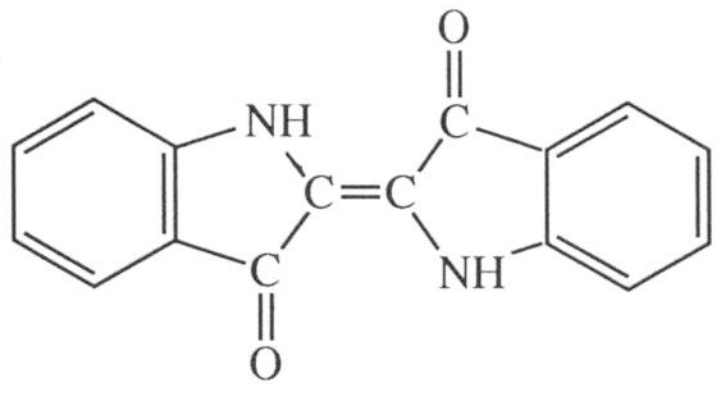

O
‖
NH C
C=C
C NH
‖
O

indigo

ing materials through reactions that still bear their names: (Hermann) Kolbe, (Charles-Adolphe) Wurtz, (James Mason) Crafts, (Stanislas) Cannizzaro, (Charles) Friedel, (Sergius) Reformatsky, (Traugott) Sandmeyer, (Ernst Otto) Beckmann, (Paul) Walden, (François Auguste Victor) Grignard.

Fresh Demand for Coal-Tar By-Products

The development of synthetic dyes sparked a demand for aromatics. Those aromatics could be produced through tar distillation. Originally wood-tar was used to protect ship bottoms and ropes. As early as the beginning of the nineteenth century, however, the gaseous products of coal distillation were being used for street lighting. And by 1820, gas factories were being built in increasing numbers, particularly in England. It soon became hard to dispose of the tarry residues, left when coal gas was manufactured.

True, Charles MacIntosh was using light oil (called benzol or naphtha) as a solvent to manufacture his waterproof material, and tar was to find a market both as a road binder and, in the form of creosote oil, as a wood preservative, especially for railroad sleepers. But such demand was incapable of absorbing the excess production of tar. The matter was solved only when the dye industry opened a new market.

Chemists had, indeed, learned how to isolate from coal tar, at least on a laboratory scale, naphthalene (Alexander Carden and Kidd in 1819), anthracene (Jean Baptiste André Dumas and Auguste Laurent in 1832), aniline and phenol (Friedlieb Ferdinand Runge in 1834), and finally benzene (August Wilhelm von Hofmann in 1845). The distillation of coal began to provide the dye industry in England with the hydrocarbon aromatics they required, and from 1865 began satisfying the needs of the German plants through their exports. Firms such as *Butler, Carless, Blagden & Co.*, and *R. Graesser* applied fractional distillation technology, which had been improved by Charles Mansfield, a pupil of Hofmann's, and which benefited from cheap coal supplies.

By 1880, the new industrial and home uses that were developing, such as incandescent gas burners and cooking ranges, required gas with stronger heating power. The higher gasification temperature this entailed reduced tar production. Also, as their needs grew, the Germans tried to shake off their dependence on imported English tar. They managed to do so by recovering the benzol produced by the coking furnaces of steelworks, thanks to an improved process developed in 1889 by Heinrich Von Brunck at BASF.

By losing her synthetic-dye manufacturing dominance to the Germans, who very soon sought to produce their own feedstocks, England was putting her whole chemical industry at risk, deprived as it would be of some of its main markets. Indeed, each ton of dye required from three to five tons of base chemicals to produce.

The Birth of Switzerland's Chemical Industry

Switzerland's chemical industry also developed from a dye base, within four companies in the Basel area admirably located to receive their feedstock from German Rhineland factories, and to supply the neighboring textile industry with dyestuffs. The industry received a boost in its early days from the policy of high prices demanded by the Renard firm in Lyon. This encouraged fuchsin production to migrate to Switzerland. It was a Basel dye manufacturer, Alexandre Clavel, who first produced fuchsin in 1860, initially for his own needs, then for other dye manufacturers. Subsequently he introduced other aniline dyes to his range of products. After his death, his company was to become Switzerland's leading dye manufacturer, known since 1884 as Gesellschaft für Chemische Industrie Basel (CIBA).

Another fuchsin production unit was set up in Basel in 1860 by J. J. Müller, who, until then, had been selling imported products on behalf of the Geigy family. *Johann Rudolf Geigy*, a member of that family, took over the business in 1862, focusing production on azo dyes after closing down the fuchsin unit in 1872. The firm of *Durand & Huguenin* dates back to 1862, when the Gerber-Kellers process was introduced for fuchsin production by Mulhouse's J. G. Dollfuss. The fourth company was *Sandoz AG*, which grew out of the association in 1886 of a chemist, Alfred Kern, with Edouard Sandoz, who had worked at Durand & Huguenin's.

The Swiss companies were soon dependent on Germany's chemical industry for their synthetic intermediates as well as their base chemicals. Thus they could not hope to compete successfully with their powerful neighbor in the field of common dyestuffs like alizarin or indigo. They directed their efforts, therefore, to the development of special high added-value dyes that gained them high worldwide repute and sparked the prosperity of Basel's chemical industry.

The Early Days of Chemotherapy

At the time the first synthetic dyes were being discovered, there was no drug industry to speak of. Apothecaries merely prepared prescriptions drawing on recipes derived from plants or even at times on inorganic substances such as Glauber's salt, mercury derivatives, or sulfur. The preparation of quinine to treat malaria was introduced into England by Thomas Morson in 1821. It was the prelude to the isolation and purification of a series of *alkaloids*.

The anaesthetic properties of nitrous oxide were, as we have seen, one of Humphry Davy's discoveries. His pupil Michael Faraday revealed the properties of ether (1818), while Liebig prepared chloroform from acetone and chlorine in 1831. But it was not before 1847 that chloroform began to be used as an *anesthetic* in England's hospitals. England's Joseph Lister was the first to use coal tar derivatives to fight the germs of infection. In 1867 he introduced carbolic acid (phenol) as an *antiseptic* in surgery. Other coal-tar derivatives that had the power of bringing down fever were developed in Germany from chance discoveries. These were followed up by remarkably systematic research.

Salicylic acid, synthesized by Hermann Kolbe from phenol in 1853, was manufactured on an industrial scale by his pupil Friedrich von Heyden twenty years later because during that time, besides its effectiveness as an *antipyretic*, it was found to have food-preservative properties. The mistaken use, in a Strasbourg clinic, of acetanilide, formed when aniline reacts with acetic acid, revealed that this dye intermediate had remarkable antipyretic properties. Kalle & Co. immediately began to manufacture it at Biebrich under the trademark of Antifebrin.

Carl Duisberg, who had joined Bayer at Elberfeld in 1884, directed the firm's research toward antipyretics in a similar range but devoid of side effects. Thus were developed first *phenacetin* in 1888 and then most notably in 1898 acetylsalicylic acid, known worldwide as *aspirin*. Hoechst, advised by Ludwig Knorr of Erlangen, was studying the family of pyrazolones, putting on the market two *painkillers, antipyrine* (1888), soon followed by *Pyramidon*. However remarkable, these products reduced only the effects of the illness; they did not attack its causes.

To go further into what was to become *chemotherapy*, it was necessary to understand by what process infection developed and bring out products that were likely to halt its course. This was to be the work of the French chemist Louis Pasteur and of three German physicians, Robert Koch, Paul Ehrlich and Emil von Behring, the true founders of *bacteriology*.

Louis Pasteur (1822–1895), the son of a tanner of Dôle in the Jura, had studied chemistry at the École Normale Supérieure. He was noticed by Antoine-Jérôme Balard, who took him on as his assistant and had him work with Auguste Laurent. He attracted attention when, at twenty-eight years old, he studied potassium acid tartrate crystals that form during wine fermentation. This led him to develop the concept of *optical isomerism*, based on the differing optical rotations observed under polarized light and to distinguish levo-rotatory, dextro-rotatory and racemic tartrate forms (from the Latin *racemus*, meaning bunch of grapes). When he became dean of Lille's Faculté des Sciences in 1854, Pasteur had the opportunity to pursue these studies further on fermentation in the alcohol distilleries of the area.

Louis Pasteur (1822–1895).

In 1857, he began identifying the microorganisms specific to each of various fermentations (alcoholic, butyric, lactic). Observing these microorganisms under a microscope and cultivating them, Pasteur proved that they developed like any other living organism and thus helped silence the remaining advocates of "spontaneous generation." Finally he discovered that harmful microorganisms could be destroyed by heating at between 45° and 60°C. *Pasteurization* is now common practice in the food industry.

Recommended by Dumas, Pasteur was commissioned, from 1865 to 1871, to study a silkworm disease threatening the livelihood of silkworm breeders in the south of France, and, therefore, also of Lyon's silk manufacturers. He discovered the root of the disease in small black corpuscles that had infected the female silkworms. He succeeded in destroying the contaminated eggs, thus halting the epidemic.

Encouraged by these early achievements, Pasteur was ready to tackle infectious germs in animals and then in humans. Using the pattern conceived in 1796 by Edward Jenner, who had invented the smallpox vaccine, Pasteur thought it would be possible to inject an attenuated form of the bacteria into live bodies, which would then be immunized against the disease these bacteria were causing. He proved his assumption right in 1881 by treating a flock of sheep attacked by anthrax. Thus it was quite deliberately that on July 5, 1885, he vaccinated the young Joseph Meister, who had been bitten by a mad dog, and saved his life.

Robert Koch was a district doctor at Wollstein in Poznan, who in 1872, at twenty-eight years of age, became interested in infectious diseases. He used different aniline dyes to color the bacteria selectively that he would observe with his microscope. He also succeeded in fixing them in a solid culture medium, whereas in a liquid medium they moved around rapidly.

He isolated in this manner the rod-bacillus of anthrax in the blood of the sick animals. In 1882, while working at the tuberculosis Charité clinic in Berlin, he isolated the bacillus of tuberculosis. Although the tuberculin culture extract he developed two years later—selling the sole rights for it to Hoechst for DM 1 million—did not prove to be the hoped-for remedy for tuberculosis, it proved useful in diagnosing the disease in both humans and animals. In 1905, Koch was awarded the Nobel Prize for Physiology and Medicine.

While still a medical student at Breslau, Paul Ehrlich (1854–1915) was drawn to study the behavior of dyes in living tissues at his cousin's, the histologist Carl Weigert, who taught him the technique of staining bacteria. He was thus able to help Robert Koch, whose assistant he had become in 1889 in Berlin. More significantly, he very soon acquired the notion that it would be possible not only to detect but also to destroy bacteria with the right dye. He demonstrated this in 1891 with methylene blue acting on the malaria parasite.

Chemotherapy was born. It now remained to develop organic compounds that, acting selectively, would not cause side effects in the body. Having no success with azo dyes, Ehrlich thought he would start off with atoxyl, an organic derivative of arsenic prepared in 1863 by A. Béchamp. In 1905, it was found to be effective against trypanosome diseases such as sleeping sickness.

After a number of systematic tests were carried out by his assistants,

Paul Ehrlich (1854–1915). Courtesy Edgar Fahs Smith Memorial Collection.

Ehrlich discovered in 1909 that number 606 of the series of compounds studied, an arsenobenzene,[9] was effective not only against the trypanosomes of sleeping sickness, but also against syphilis. In December of that same year Hoechst, for whom Ehrlich had been working since 1907 in the field of chemotherapy, marketed his compound 606 under the name of Salvarsan, a name that encompassed both the drug's curative purpose and its chemical origin. Syphilis was close to being conquered.

Ehrlich had also worked in Berlin with Emil von Behring in 1902 at the time the latter had discovered the diptheria antitoxin, and he played an active part in preparing the *antidiphtheria serum* which Hoechst was to market. In 1908 Paul Ehrlich was awarded the Nobel Prize for Medicine and Physiology for his work on immunology.

An offshoot of dye chemistry, the birth of chemotherapy was due to Pasteur's unique genius, then to the remarkable research carried out by three German physicians, each adding a stone to the edifice, and working in close cooperation with the German chemical industry.

Explosives and Alfred Nobel's Discovery

In five hundred years the formula of gunpowder has hardly changed.[10] Berthollet did attempt to replace saltpeter by *potassium chlorate*, but a violent explosion in 1797 during its preparation postponed its widespread use for a century. Meanwhile, after Pelouze had demonstrated that cellulose could be nitrated, a chemistry professor in Basel, Christian Schönbein, succeeded in 1846 in preparing *nitrocellulose* by having concentrated sulfuric and nitric acids react with cotton. That same year, Italy's Ascanio Sobrero produced *glycerol trinitrate* (nitroglycerin) by treating glycerin with the same mixture of acids. These were the two first synthetic explosives to be discovered. Although they were several times more powerful than black gunpowder, they presented serious drawbacks owing to their instability. Nitroglycerin, moreover, was a liquid and not a powder. In 1848 two workshops where nitrocellulose was being manufactured were destroyed by an explosion with

[9] It was 3,3′-diamino-4,4′-dihydroxyarsenobenzene with the formula

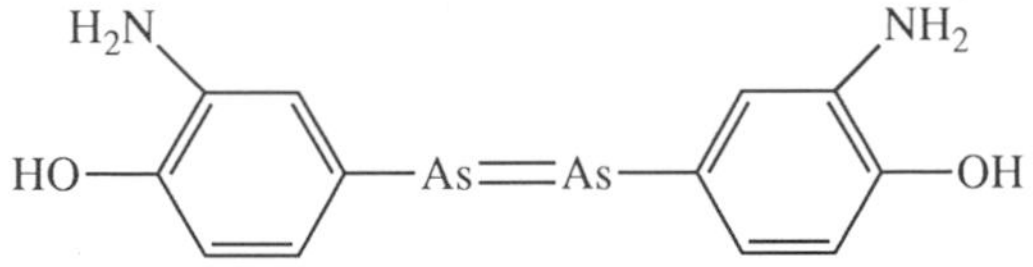

H_2N NH_2

HO— —As=As— —OH

Salvarsan

[10] Gunpowder is a mixture of about 1.8 parts sulfur, 2 parts carbon, and 10 parts potassium nitrate, serving as oxidant.

Robert Koch (1843–1910), Paul Ehrlich (1854–1915), and Emil Adolph von Behring (1854–1917). Courtesy Société Française Hoechst.

considerable loss of life, one at Faversham in Kent and the other at Le Bourget near Paris. Afterwards, no government dared become involved with the new explosive.

In 1863, however, Frédérick Abel, who had been one of Hofmann's early pupils at the Royal College of Chemistry, demonstrated that nitrocellulose could be stabilized by compressing the cake, obtained after nitration, in order fully to eliminate residual acid. Thereafter the explosive, known as *smokeless powder*, was used by the Royal Navy in torpedoes and undersea mines.

Nitroglycerin remained a laboratory curiosity until the Nobel family started working on it in their Stockholm workshop in 1863. Emmanuel Nobel was a Swedish industry leader who had become interested in explosives during the Crimean War. He had afterwards successfully developed oil fields in the Caucasus, with his four sons. Leaving the two older sons in Baku in 1859, he returned to Sweden with the younger two. One of them, the third son, was Alfred Nobel, who was to become world-famous.

Young Alfred met Ascanio Sobrero, the inventor of nitroglycerin, when working in 1850 in Pelouze's laboratory. Nitroglycerin was an unstable liquid. Even so, the Nobel family chose to use it for their studies because, as a by-product of soapmaking, it was easily available. A violent explosion occurred in Stockholm on September 3, 1864, killing the youngest son as well as four assistants. Undaunted, Alfred Nobel continued his work, merely shifting it to a canal boat anchored not far from the Swedish capital. In 1866 he succeeded in stabilizing nitroglycerin by absorbing it on kieselgur, a siliceous earth. He named the new explosive *dynamite* and took out patents for it in various countries of the world. He also took out a patent for the mercury fulminate detonator that was now needed to set off the explosion.

Nobel's discovery turned out to be timely because powerful explosives were needed for mining and major public works, in particular to blast tunnels for the expanding railway networks. Between 1867 and 1872, world production of nitroglycerin soared from 11 to 1,350 tons.

By 1873 Alfred Nobel had a stake in fifteen dynamite manufacturing companies, thirteen in Europe and two in the United States. Much of his income was devoted to further research. In 1875 he developed blasting gelatin, a controlled combustion mixture of nitroglycerin and nitrocellulose, and in 1887, a smokeless powder, *ballistite*, for which there was a ready military market. Two years later, Abel and James Dewar based the *Cordite* they offered British arsenals on such a mixture. In 1884 Nobel associated Dynamit Nobel AG with three other nitroglycerin manufacturers in Germany, and two years later he set up one of *the first multinationals* with two holding companies that controlled his interests worldwide: The Société Centrale de Dynamite in Paris, encompassing France, Switzerland, and Italy, and the Nobel Dynamite Trust Co. in London, which covered England, Spain, Germany, and America.

By the end of his life, Alfred Nobel had amassed a large fortune. Besides

Alfred Nobel (1833–1896).

the revenues from dynamite, his fortune also came from his Baku oilfield interests and from the royalties accruing from the 350 patents he had taken out worldwide. Nobel's will provided an endowment of 32 million Swedish Kronen for the annual award of five highly distinguished prizes that were to immortalize his name far more than his invention of dynamite.

Other explosives were being discovered while nitroglycerin use was developing. Paul Vieille of the École Polytechnique had conceived of stabilizing nitrocellulose by dissolving a partly nitrated cellulose, *pyroxylin*, in an ether and acetone mixture. From the resulting collodion he extruded smokeless-powder sticks, *B Powder*. *Picric acid* (trinitrophenol), which had served since

1850 to dye silk and wool yellow, also turned out to be a powerful explosive when *A. Turpin* in 1886 devised means to detonate it. The French armies used it under the name of *melinite*. In England, picric acid was first tested at Lydd in Kent and manufactured under the name of *Lyddite* especially by Read Holliday for use in the Boer war (1899–1901). *TNT* (trinitrotoluene), which was developed in Germany, came into its own after 1914.

Because of these developments, the 1870–71 Franco-Prussian war was to be the last European conflict in which gunpowder would still be used. In the United States, smokeless powder was widely adopted by military authorities after the war with Spain (1898).

Dynamite for strictly civilian use was first manufactured in California. In 1880, Repauno Chemical Co., a joint association of Du Pont with Laffin & Rand, began its production at Gibbston, New Jersey. Previously, gunpowder producers in the United States had set up an association in 1872 to ward off a price war. But the agreement did not last, and Du Pont gradually wiped out all competition. The federal government broke up its monopoly on June 13, 1913, ordering the formation of three separate entities: Du Pont, Hercules Powder Co., and Atlas Powder Co.

The development of synthetically produced *nitrate explosives* significantly affected the life of nations. It boosted major public works and mining programs throughout the world. It provided the armies and navies of different countries with increasing striking power. It created fresh demand for a number of important feedstocks such as concentrated sulfuric and nitric acids, glycerin, wood cellulose and cotton linters (residues left after ginning), as well as aromatic derivatives such as toluene and phenol. It prompted the birth, under Nobel's aegis, of one of the first multinationals and established E. I. du Pont de Nemours' leadership in the United States, in explosives and, subsequently, in chemicals. Finally, it paved the way for new applications of nitrocellulose and other cellulose substances.

Soda Plants: The Solvay Ammonia Soda Process—Electrolytic Alkali

The Leblanc process for sodium carbonate production had serious environmental drawbacks, such as hydrochloric acid fumes and accumulation of calcium sulfide (black ash) waste on factory floors. Under the 1863 Alkali Act, it became compulsory for English soda manufacturers to recover most of the hydrochloric acid produced in their plants. Weldon's process in 1866, then Deacon's in 1868, improved in 1880 by one of Bunsen's pupils, F. Hurter, converted the acid to chlorine by air oxidation over a catalyst.[11] Chlorine then acted on lime to produce bleaching powder. But in 1867, English soda plants

[11] By the reaction: $4HCl + O_2 \longrightarrow 2H_2O + 2Cl_2$

were still emitting 45 percent of their acid into the atmosphere because the market for chlorine was not sufficiently large.

This waste of acid was curtailed when it was found that bleaching powder could be used to process North Africa's esparto grass, used in the manufacture of paper. Production of bleaching powder in England then soared to 150,000 tons by 1880. The recovery of sulfur from black-ash waste was achieved in Widnes, in Cheshire, England, by Ludwig Mond and then improved by A. M. Chance in 1882. The equipment used in the Leblanc process was also improved, especially for the preparation of sulfuric acid (larger lead chambers, Glover and Gay-Lussac Towers). But despite these improvements, Leblanc soda was faced with a formidable challenge when the *ammonia-soda process* appeared.

Many scientists, chief among them the physician Auguste Fresnel in 1811, had thought of turning salt into sodium bicarbonate using ammonium bicarbonate.[12] All the chlorine could then be absorbed by lime, the only residue being calcium chloride.[13] James Muspratt at Saint-Helens, followed by Henry Deacon failed in their attempts because of technological difficulties. Commercial use of the process was achieved only when gas plants provided sufficient ammonia liquor at reasonable prices.

Two Frenchmen, the chemist T. Schloesing and the engineer R. Rolland, succeeded in 1854 in setting up at Puteaux, near Paris, a continuous production plant for ammonia-soda. But the French administration refused to give up its salt tax, thus rendering the process uneconomical. The factory was forced to shut down in 1858 after all the capital had been exhausted. It was not until 1872, when Parliament voted to abolish the excise tax, that Daguin was able to apply the Schloesing and Rolland process in a plant located at La Madeleine near Nancy.

Meanwhile, Belgium's Ernest Solvay obtained a tax exemption in his country on the salt used in the ammonia process he had patented April 15, 1861. Solvay, although a self-taught man, had acquired some practical experience in this area through his father who was a saltmaker and his uncle, Louis Semet, who was the manager of a gas plant near Brussels. With the help of his brother, Alfred, who afterwards was to work with him, and with his family's financial backing, he set up a plant in Couillet, near Charleroi, and began producing ammonia soda in 1864. During the early years Solvay's perseverance was sorely tried. But by applying his practical mind to the design of new equipment (the Solvay gas-liquid counter-flow mixing pipe, an improved lime

[12] The reactions can be represented this way:

$$NaCl + NH_4HCO_3 \longrightarrow NaHCO_3 + NH_4Cl$$

$$2NaHCO_3 \longrightarrow Na_2CO_3 + H_2O + CO_2$$

[13] $2NH_4Cl + Ca(OH)_2 \longrightarrow CaCl_2 + 2H_2O + 2NH_3$

Ernest Solvay (1832–1922). Courtesy Solvay.

oven, a rotating filter) and by watching with painstaking care the different reaction sequences, he managed perfectly to master the production process.

World production of sodium carbonate, amounting to 150,000 tons in 1863, entirely from the Leblanc process, was to soar to 1,760,000 tons in 1902, of which 1,616,000 tons were from the ammonia-soda process. The price of soda fell by two-thirds in the interval because of the fuel savings and the continuous production inherent in the ammonia system.

In 1873, a Solvay plant was built in Dombasle near Nancy. The previous year, Ludwig Mond, who, although lacking diplomas, had worked with Bunsen in Heidelberg before settling down in Widnes, had started negotiating with the Solvays. Associating with a Swiss accountant, J. Brunner, from Hutchinson Chemical Works, which was using the Mond process for sulfur recovery from black-ash waste, he started producing Solvay soda near Widnes in Cheshire in 1874. By the end of the century, Brunner, Mond & Co. bought up and closed down the plants of all their English competitors in the field. The era of Leblanc soda works was coming to an end. Before admitting defeat, however, Leblanc soda producers set up in 1890 the United Alkali Co., a merger of forty-eight plants of the Tyne, Scotland, Ireland, and Lancashire totaling 12,000 workers.

Despite its size, the United Alkali Co. soon ran into trouble. Cellulose pulp manufacturers refused to deal with a single bleaching powder supplier. The main export market faded away as protectionist tariffs were set up in the United States. Competition from ammonia-soda barely allowed for profit margins and, accordingly, prevented the upgrading of already obsolete soda plants. The causticizing and production of chlorine that were to improve the running of Leblanc plants were jeopardized by the development of electrolytic processes that produced both chlorine and very pure caustic soda. And the development in Germany of the contact process hindered sales of sulfuric acid produced in lead chambers.

The merger of soda plants within United Alkali, in fact, only postponed what was an unavoidable decline. England suffered from this decline more than other industrial nations because of the importance which Leblanc complexes had acquired in the country's chemical industry. Germany had not invested heavily in this type of soda plant and so managed very easily to direct its activities towards the production of other inorganic chemicals, more particularly towards sulfuric acid and ammonia-soda using the Hönigmann and Solvay processes, and toward electrolytic soda, which was finding growing markets in the organic chemicals industry then in full swing.

The same developments took place in the United States, where industry had long relied on natural alkalis or on imported soda and easily adapted to the new technologies. The first ammonia-soda production was brought on stream in Syracuse, New York, by the Solvay Process Co. in 1884. The Solvay brothers had joined up with the Americans William Hazard and Rowland Cogswell for the purpose.

Eight years later, two competitors emerged with the help of English technology. Michigan Alkali Co. was founded in Wyandotte, Michigan, by J. B. Ford to provide his two plate glass factories with adequate soda supplies. At the end of the century, the founder of the Pittsburgh Plate Glass Co., John Pitcairn, also decided to integrate soda production into his company and set up the Diamond Alkali Co. in Barberton, Ohio. By 1910 he was supplying other glass manufacturers.

With the help of his uncle Louis Semet, Ernest Solvay had also set about supplying his soda plants with ammonia. He had developed a system of coke ovens making it possible to recover ammonia liquors as well as tar distillation by-products. In the United States the Semet Solvay Corp., which was to become part of Allied Chemical Co., developed these ovens and by 1913 was treating up to 10 million tons of coke a year.

The development of *electrolytic soda*, which completed the ruin of Leblanc soda plants, was essentially the work of the Germans and Americans. As we have seen, as early as 1807, Davy had isolated sodium by passing an electric current through molten soda, while Faraday had established the quantitative laws of electrochemistry. To electrolyze a salt solution in an economically acceptable manner, however, two conditions had to be fulfilled. The first was to have cheap current available; this became possible through Germany's Werner Siemens' work on electromagnetism and the invention of the dynamo by Belgium's Zenobe Gramme between 1866 and 1872. The second condition was to prevent the caustic soda that was formed at the cathode (together with hydrogen) from reacting with the chlorine that appeared on the anode or with the electrolyte. The Duisburg firm of Matthes & Weber solved the problem by placing a porous membrane between the electrodes, thus preventing the mixing of the two products. Their diaphragm cell was patented in 1886.

A chemist, Ignatz Stroof, manager of the Griesheim plant in Frankfurt, fearing competition in the dye field from neighboring Hoechst, formed an association with Matthes & Weber in order to diversify, bought up their patent, and developed a continuous electrolysis process which he started at Bitterfeld in 1894. Potassium lye was the first alkali he produced, for it gave higher yields than caustic soda and could be sold at a higher price, especially for indigo synthesis. Griesheim Elektron, as the company was to become in 1891, was soon to grow into Germany's foremost electrochemicals producer.

While Ernest A. LeSueur was improving the diaphragm cell in the United States, another American, Hamilton Castner, was developing in 1892 in Europe a mercury cathode cell in which the sodium was trapped in an amalgam that subsequently reacted with water to produce caustic soda. Castner, who was manager of Aluminium Co. in Oldbury near Birmingham, also invented the graphite electrode. Austria's Carl Kellner, in a separate move, came out with a mercury cell at about the same time. Solvay could not remain indifferent to such developments, and in 1895 he was the driving force behind the Castner-Kellner association in which he had a share. Production of electrolytic soda by the mercury process could then begin in England in a unit of the Aluminium Co. at Runcorn in 1897, later in Germany in Deutsche Solvay's Osternienburg plant. The joint production of chlorine and high purity caustic soda by electrolysis went directly against the interests of United Alkali, the more so as each ton of chlorine produced electrolytically used up 1.6 tons of sodium carbonate produced in a Leblanc soda factory.

Industry leaders in the United States had perceived very early the advantages of electrolysis. Thomas Mathieson set up a cell based on the Castner process at Saltville, while Herbert Dow in Midland, Michigan, was producing bromine and electrolytic soda for his own needs, without any outside technical help, setting the stage for his founding of the Dow Chemical Company.

Taking advantage of cheap hydroelectric energy, Mathieson set up electrolysis units at Niagara Falls in 1897. In 1906, John Hooker, founder of the Hooker Electrochemical Co., followed suit. At the same time Pennsylvania Salt Manufacturing (Pennsalt) was setting up diaphragm cells in Wyandotte in 1908 with the aid of England's Arthur E. Gibbs.

The success of brine electrolysis can be explained by the many advantages it offered: the necessary investments were low and units could easily be tailored to customer requirements. There were no by-products, and very pure caustic soda was produced. Also, chlorine demand continually increased as new uses for it were found in such areas as paper making and water treatment.

Phosphorus and Its Use for Matches

Phosphorus as an element was isolated for the first time in 1669 by Hennig Brandt, a Hamburg alchemist, from phosphates contained in urine. The phenomenon of phosphorescence, and the ability of white phosphorus to flare up spontaneously when in contact with air, were matters of wonder in olden times. But it was only in 1785 that Sweden's Carl Scheele published a method for preparing this element by reducing with coal the calcium phosphate contained in bone ash in a proportion of 27% to 37%.[14]

The "phosphorus candle" appeared in France in 1780. It consisted of a sealed glass vial containing a strip of paper coated with phosphorus that ignited when the container was broken. The first Lucifer strike-anywhere *matches* were developed by Charles Sauria, a Frenchman, who failed, however, to take out a patent for his invention. His matchheads contained white phosphorus, sulfur, and potassium chlorate.

White phosphorus was produced industrially as early as 1838 by the firm of Veuve Dupasquier et Coignet in La Guillotière near Lyon, who were already treating fresh bones to extract the gelatin. Subsequently they manufactured matches until a state monopoly was established in 1872.

England, however, which imported phosphorus from the continent but had available resources of cheap coal, was in a good position to manufacture

[14] The reaction involving sulfuric acid is as follows:

$$2Ca_3(PO_4)_2 + 6H_2SO_4 + 10C \longrightarrow P_4 + 6CaSO_4 + 10CO + 6H_2O$$

matches. In fact, a Quaker, Arthur Albright, who since 1842 had been associated with the Sturges Brothers in the production of potassium chlorate in Birmingham, persuaded his partners to open a white phosphorus workshop near Oldbury. They began production in 1851, using bone ash or mineral phosphates. The unit was ideally located, as its coal supplies and acids were provided by the Chance Brothers' adjoining factory.

The Sturges, who were not confident about the future of phosphorus, separated from Albright three years later. With his Quaker friend, John Edward Wilson, Albright then set up in 1855 the firm of Albright & Wilson, which was to play a major role in its chosen field. Its production of white phosphorus soared from 54 tons in 1855 to 446 tons in 1881, and 4,500 tons in 1892.

Although the white phosphorus in a matchhead did not exceed 5 percent, the substance could cause necrosis of the jaw when handled. A workwomen's strike which broke out in 1888 at the Bryant & May match factory over the inherent danger of necrosis has remained famous in the social annals of England.

In 1841 Berzelius, who had described *allotropy* as the phenomenon of an element's being capable of existence in different forms, observed the transformation, under heat, of white phosphorus into an amorphous and stable *red phosphorus*. Four years later, Anton von Schrötter from Vienna developed a process for preparing red phosphorus by gradually heating white phosphorus in a cast-iron pot in an inert environment. Albright acquired rights on the process, improved it by reducing the dangers of explosion, and even licensed it out to Coignet.

Following research carried out by Sweden's A. Pasch and Germany's W. Bottger, the Lundstrom brothers from Jönköping, Sweden, succeeded in 1855 in producing a "safety match," which contained no phosphorus and which ignited when scraped over a surface coated with red phosphorus, crushed glass, and glue. These boxes of *Swedish matches* provided red phosphorus with its first market. At the same time, white phosphorus was used less and less until it disappeared almost totally on the eve of the 1914 war in favor of phosphorus sesquisulfide (P_4S_3) which J. Sevenne and G. Cahen from Paris had patented in 1898 for incorporation in matchheads.

When the dynamo and cheap electricity became available, phosphorus production took a new turn. Three Englishmen from Wolverhampton, W. Readman, G. Parker, and R. Robinson, patented in 1888 a process which used an electric oven to produce phosphorus in a continuous and particularly economic manner.[15]

[15] According to the reaction:

$$2Ca_3(PO_4)_2 + 6SiO_2 + 10C \longrightarrow P_4 + 6CaSiO_3 + 10CO$$

Faced with this new competition, Albright & Wilson bought up the company using the process and set up a phosphorus oven in Oldbury. Production, which hovered around 200 tons a year in 1902, was mainly geared to making matches and was halted only in 1920. Albright & Wilson had also set up a subsidiary in 1897 at Niagara Falls to take advantage of cheap electric current. What used to be the Oldbury Electro-Chemical Company was subsequently bought up by Hooker. But it was to be more than thirty years later that the electric oven process really took off, when fresh uses for phosphorus derivatives appeared.

Chemistry at the Frontiers of Metallurgy: The Adventure of Aluminum

Because of its great affinity for oxygen, aluminum cannot be obtained by reducing its oxide, unlike other metals such as iron, copper, lead, or zinc. When Hans Christian Oersted from Denmark isolated aluminum in 1825, followed two years later by Friedrich Wöhler, it was by reacting anhydrous aluminum chloride ($AlCl_3$) with potassium. One of Louis Thénard's students, Henri Sainte Claire-Deville, was the first to devise an industrial process based on this principle. Graduating at twenty-five with a doctorate in both medicine and science, he worked on the subject at the École Normale Supérieure, where he succeeded Balard as chemistry professor in 1851. In 1854 he managed to improve the Wöhler process, by replacing potassium with less-costly sodium and $AlCl_3$ by a double salt ($NaCl$-$AlCl_3$) that made the separation of aluminum easier. Sainte-Claire Deville obtained the alumina required to prepare the $AlCl_3$ by treating bauxite ore[16] with a 50 to 65 percent Al_2O_3 content with sodium carbonate in high temperature furnaces.

Compagnie des Produits Chimiques d'Alais et de la Camargue became interested in aluminum obtained by this process. The company had been founded in 1855 by Henri Merle in partnership with the inventor of ultramarine blue, Jean Baptiste Guimet. On the advice of J.B.A. Dumas, Henri Merle, who was from Alais in the Gard region of France, had established a Leblanc soda plant at Salindres, not far from the Camargue salt marshes and from sources of coal and limestone. As there was also bauxite ore close by, in 1860 he brought on line the Sainte-Claire Deville Processes for alumina and aluminum.

The high price of sodium still made the new metal expensive. Despite backing from Napoleon III, who introduced aluminum cutlery in the Imperial Court, its use remained limited to luxury items for thirty years, and production was restricted to two sites, Salindres and then Oldbury, where Castner, the American, was to work from 1888 for the Aluminium Company in England.

As it had done for phosphorus, cheaper electricity set off a revolution in

[16] So called because it was discovered in the Baux de Provence region.

aluminum manufacture by bringing down production costs. Almost simultaneously, in 1886, a Frenchman Paul Héroult and an American Charles Hall, aged 23 and 22 respectively and both destined to die the same year (1914), invented and patented, each in his own country, the electrolytic process that is still used today to produce aluminum. The method consists in passing a current of several hundred amperes through a vat containing purified alumina dissolved in cryolite.[17] The molten aluminum forms a deposit on the graphite coating of the steel vat, which acts as a cathode. At the same time oxygen is formed on the anode made of carbon blocks which are burned away in this process.

In a parallel move, the German chemist K. J. Bayer, who worked in a Russian factory, patented in 1887 a process to purify alumina by caustic soda acting at high temperature on bauxite, and $A1_2O_3$ precipitating from sodium aluminate solutions, a process that was to prove far more economical than Sainte-Claire Deville's.

Oddly enough, the Compagnie des Produits Chimiques d'Alais et de la Camargue was not immediately interested in such significant technical progress. After Merle's death in 1877, the company was managed by Alfred Rangod. It later became A. R. Péchiney et Cie. (Péchiney was Rangod's father-in-law.) Rangod, who had been a pupil of Pelouze, was a demanding boss. He was feared but held in high regard. Faced with competition from Solvay soda, he diversified his business into chemicals. Besides alumina, he produced caustic soda and chlorates, as well as copper sulfate, which was being increasingly used as a fungicide for grapevines.

Unable to convince Rangod, Héroult licensed out his process first to Société Métallurgique Suisse, from which was to emerge Aluminium Suisse, established in Schaffouse in 1887; then to Société Métallurgique de Froges the following year, and finally to British Aluminium, which set up a plant in Foyers, Scotland. Once initial difficulties were overcome, world production of aluminum by electrolysis grew to 1,000 tons in 1893 and its price fell from 87.50 F/kilo in 1886 to 3.95 F/kilo in 1895.

Péchiney, accordingly, was forced to cease production at Salindres in 1890, and Aluminium Co. had to turn to other kinds of production, such as cyanide (NaCN), used for the flotation of ore in the Rand's gold mines. As for Charles Hall, he was involved in setting up the Pittsburgh Reduction Co. in 1888, which was to become Aluminum Company of America and make him a millionaire.

In 1897, Péchiney finally made up his mind to make a comeback on the aluminum market, buying back the Calypso unit in the Maurienne Valley from Société Industrielle de l'Aluminium, which was operating under the Hall license. But it was not until 1907 that the Salindres plant started producing

[17] A sodium aluminum fluoride-based mineral found in Greenland.

alumina purified through the Bayer process. It continued to use also the Sainte-Claire Deville process, which was abandoned only in 1923.

As demand grew, all five producers of aluminum by electrolysis—an Englishman, an American, a Swiss, and two French—had gotten together in 1901 to divide the market. Their agreement did not survive the 1908 slump, when aluminum prices fell from 4 F to 1.50 F/kilo. From then on, only the French producers managed joint sales from a single establishment set up in 1912, called L'Aluminium Français.

The Electric Furnace

As has been seen, electricity was applied to chemicals in two different ways. Through electrolysis, components of a molten or dissolved substance could be dissociated and sodium, aluminum, chlorine, soda or caustic potash, for instance, extracted. Alternatively, electricity made high temperature fusion of certain components possible with no intervention in the chemical reaction. These components could then interact, as in the dry preparation of phosphorus.

The furnace developed by Héroult in 1887 extended the field of electrothermal applications. With the electric furnace, both a Frenchman Henri Moissan and a Canadian T. L. Willson, produced *calcium carbide* from coal and lime in the same year, 1892. Moissan had studied pharmacy before becoming Edmond Frémy's student at the Musée d'Histoire Naturelle. Moissan not only experimented with calcium carbide, but also discovered, in 1886, a method to isolate the element *fluorine* through electrolysis of potassium fluoride solutions in anhydrous hydrofluoric acid, which won him the Nobel chemistry prize in 1906.

Willson was the one who perceived most clearly the industrial potential of calcium carbide. In 1895, he sold his process and patents to Acetylene Illuminating Co., an English firm whose plant near Foyers, close to the British Aluminium Company, soon produced as much as 500 tons per year of calcium carbide. The reaction of CaC_2 on water, observed by Wöhler in 1862, produced *acetylene*. Acetylene quickly proved its usefulness for lighting purposes (miners' lamps, bicycle and automobile lights) as well as for welders' oxyacetylene torches.

In 1898, Germany's Fritz Rothe, working for Adolf Frank and Nikodem Caro, who patented the process in 1902, demonstrated that nitrogen passed over calcium carbide in an electric furnace heated to 1,000°C, produces *calcium cyanamide*.[18] The importance of calcium cyanamide is that it was the first fully synthetic fertilizer ever achieved.

[18] According to the reaction:

$$CaC_2 + N_2 \longrightarrow CaCN_2 + C$$

The first Héroult furnaces for aluminum production at Froges, 1890.

Unlike the Germans, the English had thought that coal would not be competitive with hydroelectric power for the production of cheap electricity. Hence they ceased producing calcium carbide by 1914, choosing instead to receive their supplies from Scandinavia. This caused them to fall very far behind their German competitors in the emerging and important field of acetylene derivatives (butadiene, vinyl acetate, vinyl chloride, acetaldehyde).

In 1891 the electric furnace served industry in many other ways when Edward G. Acheson managed to manufacture *silicon carbide*, known for its abrasive properties under the trade name Carborundum. He also developed graphite electrodes.

Chemicals and Photography

The effect of light on silver halides had been observed since 1727; but it was only in 1816 that the French physicist Nicéphore Niepce, experimenting with silver chloride, succeeded in fixing a negative image in a dark room. His partner, the theater producer Louis Daguerre, improved the process in 1839 by exposing a silvered copper plate to iodine vapor. The latent image, and, together with it, photographic art were born.

With ''daguerreotypes'' only a single, fragile print could be made. With the Calotype invented by England's Fox Talbot it was possible to obtain several positive prints from a single negative. Increased speed was obtained in 1850, when a glass plate coated with a nitrocellulose collodion was introduced. But the decisive step forward was made with the gelatin-silver bromide dry plate which produced instant prints and opened up photography to the public. It still remained to proceed from a plate to a film.

England's Alexander Parkes had developed a semisynthetic plastic in 1855 consisting of nitrocellulose and camphor. What was to be known as celluloid was used by Goodwin to prepare a photographic film in 1887. Two years later in the United States, George Eastman, who had started producing dry plates in his Rochester, New York, workshop in 1880, brought out a nitrocellulose-based transparent film on a roll and, at the same time, put on the market his portable Kodak camera. Thus it was that fifty years after the daguerreotype was invented, photography turned into a popular art. Chemistry was to find growing outlets for its products in the photographic trade, both for supporting media and for the photosensitive surfaces themselves.

Pierre Wittmann, for one, had bought a drugstore in 1845 in Paris' Marais area. Thirteen years later, he handed it over to his son-in-law Étienne Poulenc, an apothecary, who set about making photographic chemicals such as collodion, gelatine, bromine plates, sodium thiosulfate, silver iodide, and silver bromide as well as iron salts, which his company manufactured in Ivry. In 1900 it became the *Établissements Poulenc Frères*. At the same time, Étienne Poulenc was selling salts for pharmaceutical uses, such as bismuth salts, citrates, valerianates, and calcium phosphates.

A number of German firms that also dealt in both photographic and pharmaceutical products were themselves in line for a brilliant future. Such was the case of *Riedel de Haen*, an offshoot of the company founded in 1861 in Hanover by Eugen de Haen, who had been a pupil of Karl Remigius Fresenius and of Robert Bunsen; also of *Schering AG*, which was set up in 1851 by Ernst Schering under the sign of "The Green Drugstore."

After W. H. Vogel of Berlin had shown in 1873 that the presence of dyes in the silver halide emulsion improved picture reproduction, *AGFA* set about making organic products for the development of photographic plates and films, then, in 1909, set up the Film Fabrik in Wolfen near Bitterfeld, contributing to Germany's fame in that area. In the United States, well before *Eastman Kodak* had won an international reputation for its technology and marketing expertise, a firm founded in 1867 by the three *Mallinckrodt* brothers in Saint Louis had been supplying mineral salts to the fledgling photographic industry.

Further progress was made in photography around the turn of the century. In 1908, cellulose acetate film began replacing nitrocellulose film, which was too flammable. At the same time, two Frenchmen, Auguste and Louis Lumière, the movie pioneers, were producing their "autochrome" plate of color photography in the Lyon-Monplaisir plant of *S.A.A. Lumière Fils*. For his part, Belgium's bakelite inventor, Léo Baekeland, had been manufacturing since

George Eastman (1854–1932) and Louis Lumière (1864–1948).

1893, at the *Nepera Chemicals Co.* in Yonkers, his Velox paper which made it possible to develop film in artificial light. George Eastman acquired the rights to it for one million dollars in 1899.

The Birth of the Plastics Industry

As often is the case in chemistry, plastics too were first natural, then artificial, that is semisynthetic, before finally becoming fully synthetic. For a long time natural horn was used to make combs and fancy goods; *guttapercha*, a gum from the Malayan peninsula, was used for insulating; and reinforced *shellac* served to make daguerreotype boxes and phonograph records. It was *nitrocellulose*, however, already found useful as gun cotton in explosives and as collodion in photography that, thanks to the experiments of England's Alexander Parkes, formed the basis of the first semisynthetic plastics to be made. Although he was not a chemist but worked as an apprentice in Birmingham's steelmaking industry, Parkes had already invented the cold vulcanization process for rubber when he started showing an interest in nitrocellulose. In 1862 he presented at London's International Exhibition, items molded from a mixture of nitrocellulose, alcohol, camphor and vegetable oils, trade-named Parkesine or Xylonite. Parkes was no businessman, though, and despite the efforts of his partner, Daniel Spill, Parkesine's early days were hard ones.

In the United States, John Wesley Hyatt, who was looking around for a cheap substitute for ivory for making billiard balls, heard of the work being done in England by Parkes and Spill. He developed a formula that contained only nitrocellulose and camphor and, in partnership with his brother Isaiah, who named the resulting product *Celluloid*, set up in 1872 the *Celluloid Manufacturing Company* in Newark, New Jersey. Hyatt was a mechanic; his successful development of the new plastic was in large part owing to the improvements he brought about in the molding and manufacturing processes. His experience was to be of service to the English inventors. *The British Xylonite Company*, which Daniel Spill founded in 1877 with an American partner, L. P. Merriam, and which was to become subsequently BX Plastics, did very well, employing 1,160 people by 1902. Celluloid, which was used for collars and cuffs, knife handles, photographic and then cinematographic films, and billiard balls, experienced its peak years around 1920. Today it has completely disappeared. The main reasons for the decline of celluloid were its highly flammable nature and the high cost of camphor, which for a long time remained a Japanese monopoly.

In the United States, celluloid was the starting point of many a large chemical firm's development in plastics. Hyatt's main rival, Arlington's Cellonite, was bought by Du Pont in 1915, whereas Celluloid Manufacturing Co., which became Celluloid Co., was taken over by Celanese in 1927 and Fiberloid Co. was absorbed in 1938 by Monsanto.

Bakelite, which emerged on the American market in 1909, was the first truly synthetic plastic. Its inventor, a native Belgian, Léo Hendrik Baekeland, had obtained a Ph.D. in chemistry at the University of Ghent, where he had taught in 1887 before emigrating to the United States. He worked for a while for an American supplier of photographic articles, then set up his own business, developing Velox paper. With the large sum of money he netted from its sale to George Eastman, he applied himself to studying in his Yonkers, New York, laboratory the substitution of shellac by a synthetic resin produced from the polycondensation of phenol on formaldehyde. This type of reaction had already been described by von Baeyer and ter Meer. In 1900, England's James Swinburne, who was trying to obtain insulating material, was offering phenol-formaldehyde resins prepared by his company, Damard Lacquer Co., for want of the moldable resin he was unable to produce.

Léo Hendrik and Céline Baekeland with their children, Nina and George, on an outing in Yonkers, New York, about 1900. Baekeland was a keen motorist. Courtesy Leo Baekeland Collection, Archives Center, National Museum of American History, Smithsonian Institution, photo no. 84-11359.

Baekeland's merit was to determine under what conditions of temperature and pressure, and with what catalyst, phenol-formaldehyde liquid resin—which he named Novolak—turned, in the autoclave when the appropriate charges were added, into a thermosetting molding powder, which naturally he called bakelite. The General Bakelite Company, established in Perth Amboy, New Jersey, in 1910, began producing the new resin the following year. After buying the rival company, Redmanol Products, it changed its name to Bakelite Corporation in 1922. In 1931 it set up its plant in Boundbrook, New Jersey, and became a division of Union Carbide Corporation eight years later. Meanwhile, Bakelite Corporation had expanded in Europe, particularly in England, where Bakelite Ltd. bought up Damard in 1927; in Germany, where well before 1914 Bakelit GmbH was functioning as a subsidiary of Rütgerswerke; and in France, where the Bakelite Française plant was installed in Bezons, near Paris. Although opaque and dark-colored, the new plastic was easily molded and had insulating properties; so it found ready markets in the growing telephone, automobile, and wireless industries.

In contrast to George Eastman, the self-made man who had created and built up the powerful Eastman Kodak Company through his brilliant business acumen, then had deliberately ended at seventy-eight a life that had been austere and dedicated to good works, Léo Baekeland was an inventor attracted more by discoveries than by industrial enterprises. He fully enjoyed the fortune gained through his innovations and until his death at eighty, led a sportive and sumptuous life.

Artificial Fibers

Like plastics, textile fibers were for long exclusively natural products. Artificial silk, produced from cellulose,[19] only appeared at the end of the nineteenth century; and fully synthetic fibers finally came into their own, with Nylon and Perlon, on the eve of World War II.

While trying to prepare a carbon filament for electric lightbulbs with his fellow countrymen C. H. Stearn and C. F. Topham, England's Joseph Swan developed and patented in 1883 a process to produce nitrocellulose filaments by coagulating collodion threads. But it was the Frenchman Louis Marie Hilaire Bernigaud, Comte de Chardonnet, who applied the discovery to textiles. A former pupil of the École Polytechnique, Chardonnet was a passionate experimenter. His interest in textiles dated from the time when Pasteur had studied the silkworm disease that threatened Lyon's silk industry. At the 1889 Paris Exposition Universelle, he exhibited the first dress made from *nitrocellulose* thread. Having devised a way to partially hydrolyze cellulose nitrate to make it less flammable, he began producing in 1891 artificial silk (*soie Chardonnet*)

[19] The term "rayon" to designate cellulose-based artificial fibers was to be in general use from 1935 on.

Louis Marie Hilaire Bernigaud, Comte de Chardonnet (1839–1924).
Courtesy Bettman Archives.

in a factory he had set up in Besançon. Société Tubize later acquired sole rights to his process.

England's Edward Weston was also looking for a filament for light bulbs when he dissolved cellulose in an ammoniacal copper hydroxide solution (Schweitzer solution) and precipitated it with sulfuric acid. Here again, application to textiles did not take place until France's L. H. Despeissis had published his work in 1890 and, especially, until the two Germans, Hermann Pauly and Bronnert, had brought about their improvements. Industrial production of *cuprammonium fiber*, known as "Bamberg silk," began in Germany in 1898.

A third bulb-filament process was developed in 1892 by Charles Cross and Edward Bevan in England. They called *viscose* the syrup obtained by treating cellulose xanthate[20] with dilute caustic soda. The viscose process became industrially viable only in 1905, after Topham had made a great number of improvements, both chemical, in the aging of the solution, and mechanical, in extrusion-drawing technology. In the same year Samuel Courtauld's textile firm, founded in 1816, acquired the English rights to the Bevan and Cross process and began making the first viscose threads in Coventry. An American subsidiary in Marcus Hook, Pennsylvania, under the name of American Viscose Company, followed in 1911. The English company became Courtaulds Ltd. in 1912, and after World War I used its patents to set up viscose production in France, Germany, and Italy as well as in Canada.

Inspired by Hilaire de Chardonnet, French industry leaders became involved in artificial fibers very early. Ernest and François Carnot, descendants of Lazare Carnot, founded the Viscose Suisse in Emmenbrück in 1905 while at the same time setting up a spinning mill at Arques la Bataille, near Dieppe. Edmond Bernheim brought his own production on stream in Givet in 1906. As for Joseph Gillet's sons, Charles, Edmond, and Paul, who were established in Izieux and whose dye and finishing business at Quai de Serin in Lyon had prospered, they joined with the Carnots and E. Bernheim to establish the Comptoir des Soies Artificielles in 1911, the basis of what was to become after the war the Comptoir des Textiles Artificiels (CTA).

The acetylation of cellulose with acetic anhydride had been studied by France's B. Naudin and Paul Schützenberger as early as 1865. But to Charles Frederick Cross and Edward John Bevan must be given the credit for patenting in 1894 the first *cellulose acetate* industrial production process.

In the United States, George Miles succeeded in 1903 in dissolving cellulose acetate in acetone rather than chloroform by partially hydrolyzing it. This made it easier to use. The previous year, Arthur D. Little in Boston, Massachusetts, who was already using cellulose acetate to insulate electric wire, had taken out a patent for its application as a textile fiber.

The industrial success of this new artificial fiber, which is less hygroscopic than viscose, was largely due to the two Basel brothers, Henri and Camille Dreyfus. Using Miles's acetone process, they had been producing since 1910 the acetate photographic film which they supplied to Pathé Frères. They began in 1916, in their British Cellulose and Manufacturing plant in Spondon, England, to make acetate lacquers that were used to coat aviation

[20] Made from carbon disulfide reacting with alkali cellulose.

cloth. When the war ended demand dropped and alternative uses for acetate fiber were strongly promoted by British Celanese, which succeeded British Cellulose. (Courtaulds Ltd. took it over in 1957.) World production of rayon fiber peaked in 1937 at 536,000 tons. By contrast, only 12,000 tons had been produced in 1913, Chardonnet's nitrate silk accounting for half that amount, cuproammonium Bamberg silk for one third, and viscose for the balance. One-third of the total was made in Germany, particularly by the large Elberfeld plant of Vereinigte Glanzstoff Fabriken.

Viscose, which was cheaper to produce, soon drove nitrate silk and cuproammonium silk out of the market. By 1924 they only accounted for 8 percent and 1 percent, respectively, of artificial fiber production. The popularity of viscose and acetate rayon pushed up consumption of carbon disulfide and acetic anhydride and, of course, of wood pulp and cotton linters,[21] which were also in increasing demand from the paper, explosives and photographic industries.

The Invention of Cellophane

As we have seen, it was possible to produce from nitrocellulose and cellulose acetate both films and fibers. C. H. Stearn, who with F. Topham had contributed to the industrialization of the viscose process, took out a patent in 1898 covering the preparation of films from regenerated cellulose. But it was Switzerland's Jacques Edwin Brandenberger who developed in 1911 a machine for the continuous production of these films.

Brandenberger was a textile chemist who worked in a dyeing and finishing factory in Thaon les Vosges which the Gillet brothers from Lyon later merged in the Société Gillet-Thaon. What Brandenberger wanted was to protect tablecloths from stains and dust by coating them with collodion. He prepared a transparent coating which he applied to the cloth and in the process became interested in the viscose film itself which he called *cellophane*. With Gillet's participation, the Société la Cellophane was set up in Bezons and started production in 1920. Three years later it sold its rights in the United States to E. I. du Pont de Nemours, and the first cellophane sheet came out of their Buffalo plant in 1924. The spectacular development of cellophane did not really take place until Du Pont's W. Hale Charch managed to make it moistureproof in 1927, making it suitable for cigarette and food packaging.

Paints, Pigments, and Varnishes

Paints, around 1850, still contained linseed oil mixed with white lead and diluted with oil of turpentine. They were more often made by housepainters

[21] Short fibers.

than by actual paint manufacturers. *White lead* (basic lead carbonate) was prepared by oxidation of lead according to an ancient Dutch process. In 1853, England exported 3,000 tons of it, half of it to the United States. In 1869 in England, H. I. Thomas and R. W. Griffiths developed a less toxic pigment that, in contrast to white lead, did not turn yellow in a sulfurous atmosphere. J. B. Orr from Glasgow began manufacturing this *lithopone*[22] in 1872. The new white pigment was used not only for indoor paints but also in blends based on rubber, as well as in linoleum and oilcloth.

Blue pigments were usually either of Prussian Blue or Ultramarine Blue. *Prussian Blue* (ferric ferrocyanide), which a Berlin colorist had accidentally discovered in 1710, was brought to England in 1756 by Frankfurt's L. Steigenberger. His London firm was the starting point of the Lewis Berger & Sons paints factory.

Ultramarine Blue, which was extracted at great expense from lapis lazuli, became an industrial product when J. B. Guimet managed to prepare it in 1828 by mixing kaolin, sodium carbonate, coal, sulfur, and silica at a high temperature, and began producing it in Fleurieu near Lyon. To make sure that his flourishing business was kept properly supplied with sodium carbonate, Guimet entered into an association with Henri Merle in Salindres in 1852. Three years later he became one of the main shareholders of Compagnie des Produits Chimiques d'Alais et de la Camargue.

Another pigment, *chrome yellow* (lead chromate), was brought into use in the early part of the last century. Andrew Kurtz was producing it in Liverpool in 1835.

For a long time *carbon-black* remained a specialty of Germantown in Pennsylvania where it was produced from 1893 onwards through incomplete combustion of natural gas. Thanks to Godfrey Cabot's method for its preparation and Edwin Binney and Harold Smith's ways of incorporating it in other materials, the town retained a true supremacy in this area. The market for carbon-black, moreover, expanded considerably when English tire manufacturers, who had been blending it in tires to give them a pleasant black sheen, discovered in 1910 that it spectacularly increased their resistance to wear.

Iron oxide pigments became popular at the end of the nineteenth century because they were used in primers to protect metal structures such as bridges and ship bottoms.

Progress proved to be slower for binding agents. The heating of drying oils to produce standoil, blown Linseed oil, was a long process entailing considerable loss of substance. Natural resins (copal, kauri, shellac, and rosin), which were used to harden the paint surface, were consistent neither in quality nor in price. Several days were required to dry films even when litharge was used. Solvents were either alcohol or oil of turpentine. There was some prog-

[22] A complex of zinc sulfide and barium sulfate.

ress when an American chemist with the Celluloid Company produced in 1882 a less volatile and less hygroscopic nitrocellulose collodion through use of *amyl acetate* as a solvent. Then in 1910, *water-based paints* for home decorating were introduced in the United States based on casein powder.

Nitrocellulose and cellulose acetate coatings significantly increased their market only in 1923, when Du Pont brought out fast-drying lacquers for automobiles, based on low-viscosity nitrocellulose. Meanwhile, the readymade paint industry was developing rapidly because little capital and little technology were required. By 1900, there were no fewer than 400 paint and varnish manufacturing firms in England alone, employing 13,800 workers.

Perfumes and Fragrances

Fragrances and aromas, like many other entities, were first obtained from animals and plants; then their chemical composition was sought and their synthesis attempted. Finally, new substances were developed that could be produced at will.

Increasing amounts of raw materials provided perfumers with a wider range of compounds just when better living conditions in the Western world and new production technologies such as steam distillation and stills with greater capacities, were opening up wider markets. It was the twin growth of supply and demand that combined to turn perfumery from what had been a luxury craft into a full-fledged industry by the turn of the nineteenth century.

The use of perfumes had developed in Italy during the Renaissance period, as body odors were smothered under strong amber and musk scent. As hygiene improved at the beginning of the eighteenth century, lighter-scented waters replaced the heavy perfumes previously used. This was the time of *aqua mirabilis*, made up of 85 percent alcohol mixed with fragrant oils. The recipe, which has remained secret to this day, was obtained from an Italian by Jean Marie Farina, who was established in Cologne. This was the origin of *Eau de Cologne*. There were many fraudulent imitations of what was to become at the end of the eighteenth century the first widely sold perfume in Europe.

From then on, France, with its two poles—Grasse for the production of fragrances and Paris for perfumes—was to play a leading role in perfumery. Grasse developed in the seventeenth century, when its glovemaking industry received a boost from the fashion of perfumed gloves. This in turn greatly intensified the cultivation of fragrant plants such as carnations, roses, jasmine, and lavender. By the end of the following century, A. Chiris (1768) and J. B. Lautier (1795) were already established in Grasse as suppliers of essential oils. They were followed shortly afterwards by J. Roure, G. Bertrand, and F. Dupont in 1820.

Perfumers were also setting up shop in Paris, Europe's fashion oracle.

They included J. F. Houbigant (1775), P. F. Lubin (1798), Jean Marie Farina (1806), who was later to be taken over by Roger & Gallet, L. T. Piver (1813), P. F. P. Guerlain (1828), F. Millot (1839), and Aimé Bourjois (1840). While France was the leading country for natural perfumes and fragrances, it was outdistanced in *synthetic aromatic chemistry*. From 1870 onwards, research in that area was being conducted by chemists in other countries.

Although J. B. A. Dumas determined in 1833 the rough formula for α-pinene, the main component of turpentine oil, the study and separation of *terpenes*, from which many oils have been derived, was carried out by Sir William Tilden, Adolf von Baeyer, William Henry Perkin, and Ferdinand Tiemann. William Perkin *synthesized coumarin*[23] in 1875 before preparing in 1904 α-*terpineol* and *dipentene*.

An American student of Victor Meyer's, Francis Dodge, who was subsequently to join the family business of Dodge & Olcott, obtained citronellal in 1889 through partial synthesis of citronella from Java oil. The preparation in 1893 of α-*and* β-*ionones* from the citral in lemon grass was carried out by Tiemann, a pupil of Hofmann's. In 1888 O. Baur made a breakthrough in preparing the first *nitrated musks*. *Heliotropin* had been synthesized by oxidation of *isosafrol* in 1879, the same year that *vanillin* was obtained from isoeugenol, and *saccharin* was discovered inadvertently at Johns Hopkins University by an American, Ira Remsen, and a German, Constantin Fahlberg, while they were working on the oxidation of *o*-toluenesulfonamide.

An enterprising American, John Queeny, working with Meyer Brothers, importers of saccharin from Germany, joined forces with his friend, John Rossiter, and with Jacob Baur, a manufacturer of ingredients for soda drinks, to produce this first synthetic sweetener, with the technical help of Louis Veillon from the Swiss pharmaceutical group Sandoz. This was how Monsanto Chemical Company was born in 1902 in Saint Louis with an initial capital of $5,000. It was named after Olga Mendez Monsanto, the beautiful Spaniard John Queeny had married six years previously. In 1904 and 1905, two other Swiss nationals, Gaston Dubois and Jules Bebie, joined Veillon, and to the production of saccharin was added that of *vanillin*, then of *caffeine* (1905), of *chloral hydrate* and *phenacetin* (1907), *phenolphthalein* (1909), and *coumarin* (1914). But given the heavy competition from German firms, the business only really took off in 1914.

Although French chemists were not ahead in synthetic aromatics research, by the end of the nineteenth century they had used the new scientific breakthroughs to add to their range of natural oil products obtained through partial or total synthesis. In Paris this was done by Georges de Laire, in Grasse

[23] By the action of acetic anhydride and sodium acetate on salicylaldehyde, the Perkin reaction.

by F. Mane, P. Robertet, Lautier, de Roure, Bertrand, Dupont, and Méro & Boyveau.

In Switzerland, Léon and Xavier Givaudan, two brothers from Lyon, and Firmenich were breaking new ground in this area, making it hard for Germany's H. Frietsche, A. Fries, Friedrich von Heyden, W. Haarmann and Karl Ludwig Reimer to reign as supreme as they did in dyestuffs. Haarmann & Reimer, in which F. Tiemann and G. de Laire had a stake, was to be acquired by Bayer in 1954.

As for the long-established English firms, they were unable to measure up to French and Swiss companies. Stafford Allen & Sons (1833), W. J. Bush (1851), and Boake, Roberts & Co. (1869) remained modest in size and merged in 1964, under the aegis of Albright & Wilson, to become Bush, Boake, Allen (BBA).

French perfumers began incorporating the new synthetic aromatic raw materials in their compositions. L. T. Piver was the first to launch "Trèfle Incarnat" in 1907, which included *methyl salicylate* discovered in 1896 by the French chemist R. Darzens in his laboratory. At the time, L. T. Piver's president was Jacques Rouché, graduate of the École Polytechnique, who was to remain head of the Paris Opera House for thirty years. In his wake, François Spoturno from Corsica, better known as François Coty, who had worked with Chiris, brought out the perfume "Origan" in 1910. Right after him, Jacques Guerlain launched his "Heure Bleue" in 1913 and Houbigant his "Quelques Fleurs." Combining natural and synthetic ingredients, French perfumers gained a supremacy they maintained in the years following World War I.

Chemicals and Wood

Chemists and the industries they serve have perennially been interested in wood for the compounds that can be extracted from it and for the cellulose which lends itself to all kinds of transformation. The *tapping* of mature trees of certain species enables one to obtain oil of turpentine by distillation, and to extract rosin as well as tars which were used to protect ship bottoms, until steel replaced wood in ship construction. This "naval stores" industry supported a large number of small businesses in the United States, Portugal, and France. But the environment changed when, starting in 1909, three American companies, Yaryan Naval Stores, later to be acquired by Hercules, Newport Turpentine and Rosin, and Williamson Chemical Co. began using old wood stumps from Georgia, Florida, and Mississippi as raw materials which they processed through steam distillation followed by extraction with solvents.

The *tanning extract* industry also underwent great changes. Natural tannins were at first prepared by the tanners themselves. Then, by 1870, firms that until then had specialized in dyewoods both in Europe and in the United States began extracting these tannins from chestnut wood. As demand increased, these firms had to import wood from abroad so as not to deplete the forests in their own countries. Imports were boosted when a German tanner in Buenos Aires discovered, in 1872, the tanning properties of the *Quebracho* wood from the Chaco forests. The Établissements Dubosc in Le Havre, France, were amongst the first to develop the new source of natural tannins.

Dispersed for a long time, the tanning extract industry consolidated in the twentieth century. American Dyewood Co. was formed in 1904 through the merger of a number of medium-sized companies, while in France the Société Gillet & Fils developed to the point where Progil was accounting, after World War II, for 50 percent of France's natural tanning production in its Labruguière plant in southwestern France. As for the synthetic tannins discovered in 1911 by Germany's E. Stiasny, they were to be manufactured for the first time by BASF.

While terpene chemicals were produced from pine, wood remained for a long time the main source of *methyl alcohol, acetic acid*, and *acetone*; and wood *charcoal* was used both for reduction in metallurgy and as one of the components of black gunpowder. *Wood carbonization under airtight conditions* was a process discovered in 1799 by Philippe Lebon, a civil engineer. He advocated use of "wood gas" for city lighting and also, eventually, for heating purposes.

Charred in large airtight cast-iron cylinders, wood, heated to a dark red heat, formed a solid residue, charcoal, and volatile matters, especially hydrogen and methane gas, which were used for gas lighting. As coal gas replaced wood gas for lighting, and coke replaced wood charcoal in metallurgy, what kept the wood carbonization factories going between 1850 and 1920 was the chemicals extracted from pyroligneous liquors. Treated with lime, the aqueous layer separated from wood tar formed calcium acetate, from which acetic acid could be obtained by reaction with dilute sulfuric acid. After vaporization, methyl alcohol or "wood spirits" condensed in a coiled tube. By dry distillation of calcium acetate, acetone could also be obtained. The preparation of lead and aluminum acetates as mordants for cotton dying and textile printing opened up one of the main markets for such wood chemicals.

As Scotland's hardwood no longer sufficed to provide the amounts of mordant required by England's textile industry, a carbonization unit was set up in the state of New York by a Glasgow businessman, J. H. Turnbull, in 1856. He exported to England the acetates produced and sold the surplus acetic acid to the Boston-based Merrimac Chemical Co.

With the discovery of Methyl Green, the dyestuff industry opened up a fresh market, from 1865 onwards, for methyl alcohol. Its use, moreover, was promoted as a substitute for ethyl alcohol, which was subject to excise duties. These duties on drinkable alcohol accounted for 4 percent to 8 percent of the tax revenues of some states. Exemptions in the case of industrial methylated spirits only took place gradually, first in England (1855), then in Holland

(1865), in France (1872), in Germany (1879), and finally in the United States (1906).

Acetone requirements, for their part, increased after 1880 with utilization of nitrocellulose for smokeless powder, films, fibers, and celluloid. Wood charcoal, which was being threatened by coal in its traditional uses, found new life as a base material for the synthesis of *carbon disulfide* required to prepare cellulose xanthate and as *activated charcoal* due to its outstanding properties as an adsorbent.

On the eve of World War I, France had twenty-three factories for the pyrolysis of wood, most of them located in forest areas and close to tanning-extract workshops. The largest, fitted out with tunnel furnaces operating on a continuous basis, were in central France at Prémery where the Lambiotte family had settled since 1886 and on the Clamecy site, which the Gillet brothers were to merge with the Société Progil after it was established in 1920.

Wood processors were also to become heavy consumers of inorganic chemicals for the preparation of *cellulose pulp* used in the paper industry. It was France's A. D. Mellier who conceived the notion in 1854 of making a *chemical pulp* by dissolving straw under heat in a dilute caustic soda solution. Three years later he patented the process in the United States. Meanwhile, in 1855 near Philadelphia, the American Wood Paper Co. had started manufacturing soda pulp made from wood shavings boiled in alkali, according to a process of England's Watt and Burgess. Then in 1867 an American, Benjamin Tilghman, took out a patent for a *bisulfite pulp*, which the Rhode Island-based Richmond Paper Company was the first to manufacture in 1883.

In Germany, Alexander Mitscherlich, the youngest son of the great chemist Eilhardt Mitscherlich, had been striving since 1877 to improve the bisulfite process. By adjusting the temperature, pressure, boiling time, and the concentration of the calcium bisulfite solution, he managed to determine the optimum conditions for preparing a pulp that would be pure enough to be used as a substitute for America's cotton linters or Algeria's esparto grass for many purposes and particularly in chemical applications.

From bisulfite liquors, many chemical substances were obtained, such as ligno-sulfonates and vanillin as well as ethyl alcohol through fermentation. The *sulfate or soda process* was discovered by Germany's C. F. Dahl in 1879. It could be used both with pine and hardwood and avoided the corrosion of equipment that in the sulfite process was caused by SO_2. It produced particularly strong paper called *kraft* (meaning strong in German). From the liquors of this process, *tall oil* was extracted and subsequently became a major source of fatty acids on the one hand, and of rosin and turpentine oil on the other, rivaling naval stores sources. The production of chemical pulp, because it involved the use of chlorine derivatives for bleaching, and of sulfur, sulfuric acid and alkali for dissolving the cellulose, opened up broad prospects for the inorganic chemicals industry, while bisulfite liquors, as well as those produced by the sulfate process, provided fresh sources of organic products.

Chemicals and Household Products

Until about 1875, chemicals, although serving various industries, had stayed away from the public mainstream. The housewife was hardly aware of what they were all about.

True, America's George Lewis, who later took over Pennsylvania Salt Manufacturing Company (Pennsalt), was already selling yeast and soap to Philadelphia housewives in 1850, while ready-to-use cans of paint also made an appearance on the market. A great many converging circumstances would be required, however, before the first popular household products reached the market during the last part of the nineteenth century: emergence of a middle class with greater means, improved habits of health and cleanliness, simple and attractive packaging materials, consistent quality articles, and the development of trademark advertising. Some very clever businessmen seized the opportunity this favorable environment offered. In Europe they included Fritz Henkel from Germany and William Lever, later to become Lord Leverhulme, from England. Both focused their talents on household products.

The Civil War in the United States had brought about a shortage of rosin in the North, because this raw material extracted from pine trees in the South was under embargo. Joseph S. and Thomas Elkinton had the idea of using sodium silicate as a substitute for rosin in soapmaking and built a silicate plant for the purpose. This was the start of Philadelphia Quartz (PQ) in 1864, still known by that name today.

Young Fritz Henkel, from Hesse, Germany, was informed of this development at a time when he was selling soda ash to the Rheinische Wasserglas Fabrik in Herzogenrath. He developed a formulation based on soda ash and sodium silicate, his "Bleich Soda," which had the advantage, besides its detergent and bleaching properties, of requiring only a part of the costly imported fats that until then had been needed for soapmaking. Henkel set up business in Düsseldorf in 1876 and was soon manufacturing soap and glycerine in addition to silicates.

Although it took him twenty years to establish his industrial repute, Henkel's success soared considerably when, in 1907, he put Persil on the market. Combining in a single formula detergent, disinfectant, and bleaching properties, it was soon to become world-renowned. Henkel used a persalt for the purpose, sodium perborate, production of which had just been started by Degussa in Rheinfelden in 1908.

Meanwhile, another German, O. Lingner, had taken advantage of the favorable climate, built up around the fight against bacilli by the discoveries of Robert Koch and Emil von Behring, to bring out in 1903 a toothpaste he called *Odol*. It enjoyed rapid development, thanks to a clever advertising campaign; and, while its dental hygiene virtues were never proved, it contributed to the emerging need for beauty and convenience products.

The most spectacular success was William Lever's. At the age of twenty-six, he decided to quit his father's grocer's shop to set up his own business and began, in 1880, to import colonial goods. The Levers had been selling soap in cakes since 1874; and young William had the idea of registering and selling under the trademark Sunlight, a good-quality soap made from vegetable oils, which were half as expensive as tallow. In 1885, he started manufacturing the soap himself in a factory which he rented in Warrington. Five years later, the factory had become too small and Lever settled down in Port Sunlight near Liverpool. The glycerine produced as a by-product of the soapmaking very conveniently found a ready market in the growing dynamite industry.

Lever then looked abroad for steady supplies of plant-based oils. In 1894, he had purchased a cottonseed oil factory in the Mississippi area; and in 1903, he set up plantations in the Pacific region and in Africa south of the Equator where, owing to the climate, trees and plants have particularly high oil yields. When in 1906, imported vegetable oil prices boomed, Lever and his brother were able to buy up rival soapmakers in England who, unlike them, had not thought to integrate upstream with their own raw materials.

Previously, Lever Brothers had set up a business in Olten, Switzerland, in 1898, then had founded with the Stoolwerk brothers in Mannheim the Deutsche Sunlicht Gesellschaft. At the same time, an oil and margarine plant was being added to the English factory at Port Sunlight (1904). Indeed, it was through *margarine* that the Lever group considerably extended its range of activities after the First World War.

Wishing to economize on the butter consumed by his armies, Napoleon III had commissioned H. Mège-Mouriez to prepare a substitute butter made from beef fat. Mège-Mouriez's success in 1869 sparked off a rapid development of margarine plants both in Europe and in the United States. By 1871, Anton Jurgens was making margarine in Holland. He was followed in 1895 by the Van den Bergh brothers who set up plants both in Holland and in England. At that time, manufacturers were also attempting to replace animal fats with the less costly and more readily available vegetable oils.

This meant giving vegetable oils the same solid consistency as tallow. Paul Sabatier, who had been a pupil of Berthelot's at the Collège de France, showed, when working with his assistant, Abbé Senderens in 1897, at the University of Toulouse, that in the vapor phase, the double bonds of unsaturated olefins and oils could be hydrogenated using nickel as a catalyst. The German chemist W. Normann, working in a Herford oil mill near Bielefeld, discovered in 1901 that the hydrogenation of double bonds in oils also occurred in the liquid phase with the same catalyst producing solid fats.

Normann was unable to convince his fellow countrymen of the importance of his invention. He was more successful with the Warrington silicate manufacturers, Joseph Crosfield & Sons, which was to become one of Lever Brothers' most successful enterprises when the latter acquired their patents and began manufacturing hydrogenated oils for soapmaking in 1909. Jurgens, however, managed to buy back from Crosfield the Normann patent rights for fats to be used in foodstuffs.

The Germans, who had not believed in the Normann process, soon had to countenance the establishment in their country of both the Oel Werke Germania in Emmerich by Jurgens the Dutchman, and the margarine works of the Van den Bergh brothers in Cleves. When these two Dutch companies were merged with Lever Brothers in 1929, the Unilever group thus formed was subsequently to supply three quarters of Germany's margarine requirements.

Through the hydrogenation process, the cost of fats for soapmaking now based on coconut oil was halved, and the price of margarine was reduced to one-third of that of butter. The era of big consumer products was opening up, to the greater benefit of the Henkels and the Levers.

The Quest for Fertilizer Raw Materials

Toward the end of the nineteenth century, fertilizer needs were increasing rapidly with world population growth and the development of farming acreage made possible by increasing mechanization. Bone meal, manure, English or Norwegian coprolite, and even guano from Peru, exported in growing quantities to Europe and the United States, no longer sufficed.

A new *nitrogen* source had emerged around 1840 as sodium nitrate was extracted from the *caliche* deposits in the desert regions of Atacama, Tacna, and Tarapaca, which would fall to Chile after its victorious war against Peru and Bolivia (1879–1892). It was only when the Shanks process was applied in 1876 to the purification of caliche, however, that *Chilean nitrate* was extensively developed, essentially by English and, to a lesser degree, German interests. In 1909 they were joined by W. R. Grace & Co., already much involved in the guano trade and a year later by Du Pont, whose so-called "Officina Delaware" mine was to provide up to 50 percent of the nitrate needed to produce the nitric acid for explosives during the First World War.

By 1902, Chile was already exporting 1.3 million tons of its nitrate. With the recovery of iodine from the nitrate mother liquors at a rate of 1500 tons a year, it was also to rank as a leading world exporter of this halogen. Further nitrogen needs were covered in the form of *ammonium sulfate* produced from the ammonia (NH_3) that came in growing quantities from coke ovens and gasworks.

Sources of *phosphorus* were also beginning to diversify. In the United States, mines for ores with a high phosphate content (57 to 67 percent) were being developed in South Carolina as early as 1867. The major Tennessee and especially Florida mines began producing in 1888. America's superphosphates industry, which had been pioneered around 1850 by William Davison in Baltimore, then partly moved to the South, to Charleston, where the Virginia-Carolina Co. took root. While often secure in superphosphates, American compound fertilizer manufacturers continued to depend, until 1930, on outside sources for nitrogen and potash.

In contrast to the iron ores in the United States, those in Continental Europe were rich in phosphorus. The process England's S. G. Thomas discovered in 1875 to obtain steel with a low phosphorus content, produced *dephosphoration slag* as a by-product. This became a significant source of phosphorus when it was discovered that, unlike natural phosphates, it could be used directly as a fertilizer after grinding, without having to be made soluble. Europe's dependence on phosphorus from the United States decreased still further after 1920 when the phosphates from Tunisia and Morocco started to be developed.

Considerable progress had been made by the end of the nineteenth century in *sulfuric acid* production needed to produce superphosphates. The lead chamber process only yielded acid of 78 percent concentration. For the synthesis of alizarin, fuming sulfuric acid was needed and its only supplier was Nordhausen in Bohemia. But it was expensive ($100/ton), and production never exceeded the 4,350 tons of the peak year, 1884. Use of iron pyrites or copper and zinc ores as a source of sulfur, on the other hand, introduced impurities into the sulfuric acid.

The discovery of *elementary sulfur* in the Gulf of Mexico area in 1869 and, more especially, the development in 1890 by Herman Frasch of a convenient extraction process for this chemically pure sulfur, launched it as a widespread raw material for sulfuric acid plants. It will be recalled that, in 1831, a Bristol vinegar maker called Peregrine Phillips had patented a *contact process* to produce sulfur trioxide (SO_3) by passing sulfur dioxide (SO_2) and air over a platinum catalyst. Two drawbacks—the catalyst being poisoned by arsenic impurities and the lack of availability of suitable chemical equipment—had to be overcome before the contact process was put into commercial use by BASF in 1888. Under this process, very pure acid in all concentrations could be obtained irrespective of where the sulfur came from.

Germany played a pioneer role in this area through work carried out since 1875 by Clemens Winkler and then by Rudolf Knietsch. The first unit to be set up in the United States, that of New Jersey Zinc, came on stream in Wisconsin only in 1910, followed soon afterwards by the General Chemical Company as a BASF licensee.

Potassium, another element needed for soil fertilization, became available when the Stassfurt salt mines were discovered in 1856 in the State of Prussia. These mines were first developed for their salt over which the state had a sales monopoly. The *carnallite* ($KCl,MgCl_2.6H_2O$) they contained was considered as a by-product until 1861 when use of potassium chloride as a soil improver was discovered. Involvement of private capital gave the development of the Stassfurt deposits a new twist. Besides sodium sulfate, bromine also was extracted from the brine, and for fifty years remained the only European source of this element.

To avoid a price war for potash, a producers trade group was set up in 1904 under Prussia's state control. The Kali Syndikat was formed at the very time when sylvinite (NaCl, KCl) mines were being discovered north of Mulhouse during exploration for lignite and oil. Until the Versailles Treaty in 1919, Germany held a world potash monopoly thanks to the Stassfurt and Alsace mines. Half the production was for domestic use, contributing to the improvement of poor soils in Saxony and Silesia. It also served to boost the yield of Germany's sugar beet producers far above those of producers in northern France.

The rest of the potash production was exported, particularly in the form of potassium nitrate. In the United States, two companies alone, American Agriculture Chemical, which was later to become part of International Minerals and Chemicals Corporation, and Virginia-Carolina Chemical, were purchasing 70,000 tons a year of the product, amounting to a hefty third of total imported tonnage.

The Fixation of Atmospheric Nitrogen

Dependence on the Chilean monopoly for nitrate supplies could not fail to worry fertilizer and nitric acid producers. Bearing in mind a likely depletion of Chilean mines, the English physicist, Sir William Crookes, speaking before the British Association for the Advancement of Science, hinted in 1898 at the possibility of *fixing nitrogen*, which makes up 80 percent of the atmosphere. In 1843, France's Henri Victor Regnault had obtained *ammonia* (NH_3) by reacting three volumes of hydrogen with one volume of nitrogen in an electric arc. The first industrial production of ammonia, however, was achieved when two Germans, Adolf Frank and Nikodem Caro, hydrolyzed *calcium cyanamide* in an autoclave using superheated steam.[24] A calcium cyanamide plant was set up in 1910 in Knapsack near Cologne using lignite, a low-grade form of coal, as a source of carbon both for making the cyanamide and for electricity production. The plant was later to become Hoechst property.

Founded in 1908 by the Deutsche Bank, Bayerische Stickstoff Werke set up a cyanamide unit in Trostberg, not far from Munich, using its own hydroelectric power. It was the starting point of the future SKW. By the time

[24] According to the reaction:

$$CaCN_2 + 3H_2O \longrightarrow CaCO_3 + 2NH_3$$

the 1914–1918 war broke out, Germany was producing 25,000 tons a year of cyanamide which was used either directly as a fertilizer because of its 20 percent nitrogen content, or to produce ammonia which was oxidized to nitric acid for nitrate production.

Some Englishmen had adopted the Frank and Caro process as early as 1907 to set up a cyanamide unit in Odda, Norway. America's Frank Washburn did the same, establishing the American Cyanamid Company. Teaming up with James Duke, the tobacco magnate, Washburn set up his plant in Niagara Falls to take advantage of the cheap electric current. Other units were created in France (Société des Produits Azotés in 1906) and in Japan (Nippon Chisso in 1909). The nitrogen needed for the process was produced by the fractionation of liquid air, a process for air liquefaction having been first achieved by Karl von Linde of Munich in 1898.

The fixation of atmospheric nitrogen by combining it at a very high temperature with oxygen to form *nitric oxide* had been demonstrated by Crookes in 1892 in a Manchester pilot unit using an electric arc. It was only in 1903, however, that two Norwegians, Kristian Birkeland and Samuel Eyde, patented their arc process, under which the oxide NO was produced at a temperature of 3,000°, then oxidized by passing through air after having been cooled to below 600°C.[25] The resulting NO_2 was then sent to absorption towers where, reacting with water, it produced diluted nitric acid. Combined with limestone, the nitric acid yielded *calcium nitrate*, which was used as a fertilizer. La Société Norvegienne de l'Azote (Norsk Hydro), which was set up by Samuel Eyde with the help of French capital, had a capacity of 150,000 tons a year of nitrate. While this production required only air, water, and limestone as raw materials, it yielded only some 2 percent to 3 percent NO and was a heavy consumer of electricity. The cyanamide process was less dependent on electric energy and on the amount of capital needed per ton of nitrogen produced. This explains why the arc process enjoyed little success outside Norway. The last furnaces were, indeed, to be closed in 1940.

Nitric acid was to be produced by *oxidation of ammonia*, a process much preferred by industry. In his lectures given at Lille University in 1842, Frédéric Kuhlmann had already pointed out that nitric oxide could be obtained by passing air and ammonia over a hot platinum catalyst at high temperature.[26] In 1902 Germany's Wilhelm Ostwald examined the optimal conditions for this reaction and took out a patent for his process, which was used in plants built in Dagenham, England, in Gerthe near Bochum, Germany, and in Vilvorde,

[25] The reactions are:

$$N_2 + O_2 = 2NO \qquad 2NO + O_2 = 2NO_2$$

[26] According to the reaction:

$$2NH_3 + 7/2\, O_2 = 2NO_2 + 3H_2O$$

Belgium. The process was adopted on a large scale only when pure ammonia could be obtained at a better cost than from cyanamide.

The Haber-Bosch Ammonia Synthesis Process

By the end of the nineteenth century, much thought had been given to achieving the synthesis of ammonia from its component parts. But the laws of chemical equilibrium were yet to be fully grasped, and there was also the problem of developing equipment and materials capable of resisting high temperatures and pressure. France's Henri Le Châtelier, who studied and subsequently taught at the École des Mines, had formulated in 1888 a law on the displacement of chemical equilibrium. In 1900 he established the temperature and pressure conditions under which the synthesis of ammonia could be achieved from hydrogen and nitrogen in the presence of a catalyst.[27] He even took out a patent in 1901, but his efforts fell short because of an explosion which remained unexplained at the time.

The German Fritz Haber rounded out Le Châtelier's work. The son of a Breslau trader in chemicals, Haber had studied under Bunsen and Hofmann before pursuing a doctorate in organic chemistry at Berlin's Technische Hochschule. He first joined his father's business, then taught dye chemistry at the Karlsruhe Technical Institute, of which he subsequently became the head. He then became involved in electrochemistry and the thermodynamics of gases and, hence, in the synthesis of ammonia. In 1908 he joined BASF. With the help of an Englishman, Robert Le Rossignol, whose forebears were Huguenots, Haber managed in 1909 to produce synthetic ammonia using osmium as a catalyst. He used a laboratory reactor that could withstand a pressure of 175 atmospheres and a temperature of 550°C. This encouraging result prompted Heinrich Von Brunck, BASF manager at the time, to dispatch two of his best engineers to Karlsruhe. They were Carl Bosch, a metallurgist, and Alwin Mittasch, who was a catalyst specialist.

In the spring of 1910 BASF abandoned the Norwegian arc process and commissioned Bosch to industrialize the Haber process. The company was playing for high stakes; failure would have meant ruin. The first pilot plant was set up in 1911 in Oppau near Ludwigshafen. When Von Brunck died the following year, the plant was producing one ton of ammonia a day.

The task of Carl Bosch and his team was complicated by the risk of impurities in the gases poisoning the catalyst, by the need to use special steels with low carbon content, and by the need to choose the most efficient catalyst which turned out to be iron. On September 9, 1913, the main obstacles had been cleared, and industrial production started in Oppau with an initial capac-

[27] According to the equilibrium $N_2 + 3H_2 = 2NH_3 + 12.9$ kilocalories: formation of NH_3 is favored by increasing the pressure and lowering the temperature but a lower temperature slows down the reaction, hence the importance of a catalyst.

ity of 8,700 tons a year of NH_3 feeding a neighboring unit which produced 36,000 tons a year of ammonium sulfate. The Haber-Bosch process was to crown BASF's technical reputation, which until then had been based on its dye production. It made Haber a rich man, for he got a pfennig for every kilo of ammonia produced; it also established his scientific fame: in 1912 he was appointed director of the Kaiser Wilhelm Institute of Physical Chemistry in Berlin, where he remained until 1933.

The Germans' success enabled them to develop the Ostwald-Kuhlmann nitric acid process under economically feasible conditions at a time when the outbreak of war caused supplies of Chilean nitrate to become uncertain. The Haber-Bosch process boosted demand for nitrogen, produced by air liquefaction, as well as for hydrogen. Until then hydrogen had been obtained mainly through electrolysis. In 1910, however, Frank and Caro, together with Linde, started producing it by liquefaction of *water gas*, the components being separated by fractional distillation.[28] Conditions were ripe for synthetic ammonia to become one of the great inorganic chemicals of the postwar era.

Cyanides and the Gold Rush

In 1887, England's John S. MacArthur, with the help of the Forrest brothers, had developed a process for extracting gold from ores containing only traces of the precious metal by using a dilute solution of *potassium cyanide* (KCN) as a complexing agent. Thanks to this discovery, ores from Alaska, Australia, and especially from the Rand in South Africa could be developed. This considerably boosted both the production of gold needed for economic development and the demand for cyanides.

At the same time, the American Hamilton Castner, who was manufacturing sodium (Na) in Oldbury for the Aluminium Company, saw his market dwindling, as sodium was not needed in the new Hall-Héroult electrolytic process started for the preparation of aluminum. Looking for substitute uses for sodium, Castner succeeded in 1894 in producing *sodium cyanide* (NaCN) through ammonia's action on sodium, then reduction of the *sodamide* formed with charcoal at 650°C.[29] After Castner's death in 1899, his process was adopted by the Glasgow-based Cassel Gold Extracting Company, a licensee of MacArthur-Forrest. The sodium needed came from the Castner-Kellner Alkali Company, which had taken over the Aluminium Company in 1895.

While sodium cyanide was used for processing gold ores, sodamide was in great demand for the synthesis of indigo. The Deutsche Gold and Silber Scheide Anstalt (Degussa), a German concern in precious metals founded in

[28] Water gas consists of carbon monoxide and hydrogen ($CO + H_2$) and is obtained by passing superheated steam over red-hot coal in a coke-gas generator.

[29] The overall reaction is:

$$2Na + 2NH_3 + 2C \longrightarrow 2NaCN + 3H_2$$

1843 by E. F. Roessler in Frankfurt, took an interest in both products. Under the management of Roessler's sons, it contributed to the development of the Castner process, then, in association with Aluminum Company, it founded in 1895 the Niagara Electrochemical Company in Niagara Falls. Degussa also formed an association with Société d'Electrochimie to produce cyanides in Martigny, Switzerland, in 1904.

Thus in a switch of a form common in the chemical industry, sodium demand, which in all likelihood would have dwindled when the Hall-Héroult furnace began to be used to produce aluminum, grew apace with sodium cyanide demand for the processing of gold ore. Furthermore, while he was trying to produce more sodium and therefore more pure caustic soda, Castner was led to develop his mercury cathode for molten soda electrolysis and to enter into an arrangement with Carl Kellner from Vienna, who had simultaneously discovered the same process.

As to the Société d'Electrochimie, it was set up in 1889 by the Alsatian Henri Gall. Gall had initially worked with F. Poirrier in Saint-Denis, then had joined with the Comte de Montlaur to develop a patent for the electrolytic production of *potassium chlorate*, taken out in 1886.

Switching from electrolysis to thermal electricity after Henri Moissan had demonstrated the advantages of the electric furnace, H. Gall founded the Société des Carbures Métalliques in 1895 and began producing acetylene in

The Poirrier factory at Saint-Denis.

Notre-Dame de Briancon. Then he produced calcium cyanamide at the Société des Produits Azotés in Premont when electric lamps replaced acetylene lamps and new markets had to be found for calcium carbide. At the same time, Gall had set up a plant for the electrolytic production of sodium in 1901 in the Vallée de la Romanche. Thus it was very natural that he became involved with cyanides and, as we have seen, became associated with the Roesslers. Also used in electroplating and organic synthesis, cyanides were extended and diversified in use. By the time World War I broke out, sodium cyanide production exceeded 18,000 tons a year.

Industrial Gases and Air Liquefaction

At the end of the nineteenth century, great progress had been made in the handling of gases and in their use for other purposes than as chemical reagents, particularly in metallurgy and refrigeration. In 1845, the Englishmen James Prescott Joule and William Thomson had demonstrated that air and most other gases were cooled when expanding from a given pressure in the absence of any external work. Technological improvements were made by Werner Siemens in Germany in the area of heat exchangers (1857) and by Karl von Linde in compressors (1875). Liquefaction technology was also pushed forward through the work on the compression of gases conducted by France's Louis-Paul Cailletet, England's William Hampson, and Switzerland's R. Pictet. The latter managed to produce low temperatures through his "cascade" process.

From 1888, Knietsch's team at BASF produced liquid chlorine, thus facilitating long-distance transportation of chlorine. *Sulfur dioxide* (SO_2) and *methyl chloride* (CH_3Cl), which can be liquefied under moderate pressures, became widely used for refrigerating purposes in Germany, France, and England. *Carbon dioxide* (CO_2) was more difficult to handle because of the high pressures required. Its use as a refrigerating source developed only after the turn of the century.

Although some gases served to produce low temperatures, others, such as oxygen and acetylene, were used to produce temperatures reaching 3,000°C by means of the oxyacetylene torch.

Oxygen had been industrially produced for the first time by chemical process by France's Léon and Arthur Brin in 1880. Their process which was inspired by a discovery made by Jean Baptiste Joseph Dieudonné Boussingault in 1851, was based on the fixation of the air's oxygen by barium oxide.[30] The Brin Oxygen Company was set up in England in 1886 to develop the process for the production of oxygen used in limelight. British Oxygen, which succeeded Brin Oxygen Company in 1906, still uses the magic lantern emblem

[30] According to the equilibrium reaction:

$$2\,BaO + O_2 = 2BaO_2$$

on its oxygen cylinders. The barium peroxide obtained in the Brin process was also used in the production of hydrogen peroxide (H_2O_2) through the action of dilute acid.[31] An importer of textile chemicals into England, Bernard Laporte, used the method in 1888 to produce in Shipley, Yorkshire, diluted hydrogen peroxide and successfully promoted its use to bleach straw hats very fashionable at the time for men as well as women.

But it was the oxyacetylene torch, already foreseen in 1895 by Henri Louis Le Châtelier and Edmond Fouché, which provided oxygen with its first major markets in welding and metal cutting. From then on, it was no longer sufficient to produce it merely through the Brin process or the electrolysis of water.

Georges Claude graduated from the École de Physique et de Chimie Industrielle (EPCI) in 1886, four years after its founding in Paris. He was working as a young engineer with the Thomson Company when it dawned on him that air was the most abundant source of oxygen, providing it could be separated from its other components. Georges Claude was an inventive young man who had already patented in 1896 a process to dissolve *acetylene* in acetone. The process is still used today and makes it possible safely to store and carry the gas which is delicate to handle.

In 1897, Karl von Linde achieved the industrial liquefaction of air

Georges Claude with the apparatus with which he made the first quantities of liquid air. Courtesy L'Air Liquide.

[31] France's Louis Jacques Thénard had discovered hydrogen peroxide in 1818.

through decompression between 200 and 50 atmospheres after cooling to −30°C, using a refrigerating apparatus utilizing ammonia. Georges Claude's improvement of the Linde cycle consisted in decompressing the air contained in the cylinder of an air-compressor engine allowing it to do external work. To put this idea into practice, Claude roped in one of his old school fellows, Paul Delorme, who was also working with the Thomson Company. On May 25, 1902, after many vicissitudes, the first drops of liquid air were produced in the La Villette workshop where an air-compressor engine had been put at the disposal of the young inventor by the Compagnie des Tramways. On November 8 of that year, the Société L'Air Liquide was set up under Paul Delorme. But liquid air, as such, was of no interest unless oxygen could be extracted from it. While Linde had been the first to extract oxygen by fractional distillation of liquid air, he was able to recover only 66 percent of the oxygen contained in the liquid. By developing a double rectification column, Georges Claude managed to bring extraction yields to 93 percent. From that day, the Société L'Air Liquide, with its acetylene, oxygen and torch production, had all the ingredients for becoming a world leader in the area of industrial gases.

By the eve of World War I, the company had greatly increased the number of its production sites because the gases were expensive to transport over long distances. In France, oxygen was produced in six plants, dissolved acetylene in seven. Abroad, plants were set up in Belgium, Spain, Italy, Russia and Canada. Other producers emerged, but each one kept to its sphere of influence: British Oxygen to England and, subsequently, to the Dominions, Linde to Germany, Aktiebolaget Gaz Accumulator (AGA), founded in 1904 by Gustaf Dalén, to Scandinavia.

In the United States, Union Carbide had been formed in 1898 by taking over a number of calcium carbide units. Another company, National Carbon, had been manufacturing electrodes from petroleum coke since 1886, in association with Myron Herrick, who was to be the United States ambassador to France for many years. Myron Herrick was a friend of Linde's, and when the latter came to Cleveland in 1906, he seized the occasion to set up a syndicate which was to buy his air liquefaction process and develop it under the name of Linde Air Products.

Meanwhile, in 1904, Prest-O-Lite, which manufactured acetylene lamps for bicycles, had bought up the Claude-Hess dissolved acetylene process. The four companies, Union Carbide, National Carbon, Linde Air Products and Prest-O-Lite, joined forces in 1917 to form Union Carbide and Carbon Company, which became one of the world leaders in industrial gases.

From Ferments to Enzymes: The Birth of Biochemistry

The processes of fermentation to produce alcoholic beverages, cheeses, vinegar and yeast, go back to ancient times. In research carried out after 1792, Lavoisier had shown that sugar was decomposed through fermentation into alcohol and carbon dioxide.[32] Before then, Carl Wilhelm Scheele had managed to isolate, between 1769 and 1780, tartaric, oxalic, citric and lactic acids via their calcium salts from their natural medium (wine dregs, fruit juices, curdled milk).

Joseph Louis Proust, who formulated the law of definite proportions, had, for his part, studied the sugar juices of plants from 1799 to 1808 and had identified three different sugars: sucrose, fructose and glucose. As for the Saint Petersburg apothecary, G. S. Kirchhoff, he had been the first to observe in 1811 that, heated with sulfuric acid, starch produced glucose.[33]

Analyses carried out by Gay-Lussac and Thénard revealed that *sugars, starch*, and *cellulose* all contained hydrogen and oxygen in the same proportions as in water. Classified as saccharides by William Prout in 1827, these substances were termed *carbohydrates* by Karl Schmidt in 1844. Although substances obtained through fermentation were becoming better known, the mechanism under which the reactions took place remained unexplained. Great chemists like Berzelius, Wöhler or Liebig, who had taken such pains to free organic chemistry from "vitalism," remained impervious to any biological explanation.

Berzelius, in particular, considered yeast as an inanimate catalyst, while Liebig saw it as the result, not the cause of fermentation. Louis Pasteur refuted many of the ideas of these great scientists through the work he conducted between 1854 and 1864. Pasteur first became interested in optically active amyl alcohol which was contained in the fusel oil produced during fermentation. This led him to the concept of molecular asymmetry which he considered one of the features of life. He then discovered that specific microorganisms (bacteria, yeasts, molds) were responsible for various fermentations. Pasteur conceived fermentation as "life without air," for this reason subsequently termed *anaerobic*. These organisms could only reproduce in an oxygen-free environment. He also believed that fermentation could only take place in the presence of living cells.

The Munich-based German organic chemist, Eduard Büchner, proved Pasteur wrong in this case. In 1886, he demonstrated that absence of oxygen was not a prerequisite. *Aerobic* fermentations were possible also. But more significantly, he managed in 1897, with his brother Hans, to prepare a yeast

[32] This is the reaction described in 1813 by Gay-Lussac and now written as:

$$\underset{\text{glucose}}{C_6H_{12}O_6} \longrightarrow \underset{\text{alcohol}}{2\,C_2H_5OH} + \underset{\text{carbon dioxide}}{2CO_2}$$

[33] According to:

$$\underset{\text{starch (polysaccharide)}}{(C_6H_{10}O_5)_n + nH_2O} \longrightarrow \underset{\text{glucose (monosaccharide)}}{nC_6H_{12}O_6}$$

juice by destroying living cells by grinding them with sand and passing them through a hydraulic press. To his surprise, the juice still catalyzed fermentation of a thick sugary syrup.

This was proof enough that ferments even in the absence of cells caused transformation of carbohydrates into alcohol and carbon dioxide. The term *enzyme* had been given to yeast-extracted ferments by W. Hühne in 1878; so the enzyme discovered by Büchner, and for which he was awarded the Nobel Prize in Chemistry in 1907, was named *zymase*.

Another great German organic chemist, Emil Fischer, greatly contributed to the chemistry of natural products. He had studied under Kekulé before becoming Adolf von Baeyer's assistant in Munich. In 1892, he succeeded von Hofmann as professor of chemistry in Berlin. He first studied uric acid and caffeine, which he chemically linked with *purine*, a hitherto unknown substance. He then discovered that phenylhydrazine, which he had prepared in 1875, formed with carbohydrates bright yellow crystallized derivatives he named osazones. He was led to the conclusion that mannose, fructose, and glucose had a similar structure, and he synthesized them. He was awarded the Nobel Prize in Chemistry in 1902 for his fundamental work on purine and its by-products and on the structure of sugars, later to be confirmed by X-ray examination. His considerable work also involved between 1899 and 1906 the decomposition of complex proteins into *amino acids*, and from these the synthesis of *polypeptides*, as well as the preparation in 1903 of a new sedative, *Veronal* (5,5-diethylbarbituric acid).

The term *biochemistry* began to be applied around 1910 to the science of substances extracted from the living world and to the mechanisms leading to their formation. Emil Fischer laid the foundations for this science. It took only a few years for the new science to make its breakthrough into the industrial world. Indeed with the needs born of World War I, substances produced through fermentation acquired great significance just as improvements in biochemical technology were paving the way for mass production.

Citric acid had traditionally been extracted from lemon juice. When sea supplies to Europe began to be cut off in 1916, it was produced industrially using the sugar-solution fermentation process patented by Wehmer as early as 1893 and improved by the use of the mold *Aspergillus niger*, discovered by J. N. Currie. There was also greatly increased demand for *ethyl alcohol* for use by dyestuff and smokeless-powder manufacturers, and this boosted its production from the fermentation of molasses.

In 1910 the French chemist, A. Fernbach, who was working at the Institut Pasteur, developed a process to obtain *butyl alcohol* and *acetone* by the fermentation of starch from cereals or potato flour. The process was applied in England in 1913 by Strange and Graham. But as acetone was used as a solvent in the gelatinization of cordite, and its supply from pyroligneous sources in

America became scarce in 1916, it proved urgent to develop acetone production through fermentation.

Chaim Weizmann, a Russian-born engineer, found a solution to the problem. He had worked briefly for Fernbach and in 1915 had studied acetone-butylic fermentation at the University of Manchester. He isolated the organism *Clostridium acetobutylicum*, which gave the best yields in the production of acetone from fermentation. In 1917 acetone began to be industrially produced, first in Canada, then in Terre Haute, Indiana. For services rendered to the allied cause, Weizmann, who was a Zionist, obtained the promise from the British Prime Minister Lloyd George at the end of the war, through the Balfour Declaration, that Palestine be offered as a homeland for the Jews. It proved the prelude to the creation, in 1948, of the State of Israel.

The Periodic Classification of the Chemical Elements

Around 1850, sixty-three elements had been identified through the conventional methods of chemical analysis. To detect elements which might exist as mere traces in matter, however, more subtle analysis was required. In 1859, Robert Bunsen and Gustav Robert Kirchhoff succeeded in detecting infinitesimal quantities of products by separating the components of light, which a substance gives off under an electric arc or in a burner, by means of a prism and lens arrangement and observing the resulting spectrum. Between 1860 and 1863, new elements such as *cesium, rubidium, thallium*, and *indium* were thus discovered through the spectroscope.

Almost simultaneously in 1869, Germany's Lothar Meyer and Russia's Dmitri Mendeleev developed approaches for classification of the elements according to their properties and attempted to put some order into inorganic chemistry. Like Meyer, Mendeleev proposed a *periodic classification* that showed that when elements were arranged according to their increasing atomic weights, analogies between them recurred at regular intervals. Both left blank spaces numbered beforehand which they expected to be filled when new elements were discovered, and Mendeleev even predicted their properties. Thus, during his lifetime, Mendeleev had the satisfaction of witnessing the discovery of *gallium* (Paul-Émile Lecoq Boisbaudran, 1875), *scandium* (Lars Fredrik Nilson, 1879), and *germanium* (Clemens Winkler, 1885), with properties very close to his predictions.

A number of minerals discovered in 1794 by the Finnish chemist Johan Gadolin near the village of Ytterby in Sweden contained oxides from new metal elements known today as *rare earths*. It was difficult to separate the elements of these rare earths and to distinguish one from the other, and therefore difficult to place them in the periodic table. It was only in 1913 that the young physicist Henry Gwyn-Jeffreys Moseley, who was killed at Gallipoli

Dmitri Mendeleev (1834–1907). Courtesy Edgar Fahs Smith Memorial Collection.

two years later, determined the atomic number of the elements and their place in the Mendeleev table from the wavelengths of their X-ray spectra.

Meanwhile, improvements in physical measuring methods led to the discovery of a group of hitherto unknown elements, the *rare gases*. Between 1892 and 1894, Lord Raleigh (born John William Strutt) and William Ramsay succeeded in isolating from atmospheric nitrogen a minute quantity of a gas that was more dense and was chemically inert. It was called *argon* (meaning "without energy" in Greek). Through spectroscopic analysis, Ramsay was able to demonstrate at about the same time that upon heating a mineral called cleveite, nitrogen was freed together with a rare gas which had already been observed in the solar chromosphere by Edward Frankland who, for that reason, had called it *helium*. Using the air liquefaction technology that had been developed by his fellow countryman Hampson, Ramsay then succeeded, aided by Morris William Travers, in isolating, by 1902, three other inert gases. Their names, derived from the Greek, indicate their rarity and novelty: *neon* (new), *krypton* (hidden), and *xenon* (stranger). These five rare gases, which had no chemical affinity, could find no place in the Mendeleev table. Ramsay suggested inserting a special column for them between the halogens and alkali metals, where they found their logical place.

The French physicist Henri Becquerel's discovery of *radioactivity* in 1896 revealed the existence of other families of elements. A former student of the École Polytechnique, where he subsequently lectured after 1895, Becquerel had observed and registered on a photographic plate, that radiations similar to X-rays were given off by a double sulfate of potassium and uranium which he was using in his research on fluorescence.

Some time after this discovery, the brilliant physicist Pierre Curie and the young Polish woman he married in 1895, Marie Sklodowska, observed that the pitchblende uranium mineral they were working on at the École de Physique et Chimie, was far more radioactive than pure uranium. They inferred that it must contain other radioactive elements. After long and tedious fractionation of the pitchblende, the two scientists were able to announce in 1898 that they had isolated two new elements. Thereafter, *radium* and *polonium* (the latter named after Marie Curie's homeland) found their place near the end of the periodic table and were soon to be followed by other radioactive elements. Becquerel and Pierre and Marie Curie were awarded the Nobel Prize in Physics in 1902.

The new elements to emerge in this prolific period, through working with minute quantities and improved measuring techniques, were soon to find a number of practical applications despite the small amounts contained in the earth or the atmosphere. *Thorium* use spread rapidly following the invention of the incandescent mantle by Auer von Welsbach in 1885. The residues of the monazite sand from which thorium was also extracted formed, after conversion into chlorides and electrolysis, the *Mischmetall* that since 1902 has replaced aluminum as a catalyst for generating high temperatures in metallic oxide reduction.

Ruthenium and *osmium*, as well as platinum and palladium, were some of the rare metals used both in jewelry and as catalysts.

Because of their inert properties, rare gases also served purposes other than chemical. Nonflammable *helium* was suggested in 1917 for filling balloons and airships instead of hydrogen, and production began at the end of World War I from the natural gas resources of Texas gasfields. Although the atmosphere's content of *argon* is only 1 percent by volume, the gas was extracted by liquified-air distillation from 1909 on, as were *neon, krypton*, and *xenon*, which served as inert fillings for incandescent light bulbs. As for *radium*, it has been used since 1908 in the form of its bromide or other salt, in the treatment of cancerous tumors.

From the Study of Solutions to the Theory of Ions

As early as 1878, the French chemist François M. Raoult, who had been a professor in Grenoble, had conducted research on the freezing points of solutions, from which he worked out a method to determine the molecular weight of organic compounds. He noted, however, that his freezing point law did not apply to solutions of salts in water. This anomaly in the behavior of mineral salts, as, also, that of acids and bases in solution, paved the way for the *theory of ionization*, which in turn helped explain a great many properties of charged atoms or atomic groupings called ions.

Michael Faraday had called *ions* those electrolyte particles that dissociate under electrolysis and move towards the electrodes. But it was Sweden's Svante Arrhenius who, while working for his doctorate in Uppsala, suggested that ions formed as soon as the salt dissolved and not, as Faraday had thought, when the current flowed.[34]

Arrhenius's thesis, which he presented in 1883, received final acceptance only because it had the support of the great German scientist Wilhelm Ostwald and later of the Dutch physicist Jacobus Henricus van't Hoff. It explained anomalies concerning osmotic pressure, electrical conductivity, vapor pressure, and the freezing point of salts in solution. But, more importantly, it made a science out of *analytical chemistry* which until then had been a mere compilation of empirical recipes. In particular, according to Arrhenius' theory, the acid and basic nature of a substance could be attributed to its capacity to produce in solution respectively H^+ ions and OH^- ions. It became usual to relate the strength of acids and bases to their degree of dissociation. In 1909 S.P.L. Soerensen proposed the *pH* concept to express a solution's acidity. The great development of electrochemistry after 1890 was due as much to a better understanding of chemical phenomena through the theory of ionization as to the availability of cheaper electricity.

[34] Thus, according to Arrhenius, a solution of ordinary salt does not consist only of NaCl molecules, but of a small proportion of the latter in equilibrium with sodium Na^+ ions and chloride Cl^- ions. The greater the dilution, the greater the dissociation into ions.

Chemical Equilibria and Reaction Kinetics

While it was known in 1850 that certain chemical reactions reached an equilibrium before all reagents had been consumed, the study of the temperature and concentration effects on such equilibria was only undertaken after 1860 by France's Marcellin Berthelot and Péan de Saint-Gilles, who were studying the esterification of organic acids. The Norwegians Cato Maximilian Guldberg and Peter Waage, in a paper published in French in 1867, stated the law of *mass action*, which helped predict how to modify the concentrations of the reactants to change an equilibrium and obtain the desired substance in the best yield. But Guldberg and Waage's law did not mention the influence of heat and time on equilibria; in other words it dealt neither with thermodynamics nor with the kinetics of reactions. Now it had long been known that most chemical reactions produce heat, and in a number of Lavoisier's experiments it had been possible to measure the calorific energy thus produced. *Thermochemistry* became a science, however, only when Berthelot determined heats of reaction between 1864 and 1879 and developed his bomb calorimeter, which is still used today.

This work of a practical nature was rounded off in 1876 by the American Willard Gibbs, whose *phase rule* was based on major mathematical developments. It helped to determine the direction in which heterogeneous equilibria would change with changing conditions. Thus, between 1897 and 1899, it became possible to fix the optimal conditions for the formation of potassium chloride from carnallite solutions in the Stassfurt potash mines.

Despite the interesting concept of an "activated molecule" proposed by Arrhenius in 1889, chemical *kinetics* was slow to develop because of the complex phenomena involved when molecules collide to produce chemical reactions. Likewise the phenomenon of catalysis, described by Berzelius as early as 1835, could not be clearly explained in the nineteenth century, even though its practical significance was fully recognized in the preparation of chlorine by the Deacon process, of sulfuric acid by the lead chamber and contact methods, of nitric acid by ammonia oxidation, and in the hydrogenation of vegetable oils.

Molecular Architecture: The Birth of Stereochemistry

For a long time chemists were satisfied with the planar representation of molecules. But when, in 1848, Pasteur's work on tartrates revealed the existence of dextro-rotary and levo-rotary tartaric acid isomers with mirror image crystal form, while the racemic acid formed from a mixture of the two was optically inactive, some explanation had to be found for optical isomerism.

The answer to the problem was given in 1874 by the young Dutchman Jacobus Henricus van't Hoff and the Frenchman Joseph Achille Le Bel, who had both studied under Charles Adolphe Wurtz in Paris. They showed, indeed, that the molecular asymmetry and the optical activity frequently observed in natural substances occurred when a carbon atom was linked to four different atoms or groups. This led them to treat the atom as if it were in the center of a *tetrahedron* with its four valences pointing to the corners. This representation in space explained cases of isomerism that had thus far remained unexplained, such as those of maleic and fumaric acids, which have an identical formula apart from space orientation but differ in chemical properties, and of the two lactic acids, which have *dextro*-rotatory and *levo*-rotatory properties. Stereochemistry was born and would prove to be applicable to atoms other than carbon.

Chemical Engineering: A New Discipline

While physicists and chemists were linking up to understand the structure of matter and giving birth to *physical chemistry*, another discipline was emerging, particularly in the United States at the beginning of the twentieth century, that of *chemical engineering*. Until the end of the preceding century, the knowhow of chemists, coupled with the skills of mechanical engineers, had been sufficient to bring laboratory processes to industrial-scale production. But the emergence of *continuous operations* as the Solvay process was being developed created a need for a hybrid science requiring both the chemist's and the engineer's knowledge.

England's George Davis was the first to provide the teaching required to train a chemical engineer from 1880 onwards in Manchester. In the United States, a number of universities—MIT in 1888, the University of Pennsylvania in 1892, and the University of Michigan in 1898—began to train students in this discipline. During the same period, Arthur D. Little introduced the concept of unit operations.

Examples of the mastery acquired by American chemical engineers in their universities can be found in the catalytic benzene oxidation process for the production of maleic anhydride developed by John Weiss for Barrett Company in 1920 as well as in Jesse A. Dubbs and Carbon Petroleum Dubbs's 1915 petroleum cracking apparatus, which Universal Oil Products developed for large-scale production. But it was undoubtedly the synthesis of ammonia by BASF, successfully achieved in 1913 in Oppau, which forged the linking of chemistry with physics and engineering as it required knowledge in areas of analysis, equlibrium reactions, high pressures, catalysis, resistance of materials, and design of large-scale apparatus.

Chapter 4
Developments During the Period Spanning the Two World Wars (1914–1945)

The Situation in 1914

During the period from 1850 to 1914, chemistry had made giant strides both in scientific and in technological areas, taking advantage of the progress made in other fields even when it was not itself the driving force. But while, on the whole, political frontiers were irrelevant to such progress, the different industrialized countries had not advanced at the same rate. On the eve of World War I, the situation could hardly fail to reflect local characteristics and the genius peculiar to each people.

Germany's Undisputed Superiority

From the time Liebig had set the ball rolling, scientific education in German universities, with their strong focus on organic chemistry, and in Institutes of Technology, with their stress on applied chemistry, had provided German chemists and those foreigners allowed to take advantage of it with good solid training.

The number of Germans awarded the Nobel Prize as soon as it was established is proof enough of the quality of German scientific training: in chemistry, Emil Fischer (1902), Adolf von Baeyer (1905), Eduard Büchner (1907), Otto Wallach (1909), Richard Willstätter (1915), Fritz Haber (1918); in medicine and physiology, Emil von Behring (1901), Robert Koch (1905), and Paul Ehrlich (1908).

Close ties between university and industry were another noteworthy feature of German chemistry. Thus scientists and businessmen joined under the Emperor's aegis to set up in Berlin, in 1911, the Kaiser Wilhelm Gesellschaft, which was soon funding seven research institutes headed by famous scientists such as Haber and Willstätter.

Under the Reich, the chemical industry also benefited from favorable geographic conditions: a good rail network and the waterway of the Rhine, along which many plants were set up; easily available raw materials such as tars, which came from the Ruhr coking ovens, and the salt deposits between the Elbe and the Weser and in the Neckar valley, and the Prussian and Alsatian potash mines; plentiful energy sources with coal from the Ruhr and Silesia, the Rhineland's lignite, and the waterfalls of the Bavarian Alps. Endowed with such human and physical advantages, Germany's chemical industry was employing in 1913 as many as 180,000 workers, in addition to 30,000 workers in the potash mines.

Sulfuric acid production amounted to 1,700,000 tons, of which a good third was produced by the contact process developed by BASF's Rudolf Knietsch. As the Leblanc process had never really taken hold in the country, ammonia-based *sodium carbonate* provided the main alkali source and Deutsche Solvay Werke ensured three quarters of the Reich's needs, particularly from its Bernburg plant.

Griesheim Elektron had considerably developed its *electrolytic soda, chlorine*, and *bleaching powder* production. The availability of chlorine, sold in liquid form since 1905, prompted its use as a disinfectant for water while *hydrogen* produced by electrolysis both in Bitterfeld and in Rheinfelden, was beginning to be used for oxyhydrogen torch welding.

To ensure oxygen supplies needed for this activity, Griesheim acquired, in 1907, rights over the Claude air liquefaction process and, in so doing, became Linde's rival in Germany in industrial gases. Moving from electrochemistry to thermoelectricity, Griesheim produced electric furnace *phosphorus*, becoming a competitor of Albright & Wilson in England and of Coignet in France.

In 1905, Griesheim Elektron, which was already manufacturing the aniline and monochloracetic acid required for indigo synthesis, had acquired K.G.R. Oehler's dye factory in Offenbach, near Frankfurt, where in 1911 cotton dyes of the *Naphthol AS* family were discovered. Very early on, German industry became involved in *acetylene* chemistry. By 1913, Griesheim had set up a pilot unit that produced *acetic acid* from acetylene via acetaldehyde. At the same time, Fritz Klatte was pioneering the production of *vinyl acetate*.

Paul Duden, perceiving the significance of this new type of chemistry by which the Germans were also producing acetone from 1917 on, persuaded Hoechst to associate with Griesheim, then with Knapsack, producers of calcium carbide near Cologne since 1906, and finally with the company founded by Alexander Wacker near Munich in 1914. Knapsack had also put on stream in 1913 the first tunnel furnace for *calcium cyanamide* production, which was also being produced in Trotsberg by the Bayerische Stickstoff Werke set up in 1908 to develop the Frank and Caro process.

As BASF's Oppau plant was in the process of starting up, the main *ammonia* source, besides calcium cyanamide, was ammonium sulfate from the Ruhr's coking plants sold by a syndicate of fertilizer producers.

Unlike France with its North African phosphates or the United States with its Florida mines, Germany had no abundant supplies of natural phosphates, and ranked only third in *superphosphates*. But it enjoyed a *potash* monopoly. The Reich government had indeed established a sales cartel in 1910 to put an end to the savage competition raging among the different potash mine owners. Membership in the syndicate was compulsory. A ceiling price was fixed for the quantities sold in Germany on a quota basis that allowed for sufficient margins. But the prices set for export were distinctly higher, particularly for sales to the United States. While the cartel protected small producers, it prevented any production rationalization. Moreover, resentful foreign clients did their utmost, after the war, to shake themselves free from the German monopoly.

The great strength of Germany's chemical industry on the eve of World War I was undoubtedly its dominant position in *dye chemicals*. Eight companies with their foreign subsidiaries alone produced 140,000 tons a year of dyestuffs out of a world total of 160,000 tons a year; and 80 percent of this tonnage was sold abroad. The major firms were BASF, Bayer, and Hoechst, followed by Weiler-Ter-Meer, AGFA, Cassella, Kalle, and Griesheim Elektron.

These companies received their aromatic hydrocarbons from the distillation units of the Ruhr's coking factories; they imported naphthalene from England and bought their aliphatic products (methanol, formalin, acetone, acetic acid) from wood distillers. The main groups—BASF, Bayer, and Hoechst—had integrated upstream and produced certain essential raw materials such as sulfuric acid, soda, and chlorine. The considerable profits they had made with the major syntheses of alizarin and indigo, and then of the new families of dyes, had enabled them to diversify into other areas of chemistry.

BASF, which had developed the sulfuric acid-contact process as well as the chlorine-liquefaction process, paid attention to synthetic tannins and, at great expense, was launching into ammonia synthesis. Bayer manufactured chrome pigments and lithopone, and was becoming involved in photographic products. It also offered a number of pharmaceuticals, including *phenacetin* (1888), the famous "aspirin," launched in 1898, and *Veronal*, a sedative obtained under Merck license in 1904.

Hoechst was successfully engaged in chemotherapy, serums and vaccines, while at the same time paving the way for acetylene chemistry. Hoechst's pharmaceuticals tradition dated back to the Giessen University professor A. Laubenheimer, who became their head of research in 1887. L. Knorr, previously a pupil of Emil Fischer, had given Hoechst an antipyretic and painkiller of the pyrazolone family, *antipyrin*, which was put on the mar-

Vats equipped with portholes, used for the first industrial synthesis of indigo at the Ludwigshafen factory of BASF at the beginning of the twentieth century.

ket in 1884. It was followed in 1893 by F. Stolz's *pyramidon*. But it was Laubenheimer who established with Robert Koch, Emil von Behring, and Paul Ehrlich the collaboration that was to lead between 1894 and 1910 to the production of *tuberculin*, of *antidiphtheria serum*, and especially of *Salvarsan*, the first successful treatment of veneral diseases. *Novocaine*, still used today as an anesthetic, was discovered by Alfred Einhorn as early as 1905.

With these remarkable specialities, Hoechst was the most prosperous of Germany's "big three" in the chemical industry before war broke out. Moreover, Hoechst had become associated with two other neighboring dye manufactures, Kalle in Biebrich and Cassella in Mainkur. The latter was also a producer of pharmaceuticals since 1904, having formed a partnership with Paul Ehrlich. This association worried the other large rival companies, who feared the loss of a number of markets. It hastened the union of Bayer, BASF, and AGFA, which Carl Duisberg of Bayer and H. von Brunck of BASF had been striving to achieve. The union or *Dreibund* materialized on January 1, 1906, in the form of a common-interest group, or Interessen Gemeinschaft, which was to be called the Little *IG*.

In this way production in Germany came, in 1914, under the control of these groups: the Little IG, Hoechst and its satellites, and the independents, among which the main ones were Griesheim Elektron and Weiler-Ter-Meer in Uerdingen. Their only foreign rivals operating in Germany were the Swiss firms of Durand & Huguenin in Huningue, Alsace, and of Geigy in Grenzach in the Duchy of Baden.

Other long-established German companies were also experiencing a certain boom at the time because of their inventiveness. In 1910, the firm founded by one of Liebig's students, Hans Albert, had added, independently of Baekeland, *phenol-formaldehyde resins* to the superphosphates it was manufacturing in Wiesbaden. They were to be known as Albertols. In Frankfurt, Degussa had become one of the main national producers of *cyanides*. This had led to their production of *sodium*, and of *sodamide* used in the synthesis of indigo. F. Rütgers had set up in Erkner in 1860 the first tar distillation factory on the continent, which supplied *phenol, naphthalene*, and *anthracene* to the dye manufacturers.

Fritz Raschig, established in Mannheim since 1891, was extracting *phenol* and *cresols* from English tars for the same purpose. This inventive industrialist developed the "Raschig rings" for filling distillation columns, which he patented in 1915. He was also the first to synthesize phenol from benzene, in 1900, and to produce in 1907–1908 *hydrazine* and *hydroxylamine* on an industrial scale.

From cresols and xylenols, he also produced powerful antiseptics through chlorination between 1911 and 1913. Another talented chemist, Otto Röhm, who had worked with Eduard Büchner on enzymes, developed *Oropon* for leather processing and set up with his fellow countryman Otto Haas, the Röhm and Haas Company in 1907 to manufacture and market the product. By 1909, the company had both German and American branches.

A great many medium-sized companies had also established a strong foothold in *fine chemicals*. Some of them had started off as mere apothecary shops dating back to the previous century, such as Riedel, Schering, and Merck. Others, such as Böhringer, de Haen, Von Heyden, and Knoll, had begun with the preparation of medicines and fine chemicals. Still others, in particular Haarmann & Reimer, focused on the synthesis of aromatic products. As for foreign groups, they were mainly represented by Solvay Werke for *sodium carbonate* and by Dynamit Nobel for *explosives*.

The Slowdown of England's Chemical Industry

Despite the presence of eminent German chemists during the second half of the nineteenth century, England's chemical education system was hardly adequate for its needs on the eve of World War I. The training of chemists in English universities was far too academic. There was a shortage of money for research. Teachers were badly paid and did not enjoy the same prestige as their German colleagues. And the relationship between industry and university was practically nonexistent.

The best brains, following William Perkin's example, quit industry, where they were poorly paid, turning to research or teaching. Apart from a few exceptional men, often of foreign origin, such as John Brunner, Ludwig Mond, Ivan Levinstein, or Hamilton Y. Castner, England's chemical industry leaders lacked the drive of their predecessors and were waging rearguard actions in areas that were condemned to extinction. A typical case is that of United Alkali, founded in 1890 to protect the *Leblanc soda* producers gathered within a single syndicate. Its production of soda fell from 240,000 tons in 1891 to 50,000 tons in 1913, out of a total production by then of 700,000 tons, alkalis being produced almost totally by the ammonia process. Indeed, Brunner, Mond & Company had bought up all competition. They also maintained close ties with Castner-Kellner, producers of electrolytic soda also associated with Solvay. Only pricing agreements among the three groups and protected markets in the colonies kept United Alkali going; for its process was outdated, the installations obsolete, and diversification had not been carried out in time.

In *superphosphates* also, competition on the Continent and in America had caught up with producers in England even though they had been first in the field, selling 620,000 tons in 1900, both at home and abroad. By 1913 they were no longer exporting the fertilizer. As for the *dye industry*, which originated in England with W. Perkin's mauveine, it was steadily declining. Only two national producers of some significance remained, British Alizarine, and especially Levinstein Ltd. in Manchester. Under the circumstances, home production hardly exceeded 4,000 tons in 1913, although needs were five times as great. A number of organic *intermediates* were manufactured by Read Holliday, which also produced sulfur black, and by Clayton Aniline which CIBA acquired in 1918. But even in that area England, originally a supplier of coal tar derivatives to the Continent, had to rely on imports from Germany.

On the eve of the war, for lack of appropriate technologies and business drive, Britain's dye industry was only working at half capacity. And while

German producers had joined forces, their British competitors were at loggerheads and quarreling over their remaining market shares. England had also given up all thought of setting up a *calcium carbide* industry and had to depend on imports from Scandinavia, setting back the development of acetylene chemicals at home. There again, the field was left open for the Germans.

The *plastics industry*, which was soaring in Germany, France, and the United States following the Hyatt and Baekeland discoveries, had only two manufacturers worth mentioning in England, British Xylonite for *celluloid* and Erinoid for galalith, a casein plastic used in button manufacture. Compensation for these failures came from the successes of a few well-managed companies that were to make their mark worldwide: Lever Brothers for *soaps* and *oils*, Courtaulds for *viscose*, British Oxygen for *industrial gases*, Albright and Wilson for *phosphorus* and derivatives, and Laporte for *hydrogen peroxide*.

Long-established firms also gave clout to England's role in the areas of *prescription drugs* and *fine chemicals*. Beecham had been founded in 1820, May & Baker in 1834, Bush (an aroma chemicals company) in 1851, Boots in 1883, and Glaxo in 1899. But busy keeping up old traditions, they had failed to develop original products by the time the war broke out, and because of this their profit margins were hardly on a par with their German rivals.

The Unsatisfactory Situation in France

After the glorious period which lasted from the end of the eighteenth century to the middle of the nineteenth, France had to concede first place to Germany in chemistry. France's extremely centralized educational system remained practically closed to the demands of modern life. Higher education in the universities and the Grandes Écoles continued to give privileged treatment to the humanities on the one hand and to pure mathematics on the other, at the expense of the experimental sciences. Chemistry, as it was still taught in the universities and associated institutes, involved more theoretical courses than practical training. Laboratories were more often then not kept for the personal research of professors.

Some chemistry schools had been set up, however, with the help of local authorities and businessmen. The École de Physique et de Chimie Industrielle, which still today belongs to the Ville de Paris, was established in 1882 and produced as many high-level physicists as it did chemists. In Lyon, the École de Chimie Industrielle was founded as early as 1883 by the Chamber of Commerce and regional business circles. Nancy had its École des Industries Chimiques from 1890, set up with the help of the Lorraine soda manufacturers, where the teaching of applied chemistry could rival that of the German Technische Hochschulen.

The Paris Institut de Chimie, later to become the École Nationale Supérieure de Chimie de Paris, had had three brilliant directors since its founding in 1896—Charles Friedel, Henri Moissan and Camille Chabrié. Well-trained chemical engineers graduated from the Institut Industriel du Nord in Lille as well as from the École Supérieure de Chimie de Mulhouse. But whether because there were too few of them for recognition, or because, as in England, industrial chemistry was not regarded as essential, the chemists who graduated from these schools rarely ended up in high positions of responsibility. Chemical industry leaders, in fact, were increasingly recruited from the École Polytechnique, at times even from the École Centrale des Arts et Manufactures; and they were not required to work in a plant for any length of time, even less, to make laboratory discoveries as were Heinrich von Brunck at BASF, Carl Duisberg at Bayer, or Paul Duden at Hoechst.

Conversely, scientists never bothered about the practical applications of their discoveries. Three Nobel Prizes in Chemistry had been awarded to Frenchmen: Henri Moissan (1906), Victor Grignard (1912), and Paul Sabatier (also 1912). And yet Moissan's electric furnace, Grignard's organomagnesium derivatives, and Sabatier's catalytic hydrogenation were adopted by industry without the slightest personal profit accruing to any of the three scientists.

Except for individuals such as Georges Claude, Henri Gall or Adrien Badin, most industry leaders had lost their technical creativity. In the circumstances, France's chemical industry, not surprisingly, offered a very mixed picture. In the first place, unlike its German or English competitors, it was scattered all over the country on a large number of sites that varied in importance. The industry employed 37,000 wage earners, but only nine plants employed more than 500 people.

The Compagnie Saint Gobain-Chaury-Cirey was the single largest producer in 1914 of *sulfuric acid* and *superphosphates* with 41 percent and 47 percent of France's total production, respectively. Through its merger with Perret-Olivier in 1872, it had acquired a sulfuric acid unit in Saint-Fons and pyrite mines in Saint-Bel. Through its glass factories, it had a mostly captive market in *sodium carbonate*, produced at a rate of 100,000 tons a year in the Varangeville soda works near Nancy. Phosphates came from the company's mines in Tunisia. Saint-Gobain also produced *chlorine, nitric acid, aluminum sulfate,* and *copper sulfate*, which served to make the *bouillie bordelaise* used to treat mildew in vines.

The Établissements Kuhlmann had acquired some importance through the three plants they owned in the north at Loos, La Madeleine, and Amiens, which also produced superphosphates, copper sulfate, and sulfuric acid. Though the European leader in superphosphate production, France played only a minor part in *nitrogen fertilizers*. The country's coke furnaces did not yield any significant quantities of ammonium sulfate, and nitrates had to be imported from Chile or Norway, where French interests had a stake in Compagnie Norvegienne de l'Azote (Norsk Hydro).

Suitably modified, the Leblanc soda works were making hydrochloric acid for bone processing, sodium sulfate for the production of glass, and especially superphosphates from their sulfuric acid.

In their Dombasle plant near Nancy, Solvay accounted for two thirds of France's *sodium carbonate* production. The production of *electrolytic sodium* was still small, probably because the need for chlorine in bleaching powder for cotton and linen was not as large in France as in England.

The harnessing of waterfalls in the Alps and the Dauphiné region sparked an imaginative electrochemical and thermoelectric industry in a number of special sectors. Under the skillful management of H. Gall, the Société d' Electrochimie was producing chlorates through electrolysis, competing with Corbin in Chedde and with Saint-Gobain in Compiègne. Adapting the Moissan electric furnace technology, in 1896 the company started developing the production first of *calcium carbide*, then of *carbon* and of *graphite electrodes*.

Beginning in 1906, it supplied *calcium cyanamide* through the Société des Produits Azotés. At the same time, in association with Degussa, it was developing *sodium* and *cyanide* production in the Romanche valley and in Martigny, Switzerland. At Saint-Fons and Villers, peroxide products were being developed.

In 1907, at the time the Héroult patents became public property, the Société d'Electrochimie became involved in *aluminum* production and was awarded a 20 percent share of the national market when the Aluminium Français was set up in 1911. The consortium also included the Société de l'Arve which, besides the aluminum produced in the Pyrenees, manufactured Corbin's chlorates in Chedde. But more importantly, Aluminium Français also encompassed the Société de Froges as well as the Compagnie des Produits Chimiques d'Alais et de Camargue.

By the time war broke out, France was producing 15,000 tons of aluminum a year, thus acquiring a leading position in Europe. At that time, Aluminium Français was setting up a plant in North Carolina. Under the management of Adrien Badin, the Compagnie des Produits Chimiques d'Alais et de Camargue, successor of A. R. Péchiney, had not abandoned its chemical activities. The company produced copper sulfate near Avignon as well as in Salindres where *aluminum sulfate* was also made. Since 1903, *aluminum fluoride* and *artificial cryolite* needed for aluminum plants had been added. With closure of its Leblanc soda works, the company was supplied with sodium carbonate by Solvay which, in turn, received salt from the Camargue salt marshes.

Although France had abundant supplies of hydroelectricity, electric furnace *phosphorus* was not produced in the country before 1925. Coignet and Company in Lyon merely produced phosphorus as a byproduct of the *bone glue* and *gelatin* they manufactured.

Like England, France missed her chance in the *dye* field. The Société

Chimique des Usines du Rhône had abondoned this activity in 1905 and the only noteworthy French producer was Société Anonyme des Matières Colorantes et Produits Chimiques de Saint Denis, whose chemist Vidal discovered sulfur black. St. Denis produced some 800 tons a year of dyestuffs, which met only 10 percent of the country's needs. There was no national production of *organic intermediates* to speak of. Customs tariffs favored their import while encouraging foreign firms, mainly German and Swiss, to set up production units in France to handle the final processing stages.

Patent legislation which protected the finished product also discouraged any move to produce existing dyes through more innovative processes. But there were areas of organic chemicals where French chemists were more successful. Because of the contributions of De Laire, Roure, Bertrand, Dupont, the Givaudans, and others, and boosted by the natural products available in Grasse for new materials and by the Paris perfume houses downstream, *synthetic aromatic chemistry* held a more than honorable place.

With the French pioneers of photography and cinema, Lumière in Lyon and Pathé in Paris, a prosperous *film* industry was building up. These chemistry-related industries were, in turn, to boost the development of two companies, the Usines du Rhône in Lyon, and Poulenc Frères in Paris. The Société Chimique des Usines du Rhône was the more important of the two. The Saint-Fons plant already employed 700 workers and 27 chemists in 1900 and developed further under the careful management of J. Koetschet and N. Grillet.

Production of *ethyl chloride, phenol, hydroquinone, salicylic acid*, and *acetic anhydride* had very naturally been extended in 1902 to *acetyl salicylic acid*, sold under the trade name of aspirin by Usines du Rhône, and then to *cellulose acetate*. The company also produced *vanillin* since 1894, to which subsequently were added other base products for perfumery, such as *coumarin, acetophenone*, and *linalool*. Thus although dyes had been abandoned, the Usines du Rhône had a booming production of chemicals for the photographic, pharmaceutical, and perfume industries despite sharp German competition. In 1914, they took over the Société Normande des Produits Chimiques, founded four years previously to supply tanneries with *oxalic* and *formic acids*.

As for the Établissements Poulenc Frères, they were turning to synthetic drugs under the instigation of Camille Poulenc, Étienne's third son, who had joined the business in 1893 after studying under Moissan. Meanwhile the photographic department was still going strong.

It was Camille Poulenc who had engaged the young chemist Ernest F. A. Fourneau. Fourneau invented *stovaine*, a substitute anesthetic for cocaine, in 1906. Stovaine was synthesized from the organo-magnesium compounds discovered by Victor Grignard not long before. When war broke out, the Vitry plant was on stream and Poulenc had added other specialities such as atoxyl, antodyne, and asquinol to stovaine.

The traditional silk manufacturing industry in Lyon had beneficial effects on France's chemical industry. The Société des Produits Chimiques Gillet & Fils, later to become Société Progil, had started off with dyes and tanning extracts and had then diversified into charcoal, becoming also a successful supplier of tin chloride and other finishing products to the Lyon area textile industries. Very naturally, following the impetus which the Comte de Chardonnet had given *artificial silk*, Gillet & Fils became involved in new fibers. They had a part in setting up the Comptoir des Textiles Artificiels which by 1913 was marketing 2,000 tons of cellulose fibers made in France. They also sparked the development of *cellophane*.

But although the Société Française du Celluloid, founded by the Hyatt brothers in Stains, was to be the first in Europe to produce this artificial plastic on an industrial scale, France ranked only third in this area in 1913, behind the United States and Germany, with a celluloid production of 3,600 tons scattered among three companies. The Oyonnax region in France's Jura mountains, traditionally geared to woodworking cottage industries, took the switch to celluloid in its stride and was manufacturing a variety of plastic products.

Adhering to a tradition dating back to the Régie Royale des Poudres at Salpêtres, France's administration, unlike all other governments in Europe, maintained a state monopoly over *explosives*, including dynamite. Only because this law was not enforced did private industry manufacture both *dynamite* and *nitrocellulose* for civil uses.

Switzerland's Chemical Industry: An Exemplary Case

Switzerland had few natural resources on which to build a prosperous chemical industry. Yet, it took advantage of its hydroelectricity to develop, on the one hand, *aluminum* production with Alu Suisse, and on the other a *calcium carbide* and by-products industry with Lonza. At the same time, *electrolytic soda* was produced in Turgi and Monthey based on the salt mines of the Basel, Aargau, and Valais cantons. Two plants manufactured *sulfuric acid*, both near Basel in Uetikon and Schweizerhalle.

As for bulk *organic chemicals* and *synthetic intermediaries*, Switzerland could not hope to compete with the large units in neighboring Germany. Leaders of Switzerland's dye industry, however, cleverly established a network of mutually profitable relations with the large German chemical concerns from which they bought their intermediates. They were careful to avoid competition in the area of the larger markets of finished products (alizarin, Congo red, indigo). Instead, they focused their development on specialties that were either original or highly technical and sold them worldwide with comfortable margins. By 1913, four Basel factories were exporting 9,000 tons of various dyes. The main company was CIBA (Gesellschaft für Chemische Industrie Basel), which employed 2,900 people, including workers in overseas subsidiaries. CIBA maintained close links with BASF dating back to the time when they were developing rhodamines in common. A fair share of the company's activity was in pharmaceutical products, making antipyrine since 1887, and the antiseptic Vioform which was a great commercial success.

J. R. Geigy SA was turned into a joint-stock company in 1914 and, while remaining active in dyestuffs and tanning extracts, managed to develop special dyes particularly for wool dying, with the help of the great chemist Traugott Sandmeyer. Sandoz focused on azo dyes, but also produced sulfur-based dyes, which Boninger had synthesized. Although the smallest of the four Basel firms, Durand et Huguenin was nonetheless just as prosperous.

Another Basel company, F. Hoffmann-La Roche, had acquired an international reputation in *pharmaceuticals*. It was founded in 1894 by Fritz Hoffmann, husband of Adèle La Roche. Hoffmann-La Roche formed a partnership with Max Carl Traub, and by 1913 was employing over 900 people including 400 abroad. Other enterprising firms such as Firmenich and Givaudan in Geneva were involved in *perfume chemicals*. In addition, TNT production had been started in 1913 by the Fabrique Suisse d'Explosifs, established in Dottikon near Zurich at the Swiss administration's request.

For the education of its chemists and their training for industry, Switzerland had the world-renowned Zurich Polytechnic, founded in 1855. Professors from all over the world came there to teach, among them Switzerland's CIBA director Gnehm, Germany's future Nobel chemistry Prize winners Richard Willstätter and Hermann Staudinger, and the analytical chemist F. P. Treadwell from the United States.

Chemicals in the Other Countries of Europe

In 1913, the Belgium-based Société Solvay celebrated its fiftieth anniversary. Solvay had developed to multinational dimensions, and produced directly or in partnership with local firms some two million tons of *sodium carbonate*, accounting for 90 percent of world production. It had set up in 1908 a soda plant in Torrelavega, Spain, and was building another in Rossignano, Italy, just before the war. In Austria, Solvay had formed an association with Verein für Chemische and Metallurgische Produktion AG, in Russia an association with the Liubimov brothers.

Elsewhere in Europe, where chemistry was in the early stages of development, countries had each focused on special areas. With *sulfur from Sicily* and phosphates imported from Tunisia, Italy was producing over 900,000 tons a year of *superphosphates*. In addition, it ranked first on the Continent as a *copper sulfate* producer, and the Piano d'Orta *calcium cyanamide* unit, set up in 1905, had an annual capacity of 14,000 tons. Sweden's Nitro Nobel Company, founded in 1864 by Alfred Nobel, produced dynamite. It had become a multinational, even though British Nobel in England, Société Centrale de

Dynamite in France, and Dynamit Nobel AG in Germany were to go their different ways in their own countries. In 1902, Alfred Nobel had bought up the Bofors armaments company. Accordingly, the Nobel Kemi division started manufacturing *smokeless powder*, and Bofors Nobel earned a world-wide reputation in *nitration*.

Another important company was Stockholm's Superfosfat Fabrik, set up in 1871. It manufactured *fertilizers* and had developed *carbide* and *chlorate* productions, using its own hydroelectric power. These two groups were to merge in 1978 to form Kema Nobel AB. Since its inception in 1881, Perstorp had specialized in the manufacture of charcoal as well as in the production of carbon black and acetic acid. In 1905, it started producing *formaldehyde* and, subsequently, phenol-formaldehyde resins. With the establishment of AGA AB in 1904, Sweden was preparing to become a leading producer of *industrial gases*. Norway had broken away from Sweden in 1905 and called upon foreign capital to set up major units using the country's hydroelectric resources.

Kellner-Partington Pulp was an English equity company, later to become a branch of A/S Borregaard. By 1914, it was employing about one thousand people to produce *calcium carbide*, *ferrosilicon*, and *electrolytic soda*. Another subsidiary of a British company, North Western Cyanamid Company, used the Frank and Caro process to produce close to 80,000 tons a year of *calcium cyanamide* for export to England. Unquestionably, the main Norwegian company was Norsk Hydro (Société Norvégienne de l'Azote). The company used the electric-arc Birkeland and Eyde process and, when war broke out, was one of the world's major producers of *nitric acid* and *calcium nitrate*.

In Tzarist Russia, the difficulties of long-distance transport and the rigorous climate had not been conducive to any appreciable development of the chemical industry. Protected by high tariff barriers, chemical companies were profitable enough, however. This explains why they attracted foreign capital. Production efforts were essentially focused on alkalis and sulfuric acid. Besides the Liubimov-Solvay association in sodium carbonate in which some fifteen hundred people were employed in three soda works, a number of electrolysis units had been set up with German capital and the Griesheim cell technology.

Use of superphosphates was not yet widespread and sulfuric acid production, which had climbed to 160,000 tons by 1913, was essentially intended for the oil refineries in Baku where the Nobel brothers, who were the largest crude producers in Russia, made their own sulfuric acid using the Tentelev contact process. It was also intended for the preparation, particularly by Ushkov and Tentelev, of *sodium sulfate*, *hydrochloric acid, chlorine* by the Weldon process, and *bleaching powder*.

Just before World War I, some progress had been made in other areas of Russia's chemical industry. Furnaces set up in the Don Basin began to supply tars and *ammonium sulfate* from 1910 onwards. *By-products of wood distillation* were being supplied for export, especially by a subsidiary of Schering's.

Getting around high customs tariffs, the great German and Swiss dye companies, attracted by a 7,000-tons-a-year market, were all active in Russia transforming their intermediates by synthesis or refining the dyes imported from their mother companies.

The Striking Progress of America's Chemical Industry

In 1914, the time was long past since the young United States had been sanctuary for Joseph Priestley, chased from England because of his sympathies for the French Revolution, or for Irénée du Pont de Nemours, shocked by the death on the scaffold of his master Lavoisier into leaving a country ruled by what he saw as irresponsible revolutionaries.

In all areas, chemicals in America had made considerable progress. Practical teaching of organic chemistry, notably at the Johns Hopkins University in Baltimore under the leadership of Ira Remsen, and at the University of Chicago, was happily influenced by German university teaching since many young Americans had studied in Germany at the end of the previous century.

By 1915, admission to institutions of higher education had become more selective, studies were frequently extended to the doctoral level, and professors were easier to attract because the pay was significant. But more importantly, there were no obstacles, as too often was the case in Europe, to the establishment of a harmonious relationship between university and industry. The best scientists were willing to become involved in industrial problems.

The chemist Arthur D. Little, who introduced the teaching of unit operations for chemical engineering students at MIT, where he had himself been a student, had his own firm of consultants from 1886. In that capacity, he had a share in developing a process for the preparation of sulfite pulp, and in extending applications of cellulose acetate.

Thanks to wealthy businessmen such as Andrew Carnegie, Andrew William Mellon, and John D. Rockefeller, generously endowed foundations contributed to the funding of research centers in a large number of universities. Although the United States could not yet boast a long tradition of chemical work, a sufficiently large number of chemists and chemical engineers were being trained by the time war broke out to ensure the development of America's fledgling industry. By 1914, the American chemical industry employed 67,000 people and was self-sufficient in most basic products, the exceptions being *potash* and *nitrate* imported from Germany and Chile.

The Americans had built up a soda industry under the protection of high customs tariffs after having been forced for a long while to rely on imports of alkalis from England because they had failed to develop Leblanc soda plants. Solvay Process, with its three soda units, was dominant in *ammonia soda*. Rivals were Columbia Chemical and Diamond Alkali, associated with glassmakers. Others were Michigan Alkali in Wyandotte, which was beginning to export part of its production, and Mathieson Alkali, already established in Virginia, which was diversifying in Niagara Falls in electrolytic soda. Cheap electricity had attracted to Niagara Falls other *electrolytic soda* producers, such as Hooker Electrochemical in 1906 and Niagara Alkali in 1909. There was also Pennsylvania Salt Manufacturing Company, which owned an electrolysis unit since 1898 in Wyandotte, Michigan, close to rich salt deposits.

As to *chlorine* and *hypochlorite*, they were increasingly being used for paper pulp processing.

With 835,000 tons of sodium carbonate and 260,000 tons of caustic soda produced in 1914, the United States' alkali industry was beginning to be of concern to English producers Brunner, Mond, and United Alkali. They had already been ousted from the American market and were now being threatened in their other traditional outlets.

Sulfuric acid production, for which the United States was the world leader, received a boost from Texas and Lousiana *sulfur* obtained through the Frasch process, from the introduction of the contact method, and from its use for paraffin processing in oil refineries, on the one hand, and superphosphate production, on the other. Besides the traditional producers such as Harrison Brothers in Philadelphia, Grasselli in New Jersey and Alabama, and Merrimac in New England, twelve sulfuric acid manufacturers had associated in 1899 as General Chemical, under the impetus of the chemist William Nichols. What was the most important American chemical group at the time was starting to diversify just before the war in *aluminum sulfate, phosphoric acid, sulfites*, and *fluorine*.

Superphosphates accounted for a large share of chemical activity. Some three million tons were being produced in the States by 1913, about a quarter of the world production at the time. Half this amount was produced by three manufacturers. Although not integrated in sulfuric acid, their units were located in Florida, in South Carolina, and in Tennessee, close to their own phosphate mines and to the cotton and tobacco plantations that constituted their main markets.

Ammonium sulfate, which was extensively used in Europe, still played a minor role in the United States. On the other hand, not far from Union Carbide's *calcium carbide* unit, American Cyanamid Company, established on the Canadian side of Niagara Falls since 1909, had raised its production capacity of *calcium cyanamide* to 57,000 tons by 1914.

In a more specialized field, Philadelphia Quartz (PQ), founded by Joseph Elkinton in 1831, had given up soapmaking in 1904 and turned to the sole production of *silicates*, promoting their use as detergents and in adhesives, especially for corrugated cardboard. Other enterprising individuals had launched into various inorganic chemical ventures. One of them was Herbert Dow, who after many setbacks both financial and technical had kept afloat his Midland plant, which used Michigan brine, by perfecting his electrolysis process, overcoming competition from German bromine and English bleaching powder. From 1903, Midland Chemical, which Dow was to absorb in 1914, produced *chloroform* on an industrial scale from *carbon tetrachloride*.

On the West Coast, where borax mines had been discovered in northern California's Death Valley, Francis "Borax" Smith had sold to English interests what was to become the U.S. Borax & Chemical Company. In San Francisco, Stauffer Chemical had become involved in *sulfur* refining and in the production of *inorganic acids, superphosphates*, and *cream of tartar*. Stauffer Chemical had been founded in 1886 by John Stauffer and Christian de Guigné, the former of German origin and the latter a native of France. After the 1906 earthquake, Stauffer rebuilt his ruined units in the same spot but set up a *carbon disulfide* production unit in New Jersey in 1912.

The prospects opened up by the American market were attracting other European investors anxious to turn their technology to account. Roessler and Hasslacher, under the control of Germany's Degussa, were using the Castner process in their Niagara Electrochemical subsidiary to produce *sodium* and *cyanides* and were starting to produce *hydrogen peroxide* and *perborates*. England's Albright & Wilson were also producing *phosphorus* and *sodium chlorate* in Niagara Falls in the Oldbury Electro-Chemical Company units. Aluminium Français was trying to establish a business in North Carolina, and Air Liquide was linking up with Air Reduction in the area of industrial gases.

Increasing requirements in precision tools boosted the *abrasives* industry, and in this area American companies made promising breakthroughs. Since 1895 Edward G. Acheson had produced in his Niagara Falls electric furnace an exceptionally hard *silicon carbide* called *carborundum*. As the years went by, grindstones and various refractory products were added. Another company that was to become famous was the Minnesota Mining & Manufacturing Company (3M), founded in 1902 by Midwest investors to develop a corundum mine. They had diversified in 1907 in abrasive papers and adhesive tapes under the clever impetus of the Scottish William McKnight. The trade name "Scotch" is still today a reminder of him.

America's *organic chemicals* industry still largely centered on natural products. Wood distillation in the pine forests of the South had given the United States world leadership in *methanol* and *calcium acetate* from which acetone was derived. These products had initiated a steady current of exports. The same was true of *turpentine* and *rosin* extracted from pine gum, more especially by Yaryan, and Newport. *Cellulose derivatives* had developed due to ready access to cotton linters and wood pulp.

Since 1911, American Viscose, a subsidiary of Courtaulds, was manufacturing *rayon* in Marcus Hook, Pennsylvania. *Nitrocellulose* was finding increasing and diversified uses in celluloid, lacquer, films, and glazed cloth. Du Pont de Nemours, which produced nitrocellulose in significant quantities for *explosives*, was becoming interested in these applications. Firmly managed by Pierre S. du Pont, the company had been forced in 1912, under antitrust regulations, to divest a number of units that produced black gunpowder and dynamite and thus gave birth to two new companies, Hercules Powder and Atlas Powder. Nonetheless, Du Pont still remained the sole civilian supplier of smokeless powder and the main dynamite producer, employing some 5,000 people. In 1910 Du Pont bought up Fabrikoid Company, before acquiring Arlington Company, the largest of the four celluloid producers in the United States, five years later.

At the time World War I broke out, the United States was lagging behind Europe in *organic synthesis*. When Semet Solvay started to be concerned in 1900 about ammonium sulfate supplies for its Syracuse soda works and put the benzene, toluene and naptha solvents from its coking furnace distillation units on the market, there were still very few takers for these aromatic hydrocarbons in the United States.

True, some tar processing units had existed since 1896 at Barrett's or since 1900 at Peter Reilly's, but they had been designed for illuminating gas or for creosote production. It was not until 1910 that an association of General Chemical, Semet Solvay, and Barrett, named Benzol Products, started offering *aniline* and *nitrobenzene*.

It was the failings of America's *dye industry* that explained the low demand for synthetic intermediates at the time. Satisfied with their dye imports from Germany, textile manufacturers had not pressured the United States administration to establish patent legislation or a customs tariff that might have favored the development of national dye production. The complex reactions involved and the discontinuous nature of the required operations had discouraged the industry from this type of synthesis. For all these reasons, and despite the fact that at the time the United States and China were the largest world consumers of dyes, there were only two producers worth mentioning in the U.S., Schoelkopf Anilin and Heller & Merz. But the 3,000 tons a year they produced locally only accounted for one-eighth of the country's needs.

Even *fine chemicals* were having a hard time because of dependence on overseas intermediary products. Until 1914, Monsanto, for one, was greatly handicapped by the need to import from its German competitors the phenetedin required for phenacetin production, and the phthalic anhydride for making phenolphthalein. Roessler and Hasslacher, however, made their own formaldehyde in Perth Amboy, while Heyden made its own in Passaic, New Jersey, supplying General Bakelite, which had put its phenol-formaldehyde resins unit on stream in 1910.

There were two traditional chemical industries in the United States which were not affected by organic synthesis failings. Well-integrated in fats, *soapmakers* experienced booming times. This was particularly the case for Procter & Gamble and Emery in Cincinnati, which had been in business since 1837 and 1840, respectively, and for Colgate in New Jersey, which had brought out its Palmolive soap in 1898 and celebrated its centenary in 1906.

Still very much in the hands of their founding families, *pharmaceutical laboratories* were expanding rapidly, each in its own specialty. One of the oldest was Pfizer, founded in Brooklyn by the Germans, Charles Pfizer and Charles Erhart. While still making *anthelmintics, iodides, boric acid*, and *cream of tartar* in 1880, they began producing *citric acid* from calcium citrate imported from Italy. The laboratory founded by Edward R. Squibb, a naval surgeon, owed its early success to *purified ether* produced since 1852. Under the name of Squibb & Sons, it was sold in 1905 to private investors. Merck had initially been a subsidiary of the Darmstadt parent company. Its American development dates back to George Merck's arrival in the United States in 1891. Merck produced *chloral hydrate, bismuth salts, salicylates* and *alkaloids* in Rahway, New Jersey.

Parke-Davis, set up by Henry C. Parke and George E. Davis, had its origin in the Duffield-Conant apothecary shop in Detroit in 1866. Since 1897 it has mainly focused on *serums* and *vaccines*. Searle, which was later to become involved in biological products, was established in Chicago since 1890. Gideon Searle was one of the first to offer some of his specialties in the form of *tablets* in 1892. Eli Lilly had set up shop in Indianapolis in 1876, and produced *alkaloids* and *plant extracts*. In 1909, the company had developed capsules that were a novelty.

Initially, the firm of William E. Upjohn and his brothers had been set up in Kalamazoo, Michigan, in 1885, to prepare pills under a new process. Only in 1913 did it turn to research. The company which the three Mallinckrodt brothers had set up in Saint Louis in 1867, for its part, had extended its initial pharmaceutical wholesale business to the production of *narcotics* and special products such as *hydrogen peroxide*, and *gallic* and *tannic acids*. Safe from foreign competition, these family businessmen generally flourished. Unlike Germany's pharmaceutical companies, however, they had not yet developed research divisions likely to produce original specialities.

The Awakening of Japan

Japan was a latecomer to chemicals. Sumitomo had developed a *sulfuric acid* production based on its Beshi copper pyrites mines that was supplied, as early as 1887, to Tokyo Fertilizer, the future Nissan Chemical Industries, to make *superphosphate*. Also since 1908, Nippon Nitrogen Fertilizer had introduced the Frank and Caro *calcium cyanamide* process. But *ammonium sulfate* was still little used.

After the conquest of Formosa, Japan had become the world's leading *camphor* producer and began developing a *celluloid* industry from 1908 onwards. The Japanese were also large *iodine* producers, extracting it since 1888 from seaweed in the Hokkaido area. Sobo Marine Products had been set up in 1908 by Mori, the founder of Showa Fertilizer, later to become Showa Denko.

As early as 1898, Osaka Gas, Mitsui Mining and Mitsubishi had introduced Semet Solvay furnaces in their plants, using the coke from their coal mines and beginning to become interested in tar distillation products.

In both *alkalis* and *dyes*, Japan's chemical industry was at a fledgling stage. On the one hand, those chemical engineering graduates from universities and technical colleges—some 3,000 in 1917—who did not turn to teaching were, for the most part, employed by the textile, porcelain, or food industries and made no contribution to chemical production. On the other hand, the Japanese had a number of obstacles to clear in *alkalis*. The Solvay process was complex and hard to copy; and the Belgians, who did not grant licenses, had awarded the Japanese export market to Brunner, Mond. Markets were scattered, and both textile and soap manufacturers were still of small size. Matters were even less favorable in electrolytic soda. Various processes existed, but the high price of salt and the scarcity of hydroelectric power prevented developing them. Competition with sodium carbonate was difficult. Despite a busy textile industry, Japan depended entirely on synthetic dyes from abroad, although it had its own natural indigo production.

World War I and Its Impact on World Chemicals

Because it lasted longer than expected and quickly spread over a vast area, World War I had a considerable impact on the chemical industry of the nations at war. Strong demand for a great many products entailed significant production capacity increases and spurred the development of new manufacturing processes. The disruption of traditional trade patterns and the need to ensure raw material supplies to keep plants operating for the war effort, meant that government intervention was needed either to share out what was available or to subsidize production of substitutes. For geographic and historical reasons, the various countries drawn willingly or unwillingly into the war suffered its repercussions in different ways. For all of them, however, the first problem to crop up involved nitrogen.

Nitrogen Supplies

Until war broke out, four-fifths of world nitrogen consumption went into fertilizers in the form of ammonium sulfate and calcium cyanamide; and two-thirds of it originated as sodium nitrate from Chile. Now *nitric acid* was essential to produce explosives, and at the time practically all of it came from Chile's nitrates.

Germany's military High Command, expecting a lightning war, had not provided for large nitrate stocks. England's blockade rendered imports from overseas a risky business. The German government had to take a number of palliative measures: for the cyanamide use as a fertilizer for farmers, it financed the Trostberg and Knapsack capacity extensions and erected two plants in Piesteritz and Chorzow; for residual sulfate it shared out the quantities supplied by the gas plants and coking furnaces among farmers and manufacturers. More importantly, it considerably increased BASF's synthetic ammonia production in Oppau, then in April 1916, during the battle of Verdun, it helped BASF to build a new synthetic unit in Leuna, near Merseburg. By 1916, synthetic ammonia accounted for 45 percent of the Reich's nitrogen production (about 95,000 tons a year), and served exclusively to produce nitric acid for explosives using the Kuhlmann-Ostwald method: undoubtedly, the Haber-Bosch process made it possible for Germany to extend the war until November 1918.

While the French and English had no access to BASF technology, of course, they had other nitrogen sources. Admiral Maximilian von Spee did manage, early in the war, to hinder transport of Chilean nitrate to the Allied countries, but the defeat of his fleet in December 1914 off the Falklands improved the situation, at least until the German submarine attacks of 1917.

England also received from subsidiaries of British companies in Alby, Sweden, and Odda, Norway, about 80,000 tons a year of cyanamide for fertilizers; and France bought from Norsk Hydro the equivalent of 30,000 tons a year of nitrogen in the form of calcium nitrate. Britain's own gas factories and coking furnaces also produced significant quantities of sulfate, estimated to amount to some 90,000 tons a year of nitrogen. As there were not enough nitric acid units, the sulfate was directly turned into ammonium nitrate.

As France's ammonium sulfate supplies were not so abundant, the cyanamide installations of Société d'Electrochimie in Notre Dame de Briançon and in Bellegarde were extended, making it possible to produce additional quantities of nitric acid from the ammonia obtained. Finally, from the very start of the war, the Allies were able to obtain explosives from the United States, while the Central Powers could not.

Particularly solicited was Du Pont de Nemours, then the only company in America capable of providing the much-needed smokeless powder, especially for the 75mm gun. By the end of the war, it was delivering over 40 percent of the Allies' standard explosives and its guncotton plant in Hopewell, Virginia, employed 28,000 people. Ammonium sulfate production was boosted for the purpose, in particular through installation of the vertical furnaces of the Koppers Company, and increased from 45,000 tons of nitrogen in 1915 to 70,000 tons in 1918.

The Demand for Organic Chemicals

In 1914, Germany accounted for 90 percent of the world's synthetic dye production. The figure gives a measure of the effect its loss during the war had on other industrialized nations, which, moreover, were hardly prepared to satisfy their own needs in organic intermediates. Furthermore, a number of basic organic products were needed to make explosives, such as phenol for picric acid and toluene for TNT. The Germans had also exported before the war a large part of their chemotherapeutic derivatives.

In *England*, the government took prompt action to deal with the situation, which had become serious. The Board of Trade signed an agreement with Swiss dye manufactures to replace German suppliers under which the British supplied intermediates and received finished products in exchange.

Production of synthetic phenol was restarted at Read Holliday. Since the Boer war, the company had been producing trinitrophenol used to make Lyddite. Clayton Aniline considerably increased its TNT production. For the first time, toluene was extracted from Borneo oil. Read Holliday, having specialized until then in synthetic intermediates, was asked to widen its production range in Huddersfield in order to increase dye production. In 1915, it set up British Dyes Ltd. in association with Calico Printers and Bradford Dyers, with the help of the British government.

Ivan Levinstein, who never got along with his rivals, extended his production to intermediates and was encouraged by the British Board of Trade to buy up Perkin's old business, Claus and Ree in 1917, as well as the Hoechst indigo unit in Ellesmere Port. In 1918, British Dyes and Levinstein merged to form the British Dyestuff Corporation.

Acetone was a gelatinizer needed for the preparation of cordite (smokeless powder), made from nitroglycerin and guncotton. Until the war acetone had been imported from the United States, where it was a byproduct of the wood distillation industry. In 1915, when demand for cordite was growing, Winston Churchill was First Lord of the Admiralty. He became interested in the work being done by Chaim Weizmann in Manchester, whose fermentation process produced 37 tons of a one-part acetone–two-parts butyl alcohol mixture from 100 tons of cereals or starch. From then on, gin distilleries and breweries were fitted out for the same purpose. An industrial unit to produce acetone by fermentation was built by the Navy and came on line in 1917 at Holton Heath. Toward the end of the war, the Weizmann process was established by the English in Terre Haute, Indiana, where the Commercial Solvents Corporation was founded in 1919.

When the war broke out, *France* did not extract significant quantities of aromatic hydrocarbons from coal tar, and plants in the north were occupied by the Germans at the start of hostilities. Benzene and toluene, therefore, had to be imported from England and the United States until a better balance was achieved in 1916. Only then did the government make real efforts to increase the production of dyes and synthetic intermediates. The only French company of any importance in that area, until then, had been SA des Matières Colorantes et Produits Chimiques de Saint-Denis.

The Hoechst plant in Creil and the BASF plant in Villeneuve Saint-Georges were requisitioned; and in February 1917, the government helped fund a new company, Cie. Nationale des Matières Colorantes (CNMC), which employed 2,000 workers.

As it had for England, Swiss industry remained a valuable supplier of special dyes to the French industry throughout the war. Phenol production was boosted when the new unit of Société des Usines du Rhône came on stream in Péage de Roussillon in 1915 with a capacity that was soon to soar to 100 tons a day. Société des Usines du Rhône was to render other services to the allied cause. As early as 1915, its Saint-Fons unit met the air forces' increasing demand for cellulose acetate, which was used to coat the wings and bodies of planes. It satisfied a significant part of the army's health services' needs in salicylates and pyrazoles. In 1917, it resumed production of saccharin to make up for the sugar shortage.

The *Germans*, who were masters in the art of organic synthesis, did not experience the same problems as their enemies at the beginning of the war and even managed until 1916, through the medium of neutral countries, to exchange some of their dyes and pharmaceuticals production against essential raw materials like cotton, rare metals, or rubber. Afterwards, they were handicapped by the British blockade and had to resort to substitute products. Thus Bayer succeeded in preparing synthetic rubber from methyl isoprene (2,3-dimethylbutadiene), starting from the acetone which Hoechst and Wacker extracted from acetic acid in their respective units in Knapsack and Burghausen.

The *United States* only came into the war in April 1917, but very early on felt its effect on supplies of organic derivatives, particularly of dyes. Phenol and benzene extracted from tars were directed, on a priority basis, to the explosives industry to meet strong Allied demand in that area. And although there was no shortage of synthetic dyes at first because the agents of German firms continued to import them from parent companies or to deliver them from U.S. plants, prices began to soar.

In order to boost national production, the United States government, less interventionist than its European counterparts, had set up a consultative commission in October 1916, putting financial expert Bernard Baruch in charge of the country's raw materials. Baruch managed to get American businessmen to cooperate in a manner that turned out to be highly profitable for the country. Congress had established a five-year tariff barrier in September 1916. Accordingly, companies such as Du Pont in Deep Water Point, New Jersey,[1] Calco in Bound Brook, New Jersey, and Dow in Midland, Michigan, all launched into dye production.

In May 1917, a number of national dye producers including Schoelkopf, National Aniline, Barrett, as well as Benzol Products for part of its range, formed the National Aniline & Chemical Company, a voluntary association accounting for a total production of 16,000 tons a year, which left only Du Pont as a serious competitor.

Under the common impact of increased production and easier access to synthetic intermediates, prices that had soared fell appreciably. By 1918,

[1] With the technical help of Levinstein in England.

America's dyestuffs industry, which had started practically from scratch three years previously, employed 178,000 people and supplied the country's needs, at least insofar as quantity was concerned.

The Emergence of Combat Gases

Combat gases had been the subject of scientific studies as soon as war broke out. In Germany, Fritz Haber conducted in-depth research on the subject at the Berlin Kaiser Wilhelm Institute.

The first chemical attack was launched with chlorine by the Germans on April 22, 1915, during the battle of Ypres. Chlorine was replaced by phosgene and was used by the French at Verdun in February 1916. Then in July 1917, the Germans used mustard gas (dichlorodiethyl sulfide), or ''yperite,'' at Ypres. This blister gas attacked not only the lungs but also the skin, so that a gas mask no longer provided sufficient protection.

By the end of the war, on both sides of the front, more than a quarter of the munitions used contained chemical agents. As they all contained chlorine or chlorinated compounds, the problem of preparing and transporting liquified chlorine and turning it into asphyxiating gases became acute. The Germans were able to solve it fairly easily because the dye factories that did the job had already used chlorine as well as phosgene in their syntheses before the war; by 1893, BASF's Rudolf Knietsch had developed a chlorine liquefaction process.

The French were not so well prepared; after the first attack at Ypres they had to improvise hastily. A Service du Matériel Chimique was established by the government, electrolysis units were set up in the Alps, and the Compagnie Alais et Camargue's plant in Saint-Auban delivered its first liquid chlorine contingent in September 1916. Including the units set up by the Service des Poudres, France's total chlorine production during the war amounted to 24,000 tons. In addition, 16,000 tons of phosgene were produced in 1918 alone.

Some of the production was subcontracted by the Service du Matériel Chimique to private concerns. The Établissements Poulenc, in particular, produced phosgene in Vitry and the Usines du Rhône Yperite produced it in Saint-Fons. In England, Castner-Kellner Company had increased its chlorine liquefaction capacity from 5 to 150 tons a week in 1915 at the request of the British government; but a large part of its production was for dye synthesis as well as disinfecting water.

The Difficult Return to Peacetime Conditions

Disastrous in its irreparable loss of human life, the 1914–18 war also seriously affected the economies of the warring countries. Chemicals were no exception to the general rule.

Besides the territories it lost, Germany also lost the near-monopoly it had held over *synthetic dyes* until 1914. Its absence from overseas markets had left the field open to powerful integrated producers in England, in the United States, and, to a lesser extent, in France and in Japan. As for the Swiss, they very cleverly managed to supplant their German neighbors during the war by signing agreements with French and English industry.

Germany's absence had at the same time left a *potash* gap, particularly in the United States, where local resources of this major fertilizer were beginning to be developed, especially by American Potash Corporation in Searles Lake, California. The war had also especially sparked associations in belligerent countries that could only make the competition the large German corporations had to face much stiffer.

Concentrations had even taken place in Germany, where an association had been formed between the Little IG and Dreibund in 1916. The new IG had seized the occasion to incorporate Griesheim Elektron and Weiler-Ter-Meer. In the United States, as we have seen, National Aniline & Chemical had been formed from the merger of a number of dye producers. In 1917, Union Carbide, Electromet, National Carbon, Linde Air Products, and Prest-O-Lite joined forces to set up Union Carbide and Carbon Company.

In England where all the explosives factories were to merge under Nobel Industries, British Dyestuffs was set up in 1918. In France the Compagnie Nationale des Matières Colorantes was established as part of the Établissements Kuhlmann. In the United States, while a number of companies, until then confined to their home market, had prospered beyond all expectation, others had taken advantage of the interruption of trade with Germany to diversify their productions. Previously threatened by competition from overseas until 1914, Dow Chemical and Monsanto could now add organic products to their range, particularly phenol and its derivatives.

By increasing its production of smokeless powder from 4,500 tons to 130,000 tons a year to meet Allied requirements, Du Pont, for one, had built up cash reserves of nearly $90 million, which it would now invest in new areas.

The government in France had encouraged for strategic reasons the establishment of a large number of plants in the southern part of the country and in the Alps far from the combat zone. In so doing, it completely changed the pattern of France's chemical industry. Germany had likewise set up large chemical plants in the eastern part of the country close to the lignite fields, and they now bordered the frontier with Poland.

This geographic *dispersion* had been going on apace, with appreciable development of production on existing sites, particularly in industries closely involved with explosives and chemical gases: ammonia, nitric acid, concentrated sulfuric acid, chlorine, acetone.

Demand, boosted by war needs, collapsed when hostilities ended, and in all developed countries the chemical industries were left with *stocks* and *surplus capacities* on their hands for which there were no immediate markets. The matter of the reparations that Germany had to pay to the Allies further complicated the situation. Thus, a 20,000-ton stock of German dyestuffs confiscated by the Allies and sold directly to users at a price fixed by the Reparations Commission prevented the establishment of a balanced market at the very time that demand in Europe was at its lowest.

While the European producers were facing the direct and perverse effects of war, the United States was emerging from the conflict in good shape. As the American government had not interfered in corporate affairs, industry had organized its war effort in a rational manner. The Allied shopping list had reached it at a time of slight recession in 1914, and the firms engaged in the war effort had prospered on the strength of full order books. Also the need to fill the gap left by declining imports had provided American companies with new opportunities for development.

In Europe, on the contrary, governments had systematically practiced intervention, directing production as they saw fit when they were not taking matters over completely. Thus, although industry could be proud of what they had achieved for their countries, except in rare cases, they had hardly drawn particular benefit from the experience. Their plants had been diverted from their long-time objectives, and they were now faced with the painful problems of reconversion.

As for the new products which had emerged from the four years of war, they were not of a nature to open up encouraging prospects for the chemical industry's future. Acetone from fermentation could not vie with the acetic acid-based synthetic process. The synthetic rubber that Bayer had produced at the rate of 2,000 tons a year was too expensive to be able to compete with natural rubber once sea links had been reestablished.

One of the war's innovations, chemical gases, had happily lost all its markets. But the chlorine used to produce those gases and the caustic soda obtained as a coproduct of electrolysis were left begging for markets. This was also the case for nitrocellulose stocks, which built up on factory floors, and of cellulose acetate, which had served to coat the wings of combat planes.

While it is true that ammonia synthesis was headed for a soaring development, it did not originate during the war. BASF's Oppau unit dated back to 1913. Thus, instead of original products, the chemical industry of developed countries essentially had inherited *excess capacities* in many standard production areas. The situation was further complicated by the world economic slump that marked the immediate postwar years, particularly 1920 and 1921. In the circumstances, industry leaders took a two-pronged series of measures to get out of their difficulties: They tried to find new uses for excess products and diversified production through internal development, company mergers, or licensing. At the same time, they strove to sign agreements with their competitors to preserve price levels and market shares, and they lobbied their governments for protection against foreign competition through the establishment of trade barriers.

Considered as compensation for war damages, the patents, trademarks and technologies of vanquished Germany were coveted by the Allied nations and put to profit for their own industries.

Innovative Trends in Postwar Industrial Chemistry (1920–1930)

In the early twenties, world chemicals were still based on the four pillars of alkalis, acids, fertilizers, and dyes, while organic base products from tar distillation constituted the upstream. In addition, there was the major breakthrough of ammonia synthesis, but this was restricted to the two BASF plants in Oppau and Leuna. It fell to the Americans to become instrumental in pushing chemicals along new paths by discovering new uses for products of which there was now a surplus, or by trying to extract value from raw materials that were still undeveloped, or, again, by developing original molecules for applications that were just emerging.

There were many favorable factors to give an impulse to the United States industry. Companies had accumulated reserves that, in some cases, were considerable and crying out to be invested in worthwhile projects. At war's end, there were some 20,000 engineers and technicians trained in the different disciplines of chemistry. Besides their own research departments, chemical companies could call on firms of consultants such as A. D. Little, Inc., or on research institutes funded by foundations that had contacts with universities, such as the Mellon Institute in Pittsburgh. More important, America's chemical industry operated in an environment that was highly favorable for new developments because of the growing demands of the Detroit automobile industry and the possibilities being opened up by oil as a plentiful and cheap feedstock for chemicals.

Europe was also trying to put to good use the technologies developed before the war in the chemicals and chemical-related field, but there can be no doubt that the drive for ever-newer products during those years came from the United States.

The Revolution in Fast-Drying Lacquers

Lacquers based on nitrocellulose had already been in use before the war. Du Pont had bought the Arlington Company in 1915 with an eye to the future of lacquers. But their development was limited by the lack of availability of amyl alcohol used as a solvent and obtained from gin and whiskey distillation. Prohibition in the United States, together with the Russian Revolution, which halted fusel oil exports, made matters worse. Butanol obtained as a by-product in the Weizmann fermentation process turned out to be an excellent substitute, however.

Research in that direction intensified, and lacquers based on low-viscosity nitrocellulose were developed by Du Pont, then by Hercules. Paints for automobiles now dried in a few hours instead of the several days previously required. In 1913, Du Pont took stock in General Motors, which had been founded in 1908 by William Grapo Durant. At war's end, this stake was increased to 24 percent. It was on the strength of this connection that Pierre du Pont became president of GM, a fact that certainly paved the way, in 1913, for use of Duco lacquer on GM cars. Three years later, all American automobiles, with the exception of Ford cars, were being finished with a cellulose lacquer.

Production of this type of lacquer increased tenfold to nearly 5,000 tons by 1926, leading to higher consumption of solvents (butanol, butyl acetate) and plasticizers (synthetic camphor, castor oil) both in the United States and in Europe, where Duco lacquers were produced under license. Duco lacquers constituted a revolution that considerably boosted production on car makers' assembly lines, leading to a search for other types of fast-drying lacquers.

As early as 1919, Kurt Albert in Germany had developed a phenolic resin that was soluble in fast-drying oils,[2] producing a lacquer that dried in four days. Beck Kohler in Austria also offered a similar resin. In the United States, from 1927 on, Resinous Products, a Rohm & Haas subsidiary, and Henry Reichhold, brought out Amberols and Beckacites, respectively, which successfully replaced natural imported resins such as Copals in industrial paints.

Since 1912, General Electric had been trying to obtain electric wire coatings and oven-dried lacquers for household equipment. The firm had been carrying out research in Schenectady, New York, on resins produced by condensation of a polyol on a dibasic acid in the presence of fast-drying oils. This is how *glycerophthalic resins* (alkyds) were born in 1928 and were used on Ford cars. They were as quick-drying as cellulose lacquers, but they had three times their solid matter content, reducing the number of layers needed and, accordingly, the time spent on the production line.

By 1933, six American producers were bringing out 5,000 tons of alkyd resins, and the phthalic anhydride which they required had become one of the base products of organic chemistry.

A New Focus on Rubber

World consumption of rubber had soared from 100,000 tons in 1910 to 297,000 tons in 1920. By 1930, it was to climb to 710,000 tons, mainly because of the expansion of automobile tire production.

[2] Through introduction of alkyl or phenyl groups in the para position of phenol.

The United States, which accounted for half this consumption, had pioneered rubber-processing technology with Charles Goodyear, the inventor of the hot vulcanization system, then with G. Cabot, who had founded the carbon black industry. Another American, George Oenslager, who worked with Diamond Rubber, discovered in 1906 the first *organic vulcanization accelerator*, *aniline*. It was subsequently transformed into *thiocarbanilide*, easier to handle.

During the war, B. F. Goodrich in Akron and U. S. Rubber in Naugatuck set up their own aniline plants because their imports were threatened by the blockade. In 1921, another accelerator, *2'-mercaptobenzothiazole*, was obtained which Goodyear marketed under the trade name Captax. In 1922, Dovan Chemical Corporation patented *diphenylguanidine* and R. T. Vanderbilt began producing *thiuram disulfide*. Thus speeded up, the time needed for vulcanization of rubber was reduced from three hours to less than one hour, and resistance to elongation greatly improved.

In 1921 antioxidants were introduced to retard deterioration of the rubber by air oxidation. They came in the form of *secondary aromatic amines* following work on acrolein begun in 1917 by Frenchmen C. Moureau and C. Dufraisse.

The Beginnings of Petrochemicals

During the war, the British through the Asiatic Petroleum Company had extracted benzene and toluene from the oil of Borneo. But development of an oil-based chemical industry required plentiful and cheap products from refineries or natural gas as feedstocks. The United States had both.

The first petrochemical unit was put on line in December 1920 by Standard Oil in Bayway, New Jersey. It produced *isopropyl alcohol* from refinery propylene under a process already described by Marcellin Berthelot in 1855. But it was George Curme's work that proved the most fruitful. He had a Ph.D. in chemistry from the University of Chicago and had worked at the Mellon Institute since 1914 as a researcher for Union Carbide. Through electric-arc oil cracking, Curme had obtained *olefins* (unsaturated aliphatic hydrocarbons) particularly *ethylene* and *propylene*, besides the acetylene he sought to produce. Separation was carried out at Linde Division's Buffalo plant.

Curme subsequently resorted to thermal cracking of natural gas-light petroleum fractions, which contained ethane and propane, to produce olefins from them. A pilot unit was set up in 1920 by Union Carbide and Carbon in Clendenin, West Virginia.

A whole series of ethylene derivatives emerged form this line of experiments: *ethylene oxide* from ethylene chlorhydrin, *ethylene glycol, dichloroethane, ethyl alcohol*. Their commercial potential was so great that Union

The first petrochemical unit: production of isopropyl alcohol by Standard Oil of New Jersey in Bayway, New Jersey, 1920.

Carbide set up a subsidiary called Carbide & Carbon Chemicals to develop them.

For the first time, teams made up almost exclusively of chemists were given the hard but fascinating task of producing new oil derivatives in a laboratory, testing them in the pilot plant, manufacturing them on an industrial scale and, finally, marketing them. In this manner the first tons of *ethylene glycol* came out of the Charleston, West Virginia, plant in 1924. Glycol was sold as a substitute for glycerine in the production of dynamite before Union Carbide advocated, in 1927, the use of the glycol as an antifreeze agent in motorcar radiators. Beginning from nothing in 1924, production of ethylene glycol in Charleston had reached close to 6,000 tons four years later, and in 1930 the only synthetic ethyl alcohol unit constructed by the U.S. before World War II was launched.

The Discovery of Tetraethyl Lead

From the earliest days of the automobile, it had been noted that engine performance could be improved if compression ratios were increased. But this required the addition of an antiknock agent to the gasoline.

Thomas Midgley, who was employed by the Delco inventor Charles Kettering at the Dayton Engineering Laboratories Company, tackled the problem in 1917. Focusing his research efforts first on iodine, then on metal-organics, he discovered in December 1921 that tetraethyl lead was the most effective antiknock agent.

General Motors, which had bought up Delco in 1920, introduced the production of tetraethyl lead for gasoline in 1923 and granted Du Pont a manufacturing license that same year. By 1924, Du Pont's Deepwater plant was producing tetraethyl lead for a company jointly owned by GM and Standard Oil of New Jersey. Subsequently called Ethyl Corporation, the company was the first to market the additive through which gasoline with a high octane content could be successfully produced, and it continued to be produced until very recently. Although trained as a mechanical engineer, Midgley served the chemical industry very well with his invention. He also served the interests of refiners and motorists.

Bromine was needed to manufacture the tetraethyl-based additive.[3] This led to a growing demand for the halogen. A joint Dow Chemical/Ethyl Corporation company was formed, called Ethyl Dow, to extract bromine from sea water starting in 1925. It produced ethylene dibromide on a large scale at Baton Rouge, Louisiana.

[3] The additive, besides tetraethyl lead, contained ethylene dibromide and dichloride.

Midgley was also the inventor of Freon gas in 1936, the first nontoxic refrigerant.

Fresh Prospects for Cellulose Acetate

With the war over, new uses had to be found for the *cellulose acetate* that had served to coat airplane cloth. At the request of the Royal Air Force, the two brothers Camille and Henri Dreyfus had come from Switzerland to set up, first in Spondon, England, then, in 1917, in Cumberland, Maryland, a unit for the production of acetate lacquer.

After the Armistice, they resumed their 1913 trial production of a cellulose acetate-based fiber, and in 1925 the Cumberland unit produced the first commercial quantities of *acetate silk*. Two years later, what had become the Celanese Corporation of America had made its niche in the American market of artificial textiles. At the same time British Celanese and Courtaulds were introducing the new fiber in England.

The Usines du Rhône had also needed to reconvert their cellulose acetate production at Saint-Fons. They first developed a Celluloid substitute called Rhodoid that was more expensive but had the advantage of being less flammable. By 1925, and after a difficult start, Rhodoid had become popular with the plastics processing industry. Moulding powders as well as acetate film were being put on the market.

Meanwhile, Usines du Rhône had formed in 1922 a joint venture called Société Rhodiaceta on a fifty-fifty basis with the Comptoir des Textiles Artificiels in order to take advantage of their textile experience in developing an acetate fiber.

Much effort went into the development of acetate rayon;[4] but success really materialized when rayon thread, dyed in the bulk, was launched in 1929. While spinning was carried out at the Lyon Vaise unit, the Roussillon chemical plant, with its cotton linters bleaching units, its acetic anhydride production, and its increased cellulose acetate output, was back by 1926 to its 1,000-strong workforce.

Rhône-Poulenc was set up in 1928; that same year it granted, together with Rhodiaceta, an acetate-processes license to Du Pont de Nemours, which had already acquired from Gillet the Viscose silk process in 1920 and the Cellophane process in 1923. Such links were not forgotten when the American firm set out to manufacture its new Nylon fiber in France. Rhodiaceta subsidiaries were set up in 1929 in Germany and Brazil, and in Italy in association with Montecatini. At the same time, two new acetate-thread competitors were

[4] The generic term rayon was first used in 1924 for artificial textiles.

emerging in the United States, Courtaulds, through American Viscose, and Tubize-Chatillon the following year, with a combined production of 2,500 tons a year.

High-Pressure Chemicals

In 1919 Fritz Haber received the Nobel Prize for Chemistry awarded to him the year before for ammonia synthesis. The prize was protested by the Allies because of the crucial role Haber had played in gas warfare.

At the same time, Carl Bosch from BASF, who was subsequently to become IG Farben's top executive, was fully aware of the Allies' interest in the process he had developed with Haber. In 1919 he managed to prevent the dismantling of the Oppau plant by signing a secret agreement with the French representative Joseph Frossard under which a dyestuff technology was transferred to France's dye industry. When France occupied the Ruhr in 1923, Bosch had been warned of the move beforehand and managed to transfer to Leuna the equipment used for ammonia synthesis.

Nevertheless, BASF's monopoly in the area came to be challenged by the high-pressure processes developed by Luigi Casale in Italy and by Georges Claude in France. George Claude, in particular, had been working since 1917 on a synthesis which required pressures of around 1,000 atmospheres. His tests were carried out in the Montereau pilot unit belonging to the Société de la Grande Paroisse, founded in 1919 by Air Liquide, in which Saint-Gobain was subsequently to acquire a stake. In 1924 Air Liquide contributed to the establishment of the Société Belge de l'Azote. At the same time, Du Pont de Nemours obtained sole rights to the Claude process in the United States.

Fearing that these new developments would lead to overcapacities of ammonia, Bosch made use of BASF's high-pressure technology to produce *methanol*, on the one hand, and *synthetic gasoline* on the other.

High-pressure hydrogenation was first conceived by Russia's Vladimir Ipatieff in 1911. But as early as 1905, Paul Sabatier, working on catalytic hydrogenation, had examined how to obtain *methanol* from carbon monoxide (CO). Another Frenchman who was head of the Service des Poudres since 1919, G. Patard, took out a patent in 1921 describing the preparation of methanol by hydrogenation under pressure of CO in the presence of a catalyst.[5] But it was Germany's Pier, working with BASF, who developed the process on an industrial scale at the Leuna unit in 1924. The Germans began exporting synthetic methanol to the United States the following year at very attractive prices in order to earn foreign currency. But in the face of strong

[5] According to the reaction $CO + 2H_2 = CH_3OH$ and using Le Châtelier's principle.

pressure from wood distillers, the American government increased customs tariffs by 50 percent on the product in 1926.

Such protection defeated its purpose, however, for it induced Du Pont and Commercial Solvents to alternate methanol synthesis with synthetic ammonia in their respective units in Belle, West Virginia, and Peoria, Illinois, simply by switching reagents and catalysts. Thus, with the exception of special cases like Tennessee Eastman's Kingsport distillation unit which had been supplying Kodak since 1920, the development of synthetic methanol considerably affected "wood alcohol" production. The synthetic product was being used in an increasing variety of cases because of its greater purity, in particular for methylation reactions, for the production of formaldehyde and as an antifreeze agent in automobile radiators.

Germany, lacking oil resources, was interested in producing *synthetic gasoline* through pressure hydrogenation, at high temperature, of coal and lignite with which it was well-endowed. The German chemist Friedrich Bergius, who had been a student of Ostwald's, had been able to observe Haber's work on ammonia while he was in Karlsruhe in 1906. In 1913, he patented a process for the hydrogenation of finely divided coal particles, then two years later, set up in Rheinau near Mannheim an experimental unit designed to produce directly thirty tons a day of gasoline. For this work he was awarded the Nobel Prize for Chemistry in 1931, together with Carl Bosch.

In 1924, Brunner, Mond & Company signed up with Bergius; and two years later a hydrogenation unit was set up in Billingham, England. This prompted IG Farben, after conclusive tests carried out by Pier with lignite, to associate with Bergius for the development of his process. A production program was outlined between IG Farben, Brunner, Mond, and Standard Oil of New Jersey, which did not want to be left out of any synthetic gasoline project being planned in Europe.

In the fall of 1928, Carl Bosch, who had strongly backed the project, received a $35 million windfall from Standard Oil for sole rights to the Bergius process in the United States. With this money, IG Farben was able to complete its Leuna installation which, since 1927, was having a hard time bringing out synthetic gasoline at a rate of 20,000 tons a year. Capacity of 100,000 tons a year was reached only in 1931–32.

Drawing inspiration from BASF's studies since 1913, the Germans Franz Fischer and Hans Tropsch, who had conducted research on coal at the Kaiser Wilhelm Institute in Muehlheim, obtained in 1923 aliphatic hydrocarbons suitable as motor fuels by passing water gas over an iron oxide catalyst at 200°C under moderate pressure (15–20 atm). The Fischer-Tropsch process, which could also produce solid wax, was put in commercial development by Ruhr Chemie in 1934, freeing Germany of the need to extract aliphatic hydrocarbons from imported petroleum.

Acetylene as a Source of Chemical Derivatives

Acetylene, as noted earlier, was initially used for lighting purposes, employed to produce high temperatures by means of the oxyacetylene torch, and finally used to manufacture calcium cyanamide and hence to fix nitrogen for the production of ammonia for fertilizers.

German and Canadian chemists also paid very early attention to acetylene because of its highly reactive properties and because it is readily synthesized from coal, through calcium carbide as an intermediate, when cheap electricity is available. The Germans believed acetylene could be an ideal feedstock on which to base their aliphatic chemicals industry, in the absence of petroleum resources, and the Canadians saw it as a means of developing their abundant hydroelectric sources. Thanks to Paul Duden, who had brought Hoechst and Griesheim together, the Germans had produced, during the war, *acetaldehyde*, subsequently oxidized to *acetic acid*, starting from acetylene. Acetic acid was in turn transformed into *acetone* (the starting point, in particular, of Bayer's methyl rubber) in the Knapsack and Burghausen plants.[6]

Acetylene chemistry had also been studied in 1917 in Shawinigan, Quebec, by a local electric company in an attempt to provide Britain with acetone for the production of cordite. In 1925, a joint venture of Shawinigan Chemical, which provided the technology; Union Carbide, which supplied the calcium carbide and acetylene; and Roessler and Hasslacher was set up under the name of Niacet Chemicals Corporation. Its production unit located in Niagara Falls, New York, supplied, besides acetic acid, which was finding growing uses for cellulose acetate and cellulose solvents, acetaldehyde and also *crotonaldehyde* and *ethyl acetate*. In the same way as wood alcohol had been displaced by synthetic methanol, so was the acetic acid obtained by synthesis replacing the one extracted from that same source.

With demand for both methanol and acetic acid increasing at a rapid pace, wood distillation companies would have been unable to satisfy new requirements on their own.

Other derivatives from the same chemistry were soon to find fresh commercial uses. This was the case with *vinyl acetate*, which benefited from the work of F. Klatte at Hoechst in Germany and of Shawinigan Chemicals, and with *polyvinyl alcohol*, also developed by the Canadians.

[6] The reactions are as follows:

$$\underset{\text{acetylene}}{CH{\equiv}CH} + H_2O \longrightarrow \underset{\text{acetaldehyde}}{CH_3CHO} \qquad CH_3CHO + \tfrac{1}{2}\,O_2 \longrightarrow \underset{\text{acetic acid}}{CH_3CO_2H}$$

$$2\,CH_3CO_2H \longrightarrow \underset{\text{acetone}}{CH_3COCH_3} + CO_2 + H_2O$$

The chlorination of acetylene had been studied in 1903 and 1904 in Germany and Austria in an attempt to find new markets for the chlorine produced in electrolysis units. In 1908, Castner-Kellner Alkali, together with the German consortium that owned the technology, had set up Weston Chemical Company. In its Runcorn unit in England it produced *tetrachloroethane* and *dichloroethane*. Thus a whole range of powerful chlorinated solvents appeared on the market. Unlike carbon disulfide and petroleum ether, these solvents had the advantage of not being flammable.[7]

By the mid-1920s *trichloroethylene* sales really took off, as dry-cleaning and metal-degreasing technologies began to be used extensively; and Weston Chemical was bought by ICI.

Organic chemical manufactures were to benefit once more from acetylene after J. A. Nieuwland's 1925 discovery of divinyl acetylene, which led to the manufacture of *Neoprene rubber*, and when Walter Reppe's work at BASF on the reactions of acetylene under pressure paved the way for entirely new chemical developments.

State Intervention

State intervention in private business had proved necessary to manage a war economy. But the practice was continued well after the Armistice, and the chemical industry was an object of official national attention. Under article 297 of the Versailles Treaty, the Allies had been made free to take advantage of Germany's trademarks and patents. Negotiations for the transfer to the industry of the Allies of German technology were conducted through the medium of national administrations. Even the United States, with its open market approach, entrusted the task of sharing out patents and trademarks to a federal agency, the Alien Property Custodian. But as the sharing was not done on an exclusive rights basis, its success was only partial.

In Europe, state intervention reached even into production. The British government remained a partner of British Dyestuff Corporation until 1925, while the French Government, after negotiations with BASF, obtained approval from Parliament for the establishment of the Office National de l'Azote

[7] The reactions can be schematized as follows:

$$CH{\equiv}CH + 2Cl_2 \longrightarrow \underset{\text{tetrachloroethane}}{CHCl_2{-}CHCl_2} \longrightarrow \underset{\text{trichloroethylene}}{CHCl{=}CCl_2} + HCl$$

$$CHCl_2{-}CHCl_2 \longrightarrow \underset{\text{dichloroethylene}}{CHCl{=}CHCl} + Cl_2$$

(ONIA). ONIA started producing ammonia in Toulouse in 1927 using the Haber-Bosch process.

The Allies had suffered early in the war from their heavy dependence on Germany's chemical industry and they were determined to build up in their own countries an industry capable of providing for their nation's most pressing needs, particularly in dyes and synthetic intermediates. Thus while Germany, which had nothing to fear from imports, and Belgium, which had no dye production to protect, kept their customs tariffs low, France, Britain, the United States, and Japan resorted to the full protectionist arsenal of high customs duties and very restrictive import licenses. Britain's administration introduced the "Key Industries Duties" concept under which a 33 percent tax was levied, and it granted import permits only for those dyes which were not produced in Britain. In the United States, import duties hovered between 55 and 60 percent; and the American Selling Price[8] clause made them ever more prohibitive. The clause was only scrapped forty years later under the Kennedy Round of GATT negotiations in 1964–1967.

It is worth noting that these protectionist practices from supposedly liberal-minded countries harmed the textile industries both in England and in the United States without boosting the competitiveness of the dye industry in either country. In addition, they considerably hampered free trade, which alone could have brought about improvements in prices and technologies by enhancing competition.

The Proliferation of Cartels

The surplus production capacity available with the end of the war, the depression which prevailed during 1920 and 1921, the alarming depreciation of the German mark, and the fears aroused by the Bolshevik Revolution, all combined to worry the chemical industry's major leaders, prompting them to form alliances. These alliances appeared all the more necessary as separate efforts to increase market share had resulted in lowered selling prices and profit margins.

Price cartels had already been set up in Germany before the war among producers of inorganic acids, fertilizers, alkalis, and, of course, dyestuffs. The cartels often also involved sales quotas. The potash cartel was reestablished in 1919, even though Germany had lost its monopoly along with the Alsace mines. By 1924, pricing agreements and export quotas were being arranged between the Kali Syndikat and the Mines Domaniales de Potasses d'Alsace.

[8] For the basis of customs duties, the clause used the domestic price of American producers, usually quite elevated, rather than the cif (cost, insurance, freight) value of imported dyes.

Cartel arrangements became general among European producers between the two wars. Americans were prevented from taking any part because of antitrust legislation (the Sherman and the Clayton Acts). In 1928, the *dyestuff* trade was "regulated" after delicate negotiations among the Germans, French, and Swiss, with the English joining in three years later.

For *phosphorus*, Albright & Wilson, Griesheim, and Coignet resumed their pre-1914 agreement. In 1929, an international *nitrogen* convention was signed involving IG Farben, ICI, Norsk Hydro, and Chile. But the Chileans were concerned that the new synthetic ammonia capacity which was appearing on the market would prevent them selling on world markets the amounts they could supply. Thus, failing the establishment of production quotas, the cartel had to wait for a demand upturn that did not occur until 1935 to increase its sales prices.

For the large chemical and mining groups, cartels were the means through which they could maintain their respective positions in the market for their main productions during the troubled period between the two wars. By freezing the situation in the expectation of better times, these groups preserved jobs in their plants. But the comfortable profits they thus acquired were paid for by the great mass of consumers and by the small producers who stagnated because they were unable to compete effectively.

Birth of the Giants

The same concerns which led the large chemical corporations into cartel arrangements induced them to organize themselves into larger entities, the better to face the hardships then prevailing. For cartels to be efficient, partners had to be few and powerful.

The concentration process that had started during the war continued after the war on both sides of the Atlantic. In *Germany*, as we have seen, an enlarged IG emerged in 1916 in the shape of a federation. A number of activities of the participating companies were excluded, such as BASF's nitrate fertilizers, Hoechst's acetylene chemicals, and Griesheim's steel alloys. Despite Carl Duisberg's reluctance, in 1924, Bosch pushed through the idea of full integration, and in December 1925 the eight associated companies merged to form *IG Farben Industrie AG*. The new giant had a workforce of 67,000 people including 1,000 chemists, and accounted for one-third of Germany's chemical industry sales. Nitrate fertilizers and dyes, including synthetic intermediates and pharmaceuticals each accounted for 36 percent of this figure, and base chemicals for 20 percent, with photographic products making up the balance.

Rationalization measures put off earlier were now quickly undertaken:

production centers specialized along particular technical lines, with AGFA taking over photographic activities, BASF high-pressure chemistry, Urdingen the main organic intermediates, and Leverkusen the azo dyes. Only the big inorganic chemicals were left on their original sites, because long-haul transport would have eroded sales margins. In fact, as far as possible, IG Farben avoided competing with its traditional clients in the area of heavy chemicals, for such competition would eat up capital and be less profitable than the more specialized chemicals in which it was involved. Instead, IG Farben took stakes in the new sectors that were developing, first in Nobel Trust's explosives then, through Köln-Röttweil AG, in viscose which it produced in three different units. Bemberg's cuprammonium silk was made in Dormagen and Aceta's acetate fiber in Berlin; Kalle, which had given up dyestuffs, turned to cellophane films and "Ozalid" paper, and Griesheim concentrated on light metals.

IG Farben's worldwide activities accounted for 60 percent of its total sales. While engaged in unifying its export networks, the group set up in Basel the IG Chemie holding company, which concentrated all its participation in foreign companies such as General Dyestuff in the United States, Norsk Hydro in Norway, and others. By 1930, the reorganization was practically completed.

The creation of IG Farben accelerated the concentration process already underway in *Britain* in alkalis, explosives, and dyestuffs. Encompassing the main Leblanc soda plants, United Alkali dated back to 1890, as we have seen. Its inorganic acids, chlorine, and ammonia-soda production had been used to the full during the war, but obsolescent processes and equipment prevented any worthwhile attempts at diversification. In 1926, the government had sold back to British Dyestuffs (BDC) the stake it held in its capital, and BDC expanded its range of vat dyes when it took over Scottish Dyes. But divergence continued among the management teams of the different companies that had formed BDC. Its research was not up to scratch and, despite tight management and the beginnings of a reorganization process, the company's profitability was not satisfactory.

Very different was the state of Nobel Industries, a holding company which in 1918 had merged all the explosives manufacturers in Britain under the sponsorship of Sir Harry McGowan, the wily boss of British Nobel. Sir Harry had already reorganized Canada's explosives industry, setting up Canadian Explosives Ltd. (CXL), which was owned 45 percent by Du Pont. In Britain, he managed not only to rationalize the production of explosives, but also to carry out within Nobel Industries a successful diversification policy. As Du Pont had done, Nobel Industries extended nitrocellulose utilization in 1919 to cellulosic lacquers and to coated fabrics. Between 1920 and 1926 Nobel Industries also took a lucrative stake in General Motors.

The fourth group that, with United Alkali, BDC, and Nobel, was to take part in the creation of Imperial Chemical Industries (ICI), was Brunner Mond, which already in 1914 had been Britain's major chemical entity. Brunner Mond had associated with Castner-Kellner Alkali in 1916, then taken it over completely in 1920. With the purchase of Chance & Hunt in 1917, Brunner Mond accounted for three-quarters of Britain's alkali production.

Purchase of the Billingham plant in 1920 from the British government and an ambitious program in ammonia synthesis[9] were beginning to weigh heavily on Brunner Mond's finances when, at the beginning of 1926, Sir Alfred Mond, the future Lord Melchett, was appointed President. Unlike his father Ludwig Mond, one of the company's founders, who had been a great chemist and the discoverer of nickel carbonyl, $Ni(CO)_4$, Sir Alfred was a politician. He had a seat in Parliament and was as interested in the cause of Zionism as in the expansion of his company.

When his alliance projects with IG Farben for synthetic gasoline and with Allied in the United States did not succeed, Sir Alfred Mond welcomed the association suggested by McGowan. Talks began in September 1926 and led to the establishment that same year of *Imperial Chemical Industries* (ICI), a rapid process considering the complexity of the operations involved. Sir Alfred was appointed president and McGowan general manager, a clear indication of the share of power carved out for the two main companies in the group. Indeed, United Alkali had not emerged from its long decline, and British Dyestuffs was just coming out of the red after difficult beginnings.

Unlike the leaders of IG Farben, ICI's new bosses had no scientific training. Although they felt unconcealed admiration for the technical feats of the Germans, they had every intention of themselves pursuing the strategy of shared-out markets and concentrations on which they had thrived until then.[10] With their alkali and explosives monopoly completed in the home market, their main object, through agreements negotiated with powerful international partners, was to give the new group uncontested predominance throughout the British Empire. This plan explains the technical agreements that were pursued with Du Pont on explosives, the close relationship kept up with Solvay on Alkalis, the entry of ICI into the dye cartel arrangements in 1931 with a quota of 7 percent (which was, in truth, modest), and its participation in the International Nitrates Convention after Billingham's future in fertilizers had been assured.

Used to high-volume continuous production, Mond and McGowan had

[9] The equivalent of $15 million at the time was invested in Billingham between 1920 and 1925.

[10] The two giants ICI and IG Farben had the same $275 million equity, and each had a share of the national market exceeding 25 percent.

little understanding of the complexities of fine chemicals produced in smaller quantities and using operations in which the Germans, trained in the disciplines of organic synthesis, excelled. This no doubt explains why British Dyestuffs was for a long time treated poorly within the group and why, because of too little attention given organic chemicals, ICI remained for many years away from the mainstream of profitable developments in pharmaceuticals and photography.

Following a traumatic lawsuit by Lever Brothers, after which Brunner Mond had to pay a $5 million penalty and sell Crosfield to the English soapmaker, the group's leaders were careful to avoid direct competition with their main clients. For the same reason, unlike IG Farben, ICI stayed away from textile fibers for a long time so as not to cross Courtaulds, kept away from solvents to respect the positions acquired by Distillers, and, of course, avoided synthetic detergents for fear of alarming Unilever.

Thus despite their similar dimensions, the two giants, ICI and IG Farben, projected very different images of themselves because of the circumstances under which they originated. There was also little resemblance with the two main chemical groups in the *United States*.

American antitrust legislation and the watchdog role of the Federal Trade Commission (FTC) prevented any merger of rival companies that would have led to dominant positions. Moreover, the U.S. home market was sufficiently vast to leave room for several large companies in the same sector. Chemical companies in the United States, therefore, had no need to worry to the same extent as their European counterparts about exporting their production or taking part in cartel arrangements. In the circumstances, the concentration processes which led to the formation of Union Carbide & Carbon in 1917, or to Allied Chemical & Dyes in 1920 had been based either on complementary activities (upstream or downstream integrations) or on the wish to diversify into new areas.

Allied was the outcome of the merger of Solvay Products Company, Semet Solvay, Barrett Company, and National Aniline with the General Chemical Company. Firmly managed by the abrasive financier Orlando Weber, this company was the major chemical corporation in America at the time and was involved in sodium carbonate, coaltar derivatives, dyes, inorganic acids, and, since 1921, synthetic ammonia produced in the Atmospheric Nitrogen Company's Belle plant in West Virginia. Allied was careful not to enter into competition with its clients and thus deliberately kept to the major base products that were difficult to export. It did not seek to establish itself abroad, and took part only in those international agreements that protected its home markets.

The second largest American chemical group after Allied was *Du Pont*, which had diversified long before. After the war, Du Pont had contributed, through its research, to the development of cellulose lacquers and tetraethyl lead. Through its acquisitions of Harrison Brothers in 1917 and of Grasselli Chemicals in 1928, it had extended its range to inorganic acids and pigments for paints. The shares it had bought back from French shareholders, both in Azote and in Du Pont Rayon and Du Pont Cellophane, gave it full control over ammonia in Belle, and over viscose and cellulose film in Buffalo.

Under Lammot Du Pont's enlightened guidance, by 1929 the company had the most diversified portfolio of the U.S. chemical industry. In this it resembled IG Farben in Germany. It was involved in all sectors, with the exception of fertilizers, alkalis and pharmaceuticals. Its 30 percent stake in GM provided it, in addition, with comfortable dividends. Conducting a more open policy than Allied, Du Pont acquired processes abroad to boost its development, and established a number of bases outside the United States through the sale of Duco lacquer licenses and cooperation with ICI in explosives. The 1929 patents agreement with ICI and the extension to Canada of their common subsidiary Canadian Industries Ltd. (CIL) further reinforced links with the British group.

Besides the four giants in Germany, Great Britain, and the United States, there were only two other trusts that had spread both outside the frontier of chemicals and the boundaries of a single country to occupy comparable positions on the world scene. These were the Anglo-Dutch companies *Royal Dutch Shell* and *Unilever*.

The Royal Dutch Shell group was born in 1907 from the association of Shell Transport and Trading, a British company founded in 1897 by Marcus Samuel, who traded paraffin oil in the Far East under the "Shell" trademark, and Royal Dutch, a Dutch firm that had been developing oil fields in the Dutch East Indies since 1880. To counter competition from Standard Oil, Royal Dutch chief executive Henry Deterding had joined with Marcus Samuel in 1903 to set up the Asiatic Petroleum Company. Four years later Royal Dutch merged with Shell Transport, but each company retained its individual personality while sharing out the capital on a 60-40 basis.

After the war and the loss of its assets in Russia, the group set about diversifying in chemicals both in the United States and in Europe in order to exploit the cracking gases, which until then were flared, and the natural gas coming out of its wells. In 1928, *Shell Development* set up laboratories in Emeryville in California and a year later *Shell Chemical* was created with the twin object of utilizing natural gas as a source of hydrogen for the synthesis of ammonia at a plant near San Francisco, and of producing *secondary butyl alcohol* and *methy ethyl ketone* from the butylenes extracted from gas cracking on Shell Oil's Martinez site near Los Angeles.

At the same time, Royal Dutch had studied the fixation of atmospheric nitrogen and had produced hydrogen from coking and oven gases, in Holland, its subsidiary, the Mekog Company, started producing synthetic ammonia in Ijminden in 1929 by the Mont Cenis process, which Shell Chemical later used in California.

The other Anglo-Dutch giant was *Unilever*, an outcome of the 1929 association between the British-based Lever Brothers and the Dutch firm Margarine Union, carried out after difficult negotiations. The two groups had long been fierce rivals in the production of margarine and soap. The most prosperous, no doubt, was Margarine Union, which had a strong base in Britain and on the continent. But Lever operated over a larger area, through its plantations in Africa and Asia and its soap factories in India and the other Dominions. The two groups complemented each other; this fact overcame the resentments accumulated over half a century. They merged under two holding companies, the English Unilever Ltd. with equity in sterling, and the Dutch Unilever NV, with equity in guilders. Sir D'Arcy Cooper and Anton Jurgens were responsible for ensuring that the merger, comparable to the one carried out by ICI, produced a group that was greater than the sum of its parts.

France's Fragmented Chemical Industry

While chemical giants were being built up in Germany, England, and the United States, in France, which had emerged as a world power after the war, the chemical industry remained extremely fragmented. Indeed, the exigencies of war had added the scattering of sites to an already fragmented structure.

Some mergers were carried out, as well as some attempts at diversification made. In 1921, Froges became associated with Alais et Camargue to give birth to Compagnie Péchiney, which was to play a major role in aluminum production. In 1923, the Établissements Kuhlmann took over the Compagnie Nationale des Matières Colorantes (CNMC) by buying back the state-controlled shares. They also became the only French producers of dyestuffs by purchasing the companies that owned the St-Clair du Rhône and Saint-Denis units. But CIBA, with its Saint-Fons plant, remained a formidable rival. Compagnie de Saint-Gobain was pushing ahead with its diversification efforts which had begun when it took a 50 percent stake in the Société de la Grande Paroisse. In 1928, it set up the Compagnie des Produits Chimiques et Raffineries de Berre.

Rhône-Poulenc was born in 1928 from the merger of Établissements Poulenc Frères and Société Chimique des Usines du Rhône. Associations had also been formed in inorganic chemicals—in soda, acids, fertilizers, and chlorates—which produced in 1925 the Société Bozel-Malétra. Two years later, celluloid manufacturers, under the Société Industrielle des Matières

Plastiques, joined forces with the Société Générale de Dynamite to form the Société Nobel Française. However, such concentrations and diversifications, while they were not devoid of interest, could not compare with the ones that had given rise to IG Farbcn or ICI.

There were many obstacles in the way of establishing a large chemical industry in France. Unlike the British Empire in which the Dominions were populated with consumers enjoying high purchasing power, and African and Asian territories harbored immense natural resources such as vegetable oils, petroleum, and minerals, the French colonial empire offered no significant markets; nor did it provide any of the major raw materials needed by the national chemical industry, if one excepts phosphates from North Africa.

In addition, the only specifically chemical companies, such as the Établissements Kuhlmann and Rhône-Poulenc, were of medium size. Large groups such as Saint-Gobain, mainly involved in glassmaking, and Péchiney, which gave priority to its investments for the production of electricity and for units to produce aluminum, were interested in chemicals only as a sideline to their basic activities—fertilizers in the case of Saint Gobain, electrochemistry in the case of Péchiney.

The problem of synthetic ammonia was to complicate even further the structures of France's chemical industry. Both the George Claude and the Casale processes involved high pressures and consequently small volumes, and led to the development of a large number of medium-sized synthesis units in contrast to the gigantic Oppau and Merseburg plants.

Georges Claude's notion of using coking-furnace gas as a source of hydrogen had led various coal mining companies to become interested in ammonia synthesis. Thus the Béthune and the Aniche mines, which had close ties with the Société de la Grande Paroisse used the Claude process, while the Lens and Anzin installations worked under a Casale license.

The government, which held the Haber-Bosch patent rights after the signing of the Versailles Treaty in 1919, chose to develop them through the State-controlled Office National de l'Azote (ONIA), which it set up for the purpose in 1924 after much protracted thought, instead of entrusting the job to private interests. At considerable cost and after technical difficulties that were overcome only with German aid, ONIA managed to bring on line in 1929 a unit in Toulouse using the Haber-Bosch process. Although never profitable, it provided nearly a third of France's ammonia production. Whereas in 1929, two units in Germany and a single one in England accounted for the major part of the home production of synthetic ammonia, in France a much smaller total capacity was produced by fifteen units scattered around the country, owned by a variety of interests and using three different processes.

Two apparently contrary trends were converging in France after the war to bring about state intervention in matters of industrial policy. On the one hand, the left-wing political parties wanted to keep under control the capitalist corporations suspected of getting rich at the expense of workers. On the other hand, the right wing, heavily represented in military circles and the immediate postwar "patriotic" Parliament, wanted to avoid a repetition of the state of unpreparedness in which the country had found itself in 1914. Left and right shared the common thought that only direct state involvement in industrial affairs could make national interests prevail over private considerations.

On every occasion when the government could have called upon the knowhow of private corporations—the development of the Haber-Bosch licenses, the sequestration of the Alsace potash mines, the allocation of a 23.5 percent share in the Iraq Petroleum Company—successive governments chose to take a direct hand in industrial affairs. This led almost simultaneously in 1924 to the establishment of the ONIA and the Mines Domaniales de Potasses d'Alsace (MDPA), while, under the authority of Premier Raymond Poincaré's man, Ernest Mercier, Compagnie Française des Pétroles (CFP) was formed. From 1930 onward, the state took a 35 percent equity stake and had 40 percent of the voting rights in the national oil company.

Just as the refining industry was coming under State control through the 1928 law, so did the fertilizer industry become "organized" through state intervention. A Société Commerciale des Potasses was set up as early as 1919 to "harmonize" sales of Kali Sainte Thérèse and the MDPA products while, at the same time, French producers resumed their place in the Kali Syndikat cartel. In 1928, the ONIA signed a convention with the Comptoir Français de l'Azote to market its fertilizers, taking advantage of high cartel prices. Worried by world overproduction, superphosphate producers traded their freedom in 1933–1934 for import quotas. They agreed to have their prices set by a commission which included representatives from the Agriculture ministry whose role it was to protect the farmers.

Fertilizers are but one example among countless others of France's industries barricading themselves at home, accepting or even clamoring for state control, venturing only timidly into concentrations and rationalizations that alone could have made them competitive. Meanwhile, their German and English rivals were building up giant corporations wide open to the world and powerful enough to negotiate international agreements.

The Chemistry of Giant Molecules

Materials made of giant molecules or *macromolecules* have stimulated human curiosity since time immemorial.

When the Indians from Central America were turning the latex of the rubber tree into elastic matter, when the Chinese were preparing paper from plant fibers two thousand years ago, they were already using empirical processes to turn natural macromolecules into practical materials. In more recent times, these processes have grown more refined. From giant molecules available in nature, attempts were made to produce *artificial substances* with original properties: vulcanized rubber, cellulose nitrate for explosives, celluloid, fibers, photographic films, lacquers, chemical pulp for paper, casein-based plastics. Then, early in this century, Leo Baekeland developed thermosetting bakelite, the first synthetic resin ever to be marketed.

While the technologies for the preparation and use of these new materials were making giant strides, little progess was made in the understanding of the macromolecules themselves. True, between 1863 and 1866, Marcellin Berthelot in his work on isomerism, did describe the preparation from a *monomer*, amylene, of a *dimer*, diamylene, and he had made the assumption that *polymerization* could be pursued until the formation of *high polymers* made up of giant molecules. Subsequently, the French chemist G. Bouchardat, while studying in 1879 the constitution of rubber, had suggested that the isoprene molecules $CH_2{=}C(CH_3){-}CH{=}CH_2$ that formed it were linked together by their ethylenic double bonds to form a single linear molecule.[11]

It was Germany's Hermann Staudinger who, in work carried out from 1920 onwards, initiated the concept of *macromolecules* formed by the gradual linking of small molecules held together by normal interatomic bonds which could lead to molecular masses in the hundreds of thousands, even millions. Staudinger was an amiable giant who had worked under Johannes Thiele in Strasbourg between 1903 and 1907 before teaching chemistry at the Karlsruhe Polytechnic Institute, then succeeding Richard Willstätter in 1912 at the Zurich Polytechnic. He became interested in the chemistry of polymers while working for BASF in 1910 on the synthesis of isoprene.

To test the validity of his concepts he carried out research in the 1920s on polyoxymethylene (paraformaldehyde) and polystyrene. The fact that it was only in 1953 that he received the Nobel Prize shows how much resistance he had to overcome for his views to triumph despite his powers of persuasion.

Staudinger's ideas received an early boost from the work on X-ray crystallography conducted by Herman Mark. Mark had taken his doctorate at the University of Vienna in 1921 and in 1926 developed his conclusions on the structure of cellulose which lent support to Staudinger's hypothesis. Mark,

[11] $-CH_2-C(CH_3){=}CH-CH_2-CH_2-C(CH_3){=}CH-CH_2-$

who was one of plastic's true pioneers, worked with BASF in Ludwigshafen in 1927 on the advice of Fritz Haber. As deputy to the head of research, Kurt Meyer, with whom he collaborated in studying the diffraction of X-rays through cellulose, he developed with Carl Wulff a process to produce styrene from ethyl benzene. Like Staudinger, his research turned on the determination of polymer molecular weights by measuring their viscosity in a solution.

Other methods such as cryometry, osmotic pressure, and ultracentrifugation used by Sweden's The Svedberg helped to confirm the results obtained through measuring viscosities. Thus macromolecular chemistry, helped along by physical chemistry and endowed with its own established theories and its own analytical and measuring methods, was becoming a fully-fledged discipline of chemical science. A better understanding of the polymerization and high-polymer structure phenomena due to research carried out between 1920 and 1930 by Staudinger, H. Mark, and Kurt Meyer paved the way for the work which in the thirties was to revolutionize the chemistry of plastics, fibers and synthetic rubbers.

The Development of Synthetic Rubber

Synthetic rubber was born in Germany out of the needs of war. It was Bayer which took out as early as 1910 the first patents on the use of sodium (Na) as a catalyst for the polymerization of isoprene. But the resulting polyisoprene, while chemically similar to rubber, did not have its good mechanical properties. Besides, isoprene itself was hard to synthesize.

Arguing from analogy, Bayer chemists believed that methylisoprene or dimethylbutadiene $CH_2{=}C(CH_3){-}C(CH_3){=}CH_2$ which could be easily obtained from acetone would produce a satisfactory rubber. Some 2,000 tons of this *methyl rubber* was manufactured in Germany during the war.

With the return to peace and the collapse of natural rubber prices, the synthetic product was no longer competitive.

Because the English had restricted exports of rubber from Malaya in an attempt to stabilize its price, worried users set up plantations in Liberia and Brazil, while in the U.S. the use of reclaimed rubber was increased from 80,000 tons in 1924 to 181,000 tons in 1926.

At the same time, Germany and Soviet Russia became interested from 1925 on in new types of synthetic rubber, while France remained indifferent to the problem, believing it could rely on its Hevea plantations in Indochina. For both the Germans and the Soviets, strategic considerations superseded simple notions of manufacturing costs for rubber. They, accordingly, pushed forward their research, which involved using butadiene instead of dimethyl butadiene as a monomer. The resulting product was IG Farben's *Buna*[12] and the Soviet Union's *SKA* and *SKB*.[13]

The two Russian chemists, Kondakov and Lebedev, who distinguished themselves by producing these two rubbers, quite likely drew a large part of their inspiration from the IG Farben patents, since industrial property was not protected in the Soviet Union.

As for the Germans, they developed in the thirties butadiene copolymers with styrene (Buna S) which had higher wear resistance, and with acrylonitrile (Buna N or Perbunan) which was more resistant to oils and solvents. With a production of 80,000 tons by 1939, the Soviet Union was the largest world manufacturer of synthetic rubber, but IG Farben, subsidized by the Nazi regime, soon caught up.

The Americans, who were less worried than the Russians about being independent in this area, came upon synthetic elastomer chemistry somewhat by chance. In 1922, a freelance inventor called Joseph Patrick had obtained a nasty smelling material that was, however, elastic and resistant to solvents, while trying to prepare an antifreeze agent by having ethylene dichloride react with a polysulfide. It was to be known under the name of *Thiokol rubber*. *Polychloroprene*, which was marketed by Du Pont in 1933 under the name of *Neoprene*, was the outcome of experiments made in the early twenties by Julius A. Nieuwland at Notre Dame University on the chlorination of acetylene to produce divinylacetylene. Du Pont acquired the rights to Nieuwland's process, and one of its research teams, which included Wallace Hume Carothers, succeeded in producing 2-chlorobutadiene through addition of hydrochloric acid to monovinylacetylene[14] and to polymerize it into polychloroprene.

The new elastomer was expensive to produce but had similar properties to those of Perbunan. Du Pont considered it a specialty; and production at Deepwater Point, New Jersey, did not exceed 3,000 tons a year on the eve of World War II.

Following agreements signed with IG Farben in 1927, Standard Oil of New Jersey was, for its part, conducting research on copolymers of isobutylene with butadiene, and then with small quantities of isoprene to produce *butyl*

[12] Abbreviation for Butadiene-Natrium, meaning a polymerized butadiene, using sodium as a catalyst.

[13] Butadiene was prepared by the Russians from fermentation alcohol (SKA) and also from petroleum feedstocks (SKB).

[14] By the reaction:

$$CH_2{=}CH{-}C{\equiv}CH + HCl \longrightarrow CH_2{=}CH{-}\underset{\displaystyle Cl}{\underset{|}{C}}{=}CH_2$$

rubber. This was to be extensively used in the inner tubes of tires because of its remarkable gas-proof properties. When Pearl Harbor cut off the United States from its natural rubber supply in the Far East however, at the end of 1941, the Americans found themselves practically devoid of any substitute product.

The Discovery of New Plastics

Searching for new materials, chemists made good use of two kinds of polymerization reactions, *polyaddition*, under which double bonds or rings are opened and produce *thermoplastics* such as the polystyrene studied by Staudinger, and *polycondensation*, which produces *thermosetting* resins such as Baekeland's phenol-formaldehyde resins. Once formed, thermosetting plastics acquire their final shape, whereas thermoplastics can be reshaped on heating. Playing on the different factors likely to affect polymerization such as pressure, temperature, and choice of catalyst, researchers also selected the polymerization method best suited to their purpose: bulk, in solvent media, in water emulsion or suspension.

The Search for Transparency

The drawback of phenol-formaldehyde resins was that they were dark-colored. The Czech chemist Hans John managed to obtain a transparent resin in 1918, through polycondensation, by replacing phenol by urea and letting it react with formaldehyde. This is how the first aminoplast resin was born. Austria's Fritz Pollack improved on this type of resin in 1921. He partly replaced the urea by thiourea and the resulting product enjoyed widespread use for bottle ware and decorative laminates.

In 1928, British Cyanides Company, which was subsequently taken over by British Industrial Plastics, began marketing, under the trade name "Beetle," aminoplast molding powders which became very popular and saved the company from imminent bankruptcy. In 1929, it gave rights to the Beetle technology for the United States market to American Cyanamid. Urea and thiourea resins, however, were, in actual fact, *translucid* rather than transparent. And although they were used for a variety of products, they could not be employed as an organic chemical substitute for mineral glass. This role was to be played by *acrylic resins*.

In 1901 Otto Röhm had published a thesis on the polymerizaton of *acrylic acid*, for which he took out a first patent in 1912. He resumed work on it in 1920 with the help of the chemist Walter Bauer, who managed to develop an economic process to synthesize acrylic acid from ethylene cyanohydrin. In

Ueber

Polymerisationsprodukte der Akrylsäure.

Inaugural-Dissertation

zur

Erlangung der Doktorwürde

einer

hohen naturwissenschaftlichen Fakultät

der

Eberhard-Karls-Universität zu Tübingen

vorgelegt

von

Otto Röhm

aus Ohringen.

Tübingen.

Verlag von Franz Pietzcker.

1901.

Doctoral thesis of Otto Röhm on the polymerization of acrylic acid, 1901. Courtesy Rohm and Haas.

1926, the process was applied by Röhm & Haas in Darmstadt. Thereafter, polyacrylates, which did not turn yellow when exposed to light, could be used as a substitute for nitrocellulose to produce the safety glass that was increasingly in demand in the automobile industry.

In 1913, the U.S. subsidiary of Röhm & Haas also began producing acrylates and their polymers in Bristol, Pennsylvania; and new applications were found for these new materials in paints and in leather. Meanwhile, following work that W. Chalmers had conducted at McGill University in Montreal in 1929–30, Röhm & Haas and ICI became interested in the esters of *methacrylic acid*.

By polymerizing methyl methacrylate $CH_2{=}C(CH_3){-}C({=}O){-}OCH_3$, Walter Bauer obtained in 1931 a transparent thermoplastic which melted at 110° and which could be molded with ease or cast in a mold to produce sheets. What turned out to be the first *organic glass* was called Plexiglas. ICI, for its part, had patented in 1931 a process to make molded objects from polymethylmethacrylate (PMA) while one of its chemists, J. Crawford, developed a year later in Ardeer an elegant and economic synthesis of methyl methacrylate from acetone, sodium cyanide, methanol, and sulfuric acid. Applied in 1934, the process gave ICI a significant economic edge for the manufacture of sheets sold under the trademark Perspex. This did not prevent ICI from selling the licenses on its process the following year to Röhm in Germany for Plexiglas and to Du Pont in the United States in 1936 for what it was to market under the name Lucite. Lighter than mineral glass, this new organic glass was to be used for the portholes of American bomber planes and to become a great industrial product.

In 1931, researchers from the U.S.-based Corning Glass worked on the idea that an organic glass could be produced by polymerizing monomers containing silicon linked to carbon atoms. They remembered that in 1899, the English chemist Frederick S. Kipping, who taught at University College in Nottingham, had used the Grignard reaction to prepare a whole series of silicon organic compounds resembling alcohols. These *silanols* in turn led to *siloxanes*,[15] analogous to ethers.

Together with these new well-defined compounds, for which there were unfortunately no practical applications, Kipping had obtained tar residues that

[15] Two silane molecules $(CH_3)_3Si{-}OH$, thus formed a siloxane $(CH_3)_3Si{-}O{-}Si(CH_3)_3$ by eliminating a molecule of water.

he was unable to identify for want of proper knowledge in macromolecular chemistry. Now it was precisely these polymers formed either with long chains or a three-dimensional network of alternate silicon and oxygen atoms, spiked with organic radicals, that caught the interest of the Corning Glass researchers as well as those of General Electric working in close contact with the Mellon Institute. Called *silicones*, these polymers could not lead to production of the sought-for organic glass, but they had the unexpected virtues of remaining stable under heat, of being water repellent and having good dielectric properties, all of which made them useful in a number of areas when war broke out.

The search for transparent matter, whether film or plastic, was also the root cause of the development of *polystyrene* and *vinyl polymers*. Prepared for the first time by Marcellin Berthelot in 1869, *styrene monomer* was later used by the Germans to produce Buna-S following Mark and Wulff's industrial synthesis in 1930 for IG Farben.

Through the study of the polymerization of styrene, Staudinger also worked out his theories regarding the structure of macromolecules.

Polystyrene was first produced in the early thirties by the Naugatuck Chemical Division of U.S. Rubber, using patents taken out in 1928 by a Russian chemist, Ivan Ostromislensky, who had emigrated to the U.S. six years previously. But success came only in 1937, when Dow Chemical began marketing under the brand name "Styron" a colorless polystyrene with excellent electrical properties.

In 1835, the French chemist Henri Victor Regnault had noticed that the *vinyl chloride* $CH_2 = CHCl$, he had prepared, became an amorphous mass when exposed to light. Quite inadvertently and without knowing what it was all about, he had just produced *polyvinylchloride* (PVC). The first industrial process for the preparation of PVC,[16] however, was patented only in 1912 by F. Klatte, who was working with Griesheim-Elektron. The resulting polymer was not stable, could not be processed easily, and could not compete with cellulosics in price.

These drawbacks were overcome between the two wars, first in Germany but especially in the United States owing to several improvements. The use of plasticizers, which Ostromislensky had advocated as early as 1912, was implemented by B. F. Goodrich. E. W. Reid, who was working for Union Carbide and Chemicals (UCC) at the Mellon Institute, discovered that copolymerization with vinyl acetate made PVC easier to mold. Stabilizers improved higher-temperature processing of PVC, and the organo-tin compounds developed by UCC even made it possible to produce clear and transparent film. UCC's vinyl resins, sold under the trade name Vinylite, were first mar-

[16] By action of HCl on acetylene.

keted in 1936 for widespread uses. Other vinyl polymers also came on the market toward the end of the twenties.

Polyvinyl acetate, developed by Klatte, was produced industrially in 1927 by Canada's Shawinigan Chemicals. It was hydrolyzed into *polyvinyl alcohol*, and patents were also taken out in 1928 for *polyvinyl-acetal* and *polyvinyl butyral*. Shawinigan licensed this last to Du Pont; it was also manufactured by Union Carbide and by Monsanto, for it very soon turned out to be the best polymer for plastic sheets used as safety glass. From a 2 percent share of the U.S. market in 1936, it accounted for a 98 percent share five years later.

Efforts to produce a transparent organic material that would be lighter and less brittle than mineral glass were thus successful in the case of acrylic sheets and of safety glass based on polyvinyl butyral. But even where the desired result was not achieved, such efforts led to the marketing of a great variety of new plastics with unexpected properties and an outstanding commercial future.

The Chance Discovery of Polyethylene

In 1930, ICI's deputy research manager F. A. Freeth, who had started his career with Brunner Mond in 1907, suggested that the alkalis division take an interest in high-pressure chemical reactions. Freeth had been in contact, as early as 1919, with the thermodynamics laboratory of the Dutch scientist Kamerlingh Onnes in Leyden. With the technical assistance of A. Michels from Amsterdam, he managed to set up equipment in Winnington, in 1931, that was capable of measuring the physical and chemical effects of very high pressures on matter.

The physical chemist R. O. Gibson, who had worked with Michels in Amsterdam, and the organic chemist E. Fawcett, who had become interested in polymerizations through his friendship with Du Pont's Wallace Carothers, were both involved in this research. ICI's dyes division had also taken an interest in the Dutch research, and in 1932, one of its distinguished chemists, Robert Robinson, suggested a series of reactions that could be carried out in the Winnington laboratory. In 1933, one such reaction involving ethylene and benzaldehyde under pressure of 2,000 atmospheres left a white waxy deposit at the bottom of the tube as a result of a leak. The first minute quantities of *polyethylene* had just been produced.

Subsequent reactions using ethylene alone sparked off explosions. So pressure-polymerization tests were quickly called off. Besides, at the time transparent polymers such as polystyrene and polymethylmethacrylates were being sought, not waxy materials with improbable applications. At the end of 1935, another ICI physical chemist, Michael Perrin, resumed the work left off by Fawcett and Gibson. He noted that oxygen acted as a catalyst in the polymerization process. This explained the first polyethylene traces that had appeared in the 1933 tests when the leak had let air in. The oxygen dose was a critical one, for if it was too low there was no reaction, and if it was too high, an explosion was set off. After the optimal conditions for polymerization were determined in 1937, pilot production of polyethylene became possible.

Upon examining this new polymer, a chemist from ICI's dyes division, B. Habgood, discovered that it was a better electrical insulator than gutta percha. He suggested to his former employers, the Telegraph Construction and Maintenance Company Ltd., that it be used for undersea cables. They were to be ICI's first clients, buying up 100 tons from the industrial Alkathene unit which had come on stream in 1939 after ICI had patented its invention in 1936.

Carothers and the Nylon Adventure

By the end of the twenties, Du Pont de Nemours Company was prepared, as was ICI, to devote a part of its funds to fundamental research. The head of its chemical department, Charles Stine, engaged Wallace Hume Carothers in 1928 to head a high-level research team that was given the most advanced equipment for its laboratory work. Carothers had worked for Roger Adams at the University of Illinois before being appointed to Harvard. He had acquired very young the reputation of being a brilliant organic chemist.

Carother's first research work at Du Pont involved divinylacetylene, which had been discovered by Nieuwland and which led to polychloroprene rubber in 1931. At the same time, Carothers was pursuing in Wilmington fundamental research on the mechanisms of polymerization. But while Staudinger in Germany was working on polyadditions, Carothers was studying the reactions of *polycondensation*. This involved diacids reacting on dialcohols with removal of molecules of water to produce straight-chain aliphatic polyesters with a molecular weight of up to 12,000.

While conducting this research, one of Carothers assistants, Julian Hill, discovered the cold-stretching phenomenon whereby polyesters, first made into long fibers, extend still further to several times their original length when stretched after they are cooled. This boosted the strength and elasticity of the linear chains.

The melting point of aliphatic polyesters was too low and they were too soluble in water to become commercially interesting fibers. Carothers abandoned this first project in 1930, then replaced dialcohols by diamines to produce the more stable *polyamides* instead of polyesters. Their degree of polymerization was also easier to control. After a long series of tests, he chose reagents containing six atoms of carbon per molecule, in this case adipic acid, and hexamethylene diamine which could be easily obtained from benzene.

The resulting *6.6-polyamide* produced a fiber that was elastic, tough and water resistant. It melted at 260°.[17]

Nylon was born. But it was not before 1936 that the first threads for an experimental lot of stockings came out. Industrial production only started in 1939 in Seaford, Delaware. This was ten years after Carothers' first attempts, and Du Pont had spent $26 million before the first polyamide fibers were put on the market. William Carothers never lived to see how brilliantly his research had succeeded. In a fit of depression, he killed himself on April 29, 1937.

IG Farben could not remain aloof from this new type of fully synthetic fiber. Polycondensation through diacids and diamines was protected by Du Pont patents taken out in 1937. So the German firm had a try with aminoacids. From aminocaproic acid, they went into cyclic caprolactam, easily obtained

[17]

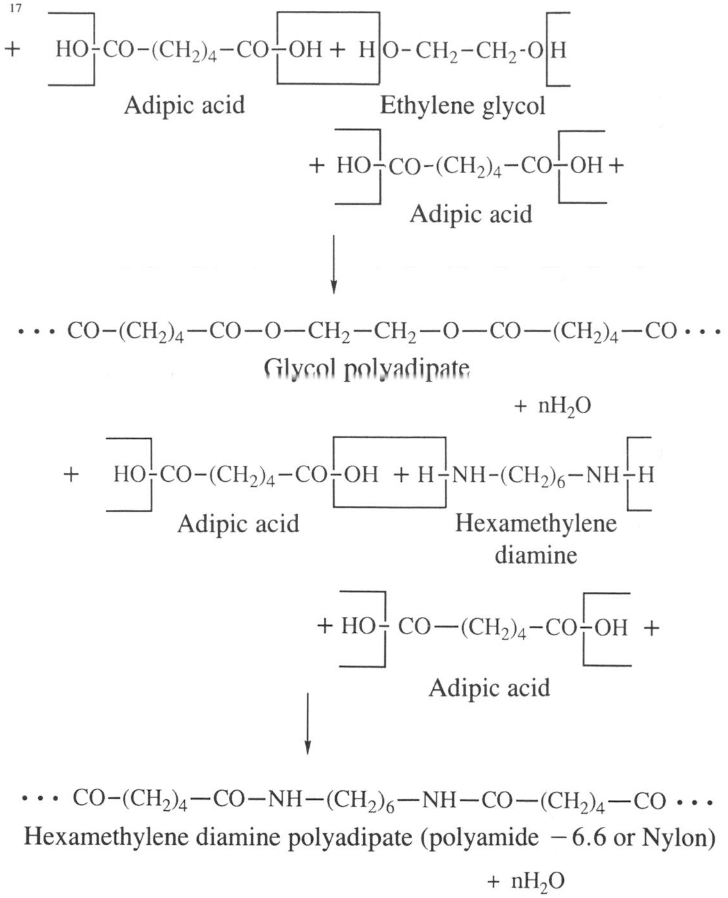

$$+ \; HO{-}CO{-}(CH_2)_4{-}CO{-}OH + H{-}O{-}CH_2{-}CH_2{-}O{-}H$$

Adipic acid Ethylene glycol

$$+ \; HO{-}CO{-}(CH_2)_4{-}CO{-}OH +$$

Adipic acid

$$\downarrow$$

$$\cdots CO{-}(CH_2)_4{-}CO{-}O{-}CH_2{-}CH_2{-}O{-}CO{-}(CH_2)_4{-}CO \cdots$$

Glycol polyadipate

$$+ \; nH_2O$$

$$+ \; HO{-}CO{-}(CH_2)_4{-}CO{-}OH + H{-}NH{-}(CH_2)_6{-}NH{-}H$$

Adipic acid Hexamethylene diamine

$$+ \; HO{-}CO{-}(CH_2)_4{-}CO{-}OH +$$

Adipic acid

$$\downarrow$$

$$\cdots CO{-}(CH_2)_4{-}CO{-}NH{-}(CH_2)_6{-}NH{-}CO{-}(CH_2)_4{-}CO \cdots$$

Hexamethylene diamine polyadipate (polyamide −6.6 or Nylon)

$$+ \; nH_2O$$

Wallace Carothers (1896–1937), inventor of Nylon. Courtesy E. I. du Pont de Nemours.

nowadays from phenol, and from there to polycaproamide fiber or *polyamide-6* which they called *Perlon*. The war was to put a stop to this study which had no strategic interest for IG Farben.

The Growth of Fluorine Chemistry

Du Pont de Nemours which had pioneered in the area of polychloroprene rubber and in synthetic fibers with its Nylon thread, also contributed to the development of fluorine chemicals through active involvement in two of the major inventions of the thirties.

The element fluorine had been isolated, as we have seen, by Henri Moissan in 1886. But its high degree of reactivity made it risky to use. As for hydrofluoric acid, *HF*, which Edmond Frémy obtained in 1856, it was hardly ever used except in its aqueous form to etch glass. Matters changed in 1930 when Charles Kettering asked Thomas Midgley to find a refrigerant both nontoxic and nonflammable for the Frigidaire division of General Motors.

Proceeding by elimination, as he had done in his research into antiknock agents, Midgley considered fluorine derivatives and noted that *dichlorodifluoromethane* best answered his purpose. Industrial production required anhydrous hydrofluoric acid as a starting point, and Du Pont began manufacturing it. In 1931, Kinetic Chemicals, a joint venture of GM and Du Pont, marketed the first fluorine-based refrigerant under the trademark of *Freon 12*. Engaged in research into the study of refrigerants at Du Pont's Deepwater laboratories in New Jersey, the chemist Roy Plunkett chanced in 1938 upon a fluorine polymer, the polytetrafluorethylene called *Teflon*, which has remarkable resistance to extreme temperatures, acids, and friction. The commercial success of these new fluorine derivatives discovered by Midgley and Plunkett gave new dimensions to the development of fluorine chemistry.

Eugène Houdry and Catalytic Cracking

The booming automobile industry of the early 1930s sparked a demand for light gasoline with a high octane content. Most oil companies had adopted the thermal cracking of petroleum, a process patented by William M. Burton of Standard Oil in 1912. What was now needed was to increase the quantities and improve the quality of the distillates produced.

An engineer from the École des Arts et Métiers in Paris, Eugène Houdry, who was also an automobile fan, had been alerted to these new needs when he took a trip to the United States in 1922. Thus far the obstacle to *catalytic cracking* had been the regeneration of the catalyst. Houdry managed to demonstrate that the alumina catalyst he was using could be regenerated at a certain temperature and in an air-and-hydrogen environment. But he was unable to convince European refiners that his discovery was worthwhile. Both the

Eugène Houdry.

The first catalytic cracking unit at Marcus Hook, Pennsylvania, 1938.

Anglo-Iranian Oil Company and Compagnie Saint-Gobain rejected his proposals.

It was in the United States, in 1930, that Houdry found the sponsors he needed to develop his invention. Through Socony Vacuum and more especially Sun Oil, he was able to set up the Houdry Process Corporation in 1931. Five years later the first "fixed bed" industrial catalytic cracker, using a blend of silica and alumina as an acid catalyst, came on stream in Marcus Hook. When World War II broke out, the Houdry process alone could produce the 97-octane gasoline required by the United States Air Force. "No other man," the president of Sun Oil, J. H. Pew, said later, "has made a greater industrial contribution to the war effort than our friend Eugène Houdry." Likewise, the spectacular development of petrochemicals would have been inconceivable without the catalytic cracking technologies initiated by this Frenchman who had commanded so little regard in his own country.

Soap Substitutes

The pupil of Gay-Lussac, Edmond Frémy, was undoubtedly the first to prepare surface-active agents through reaction of sulfuric acid on olive and castor oils to produce soap that was different from the ordinary variety. Until World War I, however, soap continued to be produced by saponification using alkalis, tallow or vegetable oils.

The growing demands of the textile industries for more efficient surfactants and the fear of a shortage of fats, which were, for the greater part, imported from overseas, had prompted the development in Germany of synthetic soaps.

In 1917, a pupil of Johannes Thiele named F. Günther, who was working with BASF's research manager Kurt Meyer, had taken out a patent for a compound obtained by combining naphthalenesulfonic acid with isopropyl alcohol which had wetting and emulsifying powers. Called *Nekal A*. it was the first fully synthetic surface-active substance to be produced, the ancestor of the *alkylaryl sulfonates*. From 1930 onwards, research in this area made great progress and new types of surfactants appeared. Boehme Fett Chemie and IG Farben, having managed to make alcohols through hydrogenation of fats under pressure and at a high temperature, over a copper catalyst, found a way to produce *sodium lauryl sulfate* industrially. Its use for shampoos was spurred by the marketing efforts of Procter & Gamble and Colgate Palmolive in the United States, and of Unilever in England.

Besides these *anionic* surface-active compounds with ionized sulfonate (SO_3) or sulfate (OSO_3) polar groups, *non-ionic* agents resulting from ethylene oxide condensation with fatty acids or alcohols were patented by BASF's M. Wittwer in 1930. Called *Igepals*, these derivatives served as models for a whole range of surfactants. Their cost was to become economically attractive when Union Carbide began manufacturing ethylene oxide through direct oxidation of ethylene in its South Charleston plant, West Virginia, in 1937.[18] That same year, in the United States, *non-ionic alkylphenol ethoxylates* not derived from fats, were marketed by Rohm & Haas under the trade name *Tritons*.

Thus from Germany's research into high performance textile auxiliaries, there emerged in the thirties both anionic and non-ionic surface-active substances. They came onto the markets of Europe and the United States and because of their excellent properties and reasonable cost gradually replaced traditional soaps, sparking a revolution in the detergents industry. Indeed, the large companies changed from traditional soapmaking to formulating powders, shampoos and detergents, using ingredients they did not manufacture and selling their products to an ever-widening public. At the same time, the dyestuff manufacturers in Germany, Switzerland, and America, were able to supply a wider range of processing agents for the leather and textile industries, based in part on these new derivatives, in order to satisfy the increasingly sophisticated demands of their clients.

The New Onslaught on Infectious Diseases

The success attained against the spirochetes of syphilis with Paul Ehrlich's Salvarsan—which made him the real father of chemotherapy—prompted the

[18] On the basis of a 1931 patent of the Frenchman Lefort.

synthesis of new molecules, particularly to fight tropical diseases at a time when European nations still possessed colonies.

During the war there had been a shortage of *quinine*, which W. H. Perkin had attempted to synthesize in 1856, because supplies of Java quinquina had been cut off. A research program conducted on antimalarials by IG Farben had successfully produced an aminoquinoline, Plasmoquin, in 1926 and more importantly in 1930 an aminoacridine, Atabrine, chemically closely related to quinine, which was still used in the Pacific during World War II.

Sulfonamides

Gerhard Domagk, who had been working since 1927 at Bayer's experimental pathology and bacteriology department in Elberfeld, was systematically screening dyes submitted to him, when he was struck by a chrysoidinesulfonamide[19] sample in 1933.

The red dye, which contained a SO_2NH_2 sulfonamide group, turned out to be particularly effective in experimental streptococcal infections in mice. In 1935, Domagk did not hesitate to use the derivative on his daughter Hildegarde, who had an attack of septicemia, thereby curing her. What was to become the first sulfa drug was thus developed under the name of *Prontosil* (or Rubiazol).

That same year, an Institut Pasteur team headed by Jacques Trefouël demonstrated that prontosil's effect was not due to the dye itself with its azo link (–N=N–), but to the colorless sulfanilamide[20] it breaks down to in the organism.

Sulfanilamide was not patentable, having already been described in 1908; so its use quickly spread under the trade name of *Septoplix*. Although it was proved that its curative properties had nothing to do with the red dye, Domagk nevertheless deserved the Nobel Prize for physiology and medicine which he won in 1939. But he was forbidden by the Nazis to accept it; so it was awarded to him only after the war.

While this research was being conducted in Germany, and then in France, C. Ewins and his team at May & Baker's in Dagenham were studying the synthesis of new sulfonamides. They produced *sulfapyridine*, which unlike

[19] The formula is

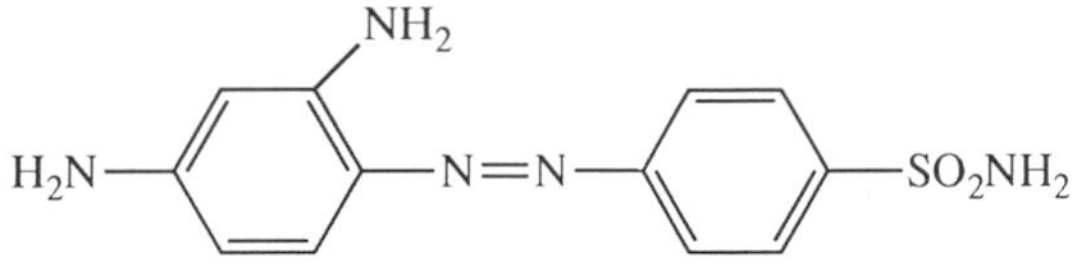

NH_2

H_2N—(benzene ring)—N=N—(benzene ring)—SO_2NH_2

[20] H_2N—(benzene ring)—SO_2NH_2

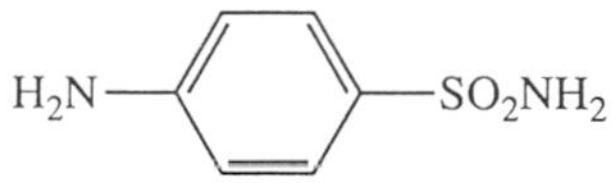

Prontosil was effective against pneumonia, and then *sulfathiazol* in 1938. While these sulfonamides were efficient against a number of bacilli, they were powerless against tuberculosis or leprosy. Other substances had to be found to fight these diseases. But meanwhile the *antibiotic* revolution had begun.

Fleming and Penicillin

In 1928, Scotland's Alexander Fleming, who taught bacteriology at London University and also worked at Saint Mary's Hospital, observed that the growth of staphylococci in a laboratory culture had been stopped by a mold that had settled on the culture. Fleming identified the mold as *Penicillium notatum* and noted that it had no toxic effect on animals or on leucocytes and he drew up a list of the germs it could destroy. But as he was no chemist, he was unable to stabilize the *penicillin* he had discovered, or to determine its structure. Matters rested there until the fall of 1939 when an Oxford team headed by the Australian Howard Florey and a German refugee Ernst Chain resumed the research. Working on a relatively small subsidy of $5,000 provided by the Rockefeller Foundation, they managed to produce penicillin in their laboratory within the next two years, to stabilize it and to demonstrate its potency through clinical tests.

The real breakthrough came when, after Florey's visit to the United States in 1941, the Department of Agriculture of Peoria, Illinois, succeeded in producing the new antibiotic through a deep-fermentation process. Penicillin revolutionized the treatment of infectious diseases taking over from sulfonamides.

In 1945, Fleming, Florey and Chain jointly received the Nobel Prize for medicine, marking recognition of their respective contributions.

Nutritional Disorders and the Emergence of Vitamins

The deficiency diseases caused by the lack in food diets of essential substances needed for the chemical functioning of the body in often minute quantities, have been known since early times.

In 1747, a naval surgeon from Scotland, James Lund, had managed to cure sailors of scurvy by giving them daily rations of oranges or lemons. Likewise, the Japanese admiral Takaki had observed in 1884 that his men no longer suffered from *beriberi* as long as their main diet of polished rice was complemented by more varied fare. Later, the English biochemist Gowland Hopkins noted in 1912 that rats fed on very pure food showed signs of *rickets* through bone decalcification. When fresh milk was included in their diet, the symptoms disappeared.

Emil McCollum in the United States discovered in 1913 that by introducing butter and egg yolk in the diet of patients suffering from xerophthalmia, a soreness of the eye's mucous membrane, the symptoms disappeared. He termed this antixerophthalmic substance, which was soluble in water and which he was trying to identify, the *fat-soluble A* factor. Meanwhile, the Polish chemist Casimir Funk was working in England and had managed in 1912 to extract from yeast a compound that was very effective against beriberi, which he called *vitamin* because of its amine content (NH_2). As this vitamin was soluble in water, McCollum called it *water-soluble B* in contrast to fat-soluble A.

England's J. C. Drummond renamed these two factors *vitamin A* and *vitamin B* in 1920. The substance which cured scurvy and which the Hungarian chemist Albert Szent-Györgyi extracted by chance from cabbages in 1928 in Gowland Hopkins's laboratory in London, was called *vitamin C*. *Vitamin D*[21] was the name given to the antirickets factor. Germany's Adolf Windaus discovered in 1926 that it came from ergosterol, a sterol extracted from rye ergot after exposure to sunlight. Two years later he was awarded the Nobel Prize in chemistry for this discovery.

Further elaboration, in the 1930s, of the factors thus identified, revealed their actual structures. Vitamin B turned out to be a complex blend of several vitamins. *Vitamin* B_1 was designated as the one that cured beriberi and that had been discovered in 1896 by the Dutch physician Christiaan Eijkman through work carried out in Java. Considering that a ton of rice husks contained only 5g of vitamin B_1, the vitamin was not isolated in pure form until 1927, and Robert R. Williams of Bell Telephone Laboratories did not establish its exact formula until 1934. Besides the amine group, the Vitamin B_1 molecule was found to contain sulfur. It was therefore called *thiamine*.

The structure of vitamin C, designated as *ascorbic acid*, was determined in 1932 by the American chemist Charles King, who succeeded in isolating it after extracting it from lemon juice.

Switzerland's Paul Karrer was the one who established in 1933 that *vitamin A* was produced by splitting the molecule of *carotene*, a colorant. This explains the yellow or orange color of food that contains it, like butter, carrots, egg yolk, and cod liver oil. By 1935, eight vitamins had been identified and characterized, including *vitamin* B_4 or *PP factor*, described since 1900 as *nicotinic acid* but whose effect as a cure for pellagra had not been recognized at the time; *vitamin E* (tocopherol); and *vitamin K*. The latter two played no role in fighting nutrition disorders.

As their structures were becoming known, chemists tried to carry out the synthesis of the principal vitamins. The Polish chemist Tadeus Reichstein, working in Switzerland, synthesized vitamin C (1938), while the American chemist Robert R. Williams synthesized thiamine (1937). At the same time,

[21] A form of this vitamin is also called *calciferol* because it plays a role in supplying bones with calcium.

C.E.K. Mees from Kodak had developed a process of intensive molecular distillation under high vacuum to obtain vitamin A from fish liver oils in concentrated doses.

As demand was growing especially for vitamin D, industry took over from scientists. Increased production of fat-soluble vitamins A and D sparked growing consumption of cod liver oil from Norway and Newfoundland. And as the price of vitamins fell, their consumption soared. Thus emerged a new industry from studies carried out since the beginning of the century by chemists and biochemists seeking to understand how deficiency diseases originated.

Chemical Messengers: Hormones

In 1850, the French physiologist Claude Bernard established the notion that a number of functions of the human body were regulated by "internal secretion centres." England's William M. Bayliss and Ernest Henry Sparling, while studying the working of the pancreas, recognized in 1902 existence of a *chemical messenger* secreted by the duodenal mucosa which stimulated the pancreatic juice. Since then, these messengers have been called *hormones*. They are produced in minute quantities by glands[22] located in different areas of the body, and are carried by the blood to other parts, boosting their physiological functions.

The simplest of these hormones is *adrenalin*. It was also the first to be isolated, in 1901, from animal adrenal glands by the Japanese chemist Jokichi Takamine working in the United States. It was synthesized in 1904 by F. Stolz. Derived from tyrosine, it is sometimes designated by its chemical name of *epinephrine*.

In 1915, Edward C. Kendall from the Mayo Foundation in Minnesota isolated from the thyroid gland another amino acid similar to tyrosine which behaved like a hormone. He called it *thyroxine*. Thyroxine contained four atoms of iodine, and it was observed that its role was to control the general metabolic rate of the body. Growth is impeded by its absence or malfunctioning. Only in 1926 was its structure established and its synthesis carried out through the work of the English researchers Charles R. Harrington and George Barger.

In 1916 the Scottish physician Edward A. Sharpey-Schafer suspected that an anti-diabetes hormone was secreted in the pancreas's Isles of Langerhans. He called this presumed hormone *insulin* (from the Latin *insula*, meaning island). Insulin was isolated in 1921 by Canada's Frederick G. Banting and Charles H. Best of Toronto University. The Nobel Prize for Medicine and Physiology was awarded to Banting in 1923 in recognition of this achievement. Although insulin was obtained in crystalline form by J. J. Abel as early

[22] Ductless glands are termed "endocrine."

as 1926, it has a complex protein structure that was fully elucidated only in 1954 by Frederick Sanger and his Cambridge team of researchers.

In 1927, two German physiologists, B. Zondek and S. Aschheim, had observed that extracts from the urine of pregnant women injected into female mice excited the mice sexually. Two years later, Germany's Adolf Butenandt and America's Edward A. Doisy succeeded in isolating the female sex hormone discovered by Zondek and Aschheim. *Estrone* was thus the first of the estrogen hormones to be discovered. Butenandt then studied male sex hormones, isolating androsterone from male urine in 1931.

It then became possible to determine the structure of these sex hormones. They resemble a *steroid*, consisting of four rings as are found in cholesterol. The Croatian chemist Leopold Ruzicka, working at the Swiss Institute of Technology, was able to describe the synthesis of androsterone in 1934.

Adolf Butenandt was awarded the Nobel Prize for chemistry in 1939 for his research on steroid hormones. But like his fellow countryman Gerhard Domagk, he only received it after the war. Leopold Ruzicka received the Nobel Prize for chemistry in 1938 for his work on polycyclic molecules.

Other steroid-type chemical messengers that were nonsexual were discovered after 1929. Adrenal glands, indeed, consist of an internal part which secretes adrenalin and an external layer, the *adrenal cortex* (from the Latin *cortex*, meaning bark) which also produces a hormone. Lack of this hormone produces Addison's disease. There are a number of *cortical* hormones, as a matter of fact, and they all belong to the steroid group. Both Tadeus Reichstein, who synthesized vitamin C, and Edward C. Kendall, who discovered thyroxin, studied the subject thoroughly, a difficult task because the small size of adrenal glands required a large number of animals to obtain sufficient cortex extracts. Thus attempts to synthesize cortical hormones were made early, but it was only after the war that a biochemist at Merck, Lewis H. Sarrett, managed to obtain in a 37-stage process the component that was to become famous as *cortisone*.

By the mid-1930s, a number of laboratories both in Europe and the United States were trying to produce the different hormones. Because they were difficult to synthesize, these firms maintained, for a long while, activities that were more biological than chemical, essentially based on recovering glands and organs from slaughterhouses.

Except for this fact, the story of hormones greatly resembles the story of vitamins. First, deficiencies leading to functioning disorders were observed; then the causes were investigated. This led to the isolation of active substances in minute quantities. Their chemical structure then had to be determined before they could be commercially developed. In either case, physiologists, biochemists and chemists working in their laboratories, had paved the way for the pharmaceutical industry.

The Problem of Crop Protection

Because plants are open to attack from fungi, insects or weeds, use of *fungicides*, *insecticides*, and *herbicides* was considered very early on to be a necessity.

Since early in the nineteenth century, elementary *sulfur* had been recommended to protect fruit trees against mildew. In 1868, an English veterinarian called William Cooper had started selling a blend of copper and iron sulfates to farmers in his area. *Copper sulfate* use became widespread when it was observed in 1882 that, mixed with lime to form *bouillie bordelaise*, it cured the Médoc vines of mildew. A botanist working at the Bordeaux Faculté des Sciences in 1885, F. Millardet, carried out a thorough scientific study of what was the first fungicide to be produced, giving it an aura of respectability.

Until 1930, England was the major producer of copper sulfate, but three years later Italy, which was the largest consumer for its vineyards, outstripped its competitors by increasing its production to 106,000 tons. Other copper salts, such as *copper oxychloride*, were used as fungicides at the time, while utilization of the extremely toxic mercury salts was becoming widespread. The discovery in 1931 by Du Pont's W. H. Tisdale and Williams of the fungicidal properties of dithiocarbamates, used until then to accelerate rubber vulcanization, greatly improved the harmlessness of fungicides to human beings as well as making them more efficient.

Research into *selective herbicides* that killed only weeds, produced *arsenic salts* and *sodium chlorate* between 1915 and 1925. In 1934, *dinitro-o-cresol* (DNOC) was discovered, but it was only in the early forties that 2,4D[23] and *MCPA*[24] provided more efficient herbicides. These selective hormones were the outcome of a study on growth control of plants which had been initiated by the discovery in 1926 of β-*indoleacetic acid* by Holland's Went, and in 1935 of α-*naphthylacetic* acid at the Boyce Thompson Institute.

The interest in *insecticides* was twofold: on the one hand, insects had to be fought because they destroyed crops; on the other hand, flies and mosquitoes that carried tropical diseases such as sleeping sickness and malaria had to be destroyed.

Nicotine was used, principally in England, to kill gnats on garden plants as soon as it could be extracted from tobacco by steam distillation. It is still used as a sulfate, although it is toxic for higher animals. *Pyrethrum* extracted by a solvent from a variety of chrysanthemum plants first came in the form of dried flowers from Dalmatia, then was imported from Japan, which became the main supplier until Kenya took over when English settlers introduced the chrysanthemum flower feverfew there in 1928. Pyrethrum has a paralyzing, not a destructive action on insects. It was mainly used in gardens and to protect

[23] 2,4-dichlorophenoxyacetic acid.

[24] 2-methyl-4-chlorophenoxyacetic acid.

cattle from flies. Its active principles were isolated by Hermann Staudinger and Leopold Ruzicka in 1924.

Other natural insecticides derived from *rotenone*, contained in certain plants from Malaya and South America, were introduced at the time despite their high price, for the same uses as feverfew. At the same time, *lead arsenate* was used increasingly to fight orchard insects, especially in the United States, where arsenates had already proved effective to fight the potato beetle as early as 1867. Americans also utilized *sodium fluosilicate*, which they imported from Italy and Denmark. But as in the case of fungicides and herbicides, spectacular progress only took place when synthetic organic insecticides were developed, foremost of which was DDT.

Dichlorodiphenyltrichloroethane,[25] or DDT, had been synthesized in 1874 by a student called O. Zeidler according to a process that was subsequently always used. It consisted of two monochlorobenzene molecules reacting with a chloral molecule in the presence of sulfuric acid. Only on September 26, 1939, however, did the chemist Paul Müller, working for J. R. Geigy in Basel, perceive the strong and persistent effectiveness of DDT as an insecticide.

Geigy had been looking, for the last twenty years, for a good insecticide. The firm had already been marketing the mothkiller, Mitin FF. Tests conducted with pyrethrum and rotenone had been abandoned because the products were affected by light. Organic chlorine derivatives led Müller to the discovery that brought him the Nobel Prize for medicine in 1948.

As early as 1942, Geigy informed the British legation in Switzerland of DDT's remarkable properties. In 1943, the world was informed that the insecticide had been the means of halting a typhus epidemic in Naples that could have caused ravages in the American army. DDT turned out to be efficient in every area of battle against insects while seeming nontoxic for higher animals. It was easy to produce, and its cost was low. It enjoyed immediate success; by 1945, production had already soared to 15,000 tons.

Thus, through a now-familiar process, the need to protect crops had led to empirical use of natural products, then to mineral salts, and finally to synthetic derivatives that were both efficient in fighting harmful crop pests and nontoxic for higher animals. A new industry had emerged from a need

[25] The formula is

H
Cl— C₆H₄ —C— C₆H₄ —Cl
CCl_3

DDT

recognized since remote times. But only in the 1930s was this need satisfied through the sustained work of organic chemists.

The Chemical Industries on the Eve of World War II

The period that had elapsed since 1920 had profoundly changed the nature of the chemical industry the world over. Mergers had been effected, resulting in the emergence of a small number of giant firms in Germany, Britain, and the United States. Several thousand shareholders held stakes in them and their shares were negotiated on the main financial markets. In this way, the general public became familiar with the activities of the chemical industry. Endowed with powerful material and human resources, the large conglomerates had developed ambitious research programs that led to spectacular industrial breakthroughs in high polymers (polyethylene, neoprene, Nylon). Such breakthroughs were made possible by high-level scientific work and spectacular technological progress, regarding both the polymerization processes and control of high-pressure reactions.

Great strides had also been made in the United States in the use of petroleum as a feedstock for the chemicals industry through better refining processes and the development of thermal and catalytic cracking. These improved the quality of distillates and, at the same time, created smaller molecules for a whole series of chemical reactions. Synthetic substitutes for natural products continued to be developed, particularly in Germany, and a whole range of surface-active agents was starting to replace soaps. Several aliphatic raw materials (methanol, acetone, acetic acid) were henceforth being produced by synthesis and no longer from wood distillation.

Most of these breakthroughs were the outcome of research carried out within industry itself. But scientists were also working in independent research institutes and, through their fundamental discoveries in vitamins and hormones, were opening up new prospects for pharmaceutical laboratories in both Europe and the United States. Such upheavals knew no frontiers. Nevertheless, they did not affect all industrialized countries in the same way.

Despite producer cartels, which created a certain degree of solidarity among competing suppliers at the expense of consumers, each national industry developed in its own natural, economic, and political environment and tried to find its own solutions to the problems it had to face.

America's Chemical Industry Emerges Strong from the Depression

The Great Depression of 1929 affected America's leading industries less than the election in 1933 of the Democrat Franklin Delano Roosevelt as President of the United States. His New Deal involved a whole series of state-inspired decisions that were at odds with the U.S. liberal tradition: establishment of the National Recovery Act (NRA), which decreed a general salary hike in the purest Keynesian spirit; forced cooperation between government and industry through pricing "codes"; passage of the Wagner Act which boosted the powers of the workers' unions; establishment of the Tennessee Valley Authority (TVA) which took an immediate interest in the production of nitrate fertilizers and superphosphates.

The chemical industry, however, was not overmuch affected by these measures. In the first place, the NRA was declared unconstitutional in 1935 and abolished. This event loosened constraints imposed on industry by Roosevelt. Industry leaders regained a free hand on prices. The Wagner Act did perhaps make life difficult for them, but labor was not a major component of the chemical industry's production costs. The TVA was, in fact, the only instance of direct state intervention in chemical production. But the American taxpayer picked up the bill. TVA losses in fertilizers in 1943, after ten years of intervention, amounted to $19,220,000.

Lacking substantial sales growth during the years of depression and unable also to increase their prices, chemical companies concentrated on improving productivity. This effort involved the use of timing, pressure-, volume- and temperature-control instruments, and the installation of reactors with a larger capacity, making use of new materials such as carbon, special steels, and Pyrex glass.

A new discipline, *chemurgy*, grew out of the need to dispose of agricultural surpluses, and companies like Corn Products, U.S. Industrial Alcohol (USI), Commercial Solvents, and Quaker Oats were active in this area.

Only a very small part of America's chemical production was exported at the time, involving some 4 percent of total sales in 1931 compared to the 27 percent exported by European groups. Had the United States been able to expand its sales abroad and apply competitive prices, there would certainly have been retaliatory measures from European producers entrenched behind their cartel habits.

Faced with Europe's arsenal of protectionist measures (high tariffs, import quotas, government subsidies, embargo threats), Roosevelt had obtained from Congress in 1934 the power to change U.S. customs tariffs and to sign, as he saw fit, bilateral trade agreements, thus replacing market forces with case-by-case arrangements. Such official intervention was a hindrance for the chemical industry, which sought to develop a base in the Southern states of the United States, where the ground was surer. At the same time, efforts were made to penetrate the foreign markets that were still open.

The Southern states were endowed with plentiful resources (cotton, wood pulp, rosin and turpentine, sulfur, salt, oil), including abundant manpower not yet organized into trade unions; they offered new markets for chemicals in the manufacture of paper, cotton and viscose rayon textiles. Following in the footsteps of Du Pont and Union Carbide, which had already settled in Belle and Charleston, respectively, other large chemical giants set up plants in the South. Production centers emerged in Baton Rouge, Lake Charles in Louisiana, Texas City in Texas, and generally throughout the region bordering the Gulf of Mexico. The rush to the South largely contributed to the chemical industry's health despite the hardships of economic depression and government intervention.

The New Look of the American Market

The emergence of new demands, the development of novel technologies and the availability of raw materials that had been little used until then, changed the nature of the United States chemical industry between the two world wars. Demand for metals, textiles, pigments, and rayon boosted *sulfuric acid* needs at a time when its production was starting to be assured by the recovery of SO_2 from sulfur minerals and of H_2S from refinery gases. Sulfur obtained through the Frasch process was supplied mainly by Freeport Sulphur and by Texas Gulf, based in Texas and Louisiana.

Hydrochloric acid (HCl) was being increasingly consumed, in particular for the synthesis of vinyl chloride and arsenious chloride, as well as for the treatment of oil wells (Dowell process). HCl was largely obtained as a by-product of the chlorination of organic compounds. *Chlorine* itself had become one of the base products of the chemical industry. In addition to its already recognized uses for the production of carbon tetrachloride (CCl_4), and then of trichloroethylene, it was now needed for the recovery of bromine from sea water, to prepare halogenated refrigerants and to synthesize organic intermediates such as chlorobenzene for phenol production (Raschig process) or ethylene chlorhydrin, forerunner of ethylene glycol, as well as of Thiokol rubber. Chlorine was also needed to produce neoprene, monomeric vinyl chloride and chlorinated rubber, the new, boom products.

Caustic soda, obtained as a by-product of electrolysis, was also being increasingly required in the Southern states by the viscose, Kraft paper, and oil industries, reestablishing a balance which the rising demand for chlorine might have threatened in the caustic soda-chlorine production units.

Nitrogen fertilizer producers, who could not help being concerned by what the Tennessee Valley Authority in Muscle Shoals was doing, were happy when Southern cotton planters adopted synthetic nitrates for their crops. This enlarged the market for ammonia, nitrate, and ammonium phosphate as well as for urea. Florida's prewar strong position in phosphate ore had eroded in Europe, which was importing phosphates from North Africa. At the same time, the Soviet Union was trying to be less dependent on its phosphate imports.

Potash was a very different picture. The Franco-German cartel had to reckon with the new sources being developed in Searles Lake, California, by American Potash & Chemical Corporation, then from 1932 onwards in Carlsbad, New Mexico, by other companies. By 1939, the United States could satisfy three-quarters of its potash needs.

With exports becoming more difficult, American companies diversified downstream to *phosphoric acid* and *phosphorus*. Furnaces were built in the Southern states, in particular by Monsanto and Victor Chemicals. Phosphoric acid found a ready market as a food acidulant while *sodium tripolyphosphate* and *tetrasodium pyrophosphate* were being increasingly used in the formulation of new detergents.

Throughout the 1930s oil processors were gradually getting closer to the chemists. The success of *tetraethyl lead* as an antiknock agent in gasoline gave impetus to dibromoethylene production, needed to remove the lead, and hence to that of *bromine*. Ethyl-Dow began extracting bromine from sea water in its North Carolina plant. Standard Oil was able to produce improved lubricants and solvents by using the Bergius hydrogenation process. The quantities of *lubricating oil* were being boosted through use of additives like Standard Oil's Paraflow, Monsanto's Santolube and Lubrizol's *o*-dichlorobenzene derivatives developed in Cleveland.

Petroleum was being increasingly regarded as a source of *aliphatic hydrocarbons*. Standard Oil had pioneered, as early as 1920, production of *isopropyl alcohol* from refinery propylene. Sharples followed with *amyl alcohols* produced from pentane chlorination. Union Carbide & Chemicals were the first to market, between 1930 and 1939, an impressive range of *alcohols, aldehydes, ketones, esters, ether alcohols, aliphatic acids* as well as *ethylene oxide* (leading to *glycols*) and *vinyl polymers* (PVC, polyvinyl acetate, polyvinylbutyral) which they produced from olefins derived from ethane, propane and butane. Research conducted by Purdue University on the nitration of methane and ethane led to *nitroparaffins*, which Commercial Solvents began producing in 1939 at its Peoria plant. Likewise, Sharples and then Rohm & Haas were marketing *alkylamines* by 1938.

This new chemistry, based on straight-chain hydrocarbons extracted from petroleum and natural gas, formed the foundation of a U.S.-born *petrochemicals* industry that was to revolutionize world chemicals after the war. American chemists established their plants closer to the Gulf of Mexico refineries for their supplies, and the oil industry integrated downstream into basic organic chemicals.

Aromatic chemicals manufacturers in the United States continued to rely on hydrocarbons like benzene, toluene, and naphthalene extracted from coal tar. As the steelworks industry was recovering around 1935, the coking furnaces built by Koppers and Semet Solvay and the gas plants which were multiplying near the towns were providing in ample quantities the aromatic derivatives needed by the chemical industry.

Creosote was used for wood protection and *tars* for road building, while the lighter products, such as benzene, toluene, xylene, and naphtha solvents, were sold directly to consumer industries. New distillation units were set up between 1925 and 1936 by Neville Chemical in Pittsburgh to produce *coumarone indene resins* and solvents, by Reilly Tar for *pyridine* and derivatives, as well as by Allied Chemical's Barrett Division which had been the first to produce, as early as 1922, *maleic anhydride* through oxidation of benzene.

As demand for *glycerophthalic resins* (glyptals) and *phthalates* for plasticizers grew, there was pressing need for *phthalic anhydride* and, consequently, for *naphthalene*. In 1935, distillation units began to extract naphthalene, which, until then, had been imported from Europe, together with *phenol*, which was used to prepare phenolic resins. Demand could no longer be met by what was produced from tar. Synthetic phenol was obtained either through benzene sulfonation (Monsanto, Barrett) or through hydrolysis of monochlorobenzene by the Raschig process developed by General Plastics. *Cresols*, which were also extracted from tar, instead of from traditional imports from England, were partly used as substitutes for phenols in phenol-formaldehyde resins.

The *naval stores* industry had developed since 1930. Hercules Powder extracted turpentine and pine oil from Southern pine trees and stumps, as well as *pale wood rosins* using a furfural-extraction process. Pine oil was used as a disinfectant and as a flotation agent for minerals. Terpene chemistry was also developing, however, and new uses were being found in perfumery, for synthesis, or as solvents, for turpentine components such as pinene, dipentene, and terpinolene, as well as for rosin in the form of hydrogenated or disproportionated rosins, and ester gums, which were serving increasingly for the formulation of inks, adhesives, and lacquers.

A process for the synthesis of *camphor* from pinene, developed by Newport Industries and taken over by Du Pont, was applied in Deepwater in 1933, making America independent of the Formosa natural camphor or of the German synthetic product for its use as plasticizer for pyroxylin. The use of *furfural* was expanding both to extract the rosin from pine tree trunks and, starting in 1933, as a selective solvent in refinery processes. Furfural was produced from oat hulls by Quaker Oats and rose from the status of laboratory product to cheap solvent.

Based on the wealth of base materials extracted from petroleum, natural gas, or agricultural byproducts, the United States had developed between the two wars an innovative *fine chemicals* industry intended to preserve the country from import contingencies. The production of *citric acid* by fermentation developed by Charles Pfizer & Company as early as 1928 freed America from the shackles of the European cartel. The United States became an exporter of the product before World War II. Pfizer also later obtained *gluconic acid* by fermentation as well as *fumaric acid*, halving its price and thus opening up new markets for the preparation of resins.

Atlas Powder announced in 1936 that it had perfected an original *sorbitol* manufacturing process by electrolytic reduction of the carbohydrates of corn syrup. Here again, the considerable drop in the production cost of the new process made it economical to use sorbitol for the synthesis of ascorbic acid (vitamin C) and to produce new surface-active agents. Cheaper sorbitol thus became a serious competitor for glycerine. Another new preparation process for *glycerine* was disclosed in 1938 by Shell Development Company using chlorination of propylene through the intermediary step of allyl alcohol. Glycerine users no longer needed to be dependent on the seesaw availability and price of the product derived from fatty acids.

Competition between natural products and their chemically produced synthetic equivalents was illustrated in 1938 when *vanillin* was produced industrially by Marathon Paper Mills Company from the lime lignosulfonates contained in bisulfite liquor. Within two years this vanillin, a byproduct of paper pulp, had taken over a third of the American market at the expense of Monsanto's synthetic vanillin.

In pharmaceuticals, while Europe originated *sulfonamides* and *tranquilizers*, as well as *germicides* using chlorinated phenols, and in 1935 the quaternary ammonium compounds discovered by IG Farben's Gerhard Domagk, *vitamin* and *hormone* production was pioneered by the Americans. Merck & Company was the first to synthesize *vitamin C* from sorbitol, with Charles Pfizer following in 1937. Robert R. Williams and J. K. Cline of Merck performed the synthesis of *Vitamin* B_1 in 1936, thereby bringing down its cost to the point where it could be incorporated in flour; its production increased to 30 tons a year by 1942.

The successful marketing of *hormones* was due to the joint effort of research teams working within independent research institutes in various laboratories belonging to pharmaceutical companies and in the largest Chicago slaughterhouses. Semi-elaborated products and glands were supplied in this way by Armour and by Wilson. *Epinephrin*, *insulin*, and the first *sex hormones* reached the market in record time, thanks to work conducted in universities such as the University of Toronto and Washington University, Saint Louis, as well as in the research centers of Eli Lilly, Parke Davis, Upjohn, and Squibb.

Some Important American Companies Between 1929 and 1939

America's chemical companies were necessarily affected by the Great Depression. Turnover of the most important group, Du Pont, fell from $214 million in 1929 to $127 million in 1932 and its working force was reduced by

10,000 employees during that time. By 1935, however, the companies had recovered; and by 1940 their sales had been boosted by the diversification efforts of industry leaders, who were simultaneously consolidating what had been acquired in the 1920s.

Du Pont bought up the stakes that its French partners had in rayon, cellophane, and ammonia subsidiaries, and took over Remington Arms, Krebs Pigments, and, in 1930, Roessler-Hasslacher. Through an ambitious research program, products as varied as tetraethyl lead, neoprene, Freon gas, Teflon, and, especially, Nylon were brought to the market. By 1937, 40 percent of the company's sales involved products that did not exist ten years before. Their popularity largely made up for the price erosion suffered by ammonia and a number of other base products. Under the firm management of Lammot du Pont since 1926, and raking in substantial dividends from its stake in General Motors, Du Pont remained the oldest, most powerful, and most diversified chemical company in the United States.

Ranking second was *Allied Chemical & Dyes*. It was the outcome of a share-exchange merger, carried out in 1920, of five different entities: General Chemical, Barrett, National Aniline, Solvay Process, and Semet Solvay. It had the structure of a holding company and, while there were obvious synergies among its various parts, each one of them developed its traditional markets, but remained faithful to the principle established by Orlando Weber of not coming into competition with client companies, by keeping to base chemicals. This explains why Allied played such a small role in international trade and had little innovative capacity in the technological area except for the ammonia unit built in Hopewell.

The reverse can be said for *Union Carbide & Carbon Company* (UCC), which had grown to a respectable size straight after the 1917 merger that created it. Between 1929 and 1939 it pursued an enthusiastic development program in aliphatic chemicals under the impulse of research conducted by George Curme at the Mellon Institute.

A pioneer in this field, UCC was the first to put on the American market synthetic ethanol, ethylene oxide, glycols, ethanolamines, acetone (from isopropyl alcohol), and acetic anhydride, all made from olefins extracted from oil and natural gas.

At the same time, and true to its origins, UCC, through its Niacet subsidiary, which had been set up in 1924 with Canada's Shawinigan Chemicals, had worked on acetylene chemistry and marketed acetaldehyde, acetic acid, vinyl acetate and, in 1936, vinyl copolymers sold under the trade name Vinylite. With this entry into resins, UCC bought Léo Baekeland's Bakelite Corporation in 1939 to enlarge its range.

Despite a more modest start, *American Cyanamid Company* was to become one of the four U.S. chemical giants. Since the death of its founder, Frank Washburn, in 1922 it was managed by an astute lawyer, William Bell, who initiated the group's diversification. The company's original operations, involving calcium cyanamide, ammonia and Florida phosphates, mainly focused on America's agriculture and took the brunt of the economic slump.

American Cyanamid did manufacture guanidine-based rubber chemicals and alkyd resins and, from 1928, had agreements with Britain's Beetle to develop aminoplast resins in the United States. But William Bell had loftier ambitions for this company. In 1929 he took over Calco, which produced dyestuffs in Bound Brook, New Jersey, and Kalbfleisch, a very old producer of acids and mineral salts.

The following year, American Cyanamid's purchase of Laboratoire Lederlé involved it in antitoxins and led it into pharmaceuticals. At the same time, it purchased an engineering company called Chemical Construction as well as an explosives company, American Powder. Internal growth was proceeding apace. Ore flotation products were developed and American Cyanamid achieved the first industrial synthesis of melamine, a new raw material which would subsequently be used to prepare melamine formaldehyde resins and decorative panels based on these resins. But it was mainly because of Bell's acquisitions policy that the group grew, in ten years, to become one of the United States' largest chemical corporations.

Dow and Monsanto were two other companies to achieve rapid growth at this time, the former because of major technological breakthroughs, and the latter because of particularly profitable acquisitions.

When Herbert Dow died in 1930, his son Willard became President of *Dow Chemical*. He was only thirty-three years old, and there were signs of hard times ahead. But the company's management team was technically competent; and the young president, who was an engineering graduate of the University of Michigan, had already proven his worth, particularly on the shop floor.

In 1932, the Dowell subsidiary was set up. Its acidification technology for oil wells led to its prosperity. The following year, Ethyl-Dow was set up under a fifty-fifty partnership to develop a remarkable process to extract bromine from seawater in Lure Beach, North Carolina. Dow, which produced both chlorine and phenol, also launched a chlorophenol research program that produced Dowicides, a line of powerful fungicides and germicides.

Dow's contribution to high polymers involved development of ethyl cellulose used in lacquers and inks, and more especially in 1937 a transparent polystyrene prepared from purified styrene, which was marketed under the name of Styron. To meet increasing demand for magnesium ore[26] required by Europe's air forces, Dow revived on a large scale a project for producing magnesium chloride extracted from seawater, a process its founder had investigated in 1910. As the U.S. military authorities were also becoming interested

[26] Magnesium production had increased from 1,000 tons in 1934 to 3,000 tons in 1938.

in the light metal before the war, Dow chose to set up a unit in Freeport, Texas in 1939; and on January 21, 1941, the first magnesium ingot extracted from the sea was produced by the Texas plant. Between 1930 and 1938, the company's sales soared from $16 million to $24 million and its workforce from 1,800 to 4,000 employees. This was mostly due to internal growth spurred by research and technology.

Monsanto's founder, John Queeny, who died in 1933, had passed on the business to his son Edgar Monsanto Queeny in 1928. He was only thirty-one years old and, although not a brilliant student, had graduated from Cornell University as a chemist. But he turned out to be a leader of exceptional qualities.

While consolidating the range of organic products that formed the core of its business (vanillin, ethyl vanillin, chlorinated phenols, aspirin), Monsanto had acquired in 1920 a 50 percent stake in Graesser Chemical, a very old Welsh firm in Ruabon which specialized in tar products. In 1929, Queeny made two major purchases, the Rubber Service Laboratories in Akron, Ohio, which propelled the company into the rubber additives sector, and the Merrimac Chemical Corporation in Boston, a firm dating back to 1853 that produced base chemicals on the East Coast, where Monsanto had no production units.

In 1936 Queeny acquired the laboratories of two well-known chemists, Charles Thomas and Carroll Hochwalt who had worked with Thomas Midgley at General Motors. They made a noteworthy contribution to Monsanto's research. Monsanto became involved in plastics when, in 1938, it completely took over Fiberloid, one of the oldest nitrocellulose production companies, which had a 50 percent stake in Shawinigan Resins. The year after, Monsanto purchased Resinox, a subsidiary of Corn Products, and Commercial Solvents, which specialized in phenolic resins. Thus, just before the war, Monsanto's plastics interests included phenol-formaldehyde thermosetting resins and cellulose and vinyl plastics (Formvar and Butvar), all located in Springfield, Massachusetts. In 1939, Monsanto also took a majority stake in the Swann Corporation in Birmingham, Alabama, then fully bought up the company two years later, thus acquiring what was to become a major position in phosphorus chemicals, and particularly in phosphates which were being increasingly used in detergents.

Other companies that were less important or less diversified than the ones mentioned were trying to develop through internal growth or acquisitions, often with success.

Hercules Powder Company had integrated upstream with cotton linters by purchasing Virginia Cellulose Company in Hopewell, Virginia in 1926. While still consolidating its explosives base and improving its rosin extraction technology, Hercules had also diversified in 1931 by acquiring Paper Makers Chemical Corporation, becoming one of the leading manufacturers of paper sizes. In 1929, it decided to build a research station near its Wilmington head office, a sign that it wished to develop its research base. By the time war broke out, Hercules had put on the market chlorinated rubber, cellulose acetate, and ethyl cellulose.

Eastman Kodak's directors had at first restricted the company's chemistry to meeting its photographic needs, then had wished to go beyond this narrow field. In 1920, they set up in Kingsport, Tennessee, the Tennessee Eastman Corporation, which extracted, from neighboring forests through wood distillation, acetic acid used for cellulose acetate films, as well as methanol. By 1930, Eastman Kodak was marketing hydroquinone, cellulose acetate, and a new cellulose acetobutyrate polymer; and the quantities sold outside the company greatly exceeded the quantities needed for internal use. Through research conducted by C.E.K. Mees, Kodak had sufficiently mastered the high vacuum distillation process to set up, together with General Mills, Distillation Products, Inc., which began by focusing its production on vitamins A and E.

The Rohm & Haas Company had managed to emerge from its initial leather-treatment business through the technical help of Otto Röhm, who had remained in Darmstadt, and through the diversification policy conducted by Otto Haas, marked by the buying out, in 1920, of Charles Lennig and the establishment of Resinous Products and Chemicals Company in 1926.

From its own research, Rohm & Haas developed proprietary products such as the non-ionic tensio-active Tritons and the Rhoplex acrylic emulsions for textiles; the Lethane thiocyanate insecticides, introduced as early as 1926; and Acryloid additives for motor oils, based on higher methacrylates. The real change in the company came with its development of Plexiglas, a sequel to research begun in 1935 in Darmstadt and transferred to Philadelphia. Plexiglas became a necessary component of American bombers during World War II.

As these chemical corporations were adding new products to their portfolios, other companies of very different origins were appearing on the chemicals market. One such example was the companies which, until then, had specialized in rubber.

U.S. Rubber, for one, which had originated in Connecticut as early as 1892, had set up in 1904 the Naugatuck Chemical Company to manufacture sulfuric acid for the reclaiming of rubber. During World War I, the company had been led to produce the aniline required for its business, then thiocarbanilide. U.S. Rubber then set out to market rubber additives, even pioneering antioxidants based on secondary amines in 1924. While pursuing major research work on latex and synthetic rubber, the company began testing, during the Great Depression years, rubber organic derivatives that it synthesized for use as fungicides. This was the start of a brilliant career in crop protection products on the part of what would become the UniRoyal Corporation.

Another very old company was *B. F. Goodrich*, which dated back to

1880. It had come to chemicals through its merger with Diamond Rubber Company in 1912. This was the company whose chemist George Oenslager had discovered in 1905 that aniline and thiocarbanilide were good accelerators. Under T. Semon's impulse, B. F. Goodrich became involved in vinyl polymers in 1925. The success of Koroseal and Geon just before the war led to construction of large PVC plants in Niagara Falls, then in Louisville, Kentucky.

Another rubber giant also came to chemicals through accelerators. In 1912, the head of *Goodyear Tire & Rubber Company*'s research department, L. Sebrell, had managed to identify Captax, an accelerator which another scientist, C. Bedford, had obtained quite by chance through sulfur reacting with thiocarbanilide. Goodyear increased the mileage of truck tires with Captax, then improved their resistance to heat by substituting rayon for the cotton lining. Goodyear became successfully involved in coatings through the development in the thirties of chlorinated rubbers trade-named Pliolite and Pliofilm. But the company kept up its rubber additive business, which was marked, in 1924, by the discovery of cheap antioxidants of the phenylnaphthylamine variety.

Firestone was a latecomer to chemicals because its main concern for a long while had been the threat to American tire makers of soaring rubber prices in Malaya imposed by Britain. The company reacted to the threat by setting up rubber plantations in Liberia.

Other American companies of different backgrounds became interested in chemicals. As early as 1920, *Commercial Solvents*, which produced butanol under the Weizmann process, had used fermentation gases (H_2, CO_2) in the production of ammonia and methanol in 1926, and also to market dry ice. Experience acquired in the thirties in high-pressure synthesis led Commercial Solvents to produce nitroparaffins from natural gas on the basis of research work undertaken at Purdue University. An outside consultant, Carl Miner, had found that the residual liquors produced in the alcoholic fermentation process of molasses in the company's Peoria plant were rich in riboflavin (vitamin B_2). So Commercial Solvents started marketing the vitamin in 1938 to be used in poultry and cattle feed. Because its fermentation expertise dated back to World War I, Commercial Solvents was one of the first U.S. companies to become involved in the production of penicillin which it set up in record time in 1943 in its Terre Haute, Indiana, plant.

Another fermentation specialist was *Charles Pfizer* of Brooklyn. It had been producing citric acid biochemically since 1923, and enlarged its range to include fumaric, gluconic, and itaconic acids. During the war, Pfizer also launched out into the production of penicillin.

Merck, a U.S. company since 1905, was established in Rahway, New Jersey. During the 1930s it had been the first American laboratory to market sulfonamides in the United States, particularly sulfapyridine and sulfathiazol.

The company also synthesized ephedrin and was a pioneer in the industrial production of vitamins (vitamin B_1, B_2, B_6, C, E, and K) thanks to a remarkable team of chemists. In due time, Merck was to become quite naturally one of the leading penicillin producers following agreements between the United States and British governments signed during the war.

The Indianapolis-based *Eli Lilly* had been the first on the biological market to produce insulin in 1922. This was the outcome of a cooperative effort with Frederick Banting and Charles Best from the University of Toronto. After 1930, *Upjohn* in Kalamazoo, Michigan, *Parke Davis*, *Squibb* and *G. D. Searle*, all working on gland extracts, focused their research, like Eli Lilly, on hormones that had recently been identified but that scientists were finding difficult to synthesize.

The Particular Nature of Britain's Chemical Industry

The economic slump which hit all industrialized countries in 1929 did not spare Britain's economy. Although basic sectors were affected, household consumption was growing for the salaries of the working population remained steady, so that as prices fell, purchasing power improved. Demand remained strong in the automobile, clothing, health, electrical appliances, and building sectors. Among the giants, those like *Unilever* or *Courtaulds*, which catered for the general public, were better off than *ICI*, for instance, which was deeply involved in heavy chemicals and very little in consumer products.

The Burden of State Intervention

After 1932, synthetic organic derivatives imported from abroad were, under the "Safeguarding Industries Act," liable to heavy customs tariffs of 33⅓ percent (Key Industry Duty). The measure had originally been intended to protect Britain's dyestuff industry. But in practice many other sectors of organic chemicals and even inorganic chemistry were affected by it. That was why *Albright & Wilson*, which had transferred its phosphorus production to its Canadian subsidiary in 1920 because it was more profitable to do so, resumed production in England in 1934, first in Widnes, then in Oldbury.

A further government measure considerably affected the development of aliphatic chemicals in Britain. As in other European countries, industrial alcohol, unlike beverage alcohol, was not taxed in Britain. But the government had gone further by subsidizing industrial alcohol suppliers since 1921 through the "inconvenience allowance," which was meant to compensate them for the red tape they had to deal with. This explains how *ethyl alcohol* became the cheapest source of aliphatic products, and why the main supplier, *Distillers Company Ltd.* (DCL), a merger of six whiskey producers established in Scotland since 1877, felt strongly encouraged to develop a chemical industry based on ethyl alcohol.

In 1928, DCL set up *British Industrial Solvents* on its Hull distillery site to turn ethyl alcohol into acetone, acetaldehyde, butyraldehyde, butanol, and acetic acid. Ethyl and butyl acetates were produced by another subsidiary, the *Methylating Company Ltd.*, in the Carshalton unit. Because of the alcohol subsidy, ICI was also tempted to produce ethylene (by a dehydration process) in Huddersfield in 1934, converting it on site into ethylene glycol and ethylene oxide. ICI also used this ethylene in 1939 for its first polyethylene unit set up in Wallerscote. ICI could very well have launched into the production of solvents for it began making synthetic methanol in Billingham in 1925 and it was obtaining butane and propane as by-products of synthetic gasoline produced on the same site under a variant of the Bergius process. However, through the "inconvenience allowance," DCL had become sufficiently strong in this area to warrant an understanding between the two firms. In 1938 they signed an agreement setting out very clearly the respective business limits of each firm.

Government intervention took place in another major area. In 1928, it levied a 4 pence per gallon tax on imported oil to encourage the production of coal-based gasoline, and in 1934 guaranteed this protection through the British Hydrocarbon Oil Act. This induced ICI to use the excess capacity of hydrogen, left over in the Billingham plant because of the ammonia slump, to set up a unit to produce synthetic gasoline through coal hydrogenation that came on stream in 1935. Later the British were to realize that neither alcohol from molasses nor from coal feedstocks could supply aliphatic hydrocarbons at prices likely to compete with oil-based products. But in the years before World War II, Britain was not really a free-trade country. Certainly the chemical industry, protected as it was by international cartel arrangements, customs duties, imperial preferences, and government subsidies, could not pretend it was operating within the frame of a market economy.

ICI's Dominant Position

After the December 1926 merger, ICI was by far the dominant chemical group in Britain. With its 40,000 employees and its sales hovering around £30 million by the end of the 1920s, it was certainly no match for IG Farben or Du Pont de Nemours. But in its own domestic market it enjoyed a true monopoly position for a large share of its production and, as its name indicated, held a preferential position within the British Empire. The cartel arrangements on alkalis and nitrate fertilizers and, from 1931 onwards, in dyestuffs, spared ICI the need to engage in price wars while at the same time guaranteeing it a share of the international markets.

Despite the strong personality of Sir Harry McGowan, who took over the group's leadership in 1931 and wielded absolute power until the rebellion of the "barons" in 1938, each division has so strong and distinct a culture that overall rationalization of operations could not take place. This offered both advantages and disadvantages. On the one hand, the autonomy of each sector was large enough for ambitious research programs to be carried out without any undue interference. This was the case with high-pressure chemistry in Billingham and the development by the dyes division of the first phthalocyanine pigment, Monastral Blue, in 1935. On the other hand, territorial conflicts became inevitable when, as in the case of polyethylene or Perspex, several divisions considered themselves to have been involved in developing a product. This made the establishment of a plastics division, after the purchase in 1933 of a majority share in the phenolic resins producer Croydon Mouldrite, all the more difficult to achieve.

Like his predecessor Lord Melchett, McGowan could boast of no technical expertise; he relied heavily on his research departments, a policy which proved fruitful. But he was personally very keen on the signing of agreements not only among international competitors but also between suppliers and clients on the home market. Thus ICI abandoned a cellulose acetate project that Rhodiaceta had proposed in 1928 because it did not wish to displease Courtaulds, which was a buyer of large amounts of its soda. McGowan subsequently made a point of associating Courtaulds in his Nylon negotiations with Du Pont, beginning in 1938. Likewise, ICI sold auxiliary products to the textiles industry but kept out of household detergents for fear of upsetting Unilever. The same spirit prompted them to associate with the Pilkington glassmakers, which were important clients for sodium carbonate, when in 1935 the project was formed to develop polymethylmethacrylate (Perspex) in transparent sheet form.

Thus just as Allied Chemical in the United States had made a point of sticking to its role of basic chemicals supplier, ICI felt bound, in prewar years, to abide by a good-conduct code because of its dominant position. This policy caused it to miss out on or delay development opportunities through not wanting to give its traditional clients any reason to complain of unfair competition.

Strong and Weak Points of Britain's Chemical Industry

While ICI had a monopoly in the key areas of alkali, chlorine, hydrochloric and nitric acids, sodium, synthetic ammonia, and nitrate explosives as well as in the major classes of dyestuffs, other corporations had managed to develop their own specialties.

Phosphorus chemistry remained the monopoly of *Albright & Wilson* which, as we have seen, had taken over production of white phosphorus at Widnes while remaining strongly entrenched in Canada through Electric Reduction. The company took advantage of increasing phosphate use in foodstuffs, which accounted for over one-third of its sales after 1934. It also supplied the detergents industry with sodium tripolyphosphate, the PVC industry with tricresyl phosphate plasticizers and the water treatment industry with sodium hexametaphosphate. From the potassium chlorate which it had long been producing for use in matches, it had gone on to sodium chlorate increasingly used to kill weeds along railroad tracks.

In Widnes, from the carbon disulfide and chlorine supplied by ICI, Albright & Wilson produced carbon tetrachloride[27] which served for such diversified uses as cleaning solvents and soil fumigants in agriculture. Since the beginning of the century sulfuric acid overproduction was endemic in England and a natural market for it was in superphosphates. In 1929, the phosphate-fertilizer sector underwent rationalization when E. Packard of Prentice Brothers merged with James and Joseph Fisons. This gave birth to *Fisons Ltd.*, which accounted in 1939 for one-third of Britain's superphosphate capacity.

Britain's Bernard Laporte had been a pioneer in peroxides. In 1948, his company was to become *Laporte Chemicals*. The use of perborates as bleaching agent in washing powders gave a boost to the firm's development. In 1932 it introduced in its Luton plant the electrolytic process for the production of hydrogen peroxide. That same year, Laporte diversified into titanium dioxide (TiO_2) through an association with National Titanium Pigments, and purchased Malehurst Barytes Company's barytes mine.

With the exception of carbon dioxide, Britain was dependent on *British Oxygen Company* for its industrial gases. The company, which had kept its name since 1906, was an offshoot of Brin Oxygen Company. Besides oxygen, nitrogen and acetylene, British Oxygen offered the whole range of rare gases by the late 1930s. With its Commonwealth units, it was one of the world leaders in its area.

Carbon dioxide had been recovered since 1929 by *Distillers* in its fermentation units and sold in the form of dry ice as a refrigerant. In the late 1930s, ICI also began marketing CO_2 produced by its hydrogen-purification process in Billingham.

England's prewar heavy organic chemicals were, as we have seen, based both on ethyl alcohol produced by fermentation from molasses, and on coal which could yield tar by distillation or be converted to carbon monoxide (CO). While *Distillers* had taken full advantage of the alcohol subsidy to develop its oxygenated-solvents chemistry, the chemistry of chlorinated solvents had originated in the excess chlorine capacities which England had been left with after World War I. Indeed, the carbon tetrachloride that Albright & Wilson produced in the Midlands as early as 1925 before transferring production to Widnes in 1933, was a by-product of sodium chloride electrolysis. The progress made in metal degreasing technology during the 1920s sparked in addition

[27] The reaction is $CS_2 + 2Cl_2 \longrightarrow CCl_4 + 2S$

a strong demand for trichloroethylene. Its consumption soared tenfold in Britain between 1928 and 1936 for the greater benefit of ICI, which had become sole supplier since its purchase of Weston Co. ICI had also found another market in 1935 for chlorine in organic chemicals in the form of Alloprene chlorinated rubber used in paints resistant to chemical agents.

Britain's coal resources had long served to develop aromatic chemicals. Before the war, tar distillation from coking furnaces and gas works were in the hands of some sixty private companies, which processed from 2,500,000 to 3,000,000 tons of tar every year. But the beginnings of concentration had taken place in 1923 when *Midland Tar Distillers* was set up in 1934, followed in 1926 by *Lancashire Tar Distillers*. Two years later *Graesser* was bought up by Monsanto Chemicals while Low Temperature Carbonization Ltd., which became Coalite and Chemical Company in 1948, was specializing in low temperature carbonization.

Fine chemicals and synthetic intermediates were also being produced from tar-based hydrocarbon aromatics by a number of medium-sized companies like *W. J. Bush & Company, Boake Roberts, W. Blythe, James Robinson* and *Hickson & Welch*. ICI also produced methanol and synthetic gasoline from coal through pressure hydrogenation in its Billingham unit.

In pharmaceuticals, England had distinguished itself when, in 1937, a group of researchers working under C. Ewins at *May & Baker*'s had discovered sulfapyridine and sulfathiazol. ICI's Pharmaceuticals Division, which discovered sulfamethazine in 1942 and the antimalarial paludrin four years later, was an offshoot of the special therapeutics research department set up in 1936 by the company's dyestuff division.

Besides the older laboratories such as *Boots Pure Drug* and *British Drug Houses* (BDH), mergers had produced large pharmaceutical groups such as *Beecham*, which had absorbed Macleans and Eno in 1938, *Glaxo*, and *Smith & Nephews*.

Copper sulfate had been introduced in England as early as 1868 by William Cooper to be used for the protection of crops. The firm of *Cooper, McDougall & Robertson Ltd.* was still leader in its branch when it associated with ICI in 1937 to form *Plant Protection Ltd.* The Vienna-born H. Ripper, who had introduced nicotine to fight insects harmful to fruit trees, founded *Pest Control Ltd.* in 1938. It was bought up by Fisons Ltd. in 1954. ICI researchers working in Jealotts Hill in 1936 discovered the role of naphthylacetic acid as a growth hormone for plants; British researchers were also to be instrumental in developing the first selective herbicides.

Three chemicals-related multinational groups originated in England: *Courtaulds Ltd.*, which had started manufacturing Viscose in England in 1905 and had set up the *American Viscose* subsidiary in the United States five years later, was producing carbon disulfide and sulfuric acid since 1916 in Stratford near Manchester. In 1928, Courtaulds diversified into cellulose acetate fibers and in 1934 took part in the establishment of *British Cellophane*. In 1940, together with ICI, it set up *British Nylon Spinners*. Only after the war, however, did Courtaulds squarely take its place in the chemical industry through its purchase of *British Celanese*.

This was also true of *Shell Chemical*, of the *Royal Dutch Shell* group, which had chemical interests, before the Second World War, in Holland and California. *Unilever*, the other Anglo-Dutch trust company, entered the chemical industry through its subsidiaries *J. Crosfield & Sons*, involved in silicates, and *Price Ltd.*, specializing in fatty acids.

Such, therefore, were the strong points of Britain's chemical industry before the war. But it also had its weaknesses. For want of calcium carbide, which it imported from Scandinavia, Britain had scarcely developed *acetylene* chemistry, so that, unlike Germany and the United States, it had no vinyl derivatives production. More generally, Britain's chemical industry was lagging in *high polymers*. Relying on Malayan natural rubber, the industry was interested neither in synthetic rubber nor in PVC, which could have served as substitutes for electrical and telephone cables. In plastics, Britain's production was relatively modest. Phenolic resins, however, were produced by *Bakelite Ltd.*, which had taken over Redmanol and Damard in 1927, and also by *Croydon Mouldrite*, in which ICI had taken a majority stake in 1933. Moreover, aminoplasts remained a specialty of *British Cyanides Company*, which became *British Industrial Plastics* in 1936.

With the exception of the old cellulosic materials manufactured in *British Xylonite* and ICI's brilliant developments in Perspex and polyethylene still in their early stages, England was far behind in thermoplastics, producing neither polystyrene nor vinyl resins. This was also the case for silicones. While they had originated through Frederick Kipping's studies at the University of Nottingham, they came to England via America, as did Nylon and Teflon. Finally, having put all its stakes in coal and sugarcane molasses as feedstock for its organic chemicals, England's chemical industry went through an agonizing reappraisal when it was confronted with the *petrochemicals* era.

Germany's Chemical Industry and the War Economy

The Germans were still reminded in the early 1930s of the supply hardships that had impaired their war effort. Their 1918 defeat had also deprived them of their African colonies, so that, unlike the French and the British, they had no overseas territories to help them cover their needs in rubber, peanut, and coconut oil. The collapse of the Ottoman Empire, moreover, had cut Germany off from Middle East oil suppliers.

In the circumstances, Germany's chemical industry remained the only hope of acquiring self-sufficiency in basic raw materials through the substitute products (*Ersatz*) it was likely to produce.

As had been the case for ICI in Billingham, IG Farben's Leuna-based program for *synthetic gasoline* production through coal hydrogenation was a costly affair and the resulting product was hardly competitive with oil-based motor fuel. By 1930, 300 million DM had already been sunk into the project; and IG Farben's management was of two minds about keeping it going.

Carl Bosch, together with Fritz Ter Meer, managed to convince his colleagues of the ongoing program's importance. When Adolf Hitler became Chancellor on January 30, 1933, IG Farben received a boost and on December 14, 1933, Bosch and Schmitz signed, on behalf of IG, an agreement with the government under which synthetic gasoline production in Leuna would be raised to 350,000 tons within the next four years. The government guaranteed the sale of the whole production at a price related to the production cost and it shared in the funding of the project. The idea of market price was superseded by the notion of strategic product. Twelve additional hydrogenation units on the Leuna model using coal and lignite were built in Germany from 1935 onwards. By 1944, their total capacity had peaked to four million tons a year.

The problem of *synthetic rubber* was raised in similar fashion. IG Farben had been forced to halt Buna rubber production in 1931–1932 because natural rubber prices on the world market had fallen too low. Besides, tire manufacturers were not very happy with the performance of Buna rubber. Hitler's government was divided as to what to do. The minister of national economy, Hjalmar Schacht, was loath to cut off Germany from overseas trade and was in favor of natural rubber imports when there existed a price advantage over synthetic rubber. Backed by his general staff, Hitler believed, on the contrary, that national self-sufficiency should outweigh economic considerations; of course, his view prevailed.

With these government assurances, Bosch set up in 1936 a large (500 ton/month)[28] Buna unit in Schkopau, close to Leuna. His closest assistant Carl Krauch, joined Hermann Goering's general staff that same year and was put in charge of raw materials and overseas trade. Thus, IG Farben became directly involved in the "four-year plan" that was to make Germany ready for war.

Other essential raw materials imported from abroad had gradually been replaced by synthetic products manufactured in the Fischer-Tropsch units. These units, which were built near Leuna, produced as much as 600,000 tons a year of the various hydrocarbons needed by Germany to synthesize the surface-active agents, solvents, and plasticizers, without which the detergents, paint, and plastics industries could not have survived.

In the textiles sector, research into fibers that could substitute for natural imported products was also pursued. The Nobel Group's old nitrocellulose

[28] In 1943, the Reich was producing 118,000 tons of synthetic rubber, including 70,000 tons from Schkopau, and the rest from Ludwigshafen and Hüls.

explosives factory, Köln-Rottweil, located in Premnitz near Berlin, had reconverted after the First World War and was producing, under the trade name "Vistra," a staple fiber which was good enough and cheap enough to find ever-incresaing markets. A program to increase artificial textile production was launched in 1935, and by 1942 the Reich was capable of producing 300,000 tons of staple fiber and 100,000 tons of artificial silk, accounting respectively for one-half and one-fifth of world production.

The Supremacy of IG Farben

IG Farben's supremacy had begun with its birth from the merger of eight companies on January 1, 1926, and the necessities of war had boosted it further. During its very first year, with sales of 1,027 million Reichsmark, IG Farben was ten times as important as the second-ranking German chemical company, Deutsche Solvay Werke; and the gap was to become wider as the years went by.

In dyestuffs, which accounted for one-third of this figure, IG Farben had a true monopoly position. This did not prevent it from carrying out rationalization measures, which improved productivity in a noteworthy fashion. In a first stage, each unit became specialized in the type of dye it could produce under the best conditions. Thus the share awarded to Ludwigshafen and, especially, to Leverkusen was increased, while Hoechst, deprived of azo dyes, saw its share reduced.

A second rationalization process began in 1930 to do away with duplication. Competition between units over a single product, which had so far been accepted by Duisberg, was henceforward banned. In other sectors, operations were shared out in relation to individual expertise: Hoechst was entrusted with *acetylene* chemicals and took on for its own use Griesheim's vinyl acetate patents and its technology; BASF retained *high-pressure reactions*, with the Oppau and Leuna plants alone employing in 1920 some 35,000 people, one-third of IG Farben's total workforce; *photographic products* were assembled under AGFA; Griesheim gave up a major share of its chemicals to concentrate on *metals* and *alloys*; Kalle, which had been forced to withdraw from dyes, took over *"Ozalid" paper* and *cellophane film*. These specializations which date back to the 1926–1930 period are still to be found in the landscape of the present-day German chemical industry.

In conjunction with its internal rationalization effort, IG Farben was also pursuing a diversification program. In the area of *textiles*, the Wolfen films unit, near Bitterfeld, had been producing viscose since 1921. Dormagen had started producing cuprammonium continuous filament in 1926, and the following year Aceta GmbH started silk acetate. Köln-Rottweil AG, with its three viscose units in Rottweil, Bobingen, and Premnitz, was absorbed by IG Farben in 1925.

With Dynamit Nobel AG, IG Farben had formed, short of absorption, a joint association in *explosives*, and with Kali Chemie marketing agreements in certain areas of inorganic chemistry. To avoid price competition in *acetic acid* and *acetone*, a joint Hoechst and Wacker subsidiary had been set up as early as 1920.

Anxious to create a good climate within the German chemicals industry, IG Farben deliberately kept away from producing certain chemicals, such as sodium carbonate, potash, superphosphates, sodium cyanide, peroxide, bromine, as well as lacquers and varnishes.

Through the medium of Bayer and Hoechst, IG Farben held a significant position in *pharmaceuticals* and basked worldwide in the fame acquired by the discoveries of these two firms. Bayer had become famous for its aspirin as early as 1876 and Gerhard Domagk's work on sulfonamides in 1935 added to its reputation. Hoechst had produced Pyramidon, Novocaine, and Salvarsan before making insulin in 1922 from pancreas glands obtained from the Frankfurt and Karlsruhe slaughterhouses. In 1939, the Hoechst laboratories brought out Dolantine, a synthetic painkiller four times as effective as Pyramidon.

Through its subsidiaries, IG Farben had developed a worldwide marketing base while maintaining overseas production at a modest level. The group focused its development mainly on exports from Germany. But through IG Chemie, a Basel-based holding company, IG had become interested in a Swiss dyestuffs manufacturer, Durand Huguenin, which it acquired in 1924. A year later, it centralized all its United States interests under General Dyestuffs Corporation. In 1927, it took a 25 percent share in Norsk Hydro, sharing in exchange its knowhow on the synthesis of ammonia.

Germany's Chemical Industry in the Environment Built Up by IG Farben

By the end of the twenties, Germany's chemical industry had a workforce of 300,000 people; and IG Farben accounted for a third of the total. It also accounted for 30 percent of the nation's chemical production and 47 percent of its chemical investments. Nonetheless, there was room for other companies to expand beside it. *The Deutsche Solvay Werke*, with its 6,300-strong workforce, provided two-thirds of the country's soda needs and was Germany's second-ranking chemical company. *Kali Chemie*, the outcome of a merger between one of the Friedrichshall potash syndicate members and Rhenania-Kunheim, was a long-time producer of heavy chemicals, principally ammonia soda. With units dispersed over sixteen sites, it was not a model of productivity; it later fell under the control of the Solvay group.

IG Farben had been clever enough to remain away from superphosphates, which it purchased merely to prepare mixed fertilizers. Because of overseas competition, Germany's superphosphate producers, like their British counterparts, were going through difficult times, and IG's trade policy did not help them. Mergers did take place, as in the association of *Pommerensdorf-Milch* with *Guano Werke* in 1927; but the sector continued to suffer from its dispersion and its overcapacities.

The company that Fritz Henkel had founded in 1876, and that had put on the market the perborate and silicate-based Persil washing powder, had grown to significant dimensions. After World War II, *Henkel* ranked among the four main chemical companies in Germany, becoming a world leader in oil-based chemicals. It is still fully controlled by the descendants of the founder, under a contract that expires in 2000. To ensure its sodium perborate supplies, Henkel had formed an association with Degussa in 1926, holding one-third of the company's preferred shares, just like IG, which had done the same for sodium. Thus sponsored, Degussa was in no danger of falling into foreign hands.

Deutsche Gold und Silber Scheide Anstalt (Degussa) had been founded by the Roesslers in 1876. The company produced sodium in Rheinfelden since 1898 and perborate since 1908. After the 1914–18 war, Degussa had taken a majority stake in Grünau and in 1929 had set up, also in Rheinfelden, an electrolytic hydrogen peroxide unit. While continuing to develop its traditional business (precious metals refining, sodium, cyanides, peroxide products) the company's leaders began diversifying in carbon black and pigments by taking interests in Wegelin, and then in Marquart, and by setting up Russ Werke Dortmund in 1936.

A number of German companies had successfully maintained their positions in certain areas in the face of IG Farben through innovative research and good technology. Founded in 1848, *Rütgerswerke* kept its leadership in tar derivatives and through improved fractionation managed to extract high added-value derivatives. *F. Raschig*, through his creative genius, successfully diversified his activities.

As previously mentioned, Raschig had developed in 1895 the so-called Raschig rings to fill distillation columns, then had carried out the first industrial synthesis of hydrazine in 1907. Between 1908 and 1914 he had built continuous tar-distillation units in Ludwigshafen and put on the market new para-chloro-meta-cresol and chloroxylenol preservatives, as well as phenol formaldehyde resins under a Baekeland license.

After the war, led by the founder's two sons, the Raschig company further distinguished itself by developing industrially an elegant phenol synthesis process via catalytic chlorination of benzene in 1935. Four years later, the U.S.-based General Plastics adopted the process.

Despite the place IG Farben occupied in fine chemicals and pharmaceuticals, there still remained a number of independent firms in these sectors.

To protect themselves against their powerful rival, *Merck Darmstadt, Böhringer & Söhne*, and *Knoll* had formed a shared interests association (*In-*

teressengemeinschaft). There was also the firm founded in 1814 in Berlin by J. D. Riedel, which had expanded by buying up Eugen de Haen's company in Seelze, near Hanover, in 1923. With its focus on fine chemicals, it employed 1,800 people, and became known in 1943 as *Riedel de Haen AG*, before becoming a subsidiary of Cassella in 1955.

Another Berlin firm that dated back to 1851 was *Schering*. After initial specialization in iodine and bromine derivatives for pharmaceutical and photographic uses, it had set up an electroplating unit in Berlin in 1908. After the war, Schering launched into pesticides and antiparasite derivatives and began research on endocrine glands which resulted in Progynone being put on the market in 1928. Its merger with Kahlbaum in 1927 reinforced the position of the two companies both in fine chemicals and pharmaceuticals. Already well known for its synthesis of camphor, by the time World War II broke out, Schering had become a very diversified group with four plants in Germany and sales in thirty foreign countries.

Despite IG Farben's preponderance, therefore, a number of German companies had managed, through their research work and clever diversification policies, to achieve a certain degree of prosperity by 1939 and enjoyed a deserved reputation abroad.

France's Chemical Industry: A Cumulation of Particularisms

Despite some concentrations, there were no groups in France by the end of the 1920s to compare, in power and diversity, with IG Farben or ICI.

Some companies had acquired distinguished positions in their special areas, like *Air Liquide* in industrial gases, but they were wary of diversification. Others had enlarged their range of products but remained outside the essential chemical sectors, like *Rhône-Poulenc* which, originating in 1928, was no longer involved in dyes but was not yet thinking of entering inorganic chemicals.

France's chemical industry still remained very scattered at the time in numerous small units all over the country. In 1929, *Saint Gobain* owned 25 chemical plants, and only Chauny and Saint-Fons had attained to a workforce of 900 people. Saint-Gobain's chemicals remained linked to sulfuric acid, superphosphates and sodium carbonate. While the company accounted for a significant share of the market in these heavy chemicals,[29] they had small added value. Production of some inorganic chemicals was started: caustic soda and chlorine through electrolysis in Wasquehal in 1927 and in Saint-Fons in 1928, nitric acid through ammonia oxidation in Chauny in 1926 and in Saint-Fons in 1929, phosphoric acid via the wet process in Chauny early in 1929.

[29] 41.05 percent, 41.25 percent, and 13.46 percent, respectively, in 1929.

Saint-Gobain also produced calcium carbide and calcium cyanamide in Modane as well as synthetic ammonia in its Grande Paroisse subsidiary which it shared on a fifty-fifty basis with Air Liquide.

Saint-Gobain's volume productions accounted for a quarter of France's inorganic chemicals. But it also had interests in other chemical sectors. It had a 20 percent stake in *Compagnie Centrale Rousselot*, which it supplied with hydrochloric acid for the glues and gelatines it produced. It had a 28 percent stake in *Société Normande de Produits Chimiques*, a Rouen company manufacturing formic acid and oxalic acid from sodium formate, HCO_2Na, and in which Progil and Rhône-Poulenc had a similar share. Besides this, Saint-Gobain bought up, between 1924 and the end of the thirties, 30.15 percent of *Société des Matières Colorantes et Produits Chimiques de Saint-Denis* to prevent it from falling into the hands of Établissements Kuhlmann.

In 1927, Saint-Gobain took a 14 percent share of *Fabriques de Produits Chimiques de Thann et Mulhouse* and was involved for 25.5 percent of the share in the establishment of *Compagnie des Produits Chimiques et Raffineries de Berre* (PCRB). Finally, to strengthen its base in Western France, Saint-Gobain had taken over *Société des Usines DIOR* in 1936, a family business that produced sulfuric acid and fertilizers in Brittany, as well as a number of fine chemicals like strychnine and rubiazol. So it was that during the 1930s, and excepting its traditional heavy inorganic chemicals productions, Saint-Gobain, whose main activity was glassmaking, seemed to be playing a holding-company role in a multitude of chemical and chemically related areas. The minority stakes it held in companies in which it was only remotely involved had been prompted either by a wish to counter a bothersome rival, or to avoid remaining left out of some promising development.

Établissements Kuhlmann, France's other chemical giant, had followed a different path to build up its future because it was essentially a chemical company. Upon the death of Frederick Kuhlmann, his son-in-law Edouard Agache had taken charge of the business in 1881. He had expanded it through the purchase in 1905 of the Wattrelos plant in Belgium and through the establishment, in that country, of the Rieme plant in 1910. When the units in northern France and Belgium were requisitioned by the Germans during World War I, Agache was forced to widen the company's geographic base in order to survive and to diversify the range of productions. Thus, he had bought up from the Rothschild family the Penarroya plant in Marseille-L'Estaque, which was languishing, and had set up units in Aubervilliers, Petit Quevilly, Bordeaux, Nantes, and Port de Bouc.

When the Service des Poudres set up the Compagnie Nationale des Matières Colorantes (CNMC) in Oissel, near Rouen, in 1916, Kuhlmann was naturally put in technical charge of the new unit. This led in 1924 to Établissements Kuhlmann absorbing CNMC with its Oissel and Villers Saint-Paul units in 1924. After 1920, the group bought up production units in Paimboeuf,

Dieuze, and Brignoud while its northern plants were being rebuilt. The company developed considerably between 1925 and 1940 under the impulse of the imaginative polytechnician Raymond Berr and the dyes specialist Joseph Frossard, while keeping to its original chemical calling, and maintaining excellent relations with IG Farben.

In 1928 Établissements Kuhlmann took a stake in *Société Coignet*, a rival of Rousselot in glues and gelatin and a long-time phosphorus-chemicals producer. Associated with French coal-mining interests, it set up, that same year also, *Marles Kuhlmann* to manufacture ethylene oxide and glycols in Choques, then founded *Courrières Kuhlmann*, which produced methanol from water gas, the following year in Harnes. Just before the war, tetraethyl lead began to be produced in Paimboeuf and Port de Bouc, leading to the founding of the *Octel-Kuhlmann* company. By then Kuhlmann was involved both in inorganic chemistry, ranking just behind Saint-Gobain in sulfuric acid and superphosphates, with access to synthetic ammonia through its coal-mine partners, and in organic chemicals, through its near-monopoly of French dyestuffs and its production of derivatives of ethylene and methanol. It came to calcium carbide, calcium cyanamide, and electrolytic chlorine through purchasing the Fredet unit in Brignoud in 1923.

By merging small glue and gelatin companies from 1917 onward, culminating in the association with Coignet in 1928, Kuhlmann became a leading producer in this area. It also became involved in bromine chemicals when it began producing antiknock agents. By 1938, Kuhlmann had achieved a more coherent development in chemicals than Saint-Gobain, because of the way it had originated, even though the units it had acquired as opportunity arose were scattered in a way that made management difficult and which no one attempted to rationalize.

A different profile again was the one presented by the Compagnie de Produits Chimiques et Electrometallurgiques, better known as *Péchiney*. It originated in 1921 from the merger of Froges with Alais and Camargue. Its focus was aluminum and ferroalloys integrated with electric energy sources. Its interest in chemicals developed through processing bauxite with caustic soda to manufacture aluminum and through acquired mastery in electric furnaces and electrolysis. It produced alumina (Al_2O_3) in Gardanne, Salindres, and Saint-Auban but had to find a market for its chlorine, of which there was excess capacity after World War I.

This is how Alais Froges et Camargue began manufacturing trichloroethylene, upgrading, at the same time, the acetylene extracted from calcium carbide by chlorinating it. Except for this, the company's chemicals, essentially inorganic, were restricted to a number of bromine and fluorine derivatives, to chlorates produced through electrolysis and to specialties like aluminum sulfate, and magnesium chloride, sulfate, and carbonate.

Directly or indirectly, other companies were also involved in inorganic chemicals. *Solvay* owned units in France that were directly managed from Brussels. There was an understanding about sodium carbonate between Solvay and Saint-Gobain concerning the neighboring soda plants in Varangeville and Marcheville-Daguin in which Saint-Gobain also had a stake. Solvay began feeling threatened in its special alkalis area when producers of electrolytic chlorine obtained caustic soda as a necessary by-product. It therefore set up an electrolytic plant in Tavaux in 1928 and became involved in chlorine derivatives.

Société d'Electrochimie was another group involved in chlorine chemicals that owed its foundation to the technology of the electric furnace which Henri Moissan had pioneered at the end of the previous century. Founded in 1889 by Henry Gall and Amaury de Montlaur to produce chlorates by electrolysis, the Société d'Electrochimie soon became involved in metal carbides, carbon electrodes, acetylene and sodium and, in 1907, in aluminum. It merged in 1919 with *Société Electrometallurgique du Giffre*, owned by Jules Barut, a producer of ferroalloys based on chromium, manganese, and molybdenum. Two years later, Paul Girod joined up with Gall and Barut. His Ugine unit in Savoie produced special electric furnace steels. The resulting Société d'Electrochimie, d'Electrométallurgie et des Aciéries Électriques d'Ugine became better known simply as *Ugine*.

Through its Pomblière and Jarrie electrolysis units—the Jarrie one initially leased from Compagnie des Matières Colorantes (CNMC)—Ugine was to become one of France's leading chlorine producers. The caustic soda obtained was used in the La Barasse alumina unit, near bauxite quarries. Ugine was an aluminum producer and had been producing cryolite from hydrofluoric acid (HF) since 1920. This is how the company integrated fluorspar (a mineral from the Massif Central and the Tarn region) upstream and developed fluorine chemicals. The Saint-Fons soda unit set up by Henry Gall as early as 1901 had been transferred to Pierre Benite in 1918 where it also produced by electrolysis hydrogen peroxide used to prepare perborates.

By the time World War II broke out, Ugine, which had started from electricity applied to chemicals and steelmaking, had a three-pronged base: ferroalloys, special steels, and aluminum-related chemicals to which it had been drawn through its other activities.

Fabriques de Produits Chimiques de Thann et Mulhouse was a very different case. Its chemical past went back to an act of association signed in 1808 between Charles Kestner and Joseph Willien for the purpose of producing sulfuric, nitric, and tartaric acids. After the Armistice, *Fabriques de Thann*, which since 1920 was managed by Auguste Scheurer's youngest brother Albert, a son-in-law of Charles Kestner, parted with the Mulhouse unit. Efforts were then focused on the reconstruction and development of the Thann site.

The company was saved through an association formed in 1922 with Joseph Blumenfeld, deputy director of the *Terres Rares* company. Blumenfeld was a Russian Jew who had graduated from the Saint Petersburg Chemical Institute. He had come to Paris in 1900 to work on a doctoral thesis on rare earths under Georges Urbain. Urbain had him appointed trustee of the sequestered branch of the German company *Auer*. Its Serquigny unit in Normandy produced cerium and thorium for the sleeves of incandescent lamps from monazite imported from the state of Travancore in south India.

This was how Blumenfeld's *Société des Terres Rares* originated. When monazite separations were carried out, large quantities of iron titanate (ilmenite) remained at the pit-head. Using the experience acquired in Urbain's laboratory, Blumenfeld set about extracting the titanium oxide (TiO_2) from ilmenite in 1921. It was a known fact from the work which France's Rossi had carried out in 1908, that TiO_2 was a pigment with a high covering power, both finer and more dense than lithopone or zinc oxide.

Blumenfeld took out patents for the preparation of this new pigment and looked around for a more appropriate site than Serquigny as well as for a source of sulfuric acid needed to decompose the ilmenite. In 1922, he granted a manufacturing license for TiO_2 to *Fabriques de Thann*, which thus became the first unit in the world to produce pure titanium dioxide. By 1930, some 11,000 tons were being produced; and process licenses were being sold to Montecatini in Italy, and then to National Pigments, which later sold its Baltimore unit to Du Pont de Nemours.

Blumenfeld, who had married Chaim Weizmann's sister, was not simply a good businessman. He was an imaginative negotiator and he cultivated international relations. He induced Urbain's brother Edouard, who held a majority share in *Société des Charbons Actifs*, to associate with Saint-Gobain in 1928, within PCRB, in order to develop in Berre a refining process granted by German interests in exchange for a stake in the new company.

Blumenfeld did Fabriques de Thann another good turn when he bought from a Czech firm the electrolysis process by which the company began, in 1926, to produce caustic potash, chlorine, and hydrogen and starting in 1930 potassium carbonate, chloride of lime, *eau de Javel* and synthetic hydrochloric acid from potassium chloride. These new productions brought about the agreement signed with Société Commerciale des Potasses to set up in 1931 a common subsidiary, *Potasse et Produits Chimiques*, pooling Fabriques de Thann, Mines Domaniales and Kali Ste. Thérèse.

Under a 1932 extension plan, Thann was able to increase its caustic potash production and start bromine chemicals. In 1936, Thann began producing zirconium oxide from Zircon, a mineral associated with ilmenite. With war looming close, the processing unit was set up in western France at La Rochelle-Pallice in 1938.

Inorganic chemicals were also the starting point of a new company set up in 1920 under the name of *Progil* (Produits Gillet) after acquiring the Pont de Claix chlorine unit set up in 1916. Société Gillet et Fils had been formed in the postwar years through a merger of the Condat, Molières, and Lyon-Vaise units. Production hinged at the time on chestnut-tree tanning extracts, silk fillers based on tin derivatives, and textile auxiliaries for dyeing and finishing. Progil became involved in silicates and phosphates through silk fillers and in 1920 acquired the Roches de Condrieu unit. Three years later it was producing carbon disulfide there; its excess chlorine was the cause of its orientation towards organic chlorine compounds.

As part of the Gillet group, Progil also became closely involved in wood products (pulp, paper, fiberboard), the purpose being to recover the chestnut-tree shavings to extract the short-fiber cellulose that was to be developed by the *Papeteries de Condat* subsidiary in 1931. It was mainly through Comptoir des Textiles Artificiels, *CTA*, and Société La Cellophane that Gillet held a dominant position in France in textiles and cellulose films.

In 1925, the year Courtaulds set up a unit in Calais, CTA accounted for 87 percent of France's artificial silk production with 10,000 tons a year scattered among twenty-five different companies. This was a far cry from the first units which had been set up in Besançon (Chardonnet silk in 1890), in Givet (Bamberg silk, Bernheim in 1902), in Arques la Bataille (viscose silk, Carnot in 1903) or in Izieux (Bemberg silk, Joseph Gillet in 1904). Usines du Rhône was just starting to develop rayon acetate and three-quarters of the production involved viscose (continuous filament and short-fiber staple).

The *Société la Cellophane*, which was set up in 1913 by the Gillet-Bernheim Group, had difficult beginnings. After licenses were sold to Du Pont in 1923 and to Kalle the following year, it began to find markets in the wrapping industry for its cellulose film which was manufactured in Mantes from 1928 onwards. This led to higher sulfuric acid, caustic soda, and carbon disulfide consumption. To ensure its cellulose pulp supplies, CTA took a 33 percent stake in a new company called *Cellulose du Pin* in which Saint-Gobain held the balance. But the unit which the company set up in Facture and which received supplies of pinewood from the Landes area, was not able to produce a completely resin-free cellulose for the manufacture of viscose or cellophane. Saint-Gobain was left with a heavy investment on its hands in an area like paper that was not really within its strategic range of business.

When after the war, the Cellulose du Pin's Tartas unit was finally capable of supplying the bisulfite pulp suitable for CTA, the rayon and cellophane markets had begun to fail as synthetic fibers and films took over.

Progil had set up in 1929 a subsidiary called *La Société des Résines et Vernis Artificiels* (RVA) with units in Lyon-Vaise and Clamecy that produced resins for varnishes and synthetic tannins. By the time war broke out in 1939, Progil formed a small, dynamic group that pursued the initial activities of its founders but was set to develop in other chemical areas when occasion arose. Maurice Brulfer, who was to become its president, had bought up the Clamecy unit in Central France in 1920. Initially centered on wood distillation, the unit expanded its range of production to include phenol formaldehyde resins and tin salts.

There were other French groups at the time that were involved in heavier chemicals that could be profitably developed with cheap hydroelectricity. Founded by Henri Gall, *La Société des Produits Azotés* (SPA) had leased its Lannemezan unit in the Pyrenees in 1921 to the Service des Poudres. There, ten years later, the most powerful carbide furnace in Europe was built and, in 1932, synthetic ammonia from electrolytic hydrogen was produced. Also for strategic reasons, another plant was set up in the Pyrenees at Soulom after World War I by Société Norvégienne de l'Azote to produce nitric acid. Set in an area that enjoyed cheap hydroelectric supplies, the unit was taken over, after the Armistice, by the Société des Phosphates Tunisiens, which later became the *Société Pierrefitte*. It produced ammonia, calcium nitrate, and ammonium nitrate, while a neighboring unit produced phosphorus, phosphoric acid, and ammonium phosphate. As to the *Société Bozel-Malétra*, it was the first in France to conduct alkaline chloride electrolysis (Socindus); since 1898, it had been producing calcium carbide and ferroalloys in the Savoie.

The Development of France's Coal-Based Chemicals

Located in the Bassin du Nord et du Pas de Calais, France's coal-based chemicals industry was restricted, until 1914, to the production of tars, ammonium sulfate, benzol, naphtha solvents, and naphthalene. In the postwar era, a number of factors boosted the development of coal-based chemicals. The gas efficiency of coking ovens improved, the production of thermal power plants increased, and the need for chemical raw materials grew. But the strongest impetus came from work carried out by George Claude, both on very high-pressure ammonia synthesis and on very low-temperature liquefaction of gases by which not only hydrogen but also methane, ethylene, ethane, and propylene could be produced. This led the mining companies of the area to associate with chemical companies like Kuhlmann, Saint-Gobain, and Air Liquide to extract value from the excess gas produced by the coking ovens and thermal power stations.

By 1939, nearly half of France's production of nitrate fertilizers came from such joint ventures. As coal gasification and hydrogenation technologies improved, synthetic methanol was produced in Mazingarbe (1926) and then in Harnes (1930), while ethylene oxide and glycols were obtained from ethylene in Choques (1930).

At the same time *Société Huiles, Goudrons, Dérivés* (HGD) was set up in 1923 by Compagnie des Mines de Lens for the large-scale processing of coke-oven tars. By 1928, HGD was processing in its Vendin le Vieil plant almost all the tar produced in the Bassin du Nord supplying pitch for electrodes, creosote for wood treatment, phenol oils, and road tar. Generally the benzols were treated with little refinement in the coking plants' distillation units. Only the Mazingarbe plant, belonging to Compagnie des Mines de Béthune, undertook the separation of benzene, toluene, xylene, and even cyclopentadiene. With the formaldehyde obtained from methanol and the phenol extracted from its phenol oils, HGD became involved in phenoplasts from 1934.

The Rise of Rhône-Poulenc and the Development of Specialty Chemicals

The merger of Établissements Poulenc Frères and Société Chimique des Usines du Rhône had produced on June 28, 1928, the *Société des Usines Chimiques Rhône-Poulenc* (SUCRP). One of the first moves of the new group that same year was to regroup under Société Parisienne d'Expansion Chimique, or *Spécia*, the pharmaceutical activities of the two partners. The Vitry plant was used for pharmaceutical and fine chemicals synthesis whilst research on prescription drugs was conducted both in Vitry on the site that in 1952 was to become Centre Nicolas Grillet, and in Dagenham, England, at May & Baker's, which Poulenc Frères had taken over in 1927. Between 1936 and 1939 these various moves generated sulfapyridine (Dagenan), sulfathiazol (Thiazamide), benzyl sulfamide (Septazine). The production of vitamins had begun earlier in 1929.

In 1936, SUCRP had set up a joint pharmaceuticals subsidiary with Montecatini in Italy, which became known under the name of *Farmitalia*. SUCRP had also inherited from Poulenc a range of rubber processing products, starting with the guanidine derivatives in 1920; then from Usines du Rhône, bulk pharmaceuticals (aspirin, pyrazoles, and salicylates), perfume chemicals (vanillin, bases for perfumes) and photographic products (hydroquinone, developers), which were produced by the Saint-Fons unit.

The Roussillon plant in the Rhône area, which was reserved for high-volume products, hinged on the acetyl group (acetaldehyde, acetic anhydride, acetic acid, cellulose acetate) and developed Rhodoid sheets, Rhodialine films, and Rhodialite cellulose acetate-based molding powders.

Through cellulose acetate chemistry Usines du Rhône had associated with Comptoir des Textiles Artificiels (CTA) to form the fifty-fifty joint venture *Rhodiaceta*. It extended to Germany as early as 1927 (Deutsche Rhodiaceta AG), to Italy with Montecatini (Rhodiaceta Italiana), to Brazil, where the subsidiary *Companhia Química Rhodia Brasileira* had been established in 1919 to manufacture the ethyl chloride used in the perfume-sprayers widely used by *cariocas* during the Rio Carnival. Rhodiaceta had sold its acetate filament process to Du Pont, obtaining in return a Nylon license in the spring of 1938. The development of this new fiber was to play a significant part in the group's postwar activities.

Thus on the eve of World War II, Rhône-Poulenc's image was that of a company involved in fine chemicals, with a competent scientific and technical staff, a company that, through its cellulose acetate chain, had upstream openings onto a thriving acetic chemicals activity and downstream onto plastic films and artificial textiles that had a highly promising future. With its still limited range of chemicals and its small overseas base, however, it could hardly compare in size and profits with the German or British chemical giants.

Other French companies in other chemical sectors where size was not a vital criterion of efficiency, such as Rhône-Poulenc, were quite honorably placed in their specialties. Such was the case of the Deux Sèvres distillers, which in 1937 took the name of *Usines de Melle*. Founded in 1872 by the industrialist Alfred Cail, this organization was converted in 1886 into a sugar-beet alcohol distillery, and from 1914 produced amyl acetate when the needs of the war and distance from the combat zone stimulated its considerable development. It started manufacturing ether for French gunpowder, acetone for English cordite, ethyl acetate to coat aircraft canvas, and furfurol to denature gasoline. After the war, the company licensed out its acquired technology in these areas, involving both biochemistry and organic synthesis.

Keeping up with the Pasteur tradition, France maintained a place of eminence in biochemicals also. *Institut Pasteur*, revived in 1887, was not an industrial concern; but serum and vaccine licenses contributed to its resources. Its staff of researchers, including Jacques Trefouel's team, which had worked on sulfanomides, maintained its reputation worldwide.

Institut Mérieux, founded in Lyon at the end of the nineteenth century by a former student of Pasteur, had focused on veterinary drugs and since 1907 had produced vaccines against bovine tuberculosis and hoof-and-mouth disease. *Institut de Sérothérapie Hémopoiétique* (ISH), had been set up in 1911 by Gaston Roussel to supply serums against anemia from horse blood. The Romainville stables had as many as one thousand horses in 1924. ISH subsequently diversified its activities setting up Union Chimique des Laboratoires Français (UCLAF), which gave birth to *Roussel-Uclaf.*

Another company, Société de Chimie Organique et Biologique, better known as *Alimentation Equilibrée à Commentry* (AEC), was organized in 1939 to to manufacture vitamins and amino acids, becoming a leading company in this field by the time it joined the Rhône-Poulenc group after the war.

France's perfume industry had taken a decisive turn in 1926, when it entered the era of fashion perfumes. *Lanvin*'s ''Arpège,'' *Chanel*'s ''No. 5,'' and *Patou*'s ''Joy'' acquired international fame. At the same time original perfumes were designed by traditional firms, such as *Guerlain* with ''Vol de Nuit,'' or *Bourjeois* with ''Soir de Paris.'' Firms in Grasse, such as *Chiris*, or in Paris, such as *Roure Bertrand Dupont*, were turning towards semisynthetic derivatives or compositions. New scents were added to the range of French perfumes by famous "noses" like Bienaimé at Houbigant's, Jean Desprez at *Millot*'s, and François Coty himself for Coty.

Hair dyes had been studied since 1907 by a young chemist, Eugène Schueller, who had set up the Société des Teintures Inoffensives pour Cheveux, which, a year later, became *Oréal*. Initial success, confirmed with the development of hairdressing salons after the war, led Schueller to design a new dye formula, the Imedia dyes. Jumping onto the bandwagon of new marketing methods through the large department stores developing in the thirties, Schueller bought up *Société Monsavon* in 1930, then in 1934 launched Dop, the first consumer shampoo.

Such was the contrasting image projected by France's prewar chemical industry. Where it was not necessary to be powerful to succeed, the wordly wisdom of a Blumenfeld, the technical ability of a Georges Claude, the enterprising spirit of a Eugène Schueller achieved success in special areas. But wherever organizational professional knowhow and good partnership relations were the keys to success, the managers of French chemical companies, with rare exceptions, were unable to assemble the indispensable conditions for creating large-scale manufacturing enterprises.

Unlike their German counterparts, France's chemical companies, with the exception of Kuhlmann and Rhône-Poulenc, did not have chemicals as their main focus and their leaders were not trained chemists. Moreover, French industry leaders preferred to remain independent rather than be linked to other firms in moves that would have achieved some coherence. This explains their marked preference for the market-sharing agreements that turned France's chemical scene into a mosaic of different companies hard to manage and hard to rationalize. Matters were further complicated by the interventions of the French government, which strove to play an active role instead of merely arbitrating among parties that were already reluctant to join forces. And, of course, when circumstances required the marshaling of forces, the weaknesses of France's chemical industry became obvious compared to its big competitors, whether in research, technology, size of units, or marketing bases abroad.

The Regrouping of Belgium's Chemical Industry

Although its management was situated in Brussels, *Solvay* was a multinational company through the stakes it had held, since its founding, in soda plants. It had significant shares in the United States-based Allied Chemicals, as well as in England's ICI and it was directly involved in a large number of subsidiaries, namely in Germany, France, and Italy. But despite defensive diversification in the electrolysis of salt and chlorine chemicals, sodium carbonate, which had been at the root of the company's success, remained its main product.

In particular, Solvay never found a way of taking a direct part in the synthesis of ammonia. It was true that in 1928–1929 Belgium's nitrogen industry was in a very scattered state. La Grande Paroisse, which was involved with Ougrée in *Société Belge de l'Azote*, was using the Claude process. The Coppée group, which, together with *Société Centrale pour la Fabrication d'Ammoniaque de Synthèse*, was the most important, was utilizing the Fauser process. Projects that were to use, respectively, the American Nitrogen Engineering Corporation and the Mont-Cenis processes, were being announced separately by *Kuhlmann* and by *Société Carbochimique*. Overall, Solvay was involved only through the stakes it held in *Union Chimique Belge*, which had only a modest ammonia production using the Casale process.

The establishment of *Union Chimique Belge* (UCB) in 1928 was the final outcome of a series of earlier concentrations incorporating, among others, the Semet-Solvay coking ovens, the Drogenbos Superphosphates and Guano chemicals, the Compagnie Franco-Belge de Colles et Gelatines and the Meurice pharmaceuticals. In 1929, UCB also absorbed Progil Belge and Produits Chimiques de Schoonarde. Thus a coherent whole was built up out of a multitude of small businesses. It was involved in base chemicals, specialties and pharmaceuticals.

In the midst of the textile crisis and under the impetus of Baron Emmanuel Janssen, another merger in artificial silk led to the creation in 1932 of *Union des Fabriques Belges et Textiles Artificiels*, called *Fabelta*. Belgium had played an important role in the development of artificial silk with *Tubize*, established in 1900, using the Chardonnet process along with Alost, Obourg, and Ninove. The units of these companies turned to viscose rayon and became part of Fabelta.

When it acquired the acetate rayon Tubize plant ten years later, Fabelta became the only Belgian producer of artificial textile thread. As for *Société Industrielle de la Cellulose (SIDAC)*, founded in 1925 to develop cellulose film, it was absorbed in 1961 by UCB, which then became deeply involved in transparent film.

Holland's Chemical Industry, Long-Established and Scattered

There are a number of Dutch chemical and chemical-related firms now grouped under *AKZO* that go back a long way. *Sikkens*, a manufacturer of lacquers and varnishes, was founded in 1792, *Ketjen* in 1835, *Noury van der Lande* in 1838. In the artificial silk sector, *AKU*, which merged with *KZO* in 1969 to form the AKZO group, was itself an offshoot of the merger of *Glanzstoff* set up in 1899 and *ENKA*, established in 1911. Likewise *KZO* (Koninklijke Zout-Organon), formed in 1967, had regrouped companies over fifty years old: *Organon*, producing bio-products, dates back to 1923, while *Zout Chemie*, one of the largest salt producers, was established in 1918.

Besides these long-established firms, Holland was traditionally open to foreign investments. While *Unilever* and *Royal Dutch Shell* were rightly at home in the Netherlands, other foreign companies had set up business in the country. Saint-Gobain had a plant in Sas de Gard, for instance, and Coppée and Montecatini had set up in 1929 the *Compagnie Néerlandaise de l'Azote*. At the time, Holland's chemical industry was essentially based on Limburg coal and since 1920 was in the hands of the Dutch coal board known as DSM (Dutch Staats Mijnen). After 1929, the main impulse for the industry was to come from ammonia. The most important producer of fertilizer was then the UKF group. On the eve of the war, therefore, there was a long way to go to achieve the concentrations that took place later. In chemicals proper, only DSM emerged as a national producer and was surrounded by small, specialized family businesses.

Montecatini and Italy's Chemical Industry

Chemicals made great strides in Italy in the 1920s, but the sites remained scattered and production unprofitable. Following the Treaty of Versailles, there were six calcium cyanamide units in the country, working at half their capacity, which was rated at 20,000 tons of nitrogen. Synthetic ammonia production, boosted by the Casale and Fauser processes, was scattered over nine different sites. Moreover, the electrolytic hydrogen used for the purpose was expensive.

With 1,200,000 tons of superphosphates produced in its plants in 1928, Italy was the third largest producer after the United States and France. It had also displaced England to become the foremost world producer of copper sulfate for the treatment of vineyards. More modestly, Italy was also involved in rayon through *SNIA Viscosa* and *Châtillon* and in phenoplasts through *Societé Industrie Resine (SIR)*, which had acquired a Bakelite Corporation license.

Dyestuff production had soared to 7,000 tons in 1928, of which a large part was sulfur dyes used to dye cotton. Despite high customs barriers, the companies involved in this area were too small and not sufficiently diversified in high-value products to be profitable. Six of them merged to form *Aziende Chimiche Nationale Associate (ACNA)*.

Greater concentration was needed to pull Italy's chemical industry out of the difficulties resulting from its dispersion and the small size of its production units. The regrouping took place through *Montecatini*. The company had started by developing pyrites mines and before the 1914 war very naturally began manufacturing sulfuric acid.

In the postwar years, Montecatini bought up various superphosphate units and soon became Italy's leading fertilizer producer. Between 1922 and 1924, it had taken control of the country's explosives industry. It afterwards became involved in ammonia, using the Fauser process, both in Italy and in

Belgium. It then purchased marble quarries and launched out into the production of aluminum.

In 1939, Montecatini took over ACNA, which was going through financial difficulties, thus becoming a diversified company involved in base chemicals through its minerals and aluminum, in inorganic chemicals through its acids and fertilizers, and in organic chemicals through its ACNA dyes and its explosives.

The Prosperity of the Basel Chemical Companies

Just before the Second World War, Basel's three big chemical companies—*CIBA, Geigy* and *Sandoz*—had not only achieved their initial aim of concentrating on high added-value specialty chemicals, but had also worked out agreements among themselves that gave them a strong negotiating position when dealing with their foreign rivals while at the same time preserving their autonomy. During the First World War, they had set up jointly in 1916 the *Schweizerhalle* SA acids unit. Then they had established a fifty-year common-interest association, under which the three companies pooled their profits and shared them according to preestablished quotas. In the framework of this agreement, Geigy and Sandoz took a share in *Clayton Aniline Company* as early as 1919. This was CIBA's Manchester subsidiary, which greatly developed its activities thereafter. The following year the three partners bought up the U.S.-based dyestuff unit of Ault and Wiborg, *Cincinnati Chemical Works*, and in 1925 they set up Société Bergamasca per l'Industria Chimica in Italy to produce sulfur dyes. In this way the three Swiss manufacturers managed to obtain a 19 percent slice of world exports within the dye cartel that was organized in 1929 and lasted until 1939.

Under rationalization measures also taken, CIBA specialized in vat dyes and Sandoz in alizarin derivatives. A total merger was rejected by Sandoz, which was of the opinion that Geigy's quota of the overall profits was too large, considering the hazardous profitability of its tannin and natural dye business.

The small Swiss common-interest association was finally broken up thirty-three years after it was established. Among other merits, it had given each partner the opportunity to develop other business areas besides dyes with the profits they had each drawn from the association. Thus, Geigy became involved in antiparasite research, marked by the development of an antimoth-worm called Mitin in 1939 and especially, during the war, by the successful DDT, while the Manchester unit focused on PVC plasticizers.

CIBA was able to engage in high-polymer research leading to the discovery of epoxy resins (Araldite). As for Sandoz, it pursued research begun in 1917 by a student of Richard Willstätter, P. Stoll, on ergotamine, an alkaloid extracted from rye ergot.

The breakup of the Swiss association did not prevent Geigy, Sandoz, and CIBA early in World War II from taking a joint majority stake in another Basel firm, *Durand & Huguenin*, nor to take part in the management of *Industrie Suisse des Goudrons*, a Pratteln-based company they had helped set up in the mid-1930s to ensure national supplies of base organic chemicals.

While remaining true to its original calling, the Basel-based *F. Hoffmann-LaRoche* had nonetheless diversified successfully in the synthesis of vitamins, becoming one of the largest world producers. *Lonza* had been set up as far back as 1897; originally it had taken advantage of cheap electricity to produce calcium carbide and hence acetylene and calcium cyanamide. It had gradually enlarged its range of products to include ammonia, nitric acid, acetaldehyde, and acetic acid.

Aware that it needed to develop a fine chemicals industry more in line with the needs of a country that, because of its size, was ill-fitted for heavy chemicals, Lonza's leaders refocused the company's business after the war on higher-value-added organic synthetic derivatives.

The same object was pursued by the leaders of *Fabrique Suisse d'Explosifs de Dottikon* set up in 1913 to make Switzerland self-sufficient in gunpowder supplies, and by the leaders of *Chemische Fabrik Uetikon*, one of the oldest concerns in the country since it was set up in 1818 to produce alkalis and inorganic acids.

The Soviet Union's Chemical Industry Lags Behind

The Bolshevik Revolution deeply affected the Soviet Union's chemical industry. The loss of Poland and the Baltic States in 1917–1918 removed important production centers from the territory. Postgraduate studies were disrupted, and the doctoral examination was reestablished only in 1934.

Laboratories suffered from a shortage of equipment. While those belonging to the Tsarist era still survived under different names, the best ones, such as the Ipatiev High Pressure Institute in Leningrad or the Karpov Institute in Moscow, were more involved in physical chemistry than in the organic chemistry that was in full swing in the West. There was also a power conflict between the Heavy Industry Commissariat, which claimed to have the upper hand in applied chemistry, and the Scientific and Technical Administration (NTU), which had authority over fundamental research institutes. More importantly, Soviet planners had given priority to the steel and electricity sectors. The chemical industry was not foremost in their minds.

During the 1920s the paper, soap, glass, and textile industries which were substantial consumers of chemicals, had not yet recovered their pre-Revolutionary level of activity; and agriculture, completely disorganized by the collectivization process, could hardly be regarded as an important outlet for fertilizers. In addition, the bureaucratic organization established by the

Soviets was not such as to adapt supply to demand. It was not surprising that in such an environment Soviet chemicals should have lagged so far behind despite the presence of competent executives trained in Tsarist times.

With a production of 145,000 tons of superphosphates and 10,000 tons of dyes in 1928, the U.S.S.R., despite its immense territory and large population, ranked thirteenth and sixth in the world respectively in these two essential areas of any developed economy. It was only in 1928 that the first synthetic ammonia unit came on line in the Soviet Union, near Gorki. It utilized the Casale process negotiated with the Italians by Ipatiev; and its capacity did not exceed 20,000 tons a year. Dyes, produced mainly in Moscow in the former German and Swiss units, had undergone little technical progress for want of recently acquired knowhow which only the Western countries could have supplied. A lack of raw materials, extracted from tar in insufficient quantities, held back production. The Russians had managed to develop a synthetic rubber industry. But, possibly for want of any biochemical tradition and a lack of competent organic chemists, the Soviet Union has remained strangely absent to this day from any significant development in areas as important as antiparasites and pharmaceuticals. Moreover, anything approaching consumer products sold through advertising campaigns by competing multinationals that are clients of the chemical industry was unknown in the Soviet Union's state-controlled economy and, therefore, foreign to the public.

The Contribution of Western Technologies to Japan's Chemical Industry

As we have seen, Japan had taken advantage before 1914 of cheap electricity to introduce European technologies using electric furnaces and electrolysis in order to manufacture calcium carbide, calcium cyanamide for ammonia production, caustic soda, and chlorine. In addition, the introduction of Semet-Solvay coke ovens had enabled a number of steelworks, as well as *Mitsui Mining* and *Tokyo Gas*, to establish the bases of aromatic chemicals from tar.

When the war cut off dyestuff imports from Germany, Mitsui Mining, as well as Japan Dyestuffs Manufacturing Company—which Sumitomo took over in 1944—were thus able to produce some of the main dyes such as indigo and alizarin. The war had also deprived Japan of the imported Brunner Mond sodium carbonate, forcing *Asahi Glass* in 1917 to attempt production of this essential product in its Makiyama unit.

Without access to Solvay technology, Asahi Glass tried to develop its own ammonia process. But it fully succeeded only in the 1930s after many financial and technical difficulties and with the help of French engineers and an American "turncoat" from Solvay Process Corporation. To help Asahi Glass withstand the price war which ICI was conducting, the Japanese government authorized the company to bypass the state salt monopoly and to import from abroad, at a lower price, the salt needed to produce sodium carbonate.

The synthesis of ammonia was carried out in Japan in several stages and in different ways. In 1921, three *zaibatsus*,[30] *Mitsui, Mitsubishi* and *Sumitomo*, in association with *Sankyo*, had set up Toyo Nitrogen Company and bought back the Haber-Bosch patents from the Japanese government. But the move turned out to be unproductive.

Private Japanese industry leaders were more daring and variously successful. This is how Noguchi introduced the Casale process in 1923 and Kaneko the Claude process that same year, whereas the Fauser process was being developed in 1924 by *Dai Nippon Fertilizer*, Japan's main superphosphate producer.

Another independent entrepreneur, Nobuteru Mori, managed to develop an indigenous technology. Mori, who had no technical background, had set up Sobo Marine Products in 1908 to extract iodine from seaweed. In 1928, he founded Showa Fertilizer Company, which was to become *Showa Denko*. As Mori could not purchase a license for European processes, he set out to develop the ammonia-synthesis method perfected by Tokyo Industrial Technology Laboratory. This is how the first unit to use Japanese technology and equipment came about.

Spurred by these captains of industry, the zaibatsus took a new interest in the problem of synthetic ammonia, the more so as under the Claude process costly electrolytic hydrogen could be replaced by hydrogen from coking oven gases. Using the American Nitrogen Engineering Corporation (NEC) technology, *Sumitomo* brought on stream in 1931 an ammonium sulfate production unit, then in 1934 a nitric acid unit, followed by a methanol unit in 1937, and a formaldehyde and urea unit in 1938. *Mitsui Mining*, which had bought back Daiichi Nitrogen Company in 1928 and set up Toyo Koatsu in 1933, used Du Pont technology. The last to enter the market was *Mitsubishi Mining*, using the Haber-Bosch process, in 1936.

The three great zaibatsus had come to chemicals through mining. Coal mines led to tar and dyes in the case of Mitsui and Mitsubishi, and copper mines to sulfuric acid and superphosphates in the case of Sumitomo. Their involvement in ammonia synthesis was what opened up for them the doors to the great chemical industry, however. This led them to set up autonomous entities to look after their interests in the area: Sumitomo Chemicals in 1934, *Mitsui Chemical Industries* in 1941, and *Mitsubishi Chemical Industries* in 1944.

Involved as it was in copper and alloys and spurred on by the duralumin orders for Japan's war effort, Sumitomo, the largest aluminum

[30] These were conglomerates, each one dominated by a family. Their origin dates back to the Meiji Restoration (1868).

consumer in Japan, had been trying since 1916 to produce it from local alumina, in the absence of bauxite supplies.

Through Nippon Denko, which, like Showa Fertilizer, became part of Showa Denko in 1939, Mori trod the same path, becoming in 1934 in fact, the first company to produce aluminum in Japan. A year later, it was the turn of Nippon Aluminum Company, an association of Mitsui and Mitsubishi with Furukawa, to do so.

With imported bauxite easily available, the three aluminum producers adopted in 1937 the Bayer process to obtain alumina. But apart from Sumitomo, which had a large captive market, they made hardly any return on their heavy investment in this area because they competed heavily with one another in a slumping world economy. Subsequent high energy costs confirmed that Japan was in no position to produce aluminum under profitable conditions.

The Japanese took an early interest in other innovations which came from Europe and America. As early as 1911, *Sumitomo Bakelite* had acquired a phenoplast license from Léo Baekeland. Through camphor used as a plasticizer, a number of Japanese companies had begun producing celluloid; and in 1919 eight of them merged to form Dainippon Celluloid Company, which became *Daicel* in 1966. Its first subsidiary, *Fuji Photo Film*, was set up in 1934 to produce nitrocellulose film. The company very naturally turned to cellulose acetate in 1939 and developed an acetic chain of products.

Artificial textiles started off a number of viscose-based filament and staple-fiber producers that gradually turned to chemicals proper. While *Kanebo*, an offshoot of Kanegafuchi Spinning, long remained faithful to its initial business, *Teijin*, in 1927, became the first to begin production of rayon in Japan. It diversified after the war in plastics and organic chemicals. *Toyo Rayon*, established by Mitsui in 1925, followed a similar path and diversified under the name of *Toray Industries* in 1970. *Shinko Rayon* was set up in 1933; taken over in 1942 by Nippon Chemical Industries, it became part of Mitsubishi Chemical Industries. *Asahi Chemical Industry*'s progress was quite the reverse. Starting off in 1931 with the synthesis of ammonia under the Casale process, it turned to other heavy chemical sectors and then began producing viscose rayon and Bamberg silk. Consequently, it is now one of the most diversified Japanese chemical groups.

Japan's oldest firms were also being forced to take into account new developments of the Western chemical industries between the two world wars. Thus *Kao Soap* and *Lion Fats & Oil*, dating back to 1887 and 1891 respectively, turned to making synthetic detergents and consumer products while remaining centered on traditional fats. Established in 1781, *Takeda* was to become Japan's leading pharmaceutical company through the foreign licenses it acquired.

Thus before World War II, Japan's chemical industry, spurred on by daring and persevering leaders and by the great *zaibatsus* that took over when a financial effort was needed, had diversified and held an honorable rank in the world. Japan's chemical industry remained dependent on foreign technologies, however, and both the proliferation of producers and the dispersion of production centers made it vulnerable to competition from the world chemical giants.

Chapter 5
World War II and Its Consequences for World Chemicals

Through the needs it created, World War II gave a spectacular impetus to the development of the discoveries made by chemists in the period between the wars.

The State of Preparedness of the Belligerent Nations

In *France*, the Service des Poudres had seen to it that the armies were well supplied with ammunition in 1939. But chemical industry leaders had done nothing about synthetic gasoline or rubber substitutes, and no serious effort had been made to develop new plastics. The French had done nothing either to win back their fellow-countryman Eugène Houdry when he went to the United States, taking his catalytic-cracking technology for aviation fuel with him. Even as the French government declared war on Hitler's Germany, the country was dependent on overseas supplies for such essentials as gasoline and rubber. Doubtless, the politicians were banking on a lightning war; and that is what they got.

England was no better prepared in the area of synthetic rubber, which she began manufacturing only in 1958. But British chemists had been interested since the 1920s in the use of coal as a source of motor fuel. Through carbonization, they produced benzol in coke ovens and gas plants; this was sold as motor fuel by National Benzole Association. More particularly, ICI had been studying since 1928 a variant of the Bergius process involving coal hydrogenation under pressure. When synthetic gasoline received protection under the 1934 British Hydrocarbon Oil Act, large-scale hydrogenation of both coal and creosote oil was carried out at Billingham. The Billingham unit was able to supply the Royal Air Force with aviation gasoline as soon as hostilities began.

ICI was also producing at the time two strategically important plastics. Perspex, which was to be used during the war for the portholes on British bomber planes, and polyethylene, which played a vital role in the development of radar.

In *Germany*, on March 16, 1935, Hitler had denounced the clauses in the Treaty of Versailles that banned the country's rearmament. The very next year Hermann Goering was put in charge of a Four-Year Plan with the avowed object of preparing the nation for war. Because of the nature of "National Socialism," German industry was driven to cooperate closely with the government, while still maintaining the forms appropriate to private commercial concerns. Cooperation was especially important for the chemical industry, which produced a number of strategic products.

As already mentioned, the early stages of the Four-Year Plan were hindered by the opposition of Hjalmar Schacht. As minister of national economy until the end of 1937, he objected to the production of costly substitutes for which he saw no economic justification, bent as he was on defending the German Mark. By the time war broke out, however, Germany had already put into practice a good part of its preparation program. Rayon production (thread and staple fiber) had soared to 270,000 tons, synthetic gasoline to 7,000–8,000 tons, while Buna rubber was being produced from 1938 in the gigantic Schkopau units. IG Farben alone accounted for three-quarters of the Four-Year Plan's chemical investment budget.

Hitler, who aimed for rubber and oil self-sufficiency at all cost, had quickly realized that he could not do without IG Farben to achieve both objectives. When one of IG Farben's managers, Carl Krauch, became Goering's assistant for the Plan's total chemical planning in 1938, there could hardly be any doubt as to how far IG Farben had become implicated in the Third Reich's war preparations. But the symbiosis was not achieved without hesitation. Carl Bosch, in particular, who had been deeply shocked by the dismissal of Fritz Haber for racial reasons in the early days of Nazism, was a worn and embittered man when he died on April 26, 1940. His successors at the head of IG Farben, C. Krauch, who became president of the supervisory board, and H. Schmitz, head of the management board, had none of Bosch's prestige, which might have made them less likely to comply with Hitler's demands had they felt so inclined.

The Scope of America's War Effort

The Americans had not remained unaffected by the defeat of France and by the danger Britain was facing. But it needed the shock of Pearl Harbor on December 7, 1941, for the idea of total war on the Axis powers to become a sudden reality that no one could avoid.

As had happened during World War I, collaboration between the Administration, scientists, and industry leaders was to work wonders. Chemicals were involved in that collaboration in a number of ways. The industry was required at the same time to do what it knew how to do, and to do it better. In traditional areas production had to increase considerably: sevenfold for aluminum and a hundredfold for magnesium, between 1939 and 1943. By 1944, oil was being produced at the unprecedented rate of five million barrels per day, and refinery capacity had grown to the point where sufficient toluene could be produced from petroleum to make 4,000 tons a day of TNT. Du Pont alone produced two million tons of military explosives between 1940 and 1945.

Together with these record production rates for established products, there was a sudden great demand for newer materials: *polyethylene* for electric insulation, *silicones* for the stable lubricants and greases required under various temperature conditions, *PVC* film for the protection of military equipment, *Nylon* for parachutes and the inner fabric of tires. Atabrine was needed for the treatment of malaria in South Pacific combat areas, while DDT made a spectacular debut in the fight against typhus in southern Italy.

To replace magnesium in incendiary bombs, a group of Harvard researchers working with Arthur D. Little had obtained, by thickening gasoline with an aluminum soap mixed with naphthenic and palmitic acids, a devastating compound, which they called *napalm*.

The United States' superiority was revealed in all its magnitude, in the implementation of four projects that proved essential to the war effort, and that were offshoots of research carried out in Europe.

The Synthetic Rubber Program

A few months before Pearl Harbor, the American Government, concerned about natural rubber supplies from Southeast Asia, had roped in a few industrial leaders to set up a small synthetic rubber production project under the Rubber Reserve Company sponsorship. Until then there were only two special kinds of synthetic rubber produced in the United States, *Neoprene* and *Thiokol*, and neither was suitable for making tires.

In 1927, Standard Oil of New Jersey had signed an agreement with IG Farben to develop the synthetic gasoline process the German firm had invented. The project lost some of its interest when the oil-rich Texas fields were discovered a year later. But the initial agreement was extended to cover butadiene production through petroleum cracking, and the possible production of synthetic rubber.

After IG Farben's chemists had obtained an isobutylene homopolymer in 1929, W. Sparks and R. Thomas in 1937 managed to produce a polyisobutylene-based elastomer using butadiene or isoprene as a co-monomer. This was the starting point of Standard Oil's *butyl* rubber, used in inner tubes. The first plant was brought on stream at the end of 1941 in Baton Rouge. Butyl, however, proved unsuitable for tires themselves; so did Buna-S, which the

United States had been importing before the war, for its thermoplastic properties were not equal to those of natural rubber.

The Rubber Reserve Program received a decisive boost with the United States declaration of war because the War Production Board in Washington targeted synthetic rubber production to increase to as much as 850,000 tons a year. The butadiene-styrene elastomer called GRS (Government Rubber Styrene) was selected. In December 1941, an agreement to pool technological knowhow was signed by the government, Standard Oil, Dow, and the four main tire manufacturers, Goodyear, B. F. Goodrich, U.S. Rubber, and Firestone. Coordination of the giant program was entrusted to W. Jeffers, then to Bradley Dewey, whose firm of Dewey and Almy later became a subsidiary of W. R. Grace. Twenty-four plants and 20,000 workers took part in the program. Over half a billion dollars was invested in equipment and machines. Styrene was produced from ethylene derived from natural or refinery gas and from coke-oven benzene. Unlike the Germans, who used acetylene-based butadiene as a starting point, the Americans resorted both to ethyl alcohol produced by chemurgy and to butylene via catalytic cracking.

The initial process, based on one used by IG Farben for Buna-S, was improved by Goodyear, which introduced continuous production. Low-temperature emulsion polymerization was made easier through use of a rosin soap (Dresinate) developed by Hercules. Thus SBR rubber came swiftly and economically into production; it was a distinct improvement, also, over Germany's Buna-S. The 8,000 tons a year of U.S. synthetic rubber production in 1941 had grown a hundredfold by the time war came to an end.

The Problem of Aviation Fuel

Gasoline demand for the internal combustion engine had been addressed by refiners as soon as automobiles came on the scene. Already by 1913, W. Burton from Standard Oil developed a process for heat cracking of heavy petroleum fractions into lighter distillates containing five to twelve carbon atoms. Universal Oil Products (UOP) subsequently improved Burton's process, and in 1920 Texaco made it continuous. While thermal cracking increased the light distillate yield, it did not improve the octane content of the gasoline produced.

The catalytic cracking process introduced by Eugène Houdry in 1936 greatly improved the yield and spectacularly enhanced the octane rating of gasoline through formation of branched hydrocarbons by molecular rearrangement. Houdry's process also reduced the reaction temperature to 500°C and lowered the pressure to 1 atmosphere. A pilot unit was set up as early as 1937 by Socony Vacuum (Mobil), followed by Sun Chemical's Marcus Hook unit, the first of industrial size. The Sun unit made it possible to produce the high yield gasoline with a high octane rating, which the U.S. Air Force required.

Houdry's "fixed bed" catalyst was quickly poisoned by carbon deposits, and regeneration was difficult. The possibilities of a "fluid-bed" catalyst had been suggested in 1929 by an independent inventor, N. N. Odell. Walter K. Lewis and his MIT research team had gone more deeply into it. Industrial scale production was first carried out by Standard Oil of New Jersey, which brought a unit on stream in 1942. In postwar years the "fluid-bed" process became universally adopted, as improved catalysts were used.

As well as producing light distillates, thermal and catalytic cracking processes also produced gases like ethylene, propylene, and butylenes, which mainly served for heating purposes. The Americans had studied conversion of these distillates into olefins through *polymerization* or *alkylation* during the 1930s. Following Ipatieff's pioneering role at UOP, Shell Oil began marketing in 1935 a di-isobutylene process using phosphoric acid as a catalyst. "Polymerized gasoline" helped satisfy during the war the ever-growing needs of aviation gasoline.

The same can be said for alkylation gasoline. There again by reacting isobutane with butylenes in the presence of $AlCl_3$ as a catalyst, Ipatieff's work in 1932 proved decisive. The first alkylation unit was put on line in 1938 by Humble Oil, a subsidiary of Standard Oil of New Jersey.

Hydrogenation technology was introduced in the 1930s by Standard Oil under its exchange-information agreement with IG Farben. By saturating olefins and by turning sulfur, nitrogen, and oxygen respectively into hydrogen sulfide, ammonia, and water vapor, the products obtained were more stable and pollution was reduced.

Another major breakthrough was Standard Oil of Indiana's *catalytic reforming* process, developed in 1940. Molybdenum-silicon catalysts are used with hydrogen to transform a straight-chain hydrocarbon such as hexane into cyclohexane through cyclization, then into benzene through dehydrogenation. Through the reforming process, aliphatic hydrocarbons could be converted into aromatics, in particular into toluene required for TNT.

By taking advantage of the discoveries of E. Houdry and V. N. Ipatieff, of university work carried out by men like W. K. Lewis, or of corporate research conducted by refinery engineers, the United States succeeded in solving the problem of aviation gasoline supplies, so vital to the war effort. It also paved the way for the new technologies which, after the war, would make the country supreme in the refining and petrochemicals sectors.

The Penicillin Saga

In 1939 an Oxford research group working under Howard Florey and Ernest Chain had managed to isolate and purify the *Penicillium notatum* Alexander Fleming had identified in 1928. But England, fighting for survival, was in no position to go ahead with large-scale production of penicillin.

Florey went to the United States in 1941 to persuade the United States administration of the project's importance. A research group from the Northern Regional Research Laboratory in Peoria, Illinois, discovered an *in-depth* way of cultivating the mold in a corn syrup and lactose environment. This three-dimensional process, which replaced the previous surface process, multiplied the yield tenfold.

A number of American laboratories were then invited to develop an industrial production process, which was achieved at a cost of $25 million. The first industrially produced penicillin was brought out by Merck in 1942. This, together with capacities put on line by Pfizer, Squibb, and Commercial Solvents, made penicillin available in sufficient quantities for the needs of combat troops. By the end of the war, U.S. production, together with Glaxo's in England, was beginning to meet civilian needs.

The Manhattan Project

During the summer of 1939, three Hungarian-born American physicists—Leo Szilard, Eugene Wigner, and Edward Teller—persuaded Albert Einstein to write to President Roosevelt to draw his attention to the potential of *uranium fission* and to the danger of the Nazis' developing before the Allies a weapon based on the principle of a nuclear chain reaction. The physicists reckoned that the explosive power of 10 grams of uranium was equivalent to 600 tons of TNT.

Only on December 6, 1941, did President Roosevelt give the signal to start development of an atomic bomb. The program went under the deliberately innocent name of the Manhattan Project. Meanwhile, under the direction of the Italian-born Enrico Fermi, a number of physicists at Columbia University were studying the means of producing sustained fission in a mass of uranium. Until then uranium had been a mere laboratory curiosity.

In natural uranium it is the fissionable *U-235 isotope* that undergoes fission. The U-238 isotope is inert but 140 times more plentiful and has the reverse effect of slowing down the chain reaction. The aim was to produce U-235 *enriched uranium*. The separation method chosen was the *gaseous diffusion* process separately developed by Harold Urey and Philip Abelson. It required conversion of uranium into gaseous *uranium hexafluoride*. It had also been observed that U-238 disintegrated into plutonium-239, just as fissionable as U-235 when absorbed by a thermal neutron.

Aided in particular by Arthur H. Compton of the University of Chicago, the chemical industry scaled up the results of laboratory processes. Du Pont was given the task of producing plutonium-239 in its Hanford, Washington, plant, Union Carbide that of separating U-235 from the bulk uranium containing mainly the inert U-238. Both companies, together with Harshaw Chemical and Hooker Electrochemical, set up fluorine units for preparation of uranium hexafluoride. Considerable progress was made in analytical chemistry because of the high purity required of the elements in reaction. Mass spectrometry, in particular, was needed for the analysis of uranium isotopes. But scientists also had recourse to extraction with solvents, via ion-exchange resins, and by chromatography.

The production of uranium-235 and plutonium-239 was taken care of in Oak Ridge,Tennessee, and in Hanford, Washington. By 1945 that production was sufficient to begin making the bomb proper. This part of the Manhattan Project was carried out in Los Alamos, New Mexico, under the supervision of the American physicist Robert Oppenheimer. On July 16, 1945, the first atomic bomb exploded in Alamagordo, New Mexico. Its explosive power was that of 10,000 tons of TNT. On August 6, 1945, a uranium bomb was dropped on Hiroshima; and three days later a plutonium bomb fell on Nagasaki.

The Manhattan Project, which was largely made possible because of the work carried out by physicists who had come from Europe, cost $1.6 billion. In its own sphere of competence, America's chemical industry had carried out the part assigned to it by the United States government.

IG Farben's Wartime Operations and How They Ended

England escaped a German invasion, thanks to its insular position and to the courage of its airmen. The English were therefore able to mobilize all their forces towards the war effort. But continental Europe, occupied by the Germans, was suffering the consequences of defeat.

As the country most coveted by the occupying power, France's situation was the most delicate, more particularly insofar as the chemical industry was concerned. Pétain and Hitler had met on October 24, 1940, in Montoire and had established the bases of Franco-German collaboration. France's industrial leaders were nonetheless caught in the turmoil of a war pursued on different fronts. The chemical industry, like all the other sectors, depended for supplies on the committees, which were in charge of sharing out what was available.

Because of their control over Compagnie Nationale des Matières Colorantes (CNMC), Établissements Kuhlmann was a particular target of IG Farben. Raymond Berr had been forced to resign because of Nazi racial laws, and Établissements Kuhlmann was now managed by René Duchemin, assisted by Joseph Frossard, the very man who had negotiated with IG Farben's Carl Bosch at the time of the 1919 Versailles Treaty.

It is hard to tell today whether the Germans exercised an irresistible pressure on him or if the friendly ties which Frossard had established in former times with IG Farben made the transactions easier. Whatever the case, an agreement was signed on November 18, 1941, by the management of the two companies under which a joint company, Francolor, was set up with IG Farben receiving a 51 percent controlling share. This move was to have serious long-term consequences for France's dye industry. Francolor was sequestered as soon as the war ended and was returned to its parent company after only an extended battle.

While IG Farben was thus compromising its French partner, it, in turn, had to comply with the increasingly formidable demands of the Nazi chiefs. Some of its leaders were even hard put to refute at the Nuremberg trial charges that they shared in the responsibility for the atrocities carried out by the regime, considering the close ties that had been woven between them and the political powers. Hitler's adventures were to cost IG Farben dearly: its plants, and in particular the Leuna unit where the synthetic gasoline was produced, were wiped out by U.S. bombing; and all its installations set up east of the Elbe fell irretrievably into Soviet hands. A large part of its overseas interests was sequestered, its trademarks confiscated, and its technologies divulged to the Occupation powers.

In August 1950, the Allied High Command decreed the liquidation of IG Farben Industrie.

Spectacular Breakthroughs in Postwar Chemicals (1945–1960)

The period extending from 1945 to 1960 was particularly prolific for the chemical process industries. The United States, far from the combat area, had considerably developed its industrial potential in the interest of the Allied cause. It was thus quite naturally at the root of the sweeping changes that were to shake up world chemicals. Europe, like Japan, was far too busy with its own reconstruction during the immediate postwar years to make a marked contribution to the renewal of chemical technologies. At the same time, America was the only country in a position to take new initiatives in the various areas where its war effort had given it undisputed mastery.

The Rise of Petrochemicals

In the 1920s and 1930s, while the rest of the world was relying on coal as the essential source of organic starting materials, the United States was blazing the trail of *oil-based chemicals*. *Carbon black*, used in the tire industry and based on oil, was an American monopoly. *Isopropyl alcohol*, obtained through propylene hydration, was produced in the Bayway unit of Standard Oil of New Jersey. In 1930 in California, Shell started to produce *hydrogen* from natural gas, then *butyl alcohols* and *methyl ethyl ketone* from n-butylenes, while in 1925 in South Carolina, Union Carbide began producing a whole series of aliphatic derivatives, especially glycols, from petroleum gases.

The spectacular development of *petrochemicals* in the *United States* after 1940 was due to two factors. On the one hand the synthetic rubber program required increasing amounts of butadiene and styrene. On the other hand, improved refining technologies to obtain aviation gasoline also opened the way to the production of other feedstock chemicals. The *ethylene* used to prepare styrene via ethylbenzene was produced either from refinery gases or from naphtha steam cracking carried out by Standard Oil of New Jersey in Baton Rouge, from 1941 on. *Butadiene* came from the dehydrogenation of cat-cracker butylene or, since 1943, from butane cracking (Standard Oil of California). In 1945, the United States produced 820,000 tons of synthetic rubber accounting for nearly 50 percent of the country's petrochemicals production. The petrochemicals industry spilled over into Canada also. Synthetic rubber units set up in Sarnia, Canada, produced 45,700 tons in 1945.

In Britain, the first unit to produce propylene and from it isopropyl alcohol and acetone was set up by British Celanese in Spondon in 1942. Before that, Shell Chemicals Ltd. had put "Teepol" production on stream early in the war. Teepol was a secondary alkyl sulfate detergent used as a soap substitute during the war. Apart from these two exceptions, there was no petrochemicals industry in Europe before the end of 1949.

The Americans themselves were of two minds about how vigorously to pursue petrochemicals. Synthetic rubber production fell below 400,000 tons in 1949; and half the GRS units had been closed down. Europeans needed to rebuild their refineries before they could think about substituting petrochemicals for their traditional coal-based chemicals. But, with the end of the U.S. slump of 1948–1949 and with new needs emerging in Europe in solvents and polymers, petrochemicals became universally recognized as an essential source of organic base products.

The North American Lead

Both in the quantities produced and in the range of products marketed, the United States had a long lead in petrochemicals. By 1950, American companies were producing some three million tons of organic derivatives from oil and natural gas. And until 1954 it was alone in having a significant market for its petroleum-based ethylene, in the polyethylene and ethylene oxide produced. Likewise its supremacy in synthetic elastomers (SBR, butyl) had opened up a wide market for its C_4 fractions. Finally catalytic reforming, first used for toluene, had extended to benzene.

With the exception of Canada, where the development of petrochemicals was boosted by the proximity of the big American oil companies already established in Sarnia and Montreal during the war and by Alberta's natural gas fields, the other industrialized countries took varying but long times to enter the petrochemicals market.

Europe's Particularisms

With the end of the war, it became clear that the "gentleman's agreement" that had been established in *England* between Distillers Company Ltd. (DCL) and ICI to share out their fields of activity could not be extended. Both for its Nylon and its polyethylene, ICI henceforward needed to be connected with oil-based feedstock.

Neither could DCL hope to perpetuate sugarcane molasses as an economical and safe raw material for the production of its oxygenated solvents. This is how DCL became associated with the Anglo-Iranian Oil Company in 1947, the future British Petroleum (BP), to form BP Chemicals Ltd. on a fifty-fifty basis. The new company's Grangemouth steamcracker in Scotland began producing ethylene and propylene in 1951 and afterwards ethanol and isopropyl alcohol. At the same time, ICI had chosen Wilton on the Tees River estuary to set up its own naphtha cracker which came on stream that same year.

Petrochemicals Ltd. was set up in 1946 to develop the "catarole" catcracker process which some Austrian refugees had perfected in England. The company, together with its subsidiaries, Styrene Products and Styrene Copolymers, was bought by Shell Chemicals in 1955. Another company, *Forth Chemicals*, was formed under a two-thirds BP Chemicals, one-third Monsanto association to produce in Grangemouth the monomer styrene needed to make polystyrene. *British Celanese* played a subsidiary role in the development of Britain's petrochemicals industry, although it had been a pioneer in olefins. It was to be subsequently bought by Courtaulds.

Germany awoke to petrochemicals much later than Britain. Under Hitler's national self-sufficiency policy, all efforts had been directed to the production of synthetic gasoline. In prewar Germany, oil refining had scarcely been encouraged. Besides this, the huge steel industry of Germany had depended on the very considerable coke production from the Ruhr coal mines. Coal-based chemicals were a logical offshoot of coke production, and the focus on those chemicals was further sharpened by the establishment of joint subsidiaries by coal-mining and chemical companies.

This was the base on which Germany had built up its feedstock supplies for its organic chemicals industry. When the war ended, it became obvious that oil could become a dangerous competitor of coal and lignite as a source of energy and raw materials. Moreover, the Leuna synthetic gasoline units were now in the Russian zone. The range of problems confronting West Germany's industrial leaders, however, and a certain natural caution, meant that it was not until 1956 that, after lengthy research, Hoechst put on stream a process to pyrolyze crude oil at 800°C, producing ethylene and propylene as "building blocks" for a petrochemicals industry.

About the same time, Hoechst began receiving methane from the Pfungstadt natural-gas fields, which were being developed by the *DEA* and which it used to prepare chlorine derivatives. In addition, Hoechst bought up in 1955 the *Anorgana* Gendorf plant built in Bavaria during the war to supply the ethylene needed to prepare ethylene oxide and glycols from carbide acetylene. Hoechst also used its pyrolysis technology to set up an original high-temperature cracking process in Gendorf in 1958 that could selectively produce ethylene from petroleum. Strangely, Hoechst changed its policy in 1960. From then on, it covered its olefin supplies through the *Caltex* refineries in Kelsterbach and through *Union Rheinische Braunkohlen Kraftstoff* (URBK) in Cologne.

Rather than develop their own cracking technology, BASF and Bayer chose to associate themselves with oil companies which supplied them with olefins. BASF and Shell set up a fifty-fifty subsidiary, *Rheinische Olefin Werke* (ROW) in Wesseling, while Bayer and BP held equal stakes in *Erdöl Chemie* in Dormagen. But production of the two refineries began only in 1955 and 1958, respectively.

One other company must be mentioned, *Hüls*, which had been formed in 1938 in Marl, Westphalia. It produced electric-arc acetylene from coal hydrogenation products and used that acetylene to carry out the synthesis of rubber. In 1944, Bentheim natural gas began to serve the same purpose. After the war, though, the Allies banned German production of synthetic rubber; so Hüls switched to producing ethylene oxide and its derivatives, as well as raw materials for detergents and plastics. Hüls subsequently became one of Germany's main olefins consumers and one of Europe's leading petrochemical companies. By 1960, Germany's petrochemicals industry had fully made up for lost time despite a slow start and the various directions taken by different groups.

Unlike Germany, *France* was able to develop its petrochemicals industry unhampered by past rigidities. Its coal-based chemical industry was not as substantial, coke for the steel industry being to a large extent imported from Germany. Moreover, France's mining companies were only involved in chemicals through the synthesis of ammonia or the production of methanol and phenol. The country's refining capacities, already substantial before the war, were rapidly rebuilt with funds from the Marshall Plan. The discovery of gas in Saint-Marcet in 1942, then of the Lacq gasfield in 1949, provided feedstock for the *ONIA* ammonia units in Toulouse, for *Société des Produits Azotés* in Lannemezan, and for *Pierrefitte* in Soulom. France was, therefore, in a good position to set up a chemicals industry based on natural hydrocarbons as both the refinery by-products and the gas from its southeastern area were available. It was through a complex network of alliances between oilmen and chemists that the purpose was achieved.

In 1949, *Produits Chimiques-Shell-Saint-Gobain* (50 percent Royal Dutch Shell—35 percent PCRB—4.9 percent Saint-Gobain) was set up to produce Teepol in Petit-Couronne in northern France and propylene-based solvents in Berre, in the south. That same year, Péchiney joined forces with Pétroles BP and Kuhlmann to establish *Société Naphta Chimie*, which came on stream in 1953, producing, for the first time in Europe, ethylene oxide through direct oxidation of ethylene (by Ralph Landau's Scientific Design process). The steamcrackers in Lavera also produced propylene, which was turned into isopropyl alcohol. By 1960 they were producing 48,000 tons of ethylene and 52,000 tons of propylene annually.

In a parallel move *Société des Dérivés de l'Acétylène* (SIDA) which had been founded by Ugine in 1929 in La Chambre, Savoie, to produce acetone and methyl ethyl ketone, built a secondary-butyl alcohol unit in Port-Jérôme, Normandy, which was put on stream in 1960. It was supplied with butylene by the neighboring Esso refinery, which had set up a steam cracker in 1959 to supply not only ethylene and propylene, but also C_4 fractions mainly for use by *Société du Caoutchouc Butyl* (SOCABU), a subsidiary it had formed with *Compagnie Française de Raffinage* (CFR), and the rubber tire manufacturers *Michelin*, *Dunlop*, and *Kléber Colombes*.

As for CFR, the refining arm of Société Français des Pétroles, it brought on stream a catcracker in Gonfreville, Normandy, in 1956 to supply olefins to the companies in which it had stakes: isobutylene for *SOCABU*, ethylene for *Manolène*, propylene tetramer for *Pétrosynthèse*. Such alliances between oil companies and chemical groups added to the already complex structure of France's chemical industry. Thus Rhône-Poulenc and Kuhlmann had joint stakes with CFR in *Manolène*, which had been set up in 1956 to produce polyethylene under a Phillips process. Likewise, *Société Pétrosynthèse*, which had been formed in 1955 to produce dodecylbenzene for detergents under an Oronite (Standard Oil of California) license, was owned jointly by CFR, Oronite and *Atlantique-Progil Electrochimie* (APEC). APEC had itself been jointly set up in 1951 by *Antar Pétroles de l'Atlantique* and *Progil Electrochimie*.

Progil Electrochimie was an offshoot of the 1950 partnership between Progil and Ugine for the purpose of manufacturing phenol in Point de Claix by the cumene-based process. The cumene was supplied by APEC from the propylene produced in Antar's Donges refinery. What rapidly became a practically inextricable maze of participation can be explained in part by the diversity of interests involved. Lack of equity, however, which restricted the financial potential of each chemical group, was also important.

The discovery of the Lacq gasfield by *Société Nationale des Pétroles d'Aquitaine* (SNPA) further contributed to the overall complexity. In 1956, Saint-Gobain, Péchiney, ONIA, and Pierrefitte jointly created *Société Aquitaine-Chimie* to produce acetylene and ammonia from Lacq gas. Two years later, Aquitaine-Chimie set up *Méthanolacq* on a fifty-fifty basis with Kuhlmann to produce methanol. Then in 1959, Pierrefitte, Rhône-Poulenc, Péchiney, Saint-Gobain and Usines de Melle established *Acétalacq* to transform Aquitaine-Chimie's acetylene into acetaldehyde. And *Azolacq* was set up in 1957 by ONIA and Pierrefitte to use Aquitaine-Chimie's ammonia to prepare ammo-nitrates and urea.

As for *SNPA*, it set out to develop the sulfur contained in large proportions in the Lacq gas[1] as well as the ethane and propane mixed with methane. Such profuse activity certainly indicated that France's chemical companies took a lively interest in petrochemicals; but such ramified interactions prevented the establishment of coherent groups capable of withstanding long-term wear and tear.

Italy began developing a petrochemicals industry in 1952, and made rapid progress because it had a very small coal-based chemicals industry. On the other hand, significant natural gasfields containing practically pure methane and no sulfur were discovered in the Po Valley after the war. It was on this easily available and plentiful raw material that *Montecatini* and *Azienda Nazional Idrogenazione Combustibili* (ANIC) established the basis of a new chemical industry that produced ammonia, acetylene, methanol, hydrocyanic acid, acrylonitrile, and SBR rubber in the Ferrare, Novare, and Ravenna units. In addition, the refining of crude oil by the oil companies made available olefins such as ethylene and propylene, so that by 1960 Italy's base petrochemical derivatives exceeded 200,000 tons. The Emilia gasfields made associations with foreign groups unnecessary; moreover, the state company ENI handled most of the production through its chemical subsidiary ANIC and through Montecatini.

Like Britain, but on a smaller scale, the *Netherlands* and *Belgium* supplied their petrochemicals industry with the feedstocks produced from the refining of imported crude oil, setting up units on harbor sites. Shell's Pernis unit near Rotterdam came on stream in 1953, and the Antwerp units of *Société Chimique des Dérivés du Pétrole* (Petrochim) in 1957.

Japan's Belated Involvement in Petrochemicals

Japan did not become involved in petrochemicals until 1957, when the Shimotsu unit belonging to *Maruzen Oil Company* came on line. Production centered on isopropyl alcohol and on butanol derivatives. There were no special obstacles in Japan to the development of an oil-based chemical industry because coal production and carbonization capacity were not sufficient for any significant coal-based chemical supply. With the exception of the Niigata natural gasfield, however, which supplies the *Nippon Gas Chemical Company* unit, Japan depends on foreign sources of hydrocarbons. For this reason, Japan's petrochemicals industry is located in the large ports close to

[1] With a content of 15.12 percent H_2S.

the refineries. The refineries are mostly owned by British and especially American interests, which supply the feedstock.

It stood to reason that the big zaibatsus like *Mitsui*, *Mitsubishi*, or *Sumitomo*, which were already heavily involved in base chemicals, would not disregard the new oil sources of raw material. By 1960, eighteen petrochemical plants had been set up in Japan, mainly located in Kawasaki and in Yokkaichi, with a production capacity of 300,000 tons a year. Despite a late start, Japan's petrochemicals industry, spurred on by the zaibatsus, had in a very short time caught up with its European rivals both regarding the extent and the range of its production. As elsewhere in the world, of course, the technologies were American. But Japanese engineers adapted them remarkably rapidly, providing their country with the grounding needed to modernize its chemical industry.

Petrochemicals, the Only Source of Propylene

While ethylene can be extracted from coking gases or from ethyl alcohol through dehydration, and butadiene can be produced from acetylene, there is no way to obtain propylene other than through the cracking of oil and gas. This makes propylene an exclusively petrochemical olefin.

Propylene is an essential base material for modern chemicals. It has served, since 1920, to prepare *isopropyl alcohol*, a major solvent and the starting material for *acetone* synthesis. In 1955, 85 percent of the acetone produced in the United States was made this way.

America's aviation fuel program during the Second World War had required the use of *cumene* to increase gasoline's octane rating. It had been prepared through alkylation of benzene with propylene. Cumene (or isopropylbenzene) had been introduced into the chemical circuit after Hock and Lang had demonstrated in 1944 that cumene[2] could be oxidized to form a hydroperoxide, which in turn was decomposed into *phenol* and *acetone*. Thus an indirect path from propylene to acetone also became available.

[2]

$$C_6H_5C(H)(CH_3)_2 + O_2 \longrightarrow C_6H_5C(OOH)(CH_3)_2 \xrightarrow{H^+} C_6H_5OH + (H_3C)_2C{=}O$$

cumene — cumene hydroperoxide — phenol — acetone

Hercules Powder and Distillers took out a patent for this major reaction, the two companies reaching an agreement to license out their process. This set off construction of cumene-based phenol units in the United States (Hercules Powder, Allied Chemical, Standard Oil of California) and, after 1953, in Europe (Distillers in Britain, Rhône-Poulenc[3] and Progil Electrochimie in France, Phenol Chemie in Germany, Société Chimique des Dérivés du Pétrole in Belgium).

Propylene also sparked an important category of synthetic detergents belonging to the alkylarylsulfonate family. Indeed, propylene's tetramer serves to alkylate benzene, and the resulting dodecylbenzene produces *dodecylbenzenesulfonate* by sulfonation. Initiated in the United States by Altantic Refining and Standard Oil, the preparation of *dodecylbenzene* for subsequent sulfonation spread to Europe in 1956. In particular, it was carried out in Britain by *British Hydrocarbon Chemicals* and *Esso Petroleum*, and in France by *Pétrosynthèse* and *Standard Kuhlmann*.

From propylene, through *allyl chloride*, Shell performed for the first time in the United States in 1948 the synthesis of glycerine. By 1955, 45 percent of the glycerine produced in the United States was obtained by synthesis. In addition, Shell used the allyl chloride obtained through propylene chlorination to produce *allyl alcohol*, a new starting material for chemical synthesis, as well as *epichlorydrin*, which constituted the start of Epoxy resins (Epikote).

From 1957 on, propylene was used as a monomer in the preparation of *polypropylene* through polymerization, both by Montecatini in its Ferrare unit and by Spencer Chemical in Orange, Texas. Propylene is also used in the *Oxo synthesis* to produce *butanol*, a major chemical for the production of solvents and plasticizers. Thus petrochemicals provided the basis for an entirely new chemical industry by supplying a way of cheaply extracting propylene from natural hydrocarbons.

The New Polymers

The war considerably boosted the production of already-known synthetic high polymers (PVC, polystyrene, phenoplasts, aminoplasts, polymethacrylates, GRS, butyl rubber, and Nylon fiber). It also sparked the development of many polymeric materials which, although described, had not yet been produced on an industrial scale.

This was the case with *polytetrafluoroethylene*, PTFE, which Roy Plunkett chanced upon in 1938. It was produced on a pilot scale by *Du Pont* in 1943, then industrially two years later. *Silicones* had likewise been exhaustively studied by Kipping between 1899 and 1937. But they became usable as lubricants and elastomers only in 1943, when *Dow-Corning* was set up to

[3] The company has developed its own process.

produce them according to the Kipping method and when *General Electric* brought on line a rival unit using a process developed in its own laboratories by Eugene G. Rochow.

An inventive American chemist, Carleton Ellis, was the first to study *unsaturated resin polyesters* in 1935. He had developed an improved process for making aminoplasts in 1933. The needs of the U.S. Army and Navy boosted the development of these materials. In 1942, *American Cyanamid*, following joint research carried out by *Owens Corning Fiberglas* and *U.S. Rubber*, set up a pilot unit to produce an unsaturated polyester resin obtained through condensation of maleic anhydride with ethylene glycol, with subsequent styrene cross-linking. By the end of the war, some 2,500 tons of these resins were being produced annually in the United States by *American Cyanamid*, *Du Pont*, *Libbey Owens Ford*, *Plaskon*, and *Marco*.

Only one small-capacity *polyethylene* unit existed in the world in 1939. It had been set up in England by ICI using its high-pressure process. The polymer proved so important for insulating radar cables that ICI disclosed its process to *Du Pont* and *Union Carbide*. Through the tubular reactors developed by Georges Felbeck, Union Carbide was able to supply in 1943 a low-density polyethylene that was superior to ICI's for radar. The next year Du Pont brought on stream a unit that used a tumbler autoclave similar to ICI's. High pressure polyethylene had thus moved from the experimental stage to that of a commercial product.

While war requirements had accelerated the development of certain high polymers, peacetime brought to the market new materials discovered by researchers on both sides of the Atlantic.

The Work of Ziegler and Natta

The drawback of high-pressure polyethylene is that its molecules are not strictly linear but form branched paraffin chains and the product loses its rigidity when heated, even below 100°C. This is too low for a number of uses, particularly in the case of objects requiring sterilization in boiling water. Through a chance discovery of Germany's Karl Ziegler, which he seized upon, a polyethylene was produced with a straight-chain structure and a melting point raised to 145°C.

Ziegler had obtained his doctorate in chemistry from the University of Marburg in 1923. His work on macrocyclic compounds in 1933 had gained him some repute; ten years later he was appointed head of the Kaiser Wilhelm Institute for Coal Research at Mülheim. It was in this research center, which was to become the Max Planck Institute, that Ziegler noticed that he had polymerized ethylene at atmospheric pressure on heating it with an organometallic catalyst, triethyl aluminum, in an attempt to prepare a fatty alcohol.

He gave up his work on fatty alcohols, and from being simply an organic chemist he became a passionate devotee of macromolecules. Together with his assistant, Holzkamp, he pursued his research on the organometallic catalysts best suited to the preparation of the *high-density polyethylene* which he had discovered.

The importance of Ziegler's work was of a nature to appeal to his longtime friend, Herman Mark, who had left Germany because of its racial persecutions, and who, since the end of the war, was head of the Polymer Institute at Brooklyn Polytechnic Institute.

Mark mentioned the matter to Robert Robinson, a consultant for Petrochemicals Ltd. That is how this young English company became one of the first to obtain a Ziegler license. Other license agreements were signed with Hoechst for Germany and Hercules Powder for the United States. Both these companies were committed naturally to share the results of their research in polyolefins, a factor which eased the path to industrial scale development. On June 18,1957, Hercules inaugurated the first American low-pressure polyethylene unit in Ziegler's presence.

It was an Italian professor, Giulio Natta, who discovered a form of polyolefin that melted at 170°C. This was *polypropylene. Natta* was a chemical engineer from the Milan Polytechnic Institute. After meeting with Hermann Staudinger in Fribourg in 1932 he turned to studying high-polymer structures. From 1938 he taught industrial chemistry at the Polytechnic, where he became the head of a small group of enthusiastic researchers in macromolecular chemistry. While Ziegler in Mülheim was able to pursue his research quite independently, Natta was forced to sign an agreement with the engineer Piero Giustiniani, general manager of Montecatini, in order to finance his work. In consequence, the marketing rights for any discovery he made were owned by Montecatini.

A three-party agreement was signed granting a license on Ziegler's patent to Montecatini, for Italy, and setting up information and research exchanges between Mülheim and Milan in the area of olefin processing. Ziegler never thought that polypropylene could be produced under his process, so that when Natta prepared the new polyolefin in 1954, he was induced by Montecatini to take out a patent in his own name, thinking that it did not interfere with his agreements with Ziegler. Ziegler, however, was highly indignant; and so ended a collaboration between two institutes that had been very fruitful, leading to the discovery of *stereospecific* catalysts in 1953. Natta had not merely prepared polypropylene. He had also demonstrated that the increase in the polymer's melting point was related to the regularity with which the propylene's methyl groups alternate in space along the carbon chain. This was the phenomenon of *stereoregularity*.

Ziegler's catalysts, combined with titanium tetrachloride, produced a regular configuration called *isotactic*, which raised the melting point after suitable stretching to direct the pattern of the molecules. In recognition of this fundamental work, which opened up new paths in high polymers, Ziegler and Natta were awarded the Nobel Prize for Chemistry in 1963.

The Success of Polyolefins

Polyethylene has different features, as we have seen, according to whether it is produced through a high- or a low-pressure process. In both cases its uses were increasingly varied: compressible containers, wrapping film, pipes, electrical insulators.

In 1952, under the antitrust move by the American administration, ICI was forced to sell the licenses for its high-pressure process to *Spencer Chemical*, *Dow*, *Eastman Kodak*, and *National Petrochemical*. At the same time BASF, which had its own tubular-reactor process, was licensing it out to *Koppers* and *Monsanto*.

Low-pressure linear polyethylene, which has a higher density and can easily be produced at atmospheric pressure, also attracted the attention of the chemical industry. Ziegler had licensed out his patents, but every licensee was left to develop the process as seemed fit. Two researchers, Hogan and Banks, working for the Bartlesville, Oklahoma-based *Phillips Petroleum Company* had also patented, in 1954, a process to produce linear polyethylene under moderate pressure by means of a heterogeneous chromium oxide catalyst.

Two years later, Phillips was producing the polyethylene industrially and licensing out a process which, unlike Ziegler's, could be used directly on a large scale and involved a cheaper catalyst. By 1970, 7.5 million tons of linear polyethylene had been produced worldwide under the Phillips process, which eclipsed Ziegler's.

Standard Oil of Indiana also developed a low-pressure method, but it fell short of success because its industrial development was put off too long.

The success of *polypropylene* was furthered by a number of factors. Because of its highly crystalline state and its well-ordered structure (isotacticity) it has remarkable thermal mechanical properties. This explains its rapid development in numerous applications: molded objects by injection, filaments, textile fibers, film. Moreover, propylene started off as a by-product of ethylene production from oil fractions. As ethylene requirements grew, markets had to be found for the propylene, which, in this context, was a cheap olefin. Finally, the discovery by Natta of isotactic polypropylene happened at a time when the chemical industry was booming and was seeking to diversify.

As for low-pressure polyethylene, *Hercules Powder* was the first to become aware of the significance of the new material and to produce it industrially in the United States under the Ziegler-Natta process. At practically the same time polypropylene patents had been taken out by Montecatini on behalf of Natta and by Phillips Petroleum on behalf of Hogan and Banks. The tough legal fight which then ensued in the United States ended only in 1982 with a High Court verdict in favor of the Americans.

Meanwhile, considerable work had been needed to make the product industrially and technologically acceptable. The Ziegler catalysts, which were delicate to handle, had to be improved; the polymerization process had to be controlled through use of hydrogen; stabilizers had to be found for the polypropylene, as its chains were less crystalline than those of high density polyethylene and, therefore, more easily affected by heat and light; and new dyeing processes had to be developed for the fibers.

By 1970, world consumption of polypropylene exceeded 1.2 million tons. Strangely enough, Montecatini, which, with Giustiniani, had very early on taken control of Natta's discovery, expended more energy on defending his polypropylene patents than on boosting production of the new polymer. True, they made considerable profits from licensing. As for Ziegler, who had invented a new chemistry, he had the rare good fortune for a scientist of raking in large royalties. Unlike Natta, he had taken out patents in his name for the work carried out under his direction at the Max Planck Institute.

Stereospecific Elastomers and Ethylene-Propylene Rubber

Ziegler's chemistry had unexpected fallout in the field of synthetic rubber. It was already known that natural rubber was an isoprene polymer in which the CH_3 groups were always arranged on the same side of the double bond (*cis*-isomer).[4] In the case of balata or gutta percha, on the other hand, the CH_3 groups are arranged on alternating sides of the double bond (*trans*-isomer). The polymer then has the properties of a hard resin and not those of an elastic material.[5] Accordingly, it was obvious that to come closer to natural rubber, the polyisoprene chain had to become *stereospecific*, that is, to be given a space conformation close to the *cis*-1,4-isomer.

One of the sponsors of the Max Planck Institute in Mülheim was *Ruhrchemie*, which had an information-exchange agreement with a fifty-fifty subsidiary of B. F. Goodrich and Gulf Oil, the *Goodrich-Gulf* Company. This company took an option on the Ziegler process in 1954 at a modest cost of \$50,000. On December 2, 1954, Goodrich-Gulf announced that its researchers had succeeded in synthesizing, besides low-pressure polyethylene, the *cis-1,4-polyisoprene*. The starting point had been a pure isoprene copolymerized with ethylene, and the result had been achieved through use of a Ziegler-Natta catalyst (aluminum alkyl + titanium tetrachloride). A group of researchers

4 $(\cdots H_2C{-}CH{=}C(CH_3){-}CH_2{-}CH_2{-}CH{=}C(CH_3){-}CH_2\cdots)_n$ *cis* 1,4-polyisoprene

5 $(\cdots H_2C{-}CH{=}C(CH_3){-}CH_2{-}CH_2{-}CH{=}C(CH_3){-}CH_2\cdots)_n$ *trans* 1,4-polyisoprene

working for *Firestone* also managed to polymerize isoprene into cis-1,4-polyisoprene following a process advocated by Ziegler in 1927 of polymerizing butadiene with lithium. Firestone announced its discovery in August 1955, a few months before *Goodyear* was also to reveal that it had produced a natural ''synthetic''rubber called Natsyn.

Shell Chemical brought the first stereospecific elastomer commercial unit on line in 1960, using lithium as a catalyst. The company had previously purchased a GRS rubber plant in Torrance, California, from the government in 1955.

Goodyear's Natsyn unit began production in 1962. Both the polymerization catalyst and the catalyst by which it produced the propylene dimer leading to the isoprene must be credited to Ziegler.

With Ziegler's catalysts, other dienes could produce stereospecific rubber. Very naturally, butadiene was first tried out. Phillips and Firestone were the first to market *cis-1,4-polybutadiene*. Its success was due to its resistance to abrasion, its good compatibility, and its low temperature flexibility. It was used to manufacture tires blended with other types of rubber. Disappointed that polyisoprene had not developed as well as expected, Shell Chemical withdrew from the field in 1967, leaving it to Goodyear and Goodrich.

This is how a chance discovery, happily seized upon by Ziegler and cleverly developed by Natta, in their research centers in Germany and Italy, not only revolutionized the chemistry of polyolefins but also, through the acumen of American oil and rubber companies, brought about industrial production of an entirely new class of stereospecific elastomers.

In 1955, using Ziegler-type homogeneous catalysts, Natta and his team succeeded in producing an ethylene-propylene elastomer. This *EPM rubber* was produced by *Montecatini* and in 1961 by *Enjay Chemical*, a subsidiary of Standard Oil of New Jersey. Like *Du Pont*, this subsidiary also marketed an improved polymer, the *EPDM* terpolymer containing 5 percent of a diene such as dicyclopentadiene or 1,6 hexanediene.

Otto Bayer and the Discovery of Polyurethanes

Otto Bayer had studied chemistry at the University of Frankfurt, the town where he was born. After obtaining his doctorate, he joined IG Farben's Mainkur plant. In 1934, when only thirty-two years old, he became research manager of Farbenfabriken Bayer in Leverkusen, one of IG's main branches. There he became interested in macromolecular chemistry and began studying the polycondensation reactions which Carothers' work in the United States had brought into fashion. The starting point he chose was the particularly reactive *diisocyanates*, which produce *urethanes* via dialcohols.

In 1937 Bayer obtained through polyaddition a viscous liquor that could be stretched to filaments similar to polyamides' behavior while he was preparing the 1,8-octane diisocyanate and making it react with 1,4-butanediol. Other diisocyanates led to polymers which could be molded or used as protective coatings and adhesives. In 1941, Bayer observed that a number of molded products contained air bubbles caused by carbon dioxide, CO_2. By adding water during the reaction, he managed deliberately to cause CO_2 to form thereby producing the first *polyurethane foams*.

Polyurethanes were used during the war in Germany for a variety of military applications. Ten years were to pass before rigid and flexible polyurethane foams were put on the market, the varieties being produced through specific selection of diisocyanates and co-reagents. Bayer set up in 1951 a *TDI* (tolylenediisocyanate) unit, then joined up with Monsanto in 1954 to form the Mobay company, which began to produce this new class of polyurethanes in the United States a year later. World production of polyurethanes, to which new aromatic components were added, jumped from 8,000 tons in 1955 to 2.5 million tons today.

Otto Bayer, who had become a member of IG Farben's Board in 1939, was named in 1964 head of Germany's Bayer, a company he had served well through his discoveries and which, by a curious coincidence, bore his name.

The Race for New Materials

Other polymers emerged in this postwar period, so rich in discoveries. A number of chemical companies became interested in the polymerization of *formaldehyde*, the subject of Staudinger's early macromolecular studies. In 1953, a group of Du Pont researchers developed the first engineering plastic belonging to the *polyacetal* family. It was put on the market six years later under the trademark of *Delrin*. A variation of this polymer based on copolymerized trioxane with a small quantity of ethylene oxide was produced in 1963 by Celanese under the name of *Celcon*.

In 1956, *Bayer* in Germany and *General Electric* in the United States simultaneously developed the *polycarbonates* on which Carothers had been working in the thirties. Polycarbonates were produced industrially in 1960. Obtained through *phosgene* reacting on diphenylolpropane or *bisphenol A*, they could be marketed in the shape of transparent sheets. Despite their high price, polycarbonates found important applications because of their specific properties. This was also the case for polyacetals, which in some cases can substitute for metals, and which form part of a whole new family of materials, the "engineering plastics."

Developing synthetic materials of great strength, or with other specific properties, is obviously highly desirable. As early as 1948, *Dow* had put on the market higher-impact-resistant *polystyrenes* which *Monsanto* had also managed to produce by changing the polymerization process. In 1950, *Naugatuck Chemical*, a subsidiary of the *U.S. Rubber* Group, produced the first ABS (*A*crylonitrile, *B*utadiene, *S*tyrene) copolymer, which was highly resistant to shock but was hard to process and was not very stable. Nine years later,

Borg Warner's chemical subsidiary *Marbon* considerably improved the properties by polymerizing the styrene and acrylonitrile in a polybutadiene-elastomer precursor. With *SAN*, an acrylonitrile-styrene copolymer marketed in 1948 in the United States, it became possible to produce a plastic that was more resistant than polystyrene to chemical agents.

Engineering plastics were to undergo remarkable development after 1960, as will be seen later. Here we shall simply mention how, in the mid-1930s, Pierre Castan, with *CIBA* in Switzerland, and S. O. Greenlee, with *Devoe & Raynolds* in the United States, had studied resins obtained through epichlorhydrin reacting with bisphenol A. This was the start of *epoxy* resins, which were remarkable for their insulating, adhesive, and mechanical properties as well as for their chemical resistance.

The successful development of these resins for coatings, glues, electrical applications, and laminated products was partly due to the epichlorhydrin that became available in 1947 through the work carried out by Shell Chemical, USA, on the chlorination of propylene, and the technical progress achieved after Castan and Greenlee had taken out their patents.

The first producers of this class of resins were, quite naturally, CIBA in Basel, the pioneers, and Shell Chemical in the United States which, as early as 1941, had worked with Devoe & Raynolds in applying them to protective coatings before developing after the war the two raw materials that served to manufacture them.

The Development of Chemical Fibers

Carothers' *Nylon* was put on the market just before the war by Du Pont. At the same time, IG Farben was selling its *Perlon* polyamide fiber, developed by Paul Schlack.

Nylon was restricted to military purposes during the Second World War. But in postwar years it was found that the fiber was not only suitable for women's hosiery but also, on slightly modifying the polymeric chain, could be used for upholstery, clothing, carpeting, tire fabric, etc. It was also found that supplied in powder-form, Nylon could be molded.

As its manufacturing cost was brought down through the introduction of new petrochemical feedstocks, Nylon became the leading synthetic fiber. In the United States alone, production, which had reached scarcely 1,200 tons in 1940, soared to 650,000 tons in 1970.

The development of *polyester fibers* was just as spectacular. Those fibers emerged in England in 1941 through research work carried out by J. R. Whinfield and J. P. Dickson. Whinfield, a Cambridge chemistry graduate, had acquired his first experience in fibers in the laboratory of C. F. Cross, known for his invention of viscose. Whinfield joined the Calico Printers Association and, in 1939, with his colleague Dickson, persuaded his employers that they should engage in research on synthetic fibers.

Carothers had initiated his studies on polycondensation by preparing polyesters from *ortho*-phthalic acid and ethylene glycol. But the polymers he obtained had melting points that were too low for the production of fibers. This was the reason why Carothers had turned to polyamides.

Whinfield and Dickson resumed research where Carothers had left off. Their starting point was *para-* or *terephthalic* acid, which by reacting with ethylene glycol produced *polyethylene terephthlate*.[6] Rough fibers could be drawn from this polymer and given a cold-stretching treatment. The resulting polyester fiber was resistant to hydrolysis and had a high melting point, features that would establish its commercial success. Thus two English researchers, working part-time in the modest laboratory of a company in no way connected with polyesters, discovered a new synthetic fiber that Du Pont, with its ten-year lead and its large teams of specialized researchers, had managed to pass by.

A patent was applied for in July 1941 but, because of the war, it was granted to Calico Printers Association only in 1946, thereby extending its period of validity. Meanwhile ICI had received in 1943 a sample of the new polymer, which was to be known as Terylene, while Du Pont began its own research on polyethylene terephtalates, from which Dacron emerged. Patent rights were sold to Du Pont in 1946 for the United States and to ICI in 1947 for the rest of the world. Both companies invested considerable sums to develop the new fiber.

After spending four million pounds in research and 15 million pounds in setting up an industrial plant, ICI brought on stream a 5,000-tons-a-year Terylene unit in Wilton in 1955. Du Pont started producing Dacron as early as 1953. While polyester, unlike Nylon, could not be turned into molding powder, it could be formed into transparent film and, of course, into textile fibers, which could readily be blended with natural fibers such as cotton. Consumption of this remarkable fiber soared in 1962 when Goodyear began using it in its tire casings.

It became possible to develop a third family of fibers, the acrylics, when

6

$$\underset{\text{ethylene glycol}}{n\ CH_2(OH){-}CH_2(OH)} + \underset{\text{terephthalic acid}}{n\ HOCO{-}C_6H_4{-}COOH} \longrightarrow$$

$$\underset{\text{polyethylene terephthalate}}{(\ldots CH_2CH_2OCO{-}C_6H_4{-}COOCH_2{-}CH_2OCO{-}C_6H_4{-}COO \ldots)_n} + nH_2O$$

a cheap process for preparing the starting material acrylonitrile, $CH_2=CHCN$, was found and when a solvent was discovered that was capable of dissolving and spinning the resulting polymer.

Otto Bayer, with the assistance of Peter Kurtz managed in 1948 to carry out the direct synthesis of acrylonitrile from acetylene and hydrocyanic acid, HCN. In 1949, a group of researchers working for Bayer in Leverkusen, succeeded in spinning a polyacrylic fiber without a break. Bayer set up a chemical-fiber unit in 1951 in Dormagen that, in addition to Perlon, soon produced 6,000 tons a year of polyacrylonitrile fiber called Dralon. Production of this fiber started in the United States in 1949 with a unit set up by Du Pont. It was followed by a *Chemstrand* (Monsanto) unit in 1952, then by *Dow* in 1958 and *American Cyanamid* in 1959.

Acrylic fiber has the feel of wool and is suitable for making knitted garments. It is weatherproof and chemically resistant, properties that have made it popular for carpeting and a large number of industrial uses. By 1973, the world capacities of this type of fiber had reached 850,000 tons.

Finally, we must note that *Hercules Powder* was the first to produce *polypropylene* monofilaments for textiles in 1961. But it took some years before polypropylene fiber was marketed on a large scale because it had little affinity with organic dyes. Mass production became possible only when *UniRoyal*, *Union Carbide*, and *Hercules* found ways of improving their dye technology. Production of polypropylene filament in the United States then soared from 6,500 tons in 1950 to 95,000 tons in 1969. Its uses diversified from ropes and jute substitutes to carpeting and upholstery fabrics.

New Deals in Big Industry

A veritable revolution occurred in the basic chemical industry over the years extending from the end of the war to the early 1960s, owing to the great steps forward made in the refining and transforming of oil and gas constituents and in the marketing of new polymers.

Progress was particularly a result of American technology in the petrochemicals sector and to American and German breakthroughs in plastics, synthetic elastomers, and chemical fibers. In particular, the main contribution in the field of high polymers came from IG Farben and the large United States chemical companies. England doubtless deserves credit for developments in polyester fibers made during the immediate postwar years. But no British company was involved in the development of synthetic rubber, and improvements in the production and use of new plastics only came much later with the acquisition of foreign licenses.

Thanks to Natta, Italy could claim the discovery of polypropylene and ethylene-propylene rubber; but for the other polymers it, too, had to resort to German and American technologies.

As for France, its industry remained strangely remote from any significant achievement in this important sector. After pioneering in the area of cellulose derivatives and, in particular, in the first artificial fibers and in cellophane, the French industry could boast of only two original fibers, albeit with limited marketing potential: Rhône-Poulenc's vinyl chloride-based *Rhovyl*, dating back to 1948, and a castor-oil derived polyamide, *Rilsan*, for which a Péchiney subsidiary, *Société Organico*, was studying development prospects in 1951.

For all other polymers, whether synthetic elastomers or plastics other than cellulose or vinyl, France had become totally dependent on foreign sources. Yet a number of French scientists, such as the university-based George Champetier and Charles Sadron, became interested in high polymers. The former established a course in macromolecular chemistry in Paris, while the latter set up a research center on macromolecules in Strasbourg.

France's administration also spent large amounts of public money on fundamental research in the framework of the Centre National de la Recherche Scientifiques, CNRS. Such research fell short of the expectations of industry, however, for there was little communication between university people and industry leaders.

It should be pointed out that the successful commercial development of the fundamental discoveries in various countries in high polymers was precisely due to close collaboration between research institutes and industry, as demonstrated by the Natta-Montecatini relationship in polypropylene, the close ties between Ziegler in Mülheim and Hoechst and Hercules or, again, by ICI's financial involvement in the development of polyester fiber on the heels of Whinfield and Dickson's research with the Calico Printers Association. It must also be said that the background and training of France's chemical leaders had not given them a bent for the new science of macromolecules. Between the two world wars, France had seen no need to adopt a policy geared to the production of substitute synthetic polymers.

The opposite was true of Germany, where the "self-sufficiency policy" of the Nazi regime had led to the production of *Ersatz* products. What had become a policy priority had been helped along by the fundamental work of scientists like Staudinger, Mark, and Kurt Meyer on macromolecules and by the close relationship prevailing between the scientists working in Germany's research institutes and industrial leaders like those of IG Farben. Those industry leaders also possessed a solid scientific background and devoted as much as 4 percent of their turnover to research, and to the development of new products.

Germany's machine-tool industry had also contributed to chemical industry's efforts. The firm of *Eckert & Ziegler*, which belonged to IG Farben, could be regarded as a pioneer for having developed, as early as 1926, the first injection-molding machine by which thermoplastic polymers could be mass produced.

The Benevolent Circle—Intermediates and High Polymers

Supporting each other, petrochemicals and high polymers greatly contributed to the postwar development of the chemical industry. The falling cost of major organic intermediates boosted new polymer production, which further pushed down the price of finished products, because of the volume effect. Demand in its turn led to new, larger, and more efficient plants to refine gasoline and produce intermediates. The benevolent circle thus engendered in the United States during the Second World War with the GRS synthetic rubber program spread its beneficent effects the world over until the first energy crisis in 1973.

The production of *elastomers* was greatly stimulated by the ethylene which the American petrochemical industry put on the market for the preparation of styrene, by the butadiene and isoprene they supplied and which was needed for the synthesis, under economically acceptable conditions, of SBR, butyl, and sterospecific rubber as well as polychloroprene (Distillers process) through 1,4-butadiene.

Chemical fibers were likewise boosted. Originally, Du Pont's starting point for *Nylon 66* had been adipic acid derived from phenol (produced from coal) that was converted to adiponitrile and hydrogenated to produce hexamethylenediamine. Beginning in 1953, Du Pont used petrochemical butadiene to synthesize adiponitrile. Coal-based phenol was also replaced by benzene derived from oil. This benzene could be hydrogenated to produce cyclohexane and, through oxidation, adipic acid. Phenol had also served as a starting point for *Nylon 6* or Perlon. The new processes developed in Europe after the war (by SNIA-Viscosa, Bayer, and DSM) were in turn based on cyclohexane for the production of caprolactam.

There was a particular problem in the case of *polyester fiber*, for while glycol was already available via petrochemically produced ethylene oxide, the other reagent, *terephthalic acid* (*TPA*), or rather *dimethyl terephthalate* (*DMT*), had never been industrially produced. In the first units set up by ICI and by Du Pont, DMT was made by nitric acid oxidation of *p*-xylene into a TPA followed by esterification with methanol to obtain DMT. The process was improved, particularly by *Witten*, a subsidiary of Germany's *Dynamit Nobel*, and licensed out to Hercules in the U.S. (the Imhausen process was independently discovered by Standard Oil of California). *Mobil Chemical* and *Amoco*, through air oxidation of *p*-xylene, managed to obtain a TPA that was pure enough to bypass DMT in the preparation of the polymer.

The emergence of polyester fibers, added to the possibility of synthesizing phthalic anhydride through oxidation of *o*-xylene instead of from naphthalene (BASF, Von Heyden), boosted the importance of xylenes as a sector of aromatics extracted from petroleum through reforming operations. In the United States alone, DMT and TPA production had soared to a total of 840,000 tons annually by 1970; the price of DMT, which had been still 32 cents a pound in 1962, fell to 14 cents a pound that same year.

The *acrylics* family of fibers also took advantage of the falling cost of the base *acrylonitrile* monomer, which became possible through process improvement. Acrylonitrile was manufactured either under the Bayer process from acetylene and HCN, or under a process used since 1940 by American Cyanamid that consisted of ethylene oxide reacting with HCN. Chemists with *Standard Oil of Ohio*, as well as a group of researchers with *Distillers* in England, developed a process in the 1950s involving propylene and ammonia. Brought on line in the United States in 1960, it turned out to be quite inexpensive because of the low cost of the raw materials used and because it also yielded HCN as a coproduct. Within ten years, acrylonitrile production by the Sohio process increased fivefold; its price dropped from 38 cents a pound to 12 cents a pound, boosting its use both for acrylic fibers (Du Pont's Orlon, Bayer's Dralon) and for ABS resins.

This close connection between the twin progress of petrochemicals and high polymers is also to be found in plastics. Indeed, the growth of *polyolefins* was furthered by the practically unlimited availability of cheap ethylene and propylene.

Vinyl polymer chemistry also underwent profound changes when petrochemical ethylene was used instead of acetylene to prepare both *vinyl acetate*, through direct acetoxylation with a palladium catalyst (ICI, Celanese, Hoechst Bayer), and *vinyl chloride*, through thermal cracking of dichloroethane. Likewise, *acetic acid* also became a petrochemical product when it came to be manufactured through the oxidation of C_4 hydrocarbons (Celanese, Hüls, BP, Distillers).

The Growth of Agrochemicals

Chemicals had already greatly contributed to the agricultural progress before the war through growing use of fertilizers and the development of the first really efficient crop-protection products to fight fungi, weeds, and insects. In the field of fertilizers, the fundamental discovery was the synthesis of ammonia. But after the war a number of not unimportant improvements were also introduced. Superphosphates with an 18 percent phosphorus content were replaced by triple superphosphate, and by *ammonium phosphate*, both containing three times as much phosphorus. The drawback of *ammonium sulfate* was that it acidified the soil. Substitutes with a higher nitrogen content, such as *ammonium nitrate*, *ammonium phosphate*, or *urea* with a 45 percent nitrogen content were used instead. *Potassium salts* were also sold in more concentrated form. The physical aspect of fertilizers was improved as granulated forms were substituted for the powdered kind that was hygroscopic and lumped during storage.

World consumption of fertilizers increased by factors of 8.2 between 1939 and 1967 for nitrogen, 4.5 for phosphorus, and 5.1 for potassium, soaring respectively to 21.3, 16.2, and 14.4 million tons. The chemical industry did its best to keep up with this spectacular rise, setting up, in particular, countless units capable of producing 1,000 tons a day of ammonia.

Together with the boosting of both the quantity and quality of fertilizers, fundamental discoveries were being made by organic chemists to protect crops against various kinds of harmful attacks. *DDT*, which had played a major role against typhus during the war, began to be used in agriculture in 1940 in Europe against the potato beetle and in 1945 in California to protect vineyards.

Another chlorinated organic insecticide, *BHC* (benzene hexachloride, i.e., hexachlorocyclohexane), had been studied in England during the early 1940s by Thomas and Stock working with ICI, in France by Dupire and Raucost, and in Spain by Ozmiz. Michael Faraday had described its synthesis in 1825, but chlorine reacting with benzene in ultraviolet light produced several BHC stereoisomers and only the *gamma*-isomer, known as *Lindane*, had the advantage of being odorless and turned out to be an active insecticide with properties resembling those of DDT.

In the framework of research on *organophosphorus compounds*, carried out in the thirties in the IG Farben laboratories, G. Schrader had identified in 1938, the insecticidal properties of tetraethyl pyrophosphate, especially against the green-fly. Subsequently research switched to nerve gases, but the Allies disclosed the results, and in 1947 American Cyanamid developed *Parathion*, followed three years later by *Malathion*,[7] which was less toxic to hu-

[7]

$$4CH_3OH + P_2S_5 + 2\left[\begin{array}{l} CH\text{-}CO_2CH_2CH_3 \\ \| \\ CH\text{-}CO_2CH_2CH_3 \end{array}\right]$$

methanol — phosphorus pentasulfide — diethyl maleate

$$\longrightarrow 2\left[\begin{array}{l} (CH_3O)_2P(=S)\text{—}S\text{—}CH\text{-}CO_2CH_2CH_3 \\ \qquad\qquad\qquad\qquad\;\; | \\ \qquad\qquad\qquad\quad\;\; CH_2\,CO_2CH_2CH_3 \end{array}\right] + H_2S$$

Malathion

mans. While Parathion and Malathion operate by *contact* with the insect, other organophosphorus insecticides, such as OMPA, which Schrader had prepared, are systemic; that is, they operate by infiltrating the plant's vascular system, killing the insects that suck the plant juices.

Although they are more expensive to produce than chlorinated products, organophosphorus insecticides, being biodegradable, are therefore useful in treatment of crops just before they are harvested, as they leave no toxic residues on the plants. Nearly a quarter of the insecticides currently used worldwide belong to the organophosphorus type.

Before being abandoned because they accumulate toxic residues in the soil, organochlorinated insecticides such as DDT and BHC were, for a while, used in conjunction with a chlorinated camphene called *Toxaphene*. It was developed by Hercules and proved very efficient against cotton pests. Other chlorinated cyclodienes were discoverd by Julius Hyman working with Velsicol Chemical Company in 1944 and were produced by a Diels-Alder synthesis.

Put on the market by Shell Chemicals in the 1950s under the names of *Aldrin* and *Dieldrin*, they proved very effective against a large number of insects. But their career was cut short by the harm they caused through their persistence and their toxicity to warm-blooded animals. They were partly replaced by the important family of *carbamates*, which had already been studied by J. R. Geigy in Basel at the end of the 1940s. The first commercial success in this series was achieved by Union Carbide in 1956 with *Sevin* or *Carbaryl*.[8] It is still widely used today because of its wide range of applications and its low toxicity.

The research on plant-growth regulators that was started in Holland between 1926 and 1934 and that had led to isolating and identifying β-indoleacetic acid, had been followed by the discovery of the property of α-naphthyl acetic acid (NAA) in 1935 at the Boyce Thompson Institute in the United States. Working in ICI's Jealott's Hill experimental station in 1940, W. G. Templeman observed that NAA had a *selective effect* on vegetation, destroying the weeds but not the crops it was intended to protect.

Templeman continued his tests with derivatives of similar compounds

[8] O
‖
O—C—$NHCH_3$

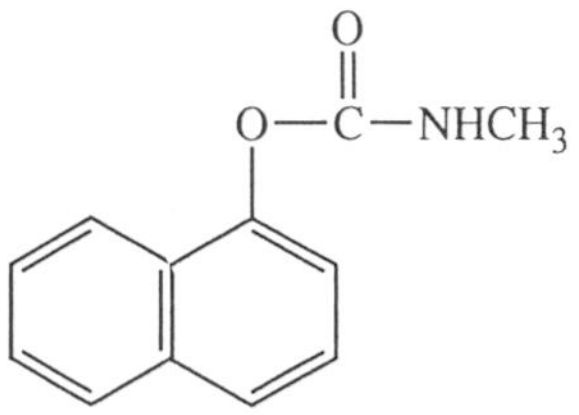

Carbaryl

provided by ICI's dye division. In 1941 he identified the sodium salt of 4-chloro-2-methylphenoxyacetic acid, or *MCPA*, as well as the salts and esters of 2,4-dichlorophenoxyacetic or *2,4-D* as selective herbicides, which were practically harmless to warm-blooded animals, were effective in small concentrations, and could be cheaply manufactured. They are synthesized from phenol, in the case of 2,4-D, or from *ortho*-cresol, in the case of MCPA, which is chlorinated and made to react with monochloracetic acid.

In England, MCPA was preferred to 2,4-D because *o*-cresol, derived from tar distillation, was available. In the United States, 2,4-D was utilized more because phenol was available in large quantities by synthesis.

American Chemical Paint (Amchem), the first company in the United States to produce 2,4-D, put 2,4-5-T (2,4,5-trichlorophenoxyacetic acid) on the market in 1948. But its use as a defoliant during the Vietnam war stirred up such controversies that it was withdrawn from the market. Other active herbicides belong to the family of *carbamates*. *IPC* (isopropylcarbamate) was developed by ICI as early as 1945, while the chlorinated derivative *chloro IPC* was marketed in 1954 by Columbia Southern, a subsidiary of PPG. In 1948, Stauffer brought out its thiocarbamates. For its part, Du Pont had become interested in *substituted ureas* as herbicides for the protection of cotton crops and began marketing *monuron* and *diuron* in the early fifties. According to the concentrations used, certain members of this group of herbicides have either a total or a selective effect.

This is also the case with *s-Triazine* derivatives, used for corn, which Geigy pioneered in 1956 with *Simazine* and particularly, two years later, with *Atrazine*. *Atrazine* was to become the most widely used herbicide, as it can easily be produced from cyanuryl chloride and is highly effective.

The development of herbicides, beginning around 1850 with the successive use of sodium chloride, ferrous sulfate, sodium arsenate, followed by sodium chlorate, took a decisive turn when selective agents such as 2,4-D and MCPA were discovered as a result of research carried out in both America and England in the early 1940s. In the United States alone, over 100,000 tons of herbicides were being used in 1966 (in terms of active principles), accounting for a third of all crop protection products.

In the fight against fungi a major breakthrough was achieved, as has been seen, with the discovery in 1934 by Du Pont's Tisdale and Williams of the *fungicidal properties* of *dithiocarbamates*, originally developed as rubber accelerators. Although sulfur and copper-based mineral products were still used because they were cheap, new organic fungicides were coming onto the market. Hester discovered a further member of the dithiocarbamates group in 1941 while working with Rohm & Haas. *Dithane D-14*, as it became known, was a sodium salt of ethylene bis-dithiocarbamate and was successfully used in Texas in 1944 on potato crops. The addition of zinc or manganese in the molecule improved the product's efficiency. A new class of fungicides, products such as *Captan* that were easily synthesized through the Diels-Alder reaction and that proved remarkably effective in treating certain fruit-tree and ornamental-plant diseases, were developed in the early 1950s by researchers working with Chevron Chemical (Standard Oil of California) and Stauffer Chemical.

By 1960, thanks to research carried out by European and especially American organic chemists, farmers had access to a whole range of crop protection products.

Medical Breakthroughs

By 1950, chemotherapists and biochemists could look back with satisfaction on the progress made in the fight against causes and symptoms of disease, progress largely due to dedicated research but often to chance as well.

Aspirin, which Felix Hoffmann developed and which Bayer put on the market in 1899, was still proving its efficacy as a painkiller, an antipyretic, and an anti-inflammatory agent to the point that thousands of tons were being produced annually throughout the world.

Barbiturates were commonly being used as sedatives. *Sulfonamides*, in the wake of Gerhard Domagk's work following the trail blazed by Paul Ehrlich for Salvarsan, and of Ernest F. A. Fourneau[9] of Institut Pasteur, had done the greatest service as antibacterials before being supplanted by penicillin.

Nutritional diseases and disorders of the internal secretory glands were successfully dealt with by isolation, identification, and industrial preparation of *vitamins* and *hormones*. But, more importantly, the era of antibiotics was ushered in through the revolutionary discovery of *penicillin* (credited to Alexander Fleming in 1929), and its development, from 1949 onwards, by Howard Walter Florey and Ernst Boris Chain. Through the joint efforts of American researchers and engineers, the amount of penicillin produced in 1945 corresponded to five million doses. Researchers then focused their attention on the total synthesis of penicillin, which was achieved in 1957 by John Sheehan at MIT, and on trying to discover new antibiotics.

Thus emerged, notably with the British-based Beecham's Ampicillin (1961), *semi-synthetic penicillins*, which were absorbed more easily than penicillin C and were effective against a great number of bacteria.[10]

Meanwhile a group of microbiologists from Rutgers University, with Selman Waksman at their head, had managed in 1945 to isolate from the soil a microorganism that produced a new antibiotic, *streptomycin*, effective against

[9] Fourneau was a pupil of Ehrlich; his group included chemist Jacques Trefouel, pharmacologist R. Nitti and physiologist Daniel Bovet.

[10] Penicillin V, for instance, can be given by mouth, as it is not destroyed by digestive juices. It is obtained by adding phenoxyacetic acid to the penicillin culture. For other semi-synthetics, the benzyl radical was introduced.

tuberculosis. In 1947, *chloramphenicol*, discovered in other soil bacteria by researchers from Yale, the University of Illinois, and the Parke Davis Company, was successful in fighting typhoid, typhus, and kidney diseases. Then in 1948, Lederle's *chlortetracyclin* and Pfizer's *oxytetracyclin* appeared on the scene. Five years later, the American Robert B. Woodward, who received the Nobel Prize in Chemistry in 1965 for his work on the structure of complex molecules found in nature, was successful in reconstituting the structure of tetracyclin.

Although they were valuable in fighting a number of infectious diseases, the antibiotics discovered in this fruitful period were no more all-encompassing than sulfonamides before them. Some microbes, which they had first successfully destroyed, even became resistant to their action. On this account, organic chemists felt the need to push further into synthetic drug research.

Germany's Gerhard Domagk, for one, famous for his discovery of Prontosil, had drawn attention in 1946 to the effectiveness of *p*-acetamidobenzaldehyde thiosemicarbazone on tuberculosis and described in 1952 the inhibiting effect of *isoniazide* on the same infectious disease. Squibb and Hoffmann-La Roche researchers also observed the same effect at about the same time. The combined action of streptomycin, isoniazide, and *PAS*, or *p*-aminosalicylic acid, was to remove tuberculosis from the list of lethal diseases.

Chemotherapy made other contributions to medical science. Even before the tuberculosis-inhibiting agents were to make sanatoria obsolete, the discovery of *psychotropes* had reduced the number of inmates in mental hospitals. Rhône-Poulenc's subsidiary, Société Specia, in the early 1950s was studying the possible use of phenothiazine derivatives as anaesthetics in surgery. In 1952, Deniker who worked in Délay's department in the Sainte-Anne mental hospital in Paris, observed that one of these derivatives, *chlorpromazine*, had a soothing effect on restless patients and wiped away delirious ideas. What turned out to be the first *neuroleptic* agent belonged to a new group of drugs that were to help patients suffering from hallucinations and delirium to return to live in their normal environment.

Through research carried out by CIBA, *reserpine*, an alkaloid extracted from *Rauwolfia Serpentina*, was recognized in 1953 as a psychotrope in the treatment of schizophrenia.

Besides neuroleptics, there were a number of derivatives structurally close to phenothiazine, like *Imipramine*, put on the market by Geigy in 1958, that turned out to be effective against depressive melancholia, another mental illness. They were tricyclic antidepressants. At about the same time, Nathan Kline of New York observed that consumption patients became slightly euphoric when treated with Hoffmann-La Roche's isoniazide. Accordingly, this antitubercular agent was also listed as an *antidepressant*. For manic-depressive patients who could not be treated with neuroleptics or antidepressants, *lithium* proved to be a major mood regulator. In milder cases of simple anxiety, patients were treated with *tranquilizers*.[11] The first one produced was propanediol carbamate or *meprobamate*, developed by the United States-based Wallace Laboratories in 1951, three years before Hoffmann-La Roche brought out *benzodiazepines*.

A wide range of chemical derivatives capable of ameliorating various brain disorders were thus developed making it less necessary to hospitalize people suffering from mental illness or to resort to electric shock treatment.

Allergy disorders such as hay fever, skin trouble, and others have been studied in France since 1939 and were attributed to histamine discharges. Two women chemists, Staub and Bovet, working with Fourneau, had discovered that year the antihistamine properties of a phenoxyethylamine derivative. Following research on phenothiazine derivatives, Rhône-Poulenc brought out in 1944 a powerful antihistamine, promethazine or "Phenergan," a type of product that induces drowsiness.

The hypoglycemic properties of *sulfonyl ureas* were discovered in 1942 by two other French scientists, the physiologists Janbon and Laubatières from Montpellier. They observed that sulfonamides belonging to the group of sulfathiodiazoles caused the glucose rate in the blood to fall. Hoechst in 1956 brought out *Tolbutamide*, the first sulfonylurea capable of treating certain forms of diabetes and one which could be administered orally, a distinct advantage over insulin. Another hypoglycemic family was that of *biguanides*, which offer the same advantage. In 1958, U.S. Vitamin Company started marketing phenethylbiguanide.

A first step in the production of *analgesics*, which, unlike morphine, would not be habit-forming, was carried out by Eli Lilly researchers with the synthesis, in 1953, of *dextro-A-propoxyphene*.

During this period, particularly intense research centered on cardiovascular diseases because of their high mortality. Chemists and clinicians sought for the active principles likely to cure them. Hypertension was treated successfully with CIBA's *reserpine*. The U.S.-based Merck, seeking for drugs to lower blood pressure, described in 1954 the effect of *methyldopa*. Shortly before the war, American and Argentine researchers had independently discovered a biological peptide, *angiotensine*, which raises blood pressure. In 1957, researchers at the Cleveland Clinic in the United States and at CIBA in Switzerland managed to synthesize this type of peptide, which was subsequently used as a vasoconstrictor in certain situations (hemorrhage, anaesthetics, hemodialysis).

The fight against *heart arrhythimia* had started in 1948 when German researchers observed that *quinidine*, like quinine extracted from the bark of the quinquina tree, was effective for this condition. But it was not until 1951

[11] The word tranquilizer was suggested by Delay.

that Squibb developed *procaineamide*, which is less toxic. Since then, a local anaesthetic, *xylocaine*, developed by Astra in Sweden, has also proved to be effective.

The administration of diuretics to cure edemas was further boosted with the discovery of *organomercury derivatives* during the 1920s, principally through work done by the Chinoin Company in Hungary. Products more powerful and less toxic that could also be taken orally were found during the 1950s in the *thiazide* class by Merck in the United States and by CIBA.

The important category of steroid *hormones* had been the subject of fundamental research since 1926 with particular credit going to the American Edward A. Doisy and Germany's Adolf Butenandt. In 1936, Tadeus Reichstein isolated cortisone from the cortex of adrenal glands using the Girard T reagent. *Cortisone* was successfully synthesized in 1946 through the joint efforts of the Mayo Clinic and Merck. It was thought that the steroid would be useful for treating Addison's disease, but in 1949 America's Philip S. Hench and Edward C. Kendall of the Mayo Clinic discovered its anti-inflammatory properties.

Since cortisone was found to be so effective in the case of rheumatoid arthritis, efforts were made to change the hormone's structure in order to improve its properties while reducing its side effects. Attempts were made to produce it from more easily available raw materials than the bovine bile originally used by Merck.

Squibb (1953) and Schering (1954) laboratories produced cortisone analogues that were more active and better-tolerated. In addition, research carried out in the early 1940s at Pennsylvania State College had shown that steroids could be produced from the sapogenins extracted from the roots of certain plants. Syntex Laboratory focused on *diosgenin* from Mexico's Barbasco plant as a source of raw material to prepare steroid hormones through hemisynthesis. Upjohn in the United States managed to simplify this preparation, while in France, Roussel-Uclaf researchers under Léon Velluz succeeded in carrying out, in 1960, the full synthesis of the *estrone* and *nortestosterone* series of hormones. They had to overcome considerable difficulties because of the complex spatial configurations of the steroid nucleus.

It had long been known that the joint action of the natural hormones, *estradiol* and *progesterone*, was capable of blocking ovulation in women, but these could not be taken orally. Legislation in France at the time prohibited French companies from becoming involved in the *contraceptives* market, so that Roussel only came to it much later.

On the other hand, in the United States, Gregory Pincus, a consultant for Searle, had been able, as early as 1956, to test a contraceptive pill on 265 women in Puerto Rico. In 1960, Searle, soon followed by Syntex, brought out a contraceptive pill containing *progestine*, a hormone similar to progesterone but able to be taken orally. Twenty years later, 55 million women throughout the world would be using this contraceptive method.

The period examined was particularly fruitful for discoveries, at times due to chance, at others the result of analogical reasoning confirmed by experience. During that time, chemistry set off a veritable revolution in all areas of medicine and gave tremendous impetus to the pharmaceutical industry as researchers working in the different laboratories closely collaborated with those in clinics and research institutes to develop and market new drugs.

Prosperity of the Chemical Industry Precedes the First Oil Crisis (1973)

The chemical industries of the free world made the most of the favorable circumstances that boosted their development in the postwar years up to the first energy crisis in 1973. The prosperity of the petrochemical industry and of its downstream clients was especially notable, as oil and natural gas were in plentiful supply and provided the obvious source of cheap feedstocks for base chemicals. A further boost was provided to petrochemicals by the growing needs for chemical intermediates owing to the emergence on consumer markets of the new high polymers, whether plastics, elastomers, or synthetic fibers.

The agricultural explosion spurred demand for fertilizers and crop protection products, which also contributed to the chemical industry's expansion. The spectacular growth of the automobile industry, a big consumer of paints, tires, and plastics, also contributed to the progress of the chemical industry. Fine chemicals were likewise carried along by the new needs of the pharmaceutical industry and of specialty producers.

The economic environment was itself undergoing great changes. *Markets* were *opening up*. After the signing of the Treaty of Rome in 1957, the European Economic Community (EEC) expanded to encompass Britain, Denmark, and Ireland. In 1973 the nine members of the EEC formed a single market of 233 million consumers.

Further negotiations, starting in 1962, called "the Kennedy Round," ended five years later with a general lowering of tariffs on industrial products within the nations that were cosignatories of the General Agreement on Tariffs and Trade (GATT). Jointly with freer trade, money was circulating more freely among countries.

This opening up of frontiers, itself a means of generating prosperity, changed the outlook of chemical leaders the world over. For Europe, foreign sales had traditionally accounted for a major share of the Gross National Product. The tendency, particularly for the Germans, still steeped in the former IG Farben's international framework, was to take advantage of lower tariffs to gain market shares by exporting more. In the United States, industry leaders responded to the prospects offered through more open markets and the general increase in purchasing power by setting up production units overseas. This was the starting point of *multinational* chemical groups.

A number of American companies, such as Eastman Kodak and Standard Oil of New Jersey, had already set up production units abroad at the beginning of the century, following a trail blazed by the Anglo-Dutch Royal Dutch Shell and Unilever multinationals. But they were exceptions, and their organizations did not relate to chemicals directly.

With units set up abroad and with boosted exports, sales of the chemical multinationals rose, in 1970, to $40 billion, 25 percent of the total worldwide sales of chemicals of $159 billion. The subsidiaries of these multinationals alone accounted for $22 billion of these sales. At first, the United States set the pace. After focusing its investments very naturally on Canada, Brazil, and Mexico, the United States chemical industry then turned to Europe, where, despite brisk local competition, prospects were promising because of the high degree of technical training of managerial staffs and the high purchasing power of the populations. Later, the flow of investments was reversed as European industry leaders set up units in ever-growing numbers in the United States.

Meanwhile, Japan, which, under the aegis of the Ministry of International Trade and Industry (MITI), had geared the reconstruction of its chemical industry to the acquisition of foreign technology within a strictly national framework, was becoming less reticent about associating with American and European partners.

There still remained areas of the world that were closed to free-trade principles. Their industries were lagging behind because they lacked the impetus supplied by the demands of a market economy.

Thus chemical industry in Latin America, while growing at a fast pace from its low starting level, based its development on an artificial policy consisting of producing substitutes for imported products. The policy disregarded the cost element, as tariff and other barriers kept out competing products and removed the need to take economic realities into account.

The countries of the Soviet bloc were even less aware of the real cost of production, living as they did in self-contained regimes that disregarded those productivity improvements that were necessarily a matter of daily concern in the chemical industry of Western countries.

Germany's Chemical Industry Rises from Its Ashes

For IG Farben the war brought about significant changes, notably the loss of all its plants located in Eastern Germany and in Poland's former German territories. Those plants were destroyed or became state-run companies. Subsidiaries operating in allied or neutral countries such as IG Chemie in Basel were sold off. Plants located in Federal Germany's territory suffered heavy war damage. The Allies took over plants in their occupation zones, and a central commission was established entrusted with the task of breaking up the group's cartel arrangements. Further, IG Farben's patents and trademarks were seized. By 1953, however, damaged units had been rebuilt, Allied controls had been lifted, the cartel broken up, and Germany's chemical industry could take off from an economic base that had been bolstered by Ludwig Erhard's monetary reform. Within this new structure, three large autonomous entities, each with its own management board and board of trustees, were set up out of the remains of the old IG. They were *BASF* in Ludwigshafen, *Bayer* in Leverkusen, and *Hoechst* in Frankfurt.

Other companies, such as *Cassella* and *Hüls*, were left with a measure of independence pending future reorganization, while still others, such as *Deutsche Solvay Werke*, *Degussa*, *Dynamit Nobel*, *Merck*, *Boehringer*, *Schering*, *Rütgers*, and *Henkel* continued their separate development as in the past.

BASF, originally set up by F. Engelhorn in 1865, was the German company most involved in base chemicals and in major organic intermediates. It was left with its Ludwigshafen and Oppau plants, which together formed the largest chemical complex in Europe. BASF's coal supplies came from its Auguste-Victoria mines in Marl-Hüls in the Ruhr. Lime and limestone for its fertilizer and carbide production were provided by its Kalkwerk Steeden subsidiary. Copper pyrites for its sulfuric acid production came from Duisburger Kupferhütte, in which BASF had a 30 percent stake.

BASF subsequently took a one-third share of Oxogesellschaft in Oberhausen which produced oxo derivatives from olefins and in which Ruhr Chemie and Henkel also had a one-third share each. It was also involved, with a 46 percent share, together with Ruhr Chemie and Goldschmidt, in the Oberhausen-based Chemische Fabrik Holten, which manufactured ethylene oxide and glycols. To round off its supplies in base chemicals, BASF set up a joint venture with Deutsche Shell called the Rheinische Olefin Werke (ROW), which brought on line in 1955 a production unit of oil-based olefins in Wesseling near Cologne, the first of its kind to be built in West Germany.

Having established its upstream supplies, BASF went ahead to develop the technologies it had perfected in the years before the war. Its expertise in high pressure catalytic synthesis had led BASF to develop a urea manufacturing process from CO_2 and ammonia and to produce, in 1942, a low-density polyethylene called *lupolen*. The experience acquired in the polymerization of styrene very naturally paved the way to expanded polystyrene, Styropor, marketed in 1951.

Likewise, BASF's knowhow in macromolecular chemistry, which had led in the 1930s to the development of acrylic dispersions, to the ''Kaurit'' formol-urea glues, to PVC and polyisobutylene, proved useful when it came to the point of developing the applications of the Nylon analog polyamide-6 or Perlon, discovered in 1940.

BASF was also able to enlarge, on the basis of work done by the great chemist Walter Reppe in 1928, the range of products obtained from acetylene and isobutylene reacting under pressure, with various aldehyde or alcohol functions. In this way major intermediates, like 1,4-butanediol, could be prepared. In 1963 an original industrial synthesis of vitamin A was successfully carried out.

Combining a wide range of integrated upstream inorganic and organic base chemicals, able to supply the nitrate fertilizer and plastics sector with an extensive and also integrated downstream synthetic intermediates sector, BASF continued to supply its traditional process chemicals, ranging from dyes to protective coatings, and leather and textile intermediates. The company diversified, at the same time, in a number of consumer products such as paints, inks, and the highly profitable magnetic tapes, in which it had become involved as early as 1935. By 1962, the group was posting a consolidated turnover of some $860 million.

This figure was to soar threefold by 1969 with the economic boom and after BASF had set up a base in Antwerp in 1964, developed cooperative ventures with Dow, and bought up Wyandotte Chemicals in the United States in 1967.

IG Farben's policy of specializing its various production units led *Bayer* to focus on very high added-value products, unlike BASF, which was more involved in basic chemicals.

F. Bayer had set up his company in Leverkusen in 1863, a name that recalls the C. Leverkus & Sons Company absorbed in 1891. But this also meant that *Bayer* found itself in the British zone of control in 1945. Under the cartel breakup arrangements, besides the main Leverkusen plant, Bayer was assigned in 1953 the original units of Elberfeld-Barmen, those of Dormagen, and the important Weiler-ter-Meer unit in Uerdingen. Stealing a march on BASF, Bayer engaged a large number of technicians who had worked with *AGFA*, whose plant was now part of the Soviet zone, and managed to put back on stream in Uerdingen and Dormagen the photographic-product and film production units, which used to be in Berlin, and the cameras, which were produced in Munich.

After taking over Cassella's acrylic fiber production, the group turned Dormagen into its main production center for chemical fibers (Bamberg silk, rayon acetate, Perlon, and Dralon). Bayer also boosted production of specialties such as rubber auxiliaries and crop protection chemicals, particularly organophosphorus products, for which there was booming demand at the time, and gradually rebuilt its dyestuffs business as well as the textile and leather auxiliaries market lost during the war. It also restored its traditional range of synthetic organic intermediates.

In pharmaceuticals, Bayer never did get back the Bayer trademark for Aspirin, which was first lost in 1917 and which had been purchased by Sterling Drug. But its research teams went to work on new discoveries and in 1952, under the impulse of the great Gerhard Domagk, produced new antitubercular agents. Bayer, which had been a synthetic rubber producer after World War I, also developed new synthetic elastomers (Perbunan). Otto Bayer's discovery of polyurethanes was the starting point of Bayer's pioneer role in the development of new high polymers both in Europe and in the United States through its subsidiary, Mobay.

Involved in the most varied sectors of fine organic chemicals, being a supplier of a wide range of artificial and synthetic fibers, utilizing to full advantage the knowhow acquired by its researchers in macromolecular chemistry to produce synthetic elastomers and new plastics such as polyurethanes and polycarbonates, Bayer did not for all that neglect the business of inorganic chemicals. Thus, while producing basic chemicals such as sulfuric acid needed for its downstream activities, Bayer manufactured inorganic pigments, more specifically titanium dioxide, chromates, enamels, and specialty products such as hydrazine, which served to prepare the azo derivatives used as blowing agents and as polymerization initiators in plastics and rubber.

Hoechst was in the American occupation zone and, like the two other big German chemical companies, resumed an autonomous existence when controls were lifted in 1953. It was assigned the Hoechst unit near Frankfurt, the neighboring units in Griesheim and Offenbach, and those in Gersthofen and Bobingen near Augsburg, as well as the former Anorgana Gendorf unit in Upper Bavaria. In addition, Hoechst gained ownership of the Knapsack-Griesheim subsidiaries in Knapsack near Cologne, Kalle AG in Wiesbaden-Beibrich, and of Behring Werke in Marburg. The group also held a 50 percent stake in the Munich-based Wacker Chemie, a 33.3 percent interest in Ruhr Chemie in Oberhausen, and a 30 percent stake in Süddeutsche Kalkstickstoff in Trostberg and in Duisburger Kupferhütte.

Under a group reshuffle which took place in 1969, Karl Winnacker, who had managed Hoechst with a firm hand since 1952, succeeded in taking over Cassella, which in the interval had become a majority holder in Riedel de Haen. The restructuring made Hoechst one of the most diversified of Germany's chemical groups. It could rely on a great pharmaceuticals tradition marked by the discovery of Pyramidon (1893) and of Novocaine (1905), by the synthesis of adrenalin (1904), by Koch's research on the tuberculosis bacillus, by the research of his colleague Emil von Behring on serums and vaccines, by the work of Paul Ehrlich, who discovered the first really effective treatment for syphilis, and by the commercial production of insulin in 1923. In 1939, Hoechst researchers synthesized Dolantine, the first morphine substitute. Then penicillin was produced in 1950, and in 1956 came the joint breakthrough achieved with Boehringer in the treatment of diabetes through sulfonylureas.

Hoechst also remained faithful to dye chemicals, through which it had achieved, between 1912 and 1924, its most successful products, Naphthol-AS and Indanthrene. A great number of dyes were put on the market by the company to meet the new requirements posed by synthetic fibers. At the same time, Hoechst became one of the major world producers of pigments. Either directly or through some of its subsidiaries, Hoechst was also strongly involved in heavy inorganic chemicals with its large production of chlorine, electrolytic soda, phosphorus and its derivatives, mineral acids, and nitrate fertilizers.

Paul Duden, who had been Hoechst's technical manager until 1932, had set up a significant acetylene-chemicals activity that produced acetone and acetic acid and was at the root of the association with Knapsack. Based on research carried out by F. Klatte on vinyl acetate in 1912, Hoechst had started producing polyvinyl acetate (the future Mowilith) and polyvinyl alcohol in 1928 in parallel with Wacker. But Hoechst lacked tradition in other polymers. It was on the lookout for a chance to get into plastics when an opportunity arose after the war to collaborate with Karl Ziegler in the development of low-pressure polyethylene, then to join with the U.S.-based Celanese in 1961 to produce polyacetal resins (Hostaform) in Germany.

Its incursion into polystyrene was less successful and proved shortlived. Hoechst, however, had been lucky to inherit from the former IG the Bobingen unit which had been producing polyamide-6 shortfibers since 1946 under the management of Schlack. This was the starting point of Hoechst's involvement in synthetic fibers, which became significant with the production of Perlon thread, then of polyesters (Trevira) in 1955. The polyacrylonitrile fiber which Cassella had developed had been assigned to Bayer.

Hoechst was also involved in films and heliography (Ozalid process) through Kalle; in crop protection products; in waxes, which were produced in Gersthofen; in fluorinated refrigerants; and in industrial gases through Griesheim. In 1964, it acquired Chemie Werke Albert in Wiesbaden and in 1968 acquired a 40 percent share in France's Roussel-Uclaf pharmaceuticals laboratory. Hoechst can, therefore, be said to be one of the world's most diversified and prosperous chemical groups.

The three major German chemical groups pursued parallel destinies. Though only offshoots of the old IG Farben, they each very soon outstripped their progenitor in size. Their management boards were headed by eminent chemists—Professors Wurster at BASF, Haberland at Bayer, and Winnacker at Hoechst—who knew one another and held one another in high esteem. Their heads were themselves assisted by extensive teams of other professional chemists. While striving to develop their hundred-year-old companies along specific lines, these leaders were careful not to compete too relentlessly with each other. They had similar activities in dyes and in textile and leather auxiliaries, in which each group had built up a long tradition of research and international trade; and their joint shares in Hüls and Ruhr Chemie had also survived.

Within the old IG Farben, there had been specific areas of cooperation with other companies, traces of which could still be found after the breakup of the cartel. These cooperative endeavors were further emphasized under the new leaders of Germany's "big three." Thus, for its supplies of oil-based feedstock, Hoechst kept to its agreement with outside refining companies, even though it had developed its own pyrolysis processes. Bayer, on the other hand, had set up a joint company with BP called Erdoel Chemie, while BASF was integrating upstream even further, both by associating with Shell in Rheinische Olefin Werke (ROW) and through purchase of the oil and natural-gas exploration/production company Wintershall.

Far behind the big three, the *Chemische Werke Hüls* nonetheless ranked fourth among West Germany's chemical companies. *Hüls* had been set up in Marl in 1938 as a joint venture of IG Farben (76 percent) and Hibernia (24 percent) to produce Buna rubber from the region's coke gases and natural gas. The company had first produced the acetylene needed to manufacture Buna rubber by an electric arc process. After the war, with its own electrolysis units, it became a major supplier of petrochemical base materials (ethylene oxide, glycols, ethanolamines, acetic acid and acetate solvents, dodecylbenzene, chlorinated hydrocarbons, vinyl chloride) and of high polymers (PVC, polystyrene, low-pressure polyethylene).

In 1955, Hüls had taken a 50 percent share in the establishment of Buna Werke Hüls, joining with Bayer, BASF, and Hoechst (making Buna cold rubber, and a *cis*-1,4-polybutadiene-ethylene-propylene terpolymer). At the same time, it extended its range of polymers to encompass polyester resins and polyamide-12.

In 1970, Hoechst sold, half to Bayer and half to VEBA, its majority stake in Chemie Verwaltung AG, the holding company that had owned 50 percent of Hüls since the cartel breakup. The company, which subsequently became fully controlled by Veba, employed 15,000 people at the time in Marl; and over 90 percent of the hydrocarbons it processed were oil-based.

Besides *Deutsche Solvay Werke*, which recovered its full independence upon the breakup of the IG cartel, there were other German chemical firms that underwent spectacular development. *Degussa*, founded in 1873, had switched, under the able management of the Roessler family, from the processing of precious metals to cyanides and ferrocyanides, to sodium and electrolysis, to peroxides and hydrogen peroxide, and finally to charcoal. It had plants all over West Germany, with the main ones in Frankfurt and Berlin. In addition, there were five carbonization units of the former HIAG. Since 1953, Degussa had been manufacturing special Aerosil silica in Rheinfelden.

Its takeover of Chemie Werk Homburg had involved it in the production of methionine and pharmaceutical derivatives. It was also involved, through its Marquart and Grunau subsidiaries, in mineral pigments and in auxiliaries for leather and textiles respectively. It had established a foothold in the production of lampblack through its 50 percent stake in Deutsche Gasruss Werke.

In 1968, Degussa set up a unit in Antwerp to produce HCN, cyanuryl chloride, and a little later, Aerosil. The following year, it joined with BASF within Ultraform GmbH to produce polyacetal resins in Ludwigshafen. Well established in its specialties, Degussa had become a prosperous company through a cleverly conducted policy of diversification.

Henkel, which was still owned by its founding family, was also a well-managed and important business. It had no plans to diversity outside of its traditional areas, however.

Together with the Feldmühle paper and cardboard business, *Dynamit Nobel* had come, in 1959, under the control of the former steel magnate Friedrich Flick, who had been sentenced in Nuremberg for collusion with the Nazis. The company's initial business had been explosives which were produced in its Troisdorf plant. After the war, Dynamit Nobel developed a number of profitable chemical products. On the Witten site it produced dimethyl terephthalate, becoming the leading supplier to Europe's polyester fiber manufacturers. It also produced caustic potash, potassium salts, and vinyl chloride—all products linked to its electrolysis activity. It also became involved in silicon and titanium chemicals, and in synthetic intermediates through malonates and orthoformates. It also produced a range of common derivatives such as fatty acid esters, PVC, and semifinished products from different plastics.

The Darmstadt-based *Röhm & Haas* company, born of the association set up in 1907 by Otto Röhm and Otto Haas to produce leather enzymes industrially, suffered badly in both world wars. The company had gained new energy in the 1930s with the success of Plexiglas, which had also boosted the fortunes of its United States sister-company, set up in Philadelphia in 1909. The links between the two companies had been severed during World War I, then re-established following the Armistice. Finally after World War II, the Haas family of Philadelphia sold its shares in the German company, creating two entirely independent entities bearing the same name. To avoid any confusion, the Darmstadt company, in which Otto Röhm's family still held a majority share, called itself *Röhm GmbH* as of January 1, 1971. Röhm GmbH remained faithful to the two production lines that had made its fortune and that had been extensively developed. These were enzymes for leather, food, and pharmaceuticals, and methacrylate chemicals for various uses (paper, textiles, plastics, additives, and lubricants).

The *Schering* company was more diversified than Röhm and also made a brilliant recovery after losing, in the wake of the war, all its assets and patents located outside West Germany and West Berlin. By the end of the 1950s,

Schering had put its German units back into shape and added, with its Bergkamen industrial products, a fourth pillar to its edifice of pharmaceuticals, agrochemicals and electroplating. Schering had already prepared in 1938 an ethinyl estradiol estrogen which was prescribed as a hormone regulator for women. Used in the formulation of contraceptive pills in 1961, this estrogen was to become one of the company's most successful products. Even before acquiring Fisons Boots, UK, in 1983, Schering held a not insignificant place in crop protection products. In particular, it had developed a range of beet herbicides.

Schering's most recent new venture has been in industrial products, based on the chemistry of fatty acids and on catalysts of the aluminum alkyl type. Although Schering's traditional electroplating business had moved with the times towards new electrolytic treatments for metals and plastics, this henceforward accounted for only 10 percent of the group's consolidated turnover.

Germany's chemical industry also remained alive in other areas where it had been successful in prewar years. While *Wacker*'s acetylene activity, so brilliantly developed, had declined, a number of traditional sectors had managed to survive. In the areas of tar derivatives, *Raschig* had lost the momentum which its founder's inventive genius had supplied. But *Rütgerswerke* was given a boost when the industry underwent restructuring and, through its subsidiaries Weyl and Thiokol, became involved in specialties. *Union Rheinische Braunkohlen (URBK)* produced synthetic cresols and xylenols from 1965 on in Wesseling, using an original methylation process of phenol, a product obtained from *Scholven Chemie*'s cumene unit.

A number of companies that were principally involved in pharmaceuticals still managed to prosper and to maintain their independence. Such was the case for *E. Merck* in Darmstadt, which was totally separate from Merck in the United States from 1919 on, for *Böhringer, Ingelheim,* and *Knoll*. But it was clear that some restructuring would be needed in this sector, for it lacked the research base to carry out an ambitious development policy.

More generally, Germany's chemical industry, having managed a brilliant recovery, still needed to carry out some frontier adjustments within the country itself in order to complete its rationalization process, and—a matter of greater difficulty—it had to expand abroad, particularly in the United States. This would give the large companies the multinational dimensions of their American counterparts.

The Peak Years of America's Chemical Industry

The United States chemical industry maintained the lead it had acquired through the war until the 1973 energy crisis, for European and Japanese companies were still busy providing for their own immediate needs, prior to forging ahead.

American industry leaders moreover, were among the first to glimpse the importance that European unification moves and the dismantling of tariff barriers within GATT would have for international trade, which, until then, had been shackled by constraints. But unlike the Europeans, who first sought to boost their exports, American industry leaders, for whom exports were not so vital, gave priority to the establishment of production units abroad, opening up the era of *multinationals*. Having both ready capital to invest and the required technologies, their task was not so hard.

Not all the major companies, of course, had this immediate global vision of new markets nor the gumption to set up international establishments. The driving force came from a small number of companies such as Dow and Monsanto that set the pace for their more reticent counterparts and, at the same time, spread the multinational concept to their European competitors.

Furthermore, this overseas expansion coincided with the American chemical industry's ambitious market development program within the United States. But a number of emerging problems demanded attention. In the first place, the world market was no longer an area shared out among the chemical giants as it had been under the prewar cartel arrangements. Competition was becoming stiffer. The federal government in Washington, in particular, wished to apply antitrust laws with renewed vigor. A second factor of major importance was the way in which environmentalists felt encouraged by the success of Rachel Carson's book *Silent Spring* in 1962. In addition, the consumer movement was shaping up under Ralph Nader; and he did not spare the chemical industry, already under attack by environmentalists. New rules, such as through the Toxic Substances Control Act (TOSCA) and the Occupational Safety and Health Agency (OSHA), were being legislated to meet this new political situation, while the powers of the Food and Drug Administration (FDA) and the Environmental Protection Agency (EPA) were being strengthened.

Economic growth in the United States was strong in the 1960s. One important boost came from the 1964 tax relief measures; but by the end of the decade growth began to slow down, and the first negative effects of inflation were felt.

Despite its size and sales (which exceeded $4 billion for the first time in 1966) *Du Pont* was not spared the hardships of this new environment. As soon as the war had ended, the Justice Department sued the company on two counts—for its Cellophane monopoly, which ended with the sale of a license to Olin following the one granted to Sylvania, and for Du Pont's share in General Motors, which had to be sold in 1965, entailing substantial loss of revenues. Du Pont was also in trouble with the antitrust laws because of its joint ventures with ICI. It was forced to dissolve the joint Latin American company, Duperial, and to break up the joint Canadian subsidiary CIL.

Du Pont was anxious to repeat its success with Nylon. It had decided to withdraw from the increasingly competitive rayon market, and it was beginning to feel the effects of competition in other areas. In 1964 it brought to market a new synthetic leather for shoes, called Corfam. The product was not a success. The Corfam project cost the company $80 million before being abandoned six years later. Both the technology and the equipment were sold to Polish interests.

Like other chemical groups, Du Pont increasingly felt the effect of restrictive regulations. In the 1970s, it was forced to give up the tetraethyl lead business and to restructure its Freon gas sector because of the suspected harmful effect of Freon on the atmosphere's ozone layer. Also, the scrapping of the American Selling Price Clause (ASP), which protected organic-derivative producers, in the framework of the general negotiations on tariffs and trade, was to have the effect of forcing Du Pont out of dyestuffs production.

In part because of these many mishaps, the Company's Board, breaking with a long tradition, decided in 1973 to appoint at its head Irving Shapiro, a lawyer who was more versed in the ways and means of the Justice Department than in the technical problems of the chemical industry. All these troubles did not prevent Du Pont from expanding abroad, however. By 1972 the size of its investments outside the United States had soared to $1.6 billion and its sales abroad to $640 million, a fifth of its total turnover.

Sales of *The Dow Chemical Company* reached the $3 billion level in 1973, as the company pursued the spectacular growth that had started during the war with its Texas petrochemicals corporations. The founder's son, Willard Dow, had died in 1949 in an airplane accident in Canada; and his brother-in-law, Leland Doan, had taken over as President. Under Leland Doan's direction, Dow's polymer business was energetically developed; by 1957 it accounted for nearly one-third of the company's turnover (styrene-butadiene latex, polystyrene, Saran film, and polyethylene). The Korean War boosted the company's sales of magnesium metal, while Dow-Corning silicones were becoming increasingly popular.

From 1959 onwards, a large part of the company's growth was generated by its units abroad, more particularly those set up in Europe, where, unlike other American companies, Dow immediately set to work to integrate production of the feedstocks for its downstream operations. In 1966 a major petrochemical unit was set up in Terneuzen, near Rotterdam, to produce styrene, polystyrene, ethylene oxide, and glycols.

Dow also set up electrolysis units in Stade in West Germany. The chlorine produced partly served to make chlorinated solvents. At the same time, Dow's leaders did not neglect the prospects that specialties could open up. Associated with Schlumberger in Dowell, Dow enjoyed the prosperity attached, at the time, to any activity related to oil exploration. Diversification into pharmaceuticals which began in 1960 with the purchase of Allied Laboratories, followed by the acquisitions of Ledoga Lepetit in Italy and of Richardson Merrell in the United States, turned out to be a fortunate move, even if there were no obvious synergies between a producer of base chemicals and the ethical drug market.

Like other major chemical companies in the United States Dow, which had succeeded so well in its industrial development and had played a pioneer role in overseas expansion, suffered from the ties which circumstances had led it to develop with the country's Defense Department. Its public image was harmed by the attacks against its role in the production of napalm ordered in 1965 by the U.S. Air Force and of the 2,4,5-T (Agent Orange) used as a defoliant in the Vietnam War.

The direction *Monsanto* took after the war was in many ways different from the path taken by Dow Chemical. Based in Saint Louis, the company was headed by the founder's son, Edgar Queeny. Monsanto became involved in the chemical-fibers sector by setting up in 1950, in joint venture with American Viscose, a company called Chemstrand to develop Acrilan, an acrylic fiber it had discovered. The following year, it acquired a Nylon license from Du Pont. In 1961, Monsanto acquired American Viscose's 50 percent share in Chemstrand and this was greatly to contribute to the company's prosperity. The hydrocarbon division subsequently set up was mainly intended to supply the textiles sector with its raw materials.

The purchase of Lion Oil in 1955 emphasized upstream integration. Its association with Bayer in 1954, to form Mobay Chemical, which was to last until 1967, paved the way for Monsanto's involvement with polyurethane foams. Monsanto also increased its polystyrene production and launched into the manufacture of vinyl resins and then of ABS. New impetus was also given to the phosphorus and phosphates sector which was to form the hard core of its inorganic division. Monsanto also started producing 2,4-D and 2,4,5-T herbicides during the fifties. This was the starting point of its involvement in the highly promising crop protection sector.

Curiously enough, however, Monsanto, whose beginnings were in fine chemicals with aspirin, vanillin, coumarin, and saccharin, now mainly focused on large-scale inorganic and organic production, leaving aside pharmaceuticals. But the research tradition acquired through men such as Thomas and Hochwalt and through the Rubber Services Laboratories was a contributing factor in the success of the specialties it developed for rubber and other organic derivatives.

The market prospects opening up in Europe could not fail to attract Monsanto's leaders. The company had been established in England since 1920, with its Ruabon and Sunderland units producing a whole range of tar derivatives. In 1962, Monsanto became a multinational by establishing the headquarters of Monsanto Europe in Brussels. The company produced chemical fibers in Northern Ireland and Luxembourg, while certain polymers like polystyrene, ABS, polyvinyl butyral sheets, and rubber specialties were produced in Britain, Belgium, France, and West Germany. When John Hanley came from Procter & Gamble in 1972 to head the company, Monsanto had taken its place among the world's chemical giants with a turnover of $2.25 billion.

The name of *American Cyanamid* was only a simple reminder, in 1945, of the company's origins: it was founded in 1907 by F. Washburn to produce calcium cyanamide in Niagara Falls, Ontario.

After World War II, the company still retained a strong position in fertilizers, its first successful venture. But under the clever management of W. Bell, who had remained president for twenty-eight years, the company greatly diversified through successive acquisitions. Through Calco, it was involved in the dyes business. Lederle had become in 1946 the group's pharmaceuticals division and produced aureomycin (chlortetracycline), acquiring precious experience in fermentation, which was boosted in 1953 through purchase of Heyden Chemical's antibiotic department. Already a supplier of liquid HCN for fumigation, American Cyanamid successfully put on the market Parathion and Malathion, thus rounding out its range of products sold to farmers. The company, which had its own phosphate mines in Florida, also supplied explosives for mines and quarries, as well as flotation products for minerals—a business that dated back to the production of calcium cyanide used for the extraction of gold.

While carrying out these acquisitions, American Cyanamid was producing original chemicals from its own research conducted in Stamford, Connecticut. It played a pioneer role in guanidines used as rubber accelerators, in dicyandiamide, in melamine, and in melamine-formol resins, a logical sequel to its urea-formol (Beetle) resins dating back to 1923. The company also contributed to the industrial production of acrylonitrile needed to implement the synthetic rubber program during the war. These chemicals naturally opened up new prospects for the company in postwar years, more especially in the high-polymer sector.

By the end of the 1950s, Cyanamid, which had bought up Formica, was thus producing melamine-based laminated sheets, polyester resins marketed under the trade name of Laminac, and a polyacrylonitrile fiber called Creslan. Although it was not involved in thermoplastics and petrochemicals and was more restrained than Monsanto and Dow in setting up units abroad, by 1973 American Cyanamid nonetheless posted a consolidated turnover of $1.472 billion. And with a product range extending from traditional heavy chemicals (inorganic acids, fertilizers, phthalic anhydride, acrylonitrile, melamine, formol) to specialties (Lederle pharmaceutical, Shulton cosmetics, crop protection products, synthetic dyes, pigments and resins, refinery catalysts, polyelectrolytes, rubber, plastics, and paper additives) to explosives and synthetic fibers, Cyanamid was one of the most diversified and prosperous American chemical companies.

Among the leading prewar chemical companies, *Union Carbide Chemicals* maintained its rank through its aliphatic chemicals inherited from Curme. Starting from olefins, it offered acids, alcohols, aldehydes, anhydrides, ketones, amines, esters, and glycols, produced both in West Virginia and in Indiana. Besides phenolic (Bakelite) and vinyl (Vinylite) resins already produced before 1940, the company sold the major thermoplastics such as polystyrene and high-pressure polyethylene as well as silicones. UCC's Canadian base was long-standing. The company had also established itself in England through Bakelite Ltd., which, in 1963, had set up, in joint venture with Distillers, the Bakelite Xylonite Company to produce PVC resins.

A specialist in a wide range of polyethylene for cables, UCC had set up during the 1950s polyethylene units in Stenungsund, Sweden; in Antwerp, where hydroxethyl cellulose was also manufactured; in India; in Australia; and in Brazil. Its international dimension also stemmed from its other activities in graphite electrodes (National Carbon Division), industrial gases (Linde Division), ferro-alloys (Union Carbide Metals Company), batteries, antifreeze agents (Union Carbide Consumer Products), all contributing to a turnover that had exceeded $1.5 billion as early as 1960.

Allied Chemical, another giant, remained, on the other hand, true to its original tradition of keeping to North America. Apart from a few inorganic chemical units located in Canada, the organic and inorganic base chemicals produced by Barrett, General Chemical, National Aniline, Nitrogen, Solvay Process, and Semet Solvay, were all focused on the home market; only a small share of their production was exported. The company's only noteworthy diversifications involved, on the one hand, caprolactam, used to prepare polyamide-6, and on the other, polyethylene waxes. Managed in a highly conservative manner, Allied Chemical carried out its development through self-financed internal growth and had never shown any interest in acquiring other companies.

The expansion of *Hercules Powder* was more spectacular. Within its traditional explosives sector, it had contributed to the war effort by running ordnance units for the Defense Department. This experience was to prove useful when propellants had to be developed for the American Navy's Polaris and Poseidon rockets. From its cotton-linters base, the company had extended its range of cellulose derivatives with production of ethyl cellulose, and then of carboxymethyl cellulose (CMC). Likewise, in 1948 it had added chlorinated camphene to its naval stores (rosin, oil of turpentine, pine oil); sold under the name Toxaphene, the chlorinated camphene proved a particularly effective insecticide against cotton pests. Its paper-sizing products based on rosin soaps provided Hercules with substantial revenues, as did its abietic acid resins for the inks, adhesives, and paints industries.

Hercules Powder also raked in substantial royalties from its process for phenol and acetone from cumene, which came on stream in Gibbstown, New Jersey, in 1954 and in association with Distillers was licensed out worldwide. In 1955, Hercules brought on stream in Burlington, New Jersey, a unit to produce dimethylterephthalate (DMT) through *p*-xylene oxidation, based on the Imhausen process acquired from Dynamit Nobel's subsidiary Witten. Hercules had been among the first companies, together with Hoechst, to become involved in low-pressure polyolefins produced with Ziegler-Natta catalysts. In 1957, it brought a Hifax polyethylene and Profax polypropylene production unit on stream in Parlin, New Jersey.

During this particularly productive period, Hercules enjoyed several successes through the work of its Wilmington, Delaware, Experimental Station. Although it subsequently gave up its pesticides sector to Boots and ceased producing phenol because it lacked the requisite size, it was shrewd enough to develop, from 1960 on, polypropylene applications in the most varied fields (fibers, films, and molded products).

From the very first, Hercules Powder had been an international company, having established itself in England and Holland after World War I to develop its cotton-linters and rosin business. This international quality was developed after 1950 when the company set up paper-sizing production units in Europe and in other parts of the world, associated with Teijin in Japan to produce DMT, and brought on stream polypropylene units in Canada, Belgium, Taiwan, and Brazil. While not always lucky in its acquisitions policy, Hercules Powder successfully carried out two good diversifications when it acquired Danish Pektin Fabriek (water soluble gums) and Polak Frutal Works (fragrances and perfumes). Later, under the management of Al Giacco, who took over from Werner Brown in 1977, Hercules Powder, under its new name of Hercules, went through further agonizing reappraisals.

Stauffer Chemical was headed in 1960 by Christian de Guigné and Hans Stauffer, grandson and nephew of the founder, respectively. Although the head office was in San Francisco, the company was managed from New York. Its range of products in inorganic chemicals reflected its origins. It was to be enlarged to encompass the phosphorus derivatives of Victor Chemicals. Through sulfur, Stauffer had become involved in chemical derivatives, such as carbon disulfide and its crop protection products which were to provide the bulk of its profits.

Stauffer had also made a fortunate choice when it took over Celanese's range of phosphorus hydraulic fluids and when it associated with Hercules in Texas Alkyls to produce Ziegler catalysts. But it was involved in types of chemicals that neither lent themselves to spectacular developments, save for the crop protection products, nor to major ventures abroad.

Quite different was the case of Rohm & Haas, managed from Philadelphia by F.O.H. Haas, one of the sons of Otto Haas, and then by Vincent

Gregory. Like its German parent company, Rohm & Haas had started from enzymes and acrylic chemicals. But in 1920 it had enlarged its base through the acquisition of Lennig and the establishment of Resinous Chemicals. It thereafter showed creativity in the most diverse sectors: Triton nonionic surfactants, Rhoplex acrylic dispersions, Lethane and Dithane insecticides and fungicides, Amberlite ion-exchange resins, Plexol lubricating additives. No doubt Rohm & Haas owed the major part of its fame and profits to Plexiglas. But riding on the wave of this popularity it managed to build up an acrylates chemistry involving methacrylates that integrated the upstream monomers and the downstream solution and dispersion resins. Through its origins and the nature of its specialties, the company was naturally inclined to expand abroad, and more particularly in Europe. This explains the size of its overseas interests and the fact that it was to rank second after Dow among U.S. chemical companies in percentage of sales abroad.

Hooker Chemical, founded by E. H. Hooker in 1907 in Niagara Falls, New York, had retained a strong position in the electrolysis of salt, its starting point. The company began manufacturing chlorinated derivatives of benzene and toluene after 1940 and afterward diversified into phosphates and phosphorus derivatives. It also bought Durez Plastics, which produced phenol-formaldehyde products as well as special resins for foundries. Well known for its electrolysis technology, the company lacked a sufficient range of original products to play an important role on the international scene.

The same may be said of *Koppers*, specializing in tar derivatives, polystyrene, and phthalic anhydride, even though it added to its range a few special products such as resorcin and BHT, an antioxidant.

A number of large American companies had come to chemicals, for which they had no particular leaning, through other activities. *Eastman Kodak*, for one, had started making its own nitrocellulose film in Rochester, New York, in 1898. It had set up a wood distillation unit in Tennessee in 1920 to provide its own source of methanol and acetic acid. Ten years later, it brought on stream a cellulose acetate unit as well as a hydroquinone production unit. It continued to make photographic products for its own needs in Rochester as well as in England, and also in France in the framework of Kodak-Pathé. But at the same time, units in Kingsport, Tennessee, were filling the needs of outside customers with cellulose acetate fibers and plastics, acetobutyrate molding powders, aliphatic solvents, plasticizers, and hydroquinone-derived antioxidants.

In this way Eastman Kodak joined the world of big chemical suppliers. The move took on a greater depth after the war when the group set up petrochemical installations in Texas. Over the years, it successfully diversified into high- and low-density polyolefins, synthetic fibers (Kodel), and new plastics (polybutyleneglycol terephthalates). Through Distillation Products—a prewar offshoot of an association with General Mills and a 100 percent subsidiary of

Eastman Kodak since 1948—the group became a producer of vitamins A and E using its molecular distillation technology.

A rival of Eastman Kodak in the area of cellulose derivatives, *Celanese* came to chemicals through rayon acetate production, which Camille Dreyfus had started in Cumberland, Maryland, in 1924. Set up under the name of the American Cellulose & Chemical Manufacturing Company by the Dreyfus brothers, the company bought the Celluloid Company in Newark, New Jersey, in 1927 and became the Celanese Corporation of America. Until the war, it focused solely on the production of cellulose acetate and its utilization in films, lacquers, and fibers. In 1940, the company began research on the preparation of the acetic acid needed for this production through oxidation of butane and propane.

In 1945, the first unit of its kind built in Bishop, Texas, produced, via vapor-phase oxidation of liquified gas obtained locally, some twenty petrochemical derivatives, including methanol, formaldehyde, acetone, acetaldehyde, and acetic acid. By the following year, the derivatives already accounted for 10 percent of the company's total sales. Half the tonnage was for Celanese captive uses.

In 1952, another unit based on liquid-phase butane oxidation technology came on stream in Pampa, Texas. Under H. Blancke who became president in 1956 after the death of Camille Dreyfus, Celanese set about diversifying from its cellulose-derivatives specialization. It took out a Phillips license involving a low-pressure polyethylene process in 1957. Celcon polyacetal resin was developed in 1961. Under agreements with ICI, Celanese became involved in 1965 in Nylon-66, in polyester fiber, and in vinyl acetate. The company proved less successful in its attempt to integrate into base materials. Its stakes in Columbia Cellulose (Canada), SIACE (Sicilian Pulp), Champlin Petroleum, and Pontiac Refining were all sold from 1968 on. Likewise, it only made a short-lived incursion into paints after buying Devoe and Raynolds in 1964.

Its European expansion was inhibited by British Celanese, headed by Henri Dreyfus. But Celanese had set up business in Mexico as early as 1944, becoming the second petrochemical producer there after Pemex. It later became associated with local fiber companies in Venezuela, Columbia, and Peru. In Canada, Celanese had set up the Canadian Chemical Company to produce petrochemicals from Alberta's natural gas. Through its polyacetal technology, it became associated in 1961 with Hoechst in Germany and in 1963 with Daicel in Japan. Thus although it had to rely on foreign licenses and had to back away from extreme integration, Celanese had developed original processes and successfully diversified in plastics and chemicals.

When *Pittsburgh Plate Glass* (PPG) was set up in 1883, it manufactured sheet glass. But in 1899, its founder John Pitcairn had the foresight to build the Columbia Chemical Company in Barberton, Ohio, to ensure its sodium carbonate supplies. His prescience went even further when he became involved in 1900 in linseed oil and in paints, which could be sold through the same channels as glass. Thus while Ford, which merged with Libbey Owens in 1930, remained a glass manufacturer, PPG's future, thanks to Pitcairn's genius, was to rest on three pillars, glass, chemicals, and paints.

During the period from 1941 to 1945, PPG was led to produce allyl diglycol carbonate (CR 39 monomer) for military purposes. It managed to develop phosgene chemicals in Barberton, in particular putting on the market in 1947 a herbicide of the carbamate type, chloro-IPC. In Natrium, Virginia, PPG had electrolysis units that provided a complement to the chlorine and soda produced by Columbia Southern Alkali, a subsidiary set up in 1951. All these activities became part of the chemicals division established in 1961 and enlarged to encompass chlorinated solvents and vinyl chloride monomer produced in Lake Charles, Louisiana, in 1964 under an original oxychlorination process. Chemical sales in 1970 already accounted for 30 percent of the group's turnover.

PPG was less fortunate when, in 1971, it took a stake in an olefin production complex located in Puerto Rico. The project was abandoned after heavy losses. Also abandoned was a titanium dioxide production project as well as one for a caprolactam unit in association with DSM.

It was through further development of its traditional chemicals business (chlorine, alkalis, phosgene, chlorinated derivatives) that PPG achieved its greatest successes, even though these base chemicals were not of a nature to lead to ambitious expansion abroad. This explains why PPG's respectable position outside its frontiers was mainly achieved through its glass and paints activities.

From Tire Manufacturing to Chemicals

Unlike their European counterparts, the United States tire manufacturers all made forays into chemicals. In 1926, Goodyear had launched its Captax accelerator on the market, the very year B. F. Goodrich developed the first vinyl chloride polymers, marketed under the trademark Geon. It was their participation in the GRS program during the war that gave *Goodyear*, *Goodrich*, and *Firestone* the opportunity, once the war had ended, to become involved in the synthetic rubber business and to penetrate on a large scale the other polymer markets, PVC and latex emulsions.

This involvement in synthetic elastomers and plastics, in addition to the production and sale of rubber additives, greatly boosted the chemicals division of the three groups. But none of them aimed to go beyond their well-defined special rubber-related chemicals sector.

U.S. Rubber followed a very different path. Set up in 1892 to produce rubber shoe soles, U.S. Rubber, with its UniRoyal trademark, was a smaller tire concern than its three competitors. But its chemicals experience dated back to 1904 when its subsidiary Naugatuck Chemical had started producing the sulfuric acid needed for tire regeneration. Cut off from its German suppliers during World War I, the company set up one of the first United States aniline units, and then developed thiocarbanilide production. It became a pioneer in the amine-type antioxidants that were to be widely used in the tire business.

Even before the war, *Naugatuck* had put to use its organic chemicals experience to synthesize fungicides and a crop protection insecticide. During the 1950s, the company brought a further range of herbicides and growth-control products on the market. During the war, Naugatuck had produced dodecyl mercaptan, used to regulate the viscosity of synthetic rubber; after the war it innovated in the sector of high polymers with Naugahyde PVC coating, nitrile rubber, ABS terpolymer, and EPDM rubber. Thus, alone among the United States tire manufacturers, U.S. Rubber managed to build up a full-scale chemical entity based on a long-time tradition, acquiring a foothold in a number of promising sectors.

The First Conglomerates

The Americans were not only the first to set up chemical multinationals, they were also the first to form *conglomerates*, that is, diversified groups involved in a number of different activities that were not necessarily related. Of these conglomerates, four were quite strongly involved in chemicals. One of the oldest had been set up in 1892 by Franklin Olin to produce explosives in a unit near Saint Louis. In 1931, Winchester Repeating Arms was acquired; and what was to become *Olin Industries* in 1944 expanded and diversified, particularly into aluminum, copper smelting, and cellophane (under a Du Pont license).

The company went into chemicals in a big way when it merged in 1954 with Mathieson Alkali Works, a concern set up in 1892 by Thomas Mathieson to produce sodium carbonate and caustic soda in Niagara Falls. With these products, Olin-Mathieson, soon to become Olin Corporation, became involved in alkalis, chlorine, hypochlorites, ammonia, and fertilizers, and acquired the major Lake Charles unit in Louisiana which Mathieson had purchased from the government in 1947. The company's portfolio then expanded to include the production of hydrazine in Lake Charles, of ethylene oxide and derivatives in Doe Run, Kentucky, and, above all, products made by the *Squibb Company*, an important pharmaceuticals firm that was already a hundred years old when it was acquired.

By 1960, sales of the chemicals division and of Squibb accounted for 31.4 percent and 15.8 percent respectively of Olin Corporation's consolidated turnover. Things were to change still further as Squibb was sold and the Olin conglomerate increasingly took on the image of a chemicals group involved both in base chemicals and in specialties (chlorocyanurics, blowing agents, and polymer additives inherited from National Polychemicals, TDI, and polyols, for polyurethanes).

W. R. Grace, followed a similar path but on a larger scale. The company had been set up in 1854 in Peru by an Irishman, William Russell Grace, to carry the guano, collected on the islands offshore from the Peruvian coastline, to San Francisco. Eleven years later, W. R. Grace established in New York a regular guano, Chilean nitrate, and Bolivian tin trade among North America, South America, and Europe. His son Joseph P. Grace took over the business and diversified it by setting up in various South American countries textile plants, sugar refineries, and paper mills, utilizing for the first time cane cellulose waste as feedstock (Paramonga in Peru). Grace also built a fleet of cargo ships to serve the western coast of South America and associated with Pan American in 1928 to create the Panagra Airlines.

When Peter Grace, age 32, took over from his father Joseph in 1945, the company he still heads today, he feared there could be danger for the group in being too concentrated on South America and in shipping. He chose, therefore, to refocus the company on the United States and to invest in the very promising sector of chemical specialties. The turning point came in 1954 with the purchase of both Dewey & Almy Chemical Company and of Davison Chemical.

Chemical engineering graduates from MIT, Bradley Dewey and Charles Almy, had founded their company in 1919 to supply mastic for sealing containers. They moved on to the manufacture of wrapping film, of latex, and various other specialties. Their business remained, even after its acquisition by Grace, a model of prosperity both in the United States and in the other countries where it was established.

Davison Chemical's beginnings date back to 1832 when William Davison built one of the first American sulfuric acid units. Following World War I, Davison Chemical became a major producer of silica gel used as a drying agent, and subsequently led Davison Chemical to become the leading supplier of catalysts to the oil industry.

In 1965 Grace bought a company called Hampshire Chemical. It had been set up seven years earlier by the prolific Bradley Dewey to develop chemicals derived from HCN. Other acquisitions, such as Chemed and its Dearborn subsidiary, turned out to be shortlived. Likewise, diversification in agriculture-related activities and in a number of service companies and consumer products was the cause of serious trouble for the group. Nonetheless, by 1982 Peter Grace could boast of having developed his conglomerate to the point of being the fifth-largest U.S. chemical company, posting a $2.6 billion turnover, 43 percent of the group's consolidated sales. More to the point, chemicals accounted for 54 percent of the profits. This was because W. R.

Grace was the world leader in chemical specialties, a particularly lucrative area when, as was the case here, it is properly mastered.

Another instance of a conglomerate combining machinery (for food packaging, pumps, military equipment) with various chemical activities was *Food Machinery and Chemical Company*, shortened to FMC in 1961. Chemicals had grown out of a series of apparently unrelated acquisitions: West Waco, which had a unit in South Charleston, West Virginia, specialized in alkalis, chlorine, and carbon disulfide as well as a natural sodium carbonate deposit (Trona) in Wyoming, and Becco, which produced hydrogen peroxide and peroxide products in Buffalo. As FMC was involved in farm machinery, it had purchased two companies which produced and formulated insecticides, Niagara Chemical and Fairchild Chemical. The group also had interests in phosphorus chemicals and in barium derivatives and developed specialties in the Nitro, West Virginia, unit (Dapon allylic resins, plasticizers, phosgene derivatives, and chlorocyanurics). The purchase of American Viscose Corporation, which became FMC's Textiles Division, ended a few years later when it was sold to its management team, for FMC had trouble grafting it onto its other activities.

FMC's chemical production was not suited to overseas expansion and, except for taking a share in Spain's Foret, the group's leaders showed no particular ambition in this area. Despite a number of assets in the shape of its crop protection products and the recent acquisition of Marine Colloids (carrageenates), FMC management seemed loath to break up the good balance achieved amongst its various activities in favor of the chemicals division.

Ethyl Corporation also could be regarded as a conglomerate that successfully made its chemical breakthrough. Its legal life began when General Motors and Standard Oil of New Jersey formed a joint venture to develop Thomas Midgley's tetraethyl lead invention and his method for preparing it from ethyl chloride. The need for bromine led to an association with Dow to form Ethyl Dow in 1930. In 1936, what was to become the largest unit for the production of gasoline antiknock agents came on line in Baton Rouge, Louisiana. After the war, Ethyl Corporation chemists developed an orthoalkylation process that led to the marketing of an original range of phenol antioxidants. Ethyl then diversified its chemicals interests by producing vinyl chloride and PVC, alkyl aluminum catalysts, sodium, bromine and bromine derivatives, as well as α-olefins.

In 1960, much to the surprise of the business community, Albemarle Paper Company, a small paper factory belonging to the Gottwald family of Richmond, Virginia, bought Ethyl Corporation. The company since then has developed in a harmonious fashion as the new owners have tried to replace sectors such as the antiknock agents, which are likely to die out sooner or later, by developing others (expansion of the Feluy unit in Belgium, purchase of Edwin Cooper and of Hardwicke Chemical, downstream realignment of the plastics sector, bromine development). As a company that started in a single product, Ethyl Corporation could well have disappeared when tetraethyl lead was threatened by a ban. But fifty years later, the talent of its chemists and of its management had turned it into a prosperous conglomerate drawing a significant part of its revenues from chemicals and plastics.

The Relationship Between Chemicals and Oil in the United States

The relationship between the chemical and oil industries in the United States was a mixed one after the end of the war. A number of chemical companies had developed processes which required well-defined petroleum fractions and had engaged in upstream integration through purchase of refining companies.

In fact, such ventures as those by Monsanto with Lion Oil and Celanese with Champlain Oil were not very successful, as we have seen. It is a little too soon to tell whether Du Pont was well-inspired when it bought Conoco.

There were other relationships established between the chemical and oil industries through associations. But some of the fifty-fifty subsidiaries did not last very long. Thus Texaco and American Cyanamid's joint subsidiary, Jefferson Chemical, set up to produce ethanolamines, glycols and ethoxylates from ethylene oxide in Texas, was fully taken over by Texaco. Likewise Petrotex, an association of FMC and Tennessee Gas Transmission to produce butadiene, butenes, maleic anhydride, and polychloroprene in Houston, Texas, was bought by Japanese interests and became Denka Chemicals before being taken over recently by Mobay.

The downstream chemical integration of oil companies was not without its own troubles, principally in the fertilizer field. It nevertheless produced powerful and diversified petrochemical subsidiaries.

Shell Chemicals, a division of Shell Oil, USA, had investigated, as we have seen, the preparation of solvents derived from *n*-butylenes as early as 1930 on the Martinez site in California. It brought on stream, in 1941, the first industrial butadiene production unit in Houston, Texas. On the same site, Shell Chemicals set up units in 1945 to produce allyl chloride and allyl alcohol by an original process of high-temperature substitutive chlorination of lower olefins. Very soon acrolein, epichlorhydrin, and finally synthetic glycerine were added to these new intermediates.

With its eleven units scattered over the United States, Shell Chemicals posted sales of $3 billion in 1982, involving a variety of products: olefins, oxygenated solvents, linear alcohols, allyl derivatives, epichlorhydrin and epoxy resins, aromatic hydrocarbons, phenol, bisphenol A, ethylene and derivatives, polypropylene, polybutadiene, Kraton, thermoplastic elastomers, insecticides and herbicides, and lube oil additives. Many of the new products and processes developed in Shell Chemicals' United States laboratories subsequently spread to the other companies of the group throughout the world, proof of the highly original research of this brilliant Shell Oil subsidiary.

Standard Oil of New Jersey also had a number of remarkable chemical achievements. The first developments, and in particular the production of isopropyl alcohol in Bayway, were made by Frank Howard, who set up the Standard Development Company in 1922 and later played a major role in the negotiations with IG Farben. Howard also took an active part in the establishment of *Ethyl Gasoline Company* in 1924 after the Bayway engineers had succeeded in producing tetraethyl lead from ethyl chloride, sodium, and lead. On the basis of technical information provided by IG Farben, Standard Development worked out in Baton Rouge, during the 1930s, the high-temperature catalytic hydrogenation processes under which isooctane for aviation gasoline and toluene for explosives were produced from oil as soon as war broke out. It was also through IG Farben that Standard was able to develop in 1934 the Oppanol-inspired oil additives produced through isobutylene polymerization (Paratone) and managed to take a foothold in synthetic rubber, from which the discovery of Butyl rubber was to emerge in 1937.

Although Frank Howard was subsequently blamed for having cooperated with IG Farben up to the declaration of war, he could point out that Standard Development had greatly contributed to pushing his group to the forefront of chemical leaders. This reputation was fully established when the subsidiary became an autonomous entity in 1966 under the name of *Exxon Chemical Company*. With sales exceeding $8 billion, Exxon Chemicals ranked fourth in 1981 among United States chemical companies, after Du Pont, Dow, and Union Carbide. It had withdrawn from synthetic fibers and reduced its fertilizer division to two sites in Holland and Canada. But it remained one of the most efficient producers of petrochemical derivatives (olefins, aromatics, oxygenated solvents) in its main plants of Baton Rouge, Louisiana, Baytown, Texas, as well as of Rotterdam and of Notre-Dame de Gravenchon in France. Its range of polymers extended from butyl and chlorobutyl rubber to polypropylene, to low density polyethylene and EPDM elastomer. Exxon Chemicals also produces plasticizers from its OXO alcohols and in particular, managed a breakthrough in specialties with its Escorez petroleum resins and its range of Paramin lube oil additives.

Other American oil companies ventured into chemicals but on a more modest scale. One of the most innovative was undoubtedly *Phillips Petroleum Company*, based in Bartlesville, Oklahoma. Already much involved in carbon black through its Texas units and its interests in subsidiaries in France, Italy, India, and South Africa, the company was also renowned for its low-pressure polyolefins and its stereospecific elastomers developed by a remarkable team of chemists. Its prosperity was, in fact, due not only to its polyethylene, polypropylene and *cis*-1,4-polybutadiene productions but also to its process licenses, sold worldwide in competition with the Ziegler and Natta methods.

Standard Oil of California began producing polybutene as early as 1935. In 1931, it had also put on the market naphthenic acids extracted from the crude oil it refined; these were to become increasinly popular as drying agents in paints. They were also used to make napalm. But it was not before 1943 that the Oronite Chemicals subsidiary was set up in Richmond. One of its first tasks was to find outlets for the xylenes which the refinery had produced since 1945 with a purity exceeding 95 percent. It was in Richmond that the original process to produce phthalic anhydride from ortho-xylene was developed. Oronite also brought on stream in 1946 a unit to produce a detergent based on alkyl aryl sulfonate.

Standard Oil of California subsequently extended its chemicals business to encompass fertilizers, crop protection products with Captan, as well as a range of lubricating oil additives that spread to Europe through fifty-fifty subsidiaries set up with British Petroleum in England (Orobis) and with Progil in France (Orogil).

Standard Oil of Indiana came to chemicals through its subsidiary *Amoco*, which successfully developed polypropylene and terephthalic acid (TPA) used as an intermediate for the production of polyester fibers. *Standard Oil of Ohio*, in which BP took a stake in 1969, developed a distinctive process for the preparation of acrylonitrile from propylene through its subsidiary *Sohio Chemical* with licenses to a number of companies worldwide. *Socony Mobil Oil* diversified very modestly into chemicals. Its subsidiary *Mobil Chemical* continued to produce olefins and aromatics in Beaumont, Texas, and paints in Metuchen, New Jersey. But it ended up selling its crop protection products division to Rhône-Poulenc because it never succeeded in building it up to proper size.

ARCO, the outcome of a merger of Atlantic Refining with Richfield Oil, kept to base petrochemicals for a long while, only taking off in more specialized chemicals when it became involved, through its Halcon process, in the propylene oxide and styrene markets and made an international breakthrough with MTBE, a gasoline additive for use as a substitute for tetraethyl lead.

In various areas, but on a smaller scale, other oil companies became involved in chemicals. *Sun Oil* produced propylene, nonene, and aromatics in Marcus Hook, Pennsylvania; *Tenneco*, which later bought England's Albright & Wilson, brought out benzene, toluene, xylenes, ethyl benzene, vinyl chloride, and vinyl acetate in its Louisiana and Houston refineries; *Texaco* also had its range of aromatics as well as olefins, and in addition produced cumene in Eagle. It also produced ethylene oxide and derivatives through Jefferson Chemical, which it took over from *American Cyanamid*.

Continental Oil Company, better known as Conoco, subsequently to be bought up by Du Pont, had a unit in Lake Charles, Lousiana, in joint venture with Cities Service, producing ethylene, butadiene, and ammonia.

While upstream oil integration generally turned out to be not very profitable for United States chemical companies and while associations between the oil and chemical industries proved transient because of different outlooks and objectives, a number of oil companies did prove they were capable of mastering chemical disciplines to their advantage.

The United States Fine Chemicals Industry

Unlike their German rivals, America's major chemical companies always had trouble adapting to the constraints of fine chemicals and specialties. Indeed, giants like BASF, Bayer, and Hoechst arose from the discovery of synthetic dyestuffs at a time when organic synthesis was a subject of honor in Germany's colleges and universities. Americans, on the other hand, had no such tradition and remained dependent on German dyes until World War I cut off their supplies and they were forced hastily to conjure up a dyestuff industry. This industry developed in the United States only behind safe tariff barriers. When tariffs were reduced in the framework of the Kennedy Round, major companies such as Du Pont, Allied, and American Cyanamid lost interest to the point that there is only a single national dye producer in the United States today, *Crompton & Knowles*.

Along with dyes, America's chemical industry also lost all the synthetic-intermediates production that was linked to them and which served to prepare pharmaceutical and agrochemical derivatives. America's large chemical companies had become prosperous through their capacity to push through wide-scale development of original processes to meet the needs of war and, in a peacetime economy, those of the vast United States market. Both engineers and sales people in these companies had become used to aiming for size. They conceived large-scale units targeted for hefty market shares.

Fine chemicals, on the contrary, whether dyes or other specialties, required multipurpose units working not on a continuous basis, but batchwise and in cycles, often short ones, involving production programs that were likely to undergo rapid changes to fit client requirements. In this type of industry, complete product ranges must be available in small volumes, and final development often required much patience.

The Swiss, as well as the Germans, perfectly understood these constraints because of a long tradition. American industry leaders, who were trained more as engineers than as chemists and who had acquired their experience within large industrial units, were not prepared, barring a few exceptions, for such a discipline. Some thought they could acquire the necessary expertise by buying companies, often family ones, involved in specialties. By trying to apply to these purchases the procedures of large companies, however, they sometimes destroyed the very flexibility and knowhow they had sought to acquire. Only the companies that accepted the principle of decentralized management gained some success in this area.

American Cyanamid was a particular case in point: Lederle in pharmaceuticals and Shulton in perfumes and cosmetics were managed in a highly satisfactory manner following their own criteria and without undue interference from headquarters.

No doubt because it is still managed by the founder's son, William Wishnick, *Witco* also respected the management styles of the different companies acquired over the years, striving to keep their top managements in place insofar as they were doing their job competently. By contrast, a prosperous specialties business like *Millmaster-Onyx* was jeopardized through being taken over by Gulf Oil. It was finally repurchased by its former owner, R. J. Milano, who returned it to profit status.

This inability of American chemical industry leaders to understand how businesses foreign to their traditional competence should be run explains, in part, the very few links established in these past few years between chemicals and pharmaceuticals in the United States. It was also true that American pharmaceutical companies, in most cases, had grown to critical size during the war and hardly felt the need to be taken over by the large groups. Thus, with the exception of Lederle, a long-time subsidiary of American Cyanamid, and of Squibb, which for a while became part of the Olin group, pharmaceutical companies prospered in total independence. Some of them had even diversified into fine chemicals, like *Abbott*, *Mallinckrodt*, and *Merck*. Others, like *Eli Lilly*, turned to agrochemicals, or like *Pfizer*, to industrial biochemistry through citric and itaconic acids. A few fermentation specialists had even struck away from their main path to follow the increasing trend toward synthetic chemicals at the expense of biochemistry.

Commercial Solvents, which had already used natural gas as feedstock between the two wars, successfully developed its own range of nitroparaffins. *U.S. Industrial Chemicals*, USI, a subsidiary of National Distillers, became a significant chemical entity. While still a supplier of fermentation ethanol, it had built up an inorganic chemicals division and also produced polyethylene. *Borden Chemical* had integrated upstream to the extent of methanol and vinyl acetate.

Thus, while the large chemical groups in the United States very rarely managed to penetrate the fine chemicals business, pharmaceutical companies and biochemical specialists adapted more readily to the disciplines of specialties and even to the production of large amounts of basic chemicals.

Foreign Investment in the United States Chemical Industry

The Europeans had been interested in the prospects offered by the American market for their chemical production and processes as far back as the turn of the century. Merck had been set up in New York in 1891 and had become an

American company after the First World War. Degussa had established Roessler & Hasslacher in 1882, sold to Du Pont in 1930. The Solvay Process Company, later a part of the Allied Chemical group, was born in 1881. Otto Röhm and Otto Haas had associated in 1907 to form a Philadelphia enterprise. Hoffmann-La Roche, which built a unit in Nutley, New Jersey, in 1928, had an import business established in New York since 1905. Other Swiss chemical giants—CIBA, Geigy, and Sandoz—jointly owned a Toms River, New Jersey, unit and the arrangement was questioned by the United States Justice Department only in 1970 when the Ciba-Geigy merger took place.

The two wars had a profound effect on European involvement in U.S. chemicals. Bayer's assets in the United States, confiscated by the American government in 1917, were sold off. Sterling Drug acquired the pharmaceuticals division while the dyes division was purchased by IG Farben in 1928, and became the following year the General Aniline Works. In 1939, IG added to its U.S. assets Agfa-Ansco and Ozalid Corporation, from which emerged General Aniline & Film (GAF).

France had a card to play in U.S. chemicals in the 1920s. Du Pont had become associated both with Air Liquide to develop Georges Claude's synthetic ammonia process, and with the Gillet textile group to produce cellophane and rayon. Edmond Gillet was on Du Pont's Board of Directors. Moreover, Air Liquide had shared in the establishment of Air Reduction, which produced liquefied gases. Du Pont, however, very soon bought up the stakes held by its French partners. Air Liquide's share in Air Reduction was sold at the end of the war because the French government had a pressing need, at the time, for dollars to pay for its purchases in the United States.

Since they were short of cash, France's chemical companies had not made many efforts to expand in the United States on a lasting basis and to make the most of the few technical assets in their possession. *Rhône-Poulenc* in 1955 had acquired Du Pont's Alamask department, and thereby owned a unit in New Brunswick, New Jersey, which produced coumarin and salicylaldehyde. But the French group's projects to produce superchlorinated PVC and terpenes ended in failure. The Deer Park plant in Texas revived only when a rare earths unit was subsequently brought on stream.

In the area of polymers, the exchange of technologies between Rhône-Poulenc and Phillips Petroleum involving Nylon in the United States and low-pressure polyethylene in Europe fell through within a few years, a victim of prevailing overcapacities.

Matters were hardly better in pharmaceuticals. Specia's Largactyl could not be developed directly for lack of sufficient means; it was licensed out to SmithKline and French, ensuring their prosperity for a long while; the association with American Home was unfortunately broken up, and the minority interest taken with Morton Thiokol ended in divorce.

Rhône-Poulenc's subsequent participation in Polychrome as well as its involvement in floppy disks were also short-lived.

The *Péchiney-Saint-Gobain* partnership was too absorbed with its interest in aluminum and glass to risk developing the PVC mass-polymerization process in the United States which its chemical subsidiary had invented. The process was merely licensed out to Dow and Hooker.

Ugine Kuhlmann was satisfied merely to set up a commercial agency in New York, but Dutch and English involvement in the United States was of a more solid nature.

There was first *Unilever*'s interest represented by the Lever Brothers units worth an estimated $116 million in 1963. The *Royal Dutch* group was involved through Shell Oil Company, which had a controlling interest in Shell Chemicals, one of the oldest petrochemical companies in the United States, set up in 1929.

ICI was to take a major step by acquiring Atlas Chemicals in 1970. It already owned Arnold Hoffman, a manufacturer of dyes and specialties for textiles, and had associated with Celanese within Fiber Industries (polyester and Nylon) as well as with U.S. Rubber within Rubicon Chemicals (TDI and aniline). *Courtaulds* had its own subsidiary in Le Moyne, Alabama, which produced rayon and Nylon 6. A British holding company, *Borax Holdings*, controlled U.S. Borax & Chemical in Boron, California, later to become part of RTZ.

The Dutch company *AKU*, after buying Glanzstoff in Germany in 1929, had taken over North American Rayon Corporation. Through its majority stake in American Enka it controlled several textile units which, besides rayon, manufactured Nylon 6 and polyester. The other Dutch group *DSM*, together with its PPG partner in their joint subsidiary Columbia Nitrogen, had launched into the production of nitrogen fertilizers, and then of caprolactam. The move, on an already congested market, was not very successful.

The only Italian chemical company to have settled in the United States in the 1960s was *Montecatini*, and it was not much luckier with its Novamont subsidiary. The Neal polypropylene unit in West Virginia suffered from technical problems and was finally sold after heavy losses.

Together with the Swiss, whose business in the United States was scarcely hindered by the war, the Germans were, of all European investors, the ones who showed the best sense of opportunity in seizing chances for gaining a foothold in the United States. Following the agreement signed with the Swiss holding company Interhandel under the Kennedy Administration, GAF was quoted on the New York Stock Exchange starting in 1965 and IG Farben's successors could not hope to regain control of it. This did not prevent German companies from seizing every opportunity likely to crop up to establish bases in the United States.

BASF, already a partner of Dow's within Dow-Badische which produced acrylic monomers, bought Wyandotte Chemicals in 1969. This was the former Michigan Alkali founded by J. B. Ford in 1890, and it provided BASF with the base products it needed to develop in the United States.

Bayer, associated with Monsanto within Mobay and with Pittsburgh Coke and Chemical within Chemagro, made a clever move when it purchased Miles Laboratories, known for its Alka-Seltzer and founded by Franklin Miles in 1880.

Hoechst took longer to find its niche. It was forced to buy out its associate Hercules in their polyester fiber subsidiary; the acquisition of Foster Grant was short-lived and Hoechst finally withdrew from polystyrene by selling out to Huntsman Chemical in Salt Lake City. But the purchase of Celanese in 1987 largely made up for these early mishaps.

Taking advantage of favorable currency rates and of technologies that could boost their United States subsidiaries, the three German giants managed to build up, in a relatively short time, a chemical base in the United States in areas which each had carefully selected. With the exception of the Swiss companies or, shortly afterward, of ICI, no European company could boast of anything approaching their achievement. Hüls' relatively modest approaches to acquire Nuodex, Henkel's too-scattered moves, Degussa's methionine investments in Mobile, Alabama, appear feeble in comparison with the ample means deployed by the three German giants.

Restructuring Britain's Chemical Industry

The structure of Britain's chemical industry had already been considerably simplified before World War II with the setting up of ICI in 1926 and the specialization of various companies in specific areas: *Distillers* in solvents, *Albright & Wilson* in phosphorus and derivatives, *British Oxygen* in industrial gases, *Fisons* in fertilizers and pesticides, *Laporte* in peroxygenated products and titanium dioxide, *Courtaulds* in artificial fibers.

Besides these companies, established as far back as the nineteenth century, others had developed through the initiative of German refugees who fled to England just before the war. *Lankro Chemicals* was founded in 1937 by F. H. Kroch to produce leather and textile auxiliaries. *Marchon Products*, named after its founders F. *Mar*zillier and F. *Schon*, was set up in 1939 and, since 1943, supplied phosphate-based detergents. It was taken over in 1955 by Albright & Wilson.

Petrochemicals Ltd., established in Carrington in 1946 to develop the Catarole process, truly played a pioneer role in Britain's petrochemicals industry. It subsequently established links with Erinoid in Styrene Products, with Pinchin Johnson and Lewis Berger in Styrene Copolymers as well as with Lankro in Oxirane, which brought out in 1949 the first tonnages of glycol ethers and ethanolamine.

Shell Chemicals acquired Petrochemicals Ltd. in 1955, together with its stakes in different subsidiaries and the rights it had acquired to Chaim Weizmann's and Karl Ziegler's patents. Reestablishing a tradition started by Liebig, Hofmann and Ludwig Mond, the Germans had thus contributed to opening up Britain's chemical industry to new development prospects.

During the postwar years, a number of gaps in the industry were also filled. To make the *acetylene* needed to produce PVC and chlorine solvents, ICI had brought a calcium carbide unit on line in 1943 in its Runcorn plant. Capacity was extended in 1958 and 1961. Through these extensions, together with the production of British Oxygen's subsidiary, Carbide Industries, located in Maydown, Ireland, which supplied the acetylene needs for Du Pont's neoprene, Britain became self-sufficient in carbide in 1965, for the first time in her history.

Synthetic rubber production only started in Britain in 1958 in the framework of a consortium set up by Dunlop, Firestone, Goodyear, North British Rubber, BTR Industries, and Pirelli under the name of *International Synthetic Rubber* (ISR). It manufactured styrene-butadiene elastomer in Grangemouth. Simultaneously, Du Pont was launching its Neoprene project in Northern Ireland, while Esso brought a butyl rubber unit on stream in Fawley in 1963.

Britain also needed to adapt to use of petroleum as a feedstock for its organic chemicals. Until the war the industry had relied on fermentation alcohol, coal, and tar derivatives. By 1965, the adaptation was achieved with 70 percent of Britain's heavy organic chemicals production based on petrochemicals, compared to only 6 percent in 1949.

The experience acquired in Billingham by ICI during the 1930s in coal hydrogenation; the pooling in 1947 of Distiller's (DCL) and British Petroleum's knowhow in a joint subsidiary called *British Hydrocarbon* set up in Grangemouth; the investments that oil groups like Shell, Esso, Phillips Petroleum, Standard Oil of California, and the American chemical companies like Union Carbide, Du Pont and Monsanto made in British chemicals—all contributed to a smooth adaptation.

While major petrochemical projects were being carried out requiring the participation of the oil and chemical industries within joint subsidiaries, each individual group was striving to order its chemicals operations around its own strong points.

ICI was to confirm its preponderance, which before the war had been achieved through its size, the variety of its products, imperial preference in the Dominions, and cartel arrangements in alkalis, nitrates, explosives. Indeed, ICI was undoubtedly a giant among postwar chemical companies. Despite the advantage it drew from its home market and its long involvement in base products, however, it henceforward had to fight the American and German giants on the world's open market; for they, too, had markets to conquer.

One of the first tasks of the company's leaders was to complete the group's upstream integration within the new environment it had to face. Through its association with Phillips Petroleum in the refinery which Phillips Imperial Petroleum owned on the Tees River, and through long-term contracts with other oil companies such as Esso in Fawley, ICI provided for its essential naphtha needs. The group had taken a 19.2 percent stake in the North Sea Ninian oil field, which began producing oil at the end of 1978. Its *petrochemicals division*, located on the Wilton and Billingham sites linked to each other through pipelines, was designed to feed the other divisions in olefins, aromatics, Oxo alcohols, and intermediates for chemical fibers (TPA and Nylon 66 salt).

From the methanol and ammonia produced in Billingham by the *agrochemicals division*, ICI makes methylamines. The methanol itself is produced from a very effective low-pressure process developed by the group and licensed out worldwide, ICI also supplying the catalysts. This methanol is also the starting point for the Proteen synthetic proteins used as animal feed. Billingham's ammonia capacity, the largest in the world, is almost exclusively directed to the production of nitrate fertilizers. ICI also brought on line in 1966 a vinyl acetate unit based on an original ethylene process, which Celanese uses in the United States.

From the ethylene supplied from Fawley, ICI produces ethylene oxide in Severnside. It is the group's strategic product used for the Lissapol detergents and to prepare the ethylene glycol intended for the explosives of the Nobel Division and for the Terylene of the Fibers Division. The *Mond division* chemicals, which still account for 15 to 20 percent of the group's turnover and personnel, remain geared, of course, to the chlorine and alkalis produced on the various northwest sites of Britain as well as in Billingham and Wilton. Fluorine hydrocarbons have been added over the years, used as refrigerants and in aerosols, as well as a very wide range of chlorine derivatives (vinyl chloride for PVC, chlorinated paraffins, chlorinated rubber, chlorinated solvents for dry cleaning and metal scouring). For historical reasons, it is also the Mond division that produces in Billingham from HCN the methyl methacrylate monomer for making Perspex as well as alkali cyanides.

ICI has regrouped in an *organic division* various products that date back to the time when it was involved in dyes, an area where it demonstrated its innovative capacities with the discovery, in 1953, of the first reactive dyes, the Procion dyes, preceded by the chance discovery in 1934 of phthalocyanine pigments such as Monastral Blue.

Over half the products in this division, which is still headquartered in Manchester, involve organic dyes and pigments. But its knowhow in organic synthesis has also served to produce intermediates and active ingredients for the specialized pharmaceutical, agrochemical (plant protection) and rubber auxiliaries (within the Vulnax subsidiary with Rhône-Poulenc) divisions. Like all dye producers, ICI had its own range of textile and paper auxiliaries. But in somewhat arbitrary fashion the organic division encompassed nitrocellulose, silicones, synthetic resins for paints, and precursors for polyurethane foam, ICI having taken part in the development of MDI in the fifties.

The *plastics division*, later to be merged with petrochemicals, has its headquarters and research and development centers in Welwyn Garden City, Hertfordshire. ICI had discovered polyethylene in 1933 and developed an industrial process for the production of polymethylmethacrylates (PMA) in 1931. The division therefore manufactures high-pressure polyethylene, which is also licensed out to more than thirty producers worldwide. In addition, it manufactures PMA, Perspex in sheets, and Diakon in molding powder, to which have been added PVC, polypropylene, and a number of special polymers such as Nylon 66, polytetrafluoroethylene, PTFE, Victrex sulfone polyether, and the Butakon butadiene copolymers. Polyethylene, polypropylene, and polyester film are also produced in this division on different sites.

When it was set up, ICI had inherited what is still the *Nobel's Explosives Division*. Most of its activities remained in Ardeer, Scotland. But in a market that is hardly developing, the division accounts, nonetheless, for no less than 20 percent of the free world's exports in its specialty.

ICI had long hesitated before getting into fibers for fear of competing with one of its main clients, Courtaulds. But in the end it has become one of the largest producers of such fibers, beginning in 1940 with the establishment of a subsidiary, British Nylon Spinners, in partnership with Courtaulds, to develop a Du Pont Nylon license. ICI subsequently bought up Courtaulds' share in the subsidiary. In 1947, ICI had acquired from Calico Printers world rights (except in the United States) for their polyester fiber. It was marketed in 1955 under the trademark Terylene after a long development program. ICI's *fibers division*, geared to the production of these two fibers, employed over 10,000 people in 1979; and its requirements in intermediate products accounted for a significant share of the petrochemicals division's production.

ICI had started its involvement in pharmaceuticals research in 1936 when it formed a specialized department within the dye division, an approach similar to that of Hoechst and Bayer. A *pharmaceuticals division* was set up in 1957. Research was carried out by the Alderley Park teams in Cheshire and had produced sulfamethazine as early as 1942. Other original products also emerged; an antimalarial, paludrine, in 1946; then a derivative Hibitane, an antibacterial in 1954; followed by a remarkable anaesthetic, Fluothane, in 1957; and in 1964, Inderal, for treating cardiovascular diseases. ICI's pharmaceuticals division, employing over 4,000 people, thus makes 80 percent of its sales from products developed by its own research teams.

The group has been equally successful with its plant protection products. As in the case of fibers, an association was first sought with a specialist in the area. ICI set up *Plant Protection* in 1937 with Cooper McDougall & Robertson, then in 1958 bought out its partner, turning the company into one of its

The first ICI polyethylene pilot plant units in Wallerscot, Cheshire, 1937.

own divisions. In the 1940s, its range included products developed in cooperation with government institutes, such as the insecticide Gammexane and the herbicide MCPA. Subsequently derivatives produced from research conducted in Jealott's Hill were added.

Plant Protection in 1960 discovered the herbicides diquat and paraquat belonging to the bipyridyl family. Original systemic insecticides, such as Menazon, from which emerged the Sayfos-type products, were also the fruit of ICI's research. ICI's involvement in paints dates back to the establishment of Nobel Chemical Finishes as a manufacturer of cellulose lacquers in 1926. The subsidiary became ICI (Paints) Ltd. in 1940. It is today the *Paint Division*, and its Dulux line was the first of its famous decorative paints to be produced.

As is to be expected from a group that is so large and so diversified, ICI has strong bases the world over. Because of Imperial Preference, ICI already had strong positions in the Dominions before the war. The group continues to be a significant producer of explosives, fertilizers, and various other products in South Africa through the *AECI* subsidiary (40 percent ICI, 40 percent de Beers), in India through Alkali & Chemical Corporation of India (ACCI), and in Australia through ICI Australia.

ICI became a producer on the European continent with the construction in 1961 of the Rozenburg unit in Holland, initially producing Perspex and polyethylene. Since then, ICI has built a plasticizer unit in Baleycourt near Verdun in France in 1962, and set up a Nylon-66 fibers unit in Oestringen, West Germany. During the 1970s, the group brought on line a polyethylene unit in Fos-sur-Mer, in southern France, while an important site was established in Wilhelmshaven in West Germany, including electrolysis facilities, vinyl chloride, and PVC units.

Under an agreement signed with Du Pont, ICI had been prevented from establishing a base in the United States before the war, even though it had settled long ago in Canada through its subsidiary *CIL*. When the war ended, ICI sought to form alliances with American firms in line with those it had signed with Celanese and U.S. Rubber. Its first major achievement, however, was the purchase of Atlas Chemicals in 1971. This was the former Atlas Powder Company of Wilmington, Delaware. A less fortunate venture was that of Corpus Christi Petrochemicals, which ICI decided to set up in 1978 in Texas with other partners to produce ethylene. The investment amounted to $600 million, and the unit was acquired subsequently by Cain Chemicals. Since then, ICI's development in the United States has mainly been geared to specialties, the management of ICI Americas being determined to pull out of petrochemicals.

Through Duperial, from which Du Pont had to withdraw following an antitrust move by the United States government, ICI has various productions in Argentina. In Brazil, it has a not insignificant base both alone, and in association with local groups.

Through the diversity of its products and its centers in all the world's strategic areas, ICI has no equal in the world. The group's leaders did indeed make inopportune decisions at times, as when, under Paul Chambers' chairmanship, they tried to acquire the fibers producer Courtaulds, or more recently when they decided upon an upstream integration in United States petrochemicals, or when they invested too late in PVC in Europe.

It is also true that, unlike Bayer and Monsanto, ICI was not able to set up a profitable business in the rubber chemicals area, and has kept to a modest place in elastomers. ICI is also absent from some interesting plastics, like polycarbonates and polyacetals. On the whole, however, the group has brilliantly succeeded in the many sectors it decided to penetrate; this was in large part due to its decentralized organization into highly competent divisions. These were self-governing, and each had research teams of great talent to judge from the number of original processes and new products developed since 1926.

The increasing involvement of *British Petroleum* (BP) in chemicals took place in two stages: first, through a common subsidiary set up with Distillers in 1951 called British Hydrocarbon Chemicals (BHC), then through the acquisition in 1967 of the stakes Distillers held in chemicals and plastics, making *BP Chemicals* the second-ranking British chemicals company after ICI. In its first stage, BP chose to associate, during the 1950s, with partners well-versed in the chemical knowhow it was lacking and able to share in the financial outlays required by the projects. This policy produced a series of joint petrochemical subsidiaries both in Britain and abroad: BHC with Distillers in Grangemouth, Scotland; Naphtachimie with Péchiney and Kuhlmann in Lavera in southeastern France; Erdoel Chemie with Bayer near Cologne.

Other alliances were formed at home involving downstream base petrochemical products: Forth Chemicals (two-thirds BHC, one-third Monsanto), which began producing styrene in 1953; Grange Chemicals (two-thirds BHC, one-third Standard Oil of California), which produced its first alkylate volumes the same year; Border Chemical (one-third each BP, Distillers, ICI), set up in 1963 to make acrylonitrile from propylene under a joint Ugine and Distillers process.[12] Standard Oil of California also became BP's partner in 1961 within BP California to supply ortho- and para-xylene, and within Orobis for lubricating oil additives.

BP's second growing stage in chemicals was achieved through acquisitions. Mobil's interests in plastics in Britain, particularly Erinoid, were bought

[12] Another joint Ugine/Distillers process enabled BP to become a 50 percent partner in Société Distugil, which produces polychloroprene in Champagnier near Grenoble, in southeastern France.

in 1965, while Distillers' joint subsidiary with Union Carbide, called Bakelite Xylonite Ltd. (BXL) since 1963, became part of the BP group through purchase of Distillers' stakes in chemicals and plastics. At a time when United States petrochemical companies were pulling out of Europe partly because of existing overcapacities, BP did the reverse by buying some of their units. It bought Monsanto's polystyrene unit in Wingles, France, and Union Carbide's petrochemicals site in Antwerp.

With its units in Britain, and its bases in France, Belgium, and Germany, BP had become one of Europe's chemical giants. A major part of its range of products, however, involved commodities: olefins, aromatics, cumene, phenol, acetone, ethanol, and acrylonitrile in Grangemouth; olefins, isopropyl alcohol, styrene, and vinyl acetate in Baglan Bay; acetic acid, acetates, and phthalic anhydride in Hull.

BP Chemicals International was also involved in Europe in polyethylene, ethylene oxide, glycols, glycol ethers, and ethanolamines, where competition remained strong. These products were based on the oil and gas resources which the parent company drew from the North Sea and which it could provide at the lowest cost. Moreover, BP Chemicals' involvement in polyethylene for cables, inherited from Union Carbide, the profits it drew from its association in Distugil's polychloroprene, the efficient processes acquired through the purchase of Distillers for cumene-based phenol as well as for acrylonitrile and acetic acid, were all favorable assets in the difficult chemical area it had chosen to develop.

Shell Chemicals Ltd., set up in 1955 when Petrochemicals Ltd. was acquired, owns two petrochemical plants in the North West, in Carrington and Stanlow, and one in Shell Haven, Essex. Carrington produces C_2, C_3, and C_4 olefins, ethylene and propylene oxides, the ethoxylates and polyols derived from them, and the major thermoplastics (low-density polyethylene, polypropylene, polystyrene).

Stanlow is geared to aromatics (benzene, toluene), to the range of solvents derived from propylene and butylenes produced on-site, to detergents, to alcohols (*n*-butyl alcohol, 2-ethylhexanol, high grade alcohols) and to phenol. In Stanlow also Shell Chemicals produces part of its specialties, lube oil additives, anticorrosion agents, Epikote resins, herbicide intermediates, high grade olefins. Shell Haven supplies the alkylate for detergents and sulfur.

With its long tradition in the discovery of new processes and special products, the Royal Dutch Shell group has directed Shell Chemicals towards the production of specialties, as it has done for its other subsidiaries worldwide. The trend was confirmed with the recent acquisition of Ward, Blenkinsop, a medium-sized company located in Widnes, which has a traditional pharmaceuticals-synthesis and fine-chemicals background.

Albright & Wilson has led a rougher life. The company dates back to 1844 when Arthur Albright began manufacturing white phosphorus for the match industry. In 1856, Albright took his friend, the Quaker J. E. Wilson, as his partner; and in 1896, Albright & Wilson set up Oldbury Electro-Chemical Company in Niagara Falls, New York, to produce phosphorus and its derivatives on the American continent.

By a strange twist of fate, it was an American group, Tenneco, which seventy-six years later became sole owner of Albright & Wilson. And yet the company had spared no effort to diversify after the Second World War. It had bought Marchon Products in 1955, Boake Roberts in 1960, W. J. Bush in 1961, and Stafford Allen in 1964. In 1964, Albright & Wilson coordinated these acquisitions under the Bush, Boake, Allen Ltd. (BBA) entity.

While remaining faithful to the phosphorus-based chemicals it produced in Oldbury, Albright & Wilson had also become involved in detergents and fatty alcohols with Marchon and its Whitehaven plant, and in fragrances and perfume bases through its BBA subsidiary. Through BBA, the company had also acquired a range of organic derivatives based on the oxidation and photochlorination of toluene produced in Widnes (benzyl and benzoyl chlorides, benzyl cyanide, benzyl alcohol, sodium benzoate).

The production of phosphorus requires 14 kwh of electricity per kilogram of phosphorus. So Albright & Wilson set up the ERCO subsidiary in Canada to take advantage of cheap electricity. But difficulties in getting the Varennes unit in Quebec on stream and the addition of a further production unit set up in 1969 in Long Harbour, Newfoundland, heavily taxed the company's financial resources. For this reason, it had to sell out to Tenneco. Tenneco subsequently got rid of BBA, returning Albright to its original chemical base, except for Marchon's activities.

Fisons's history was also a lively one. The company, which took the name of Fisons Ltd. in 1942, was one of the leading mixed fertilizer producers in Western Europe when it became involved in pesticides and horticultural products. These were grouped within a single division with headquarters in Cambridge. Fisons gained a foothold in the hydrazine and derivatives market (blowing agents, polymerization catalysts) through its purchase of Whiffens, with Loughborough and Widnes production units. Through other acquisitions carried out between 1937 and 1964 (Binger Laboratories, Genatosan, Fulford-Vitapointe) Fisons Pharmaceuticals Ltd. was formed, with a range of products that extended from toiletries to prescription drugs.

Since then, Fisons has shed its fertilizer business. The Cambridge division, after merging its plant-protection sector with Boots', was sold to the German company Schering at the same time as the industrial products department inherited from Whiffen. Fisons has now become a strictly pharmaceuticals business also involved in scientific instruments.

While Fisons was withdrawing from chemicals, another company that had started with textiles, was pursuing its efforts to go further into chemicals. *Courtaulds* began in 1815 with Samuel Courtaulds' ribbons and mourning crepe. It became involved in chemicals through the Cross and Bevan viscose process developed first in England in 1905, then in the United States with the establishment of American Viscose five years later.[13] During the twenties, Courtaulds set up rayon viscose units in France, Germany, Italy, and Canada.

A dispute arose before the war with British Celanese over acetate rayon. Courtaulds was to buy this group in 1957. In 1937, Courtaulds had taken a 75 percent share in the setting up of British Cellophane. Three years later a fifty-fifty partnership was formed with ICI within British Nylon Spinners, an association that ended in 1964. Courtaulds had started, as early as 1916, to manufacture in Stretford, near Manchester, the sulfuric acid and carbon disulfide needed to prepare viscose.

Through cellulose acetate produced by Nelson Acetate, in which Courtaulds had a 50 percent stake, the company became involved in acetic anhydride and in ketene chemicals (such as acetoacetates). It enlarged its range of cellulose and acetic acid derivatives by manufacturing monochloroacetic acid and water-soluble gums (CMC, methyl cellulose) in Spondon and Coventry. Purchase of Leeks Chemicals gave Courtaulds a foothold in acetate-type plasticizers like triacetin, used in cellulose acetate fibers and cigarette filters. One of Courtaulds' successful diversifications was the acquisition of Pinchin Johnson, which led to the setting up of International Paints. Strongly involved in acrylic fibers, the group also played a pioneer role in the promising area of carbon fibers, with polyacrylonitrile (PAN) as the precursor.

With its workforce of 50,000 in Britain and 30,000 abroad, Courtaulds is an essentially textile-oriented group. But its diversification in specialized chemicals has proved successful because of timely acquisitions, even though chemicals proper still form a modest share of its overall activities.

Croda, which was quoted on the Stock Exchange in 1964, was set up in 1925 to manufacture lanolin from the wool grease recovered from textile factories. Croda started enlarging its range of products in 1945 with fatty acids. Then under Fred Wood's successful management, it diversified into resins, paints, and food through a series of acquisitions. With over twenty subsidiaries, it became something of a small international conglomerate. Croda's involvement in chemicals proper was less successful. Its subsidiary Synthetic Chemicals which produced *p*-cresol, xylenols, and synthetic pyridine and which had been part of Midland Tar, was finally sold to Shell Chemicals.

Although *Unilever* had always shown a marked preference for consumer products, it was the sole shareholder since 1896 of Joseph Crosfield & Sons, the silicate manufacturer and in 1937 it acquired the fatty acid specialist Price.

In 1960 Unilever formed a chemical group in Britain, then set about buying up small and medium-sized national companies involved in adhesives, synthetic resins, and perfume bases. Pushing further into these three areas,

[13] Courtaulds sold American Viscose in 1941.

Unilever bought National Starch in the United States, Synrès in Holland, and Bertrand Frères in France. Since that time the Anglo-Dutch group had become a significant world force in specialty chemicals of a very specific nature. But except for the very understandable case of oleochemicals and silicates, it has stopped short of going too far upstream.

Britain was lucky to possess a *pharmaceuticals industry* that was very efficient, thanks to a few laboratories set up during the nineteenth century that had managed to remain dynamic. *Beecham*, which had bought Macleans and Eno in 1937, stood out after the war through an ambitious research program that began in 1954 and produced new synthetic penicillins. *Boots*, which had shed its plant protection business, concentrated on pharmaceuticals and cosmetics as well as toiletries. Glaxo Laboratories was the main unit of the *Glaxo Group*; it was set up in 1962 to encompass about sixty subsidiaries involved in pharmaceuticals, immunology, veterinary products, and agrochemicals. Glaxo associated with British Drug Houses (BDH) in 1965 to form a common subsidiary for the wholesale marketing of their pharmaceuticals. *Smith & Nephew*, set up as a group in 1937, also owned a large number of subsidiaries involved in prescription drugs, hospital supplies, and consumer products. These subsidiaries dealt not only in various sectors at home but also had an international dimension through former bases set up in Commonwealth countries.

One industry that had trouble surviving after the war was the tar derivatives industry, which had originated in Britain. The nationalization of the coal and steel industries did not help, although a private sector remained. It was perhaps this dual structure that made the necessary mergers difficult, such as those successfully carried out in West Germany.

With the exception of *Coalite & Chemical*, which had been wise enough to concentrate on the chlorine derivatives of phenol and which supplied through its low-temperature carbonization process the compressed coal bricks very popular for home heating, most of the companies within this sector like British Tar, Midland Tar, Yorkshire Tar, were merely trying to survive when they were not dying out. A company like *Hickson & Welch*, dating back to 1893, managed to prosper because its synthetic intermediates and its dyes were downstream from tar derivatives and it diversified in the treatment of wood and fine chemicals.

When the first energy crisis broke out, Britain's chemical industry, which, like those of all industrialized countries, had already been hit by the consequences of excess capacities in fibers, plastics, and petrochemicals, was already well on the way to being restructured. ICI emerged as the dominant group, being involved in all chemical sectors and having a good research base as well as already strong international positions.

The Shell and BP Oil companies played their part in production rationalization, appropriating in the process fair-sized market shares both in Britain and on the Continent. For firms like Fisons and Albright & Wilson restructuring practically meant dismantling, and the environment was hardly favorable for the tar-derivative industry.

There were some flourishing industries, however, particularly among the large pharmaceutical companies and in the consumer products area related to chemicals. More especially, the outlook had been considerably simplified for British chemicals. New moves could now be made by the surviving groups to improve the ways and means of measuring up to their powerful foreign competitors.

The Complexities of the French Chemical Industry

The postwar French chemical industry emphasized its particularities even further compared to the American, German and British industries. Its research was for the most part carried out through public funding by state-owned research bodies. The main one, the Centre National de la Recherche Scientifique (CNRS), was staffed by as many as 25,000 people encompassing all disciplines. The relationships between the CNRS researchers and those of the different university laboratories, on the one hand, and industry on the other, were distant ones, which explains why so little of France's chemical research had industrial consequences. Although in the specialized schools chemistry was taught by a competent staff, as it was also in the university departments, the careers open to graduates were not very exciting.

On the one hand, chemical industry leaders had come to chemicals through other industries that were regarded as more prestigious, such as glass, aluminum, steel, oil, or textiles. On the other hand, even in their own specialty, graduates from the higher chemical schools hardly ever climbed to posts of responsibility where they would have been likely to develop their leadership talents. These posts were generally given to graduates of France's Grandes Écoles, where mathematics rather than applied sciences was the leading subject. The Écoles de Chimie more often than not recruited those who had failed to pass entrance examinations for the Grandes Écoles.

In addition to these characteristics, which had prevailed since before the war, France had acquired new features that set it apart from other industrialized nations of the free world.

Planification considered by General de Gaulle as an "ardent obligation," and the need for a concerted industrial policy were accepted as gospel truths by the civil service, but also by industry leaders who often were recruited within the government administration. The role of the state was further emphasized by the duties it felt it had to assume after every war on behalf of the nation's higher interests. After World War I, the state, which was already involved in the chemical industry through the Régie des Alcools or the Monopole des Poudres, became even more so through the establishment of the ONIA and the Mines Domaniales de Potasse d'Alsace (MDPA).

After World War II, the nationalization of the coal mines gave the state a good opportunity to take a firm foothold in coal-based chemicals through the Houillères du Nord et du Pas-de-Calais (HBNPC) and the Houillères du Bassin de Lorraine (HBL). Sequestered because of its links with IG Farben, Francolor was subsequently returned to Établissements Kuhlmann after a long procedure. But although Francolor was returned to the private sector, the rise of the oil industry was to provide further occasion for public meddling in the chemical industry.

While the French government was loath to relinquish its industrial stakes, it was in no way opposed to associating with private groups, thus still further adding to the complexities of the French chemical industry. Moreover, companies that were set up for very specific purposes moved on to such different tasks as time went by that the reasons for setting them up were quite forgotten. *Charbonnages de France Chimie* (*CdF Chimie*), for instance, has practically nothing to do any more with coal-based chemicals. Likewise, *Société Nationale Elf Aquitaine, SNEA*, set up to develop the gas from Lacq and its by-product, sulfur, became, under the name ELF Aquitaine, a conglomerate encompassing oil and gas exploration worldwide, as well as refining and a large number of other activities ranging from heavy to fine chemicals through ATO and to pharmaceuticals and perfumes and fragrances through Sanofi.

Well before the 1981 nationalization measures that put the whole of France's chemical industry under state control—a unique occurrence in the free world—the industry was already under tight government supervision through economic planning, through authorizations needed for any investments made by foreign companies in France, through state-owned or joint state and private companies, and through the close ties established between government management and civil service, all of whose members came from the same Grandes Écoles.

In their operations in France, chemical companies were under two handicaps. For one thing, the cost of energy, which often had such a determining effect on market prices, depended on factors over which they had no control. Indeed, the oil industry was governed by a 1928 law that gave the government a monopoly over oil imports into France, while the gas and electricity utilities were also state monopolies. This meant that energy in France was not supplied under normal market conditions.

An aggravating circumstance was the price control maintained in France for thirty years. Rigorously applied to chemicals, it prevented the industry from building up financial reserves when it could have done so. Caught in this vise, the companies were unable to make the same profits as their overseas competitors, and they were too short of cash to finance their investments.

The practically inextricable network of subsidiaries that developed within the industry grew not only out of these financial difficulties, but also out of the need to associate with foreign partners that could provide the technologies it was unable to develop on its own. Considering all these constraints, it is amazing that the industry managed to catch up the time lost after the war during the "thirty glorious years" of economic expansion.

The only company of truly chemical origins was *Rhône-Poulenc* born in 1928 under the name of Société des Usines Chimiques Rhône-Poulenc (SUCRP), out of the merger of Société Chimique des Usines du Rhône with Établissements Poulenc Frères. Since 1935, SUCRP had been headed by Albert Buisson, who had friendly relations with Camille Poulenc. He remained president until 1956. Buisson, a pharmacist, later acquired legal and financial training that proved highly advantageous for the group during his long career there. The general manager, Nicolas Grillet, was an industrial leader who understood research problems. The pharmaceutical tradition remained strong in the group: the Spécia subsidiary dated back to 1928, Albert Buisson in 1956 brought in the Société Théraplix, and between 1963 and 1968 majority stakes were taken in Laboratoire Roger Bellon and in Institut Mérieux.

Both in pharmaceutical synthesis and in biochemistry the group has a large number of achievements to its name: sulfonamide in 1935, the first antihistamines in 1942, penicillin in 1945, the first neuroleptic in 1950, spiramycin between 1953 and 1955. Through Poulenc Frères, which had acquired a majority stake in 1927, SUCRP had inherited a subsidiary in Britain, May & Baker, which specialized in fine chemicals and pharmaceuticals.

Through Usines du Rhône, Rhône-Poulenc had also become a 50 percent shareholder of Rhodiaceta, which was involved in artificial fibers. Owing to agreements signed in 1939 with Du Pont, the range was enlarged to include Nylon. The Société Rhovyl subsidiary was set up in 1948 by SUCRP-Rhodiaceta and Saint-Gobain to develop the first vinyl fiber, invented in 1937 by a chemist with Rhodiaceta. The production unit was located in Tronville en Barois, in Northern France. Rhodiaceta also brought from ICI in 1954 the license for France of the polyester fiber marketed under the trade name Tergal.

The following year, Rhodiaceta and Comptoir des Textiles Artificiels (CTA) developed in their research centers an acrylic fiber, Crylor. In 1961, the castor oil-based polyamide fiber, Rilsan, invented in 1947 by Péchiney and produced by its subsidiary Organico, became the joint property of Rhône-Poulenc and Société Valentinoise d'Applications Textiles. That same year, Rhône-Poulenc SA was set up as a holding company and acquired CTA's entire production base both in France and abroad as well as the Société La Cellophane. In exchange for handing over CTA, the Gillet group was given a 15 percent stake in Rhône-Poulenc SA.

It made sense to encompass these different acquisitions within a company called Société Rhône-Poulenc Textiles, which was set up in 1949. As long as it enjoyed exclusive rights on the Nylon and polyester patents, the group made large profits out of its textiles division. When the patents expired and France became a member of the European Economic Community, Rhône-Poulenc was forced to undertake agonizing reappraisals despite the fact that it was well-integrated upstream on intermediates and downstream in the weaving business.

The Rilsan fiber was abandoned, as was Crylor a few years later, squeezed out of the French market by Courtaulds' acrylic fibers. Rhovyl still remained a worthwhile specialty, but its markets were restricted. The greater part of Rhône-Poulenc Textiles' efforts were henceforth to be devoted to rationalizing its Nylon and Tergal productions in order to cope with world overcapacities entailing considerable losses for the group from the seventies onwards.

Likewise the cellulose film which Société La Cellophane produced in Mantes and Bezons near Paris was abandoned because of a dwindling market, after a long career that had started in 1913.

Rhône-Poulenc's chemicals proper were also very diversified. The Saint-Fons factory had separate facilities for the production of phenol, salicylic acid, and acetic anhydride. Multipurpose units manufactured various products for pharmaceuticals, agrochemicals, and perfumes.

Rhône-Poulenc's technical services can be credited with setting up one of the first phenol-synthesis units using the cumene process. The company also pioneered in Saint-Fons the synthesis of dihydroxyphenols through hydrogen peroxide oxidation of phenol to produce pyrocatechol and hydroquinone.

Although SUCRP had a long involvement in rubber additives, it had never pushed far into the area. After an association with ICI in Vulnax, it finally pulled out of the field. The company had better luck in plant-protection products, particularly when its own technical strengths in this area were added to Pechiney-Progil's marketing skills.

SUCRP's involvement in plastics had occurred through cellulose acetate, but it also produced vinyl resins since 1937. The range was enlarged after the war through association with General Electric for silicones and with Phillips Petroleum for low-pressure polyethylene. In 1956, the company, together with Saint-Gobain, Progil and Ugine, formed the Société Dauphinoise de Fabrications Chimiques (Daufac) to produce vinyl chloride monomer in Jarrie (Isère). For lack of the necessary petrochemicals integration, however, Rhône-Poulenc finally pulled out of all the main thermoplastics, keeping to the specialties, Technyl Nylon, silicones, polyimide resins and butadiene-styrene latex. The only monomer retained was the vinyl acetate produced in Pardies.

The group had a number of significant positions abroad after Rhône-Poulenc SA was set up. To the overseas pharmaceuticals subsidiaries, textile companies were added. While it had trouble taking a foothold in the United States, Rhône-Poulenc was strongly entrenched in Brazil, where its RIQT

Rhône-Poulenc synthetics plant at Saint-Fons. Courtesy R.P.-DCNB.

subsidiary was the leading foreign group in the country. In Argentina, the Rhodia SA subsidiary remained involved in synthetic fibers and pharmaceuticals, but had pulled out of such chemicals as sorbitol and solvents. Through Rhodiaceta's former base, the group in Europe produced synthetic fibers in Switzerland (Viscose Suisse), in Germany (Deutsche Rhodiaceta AG), and in Spain (SAFA).

Thanks to the May and Baker subsidiary, Rhône-Poulenc was present in pharmaceuticals and agrochemicals not only in Britain but also in various countries of the former Commonwealth. Rhône-Poulenc's policy of association with outside partners, while more restricted than that of other French chemical groups, helped it to diversify in areas in which it would have been reluctant to tread alone.

This policy produced a number of subsidiaries that retained their original form only fleetingly: Manolène for polyethylene, Acétalacq for acetaldehyde, Daufac for vinyl chloride monomer, Sodes for synthetic ethanol, Tolochimie for TDI and isocyanates. Lautier, which produced in Grasse the basic ingredients needed for perfumery, became part of the group but was sold to Florasynth after a few disappointing years. Givaudan-Lavirotte also gained little from its association with Rhône-Poulenc. The same could be said of Prolabo, a very old subsidiary, which never managed to reach a critical size for want of any decisive backing from the parent company. Alimentation Equilibrée à Commentry which brought a precious complement of vitamins and amino acids to Rhône-Poulenc, was set up in 1939 and gradually absorbed by the group.

Just before the great restructuring of France's chemical industry, Rhône-Poulenc's image was that of a very carefully managed group financially, that had its strong points in fine chemicals through its Saint-Fons and Vitry units and its Centre des Carrières research base, and that had promising activities through some of the recently acquired subsidiaries (Mérieux, Roger Bellon, AEC). But the group was exposed to the difficulties brought about by world overcapacities in its chemical fibers sector. And it was still of two minds about integrating in upstream petrochemicals and diversifying into plastics. Finally, the group's leaders still had to decide about the future of subsidiaries such as Lautier, Prolabo, Givaudan-Lavirotte, and to work out a new policy for approaching the large United States market.

The joint venture *Péchiney-Saint-Gobain (PSG)* was formed in 1959; in 1961, both parent companies incorporated their chemical assets, including their stakes in chemical subsidiaries. Because of its origins, the company was much involved in heavy inorganic chemicals (mineral acids, chlorine, soda, ammonia) and organic chemicals (vinyl chloride, chlorine solvents, phthalic and maleic anhydrides). The joint venture had also become France's foremost producer of mixed fertilizers and was at the same time involved in a number of interesting specialties (organic fluorine derivatives, rare earths, aluminas, catalysts).

PSG could also benefit from the strong positions held by its two partners in vinyl polymers, and it produced polyester resins and latex. It had moreover inherited stakes in a variety of sectors: titanium dioxide and potash products (Thann et Mulhouse), plasticizers, solvents, and fine chemicals (Melle-Bezons), polystyrene (Plastichimie), plastic tubes (Armosig), and plant protection products (Pepro).

PSG had achieved a few successful industrial developments: mass polymerization of PVC; a "chloé" unit for the production of chlorinated solvents from ethylene; a liquid-extraction process for the separation of rare earths; and an efficient process for phosphoric and nitric acid manufacture. But by their nature, PSG's base products were not really suitable for overseas expansion and positions abroad; indeed, they were limited to the Reposa subsidiary in Spain for vinyl polymers and the Zuidchemie subsidiary in Holland for fertilizers.

PSG's industrial structure was also difficult to manage. There were seventeen plants and two mines scattered over France, and the merger of the two companies led to no rationalization. Moreover, there was no satisfactory integration of raw materials. Péchiney had retained hold of its stake in Naphtachimie, so that petrochemicals remained outside the company. And even for ammonia, partners such as APC and Cofaz had to be found to finance the 1,000-tons-a-day unit in Grand Quevilly in Normandy.

Although it ranked third among France's producers and had a 13,000-strong workforce, PSG, generally speaking, did not get from its partners the financial support needed to rationalize and develop its various sectors of activity. Neither Saint-Gobain nor Péchiney seemed inclined to invest heavily in their joint subsidiary.

The company known as *Progil* since 1920 had by that time added the disodium phosphate unit in Roches de Condrieu near Lyon to the three plants in Condat, Molières and Lyon-Vaise, also near Lyon, which until then had formed the chemicals units of Gillet et Fils. Société Le Chlore Liquide was absorbed in 1923, and the same year Progil began producing in Roches de Condrieu the carbon disulfide needed for viscose. Five years later Résines et Vernis Artificiels (RVA), was set up to produce phenolic resins and alkyds. After World War I, Maurice Brulfer, a chemical engineer who had graduated from the École des Industries Chimiques in Nancy, was appointed head of Progil. A man with true entrepreneurial spirit, he had acquired Société des Produits Chimiques de Clamecy in 1920. The dedication to chemicals which he brought to the group was kept up under his successor Jean Montet.

From its origins and its prewar acquisitions policy, Progil retained three main types of business after the war: wood and derivatives, inorganic chemicals and synthetic resins. In order to pursue its development despite its more restricted financial means, Société Progil undertook a systematic policy of associations with partners that could provide both funds and technology.

While it had acquired Salines du Sud-Est to integrate the rock salt needed for its Pont de Claix electrolysis, Progil resorted to associations for its petrochemical supplies. It took a stake in the Feyzin steam cracker together with Elf, SNPA, Solvay, and Ugine. Through its subsidiary Progil Electrochimie, it had a stake in Ugine's Pont de Claix phenol and propylene oxide units. Progil Electrochimie was itself associated with the oil company Antar for the supply of cumene that came from the Donges refinery.

The association with Ugine produced other subsidiaries: Plastugil for plastics and elastomers, Progil Bayer Ugine, PBU, for polyurethanes, Ugilor for acrylonitrile, acrylates, and methacrylates. A subsidiary that was to have a great future was Péchiney-Progil, set up to develop plant protection products.

Progil diversified its portfolio with yet other subsidiaries: Pétrosynthèse, set up with Compagnie Française de Raffinage and Standard Oil of California to produce detergents; Orogil, also with Standard Oil of California, to make lube oil additives; Diversey-France, for industrial cleaning agents; SiFrance, for silicates; Soprosoie, for leather and textile additives; Xylochimie, for the treatment of wood.

In pure French tradition and as early as 1965, a report was commissioned by the then Prime Minister Georges Pompidou from a high civil servant, B. Clappier, outlining the shape which France's chemical industry should be given. The report advocated reorganizing the state-owned companies around the two national oil companies, Elf Aquitaine and Société Française des Pétroles. In addition two "international-size" private groups should be set up.

In the framework of France's Fifth Economic Plan, tax incentives were offered to push through these mergers. One of the two private-sector poles was to be Rhône-Poulenc. W. Baumgartner, former governor of the Banque de France, and successor to Marcel Bô as the group's president, set about to achieve it. Baumgartner had no industrial experience and, of course, no experience of the chemical industry; his appointment was no doubt due to his good contacts both within the government and within the high administration, which could smooth the path to mergers.

The first merger announced was between Rhône-Poulenc and Progil on April 20, 1969. Its first consequence was to raise to 20 percent the Groupe Gillet's stake in the capital of the holding company Rhône-Poulenc SA. That same year, the Péchiney-Saint-Gobain company, which had been given the 57.2 percent share that Péchiney held in Naphtachimie, became a 51 percent subsidiary of Rhône-Poulenc SA.

Rhône-Poulenc SA increased its stake in PSG by stages, setting up Rhône-Progil in 1971, then fully absorbing Rhône-Progil and SUCRP within Rhône-Poulenc Industries in 1975.

Viewed in terms of consolidated sales, these mergers did produce an international-sized group[14] and reduced the predominance of the textiles sector while improving upstream integration. But as both Péchiney Saint Gobain and Progil were short of capital and had a debt ratio of 35 percent compared to Rhône-Poulenc SA's mere 4.8 percent at the time of the merger, the newly formed group was no longer financially healthy. Moreover, the two chemical companies absorbed had very small overseas bases, so that the share of RP.SA's consolidated sales abroad had dwindled. Finally, it required strong group leadership to carry out the rationalizations needed to combine into a sound whole such ill-assorted companies, which until then had been rivals, and to prevent the formation of cliques. On the advice of the McKinsey consulting firm, Rhône-Poulenc set up eight divisions and an Executive Committee assisted by functional departments. Nevertheless, this was by no means the end of Rhône-Poulenc's period of restructuring.

A second private chemical pole could have been set up around *Ugine Kuhlmann*, but events and men decided otherwise. The merger of Ugine and Kuhlmann, announced in May 1966, pushed the new group to second place in

[14] Consolidated turnover in 1974 amounted to 20,351 million francs, placing Rhône-Poulenc among the top ten world-chemical companies.

France's chemical industry, causing no little surprise to observers. Indeed, Ugine had chosen Progil as its partner within a large number of common subsidiaries, while Kuhlmann had shown a preference for an association with the national coal companies, so that a Progil-Ugine merger seemed more logical.

Through the play of prewar mergers, *Ugine* (Société d'Electrochimie, d'Electrométallurgie et des Aciéries Electriques d'Ugine) was involved in three sectors of activity: chemicals, a category that also encompassed aluminum; ferro alloys; and special steels. Until World War II, its president had been Georges Painvin, a polytechnician from the Corps des Mines; he regarded the range of the group's activities as a guarantee of stability.

Ever since it had been set up in 1906, Société des Produits Azotés (SPA) had maintained close ties with Ugine both through its leadership and the financial ties which had been established between the two groups. The discovery of gas in Saint-Marcet during the war, and the subsequent development of the Lacq gasfield, had turned the Lannemezan site in the Pyrenees into a major production center for SPA (ammonia, nitric acid, hydrazine, acetylene black) and for Ugine (aluminum).

Ugine had also expanded into the chemistry of halogens (chlorine, fluorine, bromine) and halogen derivatives (refrigerating fluids, propulsion gases, fireproofing agents, aluminum chloride). Through its knowhow in fluorine derivatives, Ugine had become closely involved in the nuclear industry, supplying, in particular, the uranium hexafluoride required for isotope enrichment. Ugine's other strengths were peroxides (H_2O_2, perborates) produced in Jarrie, in the Lyon area, chlorates, and sodium chlorite produced in Chedde and in Prémont in the Alps.

Also in that area, in La Chambre, Société Industrielle des Dérivés de l'Acétylène (SIDA), a subsidiary of Ugine, produced oxygenated solvents from acetone and secondary-butyl alcohol. Moreover, Progil and Ugine had numerous links through subsidiaries such as Progil Electrochimie, Plastugil, Distugil, and Ugilor.

Just before it merged with Kuhlmann, Ugine was France's leading special-steel producer and ranked second in aluminum. But chemicals accounted for only 26 percent of its turnover. *Kuhlmann*, on the other hand, was exclusively a chemical company, dating back to 1825, when it was set up in Lille as a sulfuric acid plant, by Frédéric Kuhlmann, a chemistry professor at the Faculté de Lille, and his brother. Between 1925 and 1940, Kuhlmann had developed strongly in inorganic acids, fertilizers, chlorine and derivatives, bone-processing products, silicates, barium derivatives, and dyestuffs (CFMC) under the direction of Raymond Berr and Joseph Frossard.

After the end of World War II, Kuhlmann's leadership lay in the hands of polytechnicians from the Corps des Mines (Perilhou, Desportes), who were latecomers to the chemical industry. Their main concerns had been first to recover Francolor, then gradually to take the company from coal-based chemicals to petrochemicals while diversifying through associations. Thus emerged Lorraine-Kuhlmann, to produce polystyrene at Dieuze; Dispersions Plastiques, in which Kuhlmann-CFMC was associated with BASF to produce expanded polystyrene and acrylic latex; and Société Normande de Polyéthylène, in which Kuhlmann was associated with Rhône-Poulenc and Compagnie Française des Pétroles to develop a low-pressure Phillips Petroleum process.

Kuhlmann also produced PVC from Brignoud vinyl chloride, and polyethers for polyurethanes under a Wyandotte license, thus completing its range of polymers, which began in 1930 with the Pollopas urea-formaldehyde resins produced in CFMC's Villers Saint-Paul plant. In 1960, Kuhlmann had taken over Produits Chimiques Coignet, which had a long-standing reputation as a producer of glues, gelatin, and phosphorus. Two years later, Kuhlmann was merging with CFMC and taking the name of Établissements Kuhlmann. In this new form, the group was involved in a variety of chemical sectors but had acquired a large part of its technology from outside companies. In 1966, it teamed up with Ugine to form Ugine-Kuhlmann. The merged group, together with Socíeté des Produits Azotés, which had been tagged on, could have formed another pole for the private chemical sector, the more so as it had extended its range to pharmaceuticals with the acquisition of Laboratoires Pointet Girard, of Fournier Frères and of IBF, as well as to inks through purchase of Lorilleux Lefranc, a respected company both at home and abroad. But there was no unified leadership, and no rationalization was carried out.

Accordingly, profits slumped and P. Grezel, one of the Banque Lazard managers, who had become president in 1971, was asked to outline a suitable strategy. This was when an attempt was made to form an association with Péchiney, whose president P. Jouven, had already pulled out of chemicals by cutting away from Péchiney-Saint-Gobain and Naphtachimie.

Within the new *Péchiney Ugine Kuhlmann* (PUK) group set up in 1972, chemicals under the name *PCUK* accounted for only 18.3 percent of consolidated sales. Moreover, they had hardly any complementarity with the other sectors of the group hinging on special steels, ferroalloys, copper smelting, and nuclear industry, all areas which received the management's priority attention.

Already in competition within Ugine Kuhlmann with aluminum and special steels for the funds required for their development, in the new organization of PUK, chemicals had become a kind of foreign body. Obviously, they could no longer serve as a pole of attraction for new mergers.

Another chemically oriented entity was to disappear in the whirlwind of restructuring. The origins of *Nobel Bozel* go back to 1808 when Établissements Malétra was producing sulfuric acid in the Rouen area. In 1925, the company had joined with Compagnie Générale d'Electrochimie de Bozel in Savoie, producing calcium carbide and ferroalloys since 1898, to form Bozel-Malétra.

Through other takeovers, in particular of Nobel explosives, the Nobel Bozel group became involved in plastics, electrolytic coatings and chemicals. Besides sulfuric acid, potash, and potassium carbonate, Nobel Bozel's inorganic chemicals included silicates, nickel and cobalt salts, and zirconium. From acetaldehyde, the company had developed original glyoxal and glyoxylic acid chemicals. In polymers, it supplied nitrocellulose collodions, urea-formaldehyde resins, as well as vinyl and acrylic emulsions in the Polysynthèse subsidiary it jointly owned with Hoechst. Nobel Bozel also produced chlorinated solvents, herbicides (MCPA; 2,4-D) and a range of surface-active agents.

The company was managed as a minority subsidiary of *Centrale de Dynamite*, a kind of conglomerate involved in gelatins, glycerins, paints (Duco), and compressed panels (Isorel), as well as in antibiotics through another subsidiary of Hoechst called SIFA. It was difficult to manage such a hodgepodge of activities with any coherence. Under the impulse of J. C. Roussel, one of the two sons of *Roussel-Uclaf*'s founder, an association was formed with what was France's second-ranking laboratory. Roussel-Uclaf was also partly owned by the Hoechst group from 1968.

At the end of the restructuring moves, Centrale de Dynamite came to own 50.2 percent of Roussel-Uclaf-SIFA and 43 percent of Nobel Bozel. The chemicals division was grouped with Nobel Hoechst Chimie, a 50 percent Hoechst subsidiary. This restructuring, which set pharmaceuticals apart from chemicals proper, entailed a number of consequences. It gave Germany's Hoechst a substantial stake in both the pharmaceutical and the chemical companies; it excluded Nobel Explosifs, which operated a rapproachment with Poudreries Réunies de Belgique; it focused Nobel Bozel on electro-metallurgy and detergents; and it paved the way for the separation of Duco and Isorel. Another opportunity to turn a private French group into a coherent chemical entity had been missed.

Since its founding in 1902, *L'Air Liquide*, one of the feathers in France's industrial cap, had only had two heads, Paul Delorme, who had founded it with Georges Claude, and his son, Jean Delorme, who pushed the group to rank with the world giants in industrial gases through his managing expertise combined with the fact that he was also financially prudent and technologically daring.

Ousted from the United States after the war, because the French government, short of dollars, forced it to sell its American assets, L'Air Liquide returned with a vengeance in the 1970s through the medium of its Liquid Air subsidiary, itself born of Exxon's industrial gases division and the purchase of a Chemetron division, the one-time National Cylinder Gas Company. The group had taken advantage of increased oxygen consumption in steel works, of nitrogen use in chemicals and coal storage industries, and of argon consumption for the welding industry. It also basked in the mutual respect that was a feature of the relationship among the world's leading industrial gas companies. Each one retained a distinct dominance in its home market and trod carefully in the territory of the others.

Like other French industrial leaders, those of L'Air Liquide came to chemicals either by chance or through the need to develop their products and processes. Georges Claude's very-high-pressure synthetic-ammonia process induced L'Air Liquide to set up Société Chimique de la Grande Paroisse in 1919. Likewise, because it was an oxygen producer, L'Air Liquide joined with Ugine to produce hydrogen peroxide in Jarrie within Société Oxysynthèse using the Laporte process. Growing demand for perborates in washing powders boosted production of peroxygenated products (perborates, persulfates, peroxides, peracids, epoxydized soybean oil). L'Air Liquide also seized the chance to acquire, early in World War II, a Lyon-based company, Lipha, giving it a foothold in pharmaceuticals.

Through shares taken in Ethylène-Plastiques and in Société Normande de Matières Plastiques, in the 1950s the group became involved in high- and low-pressure polyethylene. Despite the inroads made in chemicals, L'Air Liquide never worked out a deliberate policy of penetration into this area. Its leaders were interested in chemicals only to the extent that they boosted consumption of industrial gases. This is why the toehold in pharmaceuticals was maintained but not pushed further and why participations in polyethylene were sold just as was, in 1987, the majority stake in Grande Paroisse. The only remaining chemicals in the group were the peroxide sector, a lucrative one in all respects.

Pierrefitte-Auby, a subsidiary of Banque de Paris et des Pays-Bas, resulted from a merger in 1969 between Société Generale d'Engrais et de Produits Chimiques Pierrefitte and Produits Chimiques d'Auby. Pierrefitte's development was geared to fertilizers. It had built a 1,000 tons a day ammonia unit in Grand Quevilly in 1969 with PSG and APC, and had set up in 1966 Compagnie Française de l'Azote (Cofaz), with Compagnie Française de Raffinage. Auby was a more diversified business. It had been linked with Société Chimique des Charbonnages de France since 1966 to produce and sell fertilizers, but it was also involved in such interesting products as fatty amines and carrageenates. It also had a minority stake in the Labaz pharmaceutical laboratory. After the merger, Pierrefitte-Auby broke the agreements linking Auby with Charbonnages and reassembled part of its fertilizer interests around Cofaz. Specialty chemicals henceforth accounted for a quarter of the new group's sales. But its life was a short one. No doubt Pierrefitte-Auby's prospects were not so rosy for Paribas because of its fertilizer involvement. Whatever the case, it was sold off in bits to various other parties.

Besides the entities involved in the uncertainties of the restructuring process, there were a number of medium-sized companies in the private sector that were essentially hinged on fine chemicals and specialties. *Société Rousselot* pushed through the development of its gelatin sector by acquisition of rival companies both in France and abroad. It had extended its range of activities to polymers, adhesives, and animal feed when it was bought in 1977 by Elf Aquitaine, which proceeded to dismantle it. *Compagnie Française des Produits Industriels* (CFPI), founded before the war by Adrien Hess, developed Amchem's patents in herbicides and added to its range a series of specialties, principally surface-active agents as well as metal processing products. *Société Française d'Organo-Synthèse* (SFOS), acquired its name in 1956 when Laboratoire Zundel et Joliet was bought by Roger Bellon. It initially supplied the pharmaceutically active ingredients for the parent company before diversifying into additives for polymers, special monomers, and synthetic resins. In 1978, SFOS bought Rhône-Poulenc's Persan unit, after which Rhône-Poulenc Santé increased its stake in Laboratoire Roger Bellon from 51 percent to 58 percent.

Other companies, which had long remained independent, either became associated with chemical and pharmaceutical groups as was the case of *Société Anonyme pour l'Industrie Chimique* (SAIC) with Rhône-Poulenc, *Sapchim-Fournier-Cimag* with Labaz, and *De Laire* with Delalande, or developed under their own steam, such as *Produits Auxilliaires et de Synthèse* (PCAS), taken over in 1981 by Stauffer Chemical.

France was the only great industrial country in the free world deliberately to have set up a state-controlled chemical industry, one that had emerged from three different horizons, fertilizers, coal, and oil. The administration, which already owned ONIA and MDPA, merged them in 1967 within *Entreprise Minière et Chimique* (EMC), encompassing Azote et Produits Chimiques, Mines Domaniales des Potasses d'Alsace and their joint market subsidiary, Société Commerciale des Potasses et de l'Azote. While salaries increased in the Alsace potash mines to the point of accounting for over 50 percent of extraction costs, the emergence of Canada's products on world markets brought about the collapse of prices.

Things were no better for Azote et Produits Chimiques (APC). Its Toulouse unit had to compete with the East European nitrate fertilizers. The reasons that had prompted the administration to set up a "national" nitrate industry were soon forgotten. APC's sole objective became to keep the company going through diversification and public funds. Hence a series of maneuvers that ended in 1971 with APC taking over Méthanolacq at the expense of Ugine Kuhlmann and with the establishment, together with BASF, of a 60,000-ton urea formaldehyde-glue unit in Toulouse.

At the same time, EMC had set up a monomer vinyl chloride subsidiary in joint venture with Holland's state mining company DSM, in Tessenderloo, called Produits Chimiques de Limbourg. Other acquisitions involved sulfur derivatives with Société Languedocienne de Soufre et Micron Couleurs, and engineering with ONIA-Gegi. Such diversifications were not good choices, considering the overcapacities that later emerged both in methanol and in vinyl chloride, and the shortage of major turnkey contracts for ONIA-Gegi. Glues also had to be restructured when BASF withdrew, and Languedocienne's sulfur derivatives were taken over by CdF Chimie.

The only diversification that could be considered successful was the Tessenderloo subsidiary, which was involved in chlorine and which developed downstream through an agreement with Huiles, Gondrons et Dérivés (HGD).

Subsequently EMC took a stake in an animal-feed company called Sanders, and tried to set up a fine chemicals division through Tessenderloo. But its chlorocyanurics derivatives unit, well integrated with the Toulouse urea unit, was sold to CdF Chimie. At the same time, fatty amine production was halted on the site.

The involvement of the French coal mines in chemicals dates back to 1921, when Georges Claude sold his synthetic ammonia process to Compagnie des Mines de Béthune. The process used coal gas as a source of hydrogen. Mines de Béthune also developed the first industrial synthesis of ethyl alcohol in 1925. By 1926, synthetic methanol was also being produced in Mazingarbe, while one of the first European units for the production of ethylene oxide from coking-plant ethylene was set up in Chocques by Marles-Kuhlmann in 1935. France's mining companies had thus gradually developed a chemical industry close to the coking plants in order to boost coal use.

The 1946 nationalization law handed over the chemical plants set up by the former companies to the coal mines, which also received the investment these companies held in subsidiaries set up with private sector firms. This was the starting point of the inextricable network of ties between the private and the public sector that became a dominant feature of France's chemical industry. The emergence of oil as a substitute for coal feedstock further complicated matters.

A first attempt at simplification was made at the beginning of 1968, when the chemicals divisions of Charbonnages de France were merged within *Société Chimique des Charbonnages* (SCC) and a subsidiary, *CdF Chimie*, was set up to market the products. At the time, the public group thus formed was posting consolidated sales exceeding one billion francs, and ranked as France's fourth chemical company. It encompassed four production sectors: fertilizers, the basic organic intermediates, thermoplastics, and organic fine chemicals. Its main production sites were in Mazingarbe, Drocourt, Wingles, Vendin, and Douvrin in the Nord-Pas de Calais area, and in Carling, as well as in Marienau in Lorraine.

The development of a coal-based chemical industry in France after World War II was made possible because a coking process in the Lorraine coal mines provided extra volumes of tar, benzols, and gas. Moreover, in 1954 the Ethylène Plastiques subsidiary of Houillères du Nord began using the coking-plant ethylene to produce polyethylene under an ICI license. But the situation rapidly changed as demand for metallurgical coal began to stagnate.

The steelworks also required less coke as they switched to fuel oil and gas to feed the blast-furnaces. The reasons that led Charbonnages to develop a chemical industry were fast disappearing as natural gas and oil were being substituted for coal.

Both the leaders of Société Chimique des Charbonnages and the administration providing the funds felt that their main job was to ensure survival of the group in the new context rather than remaining true to their initial calling, which would have restricted their development possibilities. In the circumstances, it was no longer necessary to locate production sites near the coal mines, but rather where hydrocarbon supplies were handier. This explains the construction in Nangis, not far from the Paris area, of a 1,000-tons-a-day ammonia unit in joint venture with Elf Aquitaine and La Grande Paroisse. It also explains the association with the Saarbergwerke state coal mines within the Klarenthal refinery which supplied naphtha to the Carling site. Other associations with Saarbergwerke involved urea and ammonia.

SCC also began building steamcrackers in Carling in 1970. The subsequent construction of the Dunkirk steam cracker, in joint venture with the Qatari government, considerably worsened the already precarious financial situation of the group.

After it was set up, SCC with its partners had carried out a number of frontier adjustments involving full control of Ethylène Plastiques in 1968 on the withdrawal of PSG and L'Air Liquide; scrapping, in 1970, the fertilizer agreement with Auby; and taking over the Courrières-Kuhlmann polystyrene unit in Dieuze. As a result of all these moves, the chemical subsidiary of Charbonnages accounted for 36 percent of France's nitrogen fertilizer market, 46.2 percent of high-pressure polyethylene, and 26 percent of polystyrene. It was the leading producer of acrylonitrile, cyclohexane for polyamides, and styrene monomer supplied from Carling. In several organic intermediates (phthalic and maleic anhydrides, methanol, oxo-alcohols from the Courrières-Kuhlmann subsidiary), it held an honorable place.

It had a number of subsidiaries that were making profits, in particular Ugilor. This had been set up in joint venture with Ugine-Kuhlmann to develop acrylic and methacrylic chemicals; and it owned efficient processes, in particular one rivaling Sohio's for acrylonitrile. Also making profits was HGD, in which PSG had a stake and which, besides developing coal tar derivatives, was involved in the fine chemicals area of products derived from toluene oxidation and included phenol-formaldehyde plastics in its range of products. But with the exception of some promising areas and while technologically sound, the chemicals division of Charbonnages was generally geared to heavy chemicals requiring substantial investments and having to compete with the world chemical leaders. There was also in this area the latent threat of overcapacity, leading inescapably to the erosion of selling prices and profit margins.

The French state became involved in oil very early. *Entreprise de*

Total Chimie's petrochemicals complex in Gonfreville near Le Havre. Courtesy Total Chimie.

Recherche et d'Activités Pétrolières (ERAP) was formed through the merger in 1966 of two state-owned companies, Régie Autonome des Pétroles and Bureau de Recherche de Pétrole, dating back to 1939 and 1945, respectively. A "rival" national group to Compagnie Française des Pétroles (CFP) had been set up from ERAP-controlled companies, Société Nationale des Pétroles d'Aquitaine (SNPA) and Elf Union, which was subsequently to become Elf Aquitaine. It was *Elf Aquitaine*'s Lacq gasfield, containing both ethane and sulfur, that started the oii group's chemical activities.

Basic intermediates such as ethylene, benzene, and styrene were produced in Lacq proper and, as early as 1966, were turned into low-density polyethylene and polystyrene.

Disappointed by the Rilsan prospects, Péchiney had sold Organico to SNPA in 1963. Three years later, Aquitaine-Organico was set up with a range of products that included Aquitaine Plastique's thermoplastics, polyamide-11 (a derivative of Organico's castor oil), and a polyamide-12 produced from the lactam-12 monomer, itself a butadiene derivative.

In 1967, an industrial complex for the production of PVC and polyethylene was set up in Balan, near Lyon, while in Feyzin, also near Lyon, a steamcracker was brought on stream in which Progil, Ugine, and Solvay were associated with SNPA.

The Elf Aquitaine group had bought the Caltex refinery in Bec D'Ambès in 1960. By acquiring Antar in 1970, it added an aromatics production unit located in Donges which supplied, in particular, *o*-xylene and *p*-xylene. With Canada and the United States, SNPA was one of the world's leading sulfur producers; and it had developed an original sulfur-based chemical industry. Besides sulfur and H_2S, this included dimethyl sulfide, DMSO, thiourea, and mercaptans.

Finally, in a further downstream diversification move, SNPA had taken a firm foothold in the health sector through its 100 percent Sanofi subsidiary. From 1973 on, Sanofi acquired the Labaz, Castaigne, and Robilliart laboratories, and took shares in the cosmetics laboratories of Yves Rocher and Société Roger et Gallet.

As a state-controlled group, Elf Aquitaine had come to chemicals through natural gas and sulfur. It was, therefore, preparing for the day when depletion of the Lacq gasfield would require turning to other activities to survive as a chemical entity.

The other national oil group, *CFP*, had shown no such definite leaning towards chemicals. Directly or through its subsidiary, Compagnie Française de Raffinage, CFR did have a 22.5 percent share in SOCABU and 40 percent in Pétrosynthèse, and together with the German companies Hüls and Scholven Chemie it set up in 1968 the Société Industrielle des Polyoléfines. Moreover, a 50 percent CFP–50 percent CFR company had been established under the name of Total Chimie, regrouping all the other chemical parts of the group.

But when the two oil groups, Elf and CFP, associated to form *ATOChimie* in 1973, CFP acted more as the silent partner, while Elf Aquitaine was the poorer partner, taking advantage of its Lacq revenues to diversify deliberately into both fine and heavy chemicals. These different tendencies were subsequently confirmed when Elf Aquitaine acquired Metal and Thermit, and Texas Gulf in the United States and CFP withdrew from ATO.

Rivalries and the pursuit of different objectives managed to break up the chemical association set up between the two national oil groups, as advocated in the Clappier report.

The state-owned chemical industry comprised yet another sector, less significant in size but very active nonetheless. It involved powders and explosives. Here again France stood apart from its big partners, with its Monopole des Poudres dating back to prerevolutionary times and ending only because of the pressure exercised by the European Economic Community (EEC). It was replaced by *Société Nationale des Poudres et Explosifs* (SNPE), set up with exclusively public funds. The company was involved in military powders and explosives and in propellants, of course, but it also produced chemicals for civilian use, drawing upon its knowhow in nitration and phosgenation (industrial nitrocellulose, chloroformates, and carbamates). In order to rationalize France's dynamite market, SNPE bought Nobel PRB, leaving only one French rival in the field, the private company Nitrochimie.

For a country with a propensity for "national solutions," France has proved singularly accessible to foreign capital eager to invest in chemicals. The reasons are historical, technical, and financial. The Belgian group Solvay established itself in France as far back as 1873 when it built the Dombasle soda works near Nancy, one of the largest on the continent. Through an agreement with the administration, in 1925 Solvay selected the Tavaux site in the Jura to build its second French soda plant and brought an electrolysis unit on line there in 1930.

From then on, it was natural that the Belgian group developed after the war the productions that fitted into its base activities: sodium and calcium products and chlorine derivatives (HCl, *eau de Javel*, chlorinated solvents, allyl chloride), peroxide products launched in 1959, as well as vinylidene chloride copolymers and PVC, in association with ICI within Solvic.

The technical reasons arise from the fact that with the establishment of the Common Market, foreign firms with original technologies no longer saw the need to associate with French partners to develop them, preferring to do so through their own subsidiaries. Before the war, for example, Shell had been associated with Saint-Gobain within PCRB; but it was only when Shell Chimie was set up as a 100 percent subsidiary of Royal Dutch Shell that its chemical branch really expanded in France. In the same way, Esso, which had formed a joint venture with CFR to develop butyl rubber, thereafter reserved the processes and products that Exxon wished to develop in France to its subsidiary

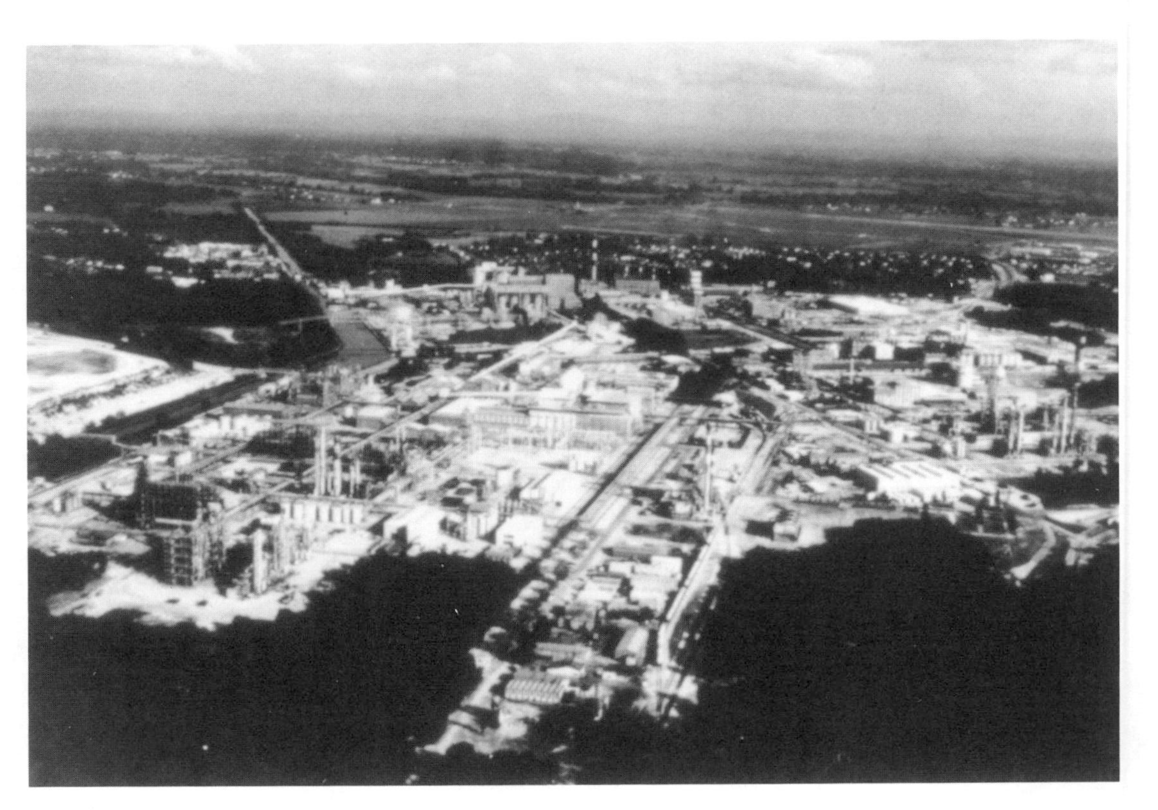

Aerial view of the Tavaux factory. Courtesy Solvay.

Esso Chimie. By 1970, Shell Chimie had developed its Berre site and Esso Chimie its Port-Jérôme units, thereby ranking sixth and eighth among the firms involved in chemicals in France.

A number of foreign investors, on the other hand, withdrew from France, leaving national chemical companies to take up the field, mainly in silicones, polyurethanes and polystyrene. Many family businesses, however, involved in pharmaceuticals, essential oils, and chemically related products, were forced to sell out to foreign investors because there were no French buyers willing or able to pay the price they were asking.

From 1963 to 1969, the number of American groups that acquired shares in France's pharmaceuticals industry rose from fifteen to twenty-two, accounting for what was believed to be sixteen percent of the market. Many reputable firms supplying the French perfumers had been taken over by Swiss, American, or Anglo-Dutch groups (Chiris, Roure-Bertrand-Dupont, Givaudan, Bertrand Frères).

At times, problems arose from estate taxes causing major ownership changes for the greater benefit of foreign groups. For instance, Hoechst had taken a foothold in the Roussel-Uclaf holding company Chimio by acquiring the stake held by one of the founder's sons, Henri Roussel. When the other son, Jean-Claude Roussel, died in 1972 in a helicopter accident, Hoechst's share already amounted to 43 percent and the French government was no longer in a position to look for a "French solution." Besides, Roussel-Uclaf was quite satisfied with its relationship with the German giant.

Following the changes that had taken place between the end of World War II and the first energy crisis of 1973, France's chemical industry had all the appearance of pieces in a puzzle, and an outside observer would have had difficulty putting them together, so dense was the tangle of private and public groups and French and foreign firms.

No doubt great strides had been made in a number of areas. Through licenses or through associations with foreign partners, France's chemical industry no longer lagged behind in products and processes. But its own research had yielded few industrial breakthroughs; and its frequent restructurings, rarely followed by rationalization, had prevented any harmonious development of its activities, which would have required steady and prolonged efforts at all levels of responsibility.

Despite some attempts at clarification, France's chemical industry was far from having achieved stable structures in 1973 as the upheavals of the next ten years would show.

The Emergence of Italy's Heavy Chemical Industry

As in France, the state exercised its influence very strongly on Italy's chemical industry in postwar years, more particularly on Montecatini's base chemicals and on the new petrochemical entities. *Montecatini*'s history dates back to 1888, when it was set up to extract copper from Tuscany's pyrites. The company started producing fertilizers very early in this century, then after World War I became involved in the synthesis of ammonia. Subsequently, either directly or through joint subsidiaries set up with foreign partners, Montecatini penetrated the sectors of aluminum; dyes, through ACNA, founded in 1931; pharmaceuticals, through stakes taken in Farmitalia and Carlo Erba, which had merged in 1978; and fibers, through Châtillon and Rhodiatoce.

Edison was an electricity utility established in Italy in 1884. Fearing it would be nationalized in the postwar political climate, it had diversified into petrochemicals in the 1950s and had become Italy's second largest chemical company. When electricity was indeed nationalized in 1962, Montecatini and Edison considered merging their chemical interests. In 1966 they formed Montecatini-Edison, called *Montedison* since 1970. The group thus set up is more in the nature of a conglomerate than a chemical group proper. Its range of activities extends from retail sales to food to insurance to finance. It was reorganized under Mario Schimberni and is now one of Italy's major chemical businesses.

Its base chemicals division, managed by *Montedipe*, supplies both the other divisions of the group as well as outside clients with olefins, aromatics, cumene, fiber intermediates, acetic acid, ethylene oxide, chlorine, and derivatives. *Montepolimeri* handles the plastics division, and on seven sites in Italy and one in Belgium (set up in joint venture with Petrofina for polypropylene) produces a full range of high polymers: polyolefins, including the isotactic polypropylene discovered by Giullo Natta, PVC, styrene and methacrylic polymers, as well as ethylene-propylene rubber.

Through its research on high-performance catalysts for polyolefins, Montedison achieved a strong enough position to be able subsequently to associate with Hercules in a joint subsidiary called Himont to become the world's leading producer of polypropylene.

The group's specialties were assembled within *Ausimont* and include fluorine derivatives extending from refrigerants to fluorine elastomers, organic peroxides, mineral pigments, including TiO_2, various products for detergents (silicates, phosphates, perborates, dodecylbenzene, and CMC), as well as a variety of metal processing agents, fuel additives, and catalysts. Resins (alkyds, aminoplasts, acrylics, and vinyls) and some of their precursors (formol, pentaerythritol, and melamine) were assembled within *Resem*.

Through *ACNA*, Montedison has continued its dyes and intermediates business. But neither productivity nor research in that area has been kept at a convincing level. Likewise, the group has drawn no reward from its fertilizer branch, taken care of by *Fertimont* since 1981. Because of price controls in Italy and overcapacities worldwide, it has remained a losing sector.

After taking control of SNIA-Viscosa, Montedison had assembled within

Montefibre, the four companies and their subsidiaries involved in fibers. Although well-integrated upstream, their range of fibers was too extended for efficiency, including as it did viscose, acetate, acrylic, polyamide, polyester, and polypropylene. Some specialization would sooner or later be required, together with some reorganization to help face up to more aggressive rivals.

The group's engineering division, called *Technimont* since 1973, had a long-established reputation, but it could not help being affected by the world slump in turnkey contracts. Pharmaceuticals were handled by the subsidiaries *Farmitalia*, in which Montedison had bought Rhône-Poulenc's 50 percent share, and *Carlo Erba*, a laboratory dating back to 1850 and enjoying a high reputation. The discovery of an anticancer drug, Adriamycin, by Farmitalia brought in substantial profits to this sector. On the whole, Montedison was diversified in its chemical activities and had carved out for itself a leading share of the Italian market.

But for a group of its size, its involvement overseas was small. Moreover, while its research had proved brilliant in certain areas (polypropylene, anticancer drugs) the group was far too involved in areas where competition was strong to achieve brilliant profits.

ANIC, which was to become Italy's second-ranking chemical group, was set up in 1936 by the state with public funds. When Ente Nazionale Idrocarburi (ENI), was established in 1953 under Enrico Mattei's sponsorship, ANIC was required to turn from refiner to chemist. It was assigned by the state the twofold mission of developing petrochemicals in Italy and of contributing to the solution of the economic and social problems which the government was having to face, mainly in the south.

The petrochemicals industry, boosted by the discovery of a methane field in the area, came on stream in 1957 with the Ravenna unit. Synthetic rubber and fertilizers were produced there as well as lampblack, in association with Phillips. In 1960, ANIC began producing PVC in Ravenna, then two years later, near another source of natural gas, set up a second petrochemical site in Gela, Sicily, fitted with an electrolysis facility. Other sites were created near methane fields, in 1964 for chemical fibers and methanol in Pisticci (Matera), in 1972 for ammonia, urea, and caprolactam in Monte San Angelo, near Manfredonia.

With its various investments, made either directly or through subsidiaries, ANIC—which in 1972 had taken over Montedison's Priolo and Brindisi steamcrackers—had set up, by the time of the first oil crisis, a group that was involved both in heavy organic and inorganic chemicals (aromatics and olefins, ammonia, mineral acids, borax, electrolysis products); in synthetic rubber (styrene, butadiene, stereospecifics, ethylene vinyl acetates, isoprene); in plastics (PVC, polyolefins, ABS, polycarbonate, polyamide-6); and chemical fibers.

The group also made efforts to diversify into fine chemicals, either through internal development (acetylene chemicals, CO chemistry) or, more particularly, by acquiring specialized firms like Bozzetto (rubber and plastics additives) and Brichima (dihydroxyphenols and derivatives). For want of internal knowhow, the involvement in pharmaceuticals was achieved only through purchase of Italian laboratories (Recordati, ISVT, Archifar, Bellco). Despite these belated efforts to get into high added-value areas, ANIC was still weighed down by its initial heavy chemical products. Most of these were to feel the full brunt of world competition and stagnant demand, the more so as their base was a purely national one.

While it had made sense to make use of the natural gas on ANIC's different petrochemical sites, the Italian administration's policy of developing units in the south that required little manpower and were likely to build up overcapacities in Europe, could neither solve the social problems of the south of Italy nor improve the financial situation of groups involved in petrochemicals. While Montedison owed its existence to the nationalization of electricity in Italy, and ANIC its importance to the size of public funds which Enrico Mattei and his successors managed to obtain for their industrial plans by convincing the government that these would improve the country's social situation, the spectacular development of *Società Italiana Resine* (SIR), can be ascribed to the skill of engineer Nino Rovelli, its leader since 1948, who succeeded in convincing successive Italian governments of the advisability of setting up a third petrochemical pole in Sardinia.

SIR had emerged in 1931 through the acquisition of Società Italiana Bakelite. It had added polyester resins in 1954, then polystyrene in 1957 to its range of phenolformaldehyde products. A major turning point was the establishment of a petrochemical site in Porto Torres (Sardinia), which began a large scale production of phenol in 1962 by the cumene process.

Either directly or through satellite companies likely to rake in public funds, Rovelli integrated increasingly upstream by bringing an ethylene unit on stream in 1965. Three years later, he was inaugurating an actual refinery, Sardoil, on the Porto Torres site. Meanwhile, SIR was adding to its range of synthetic resins, extending from phenolics to aminoplasts, polyesters, epoxies, furanes, and alkyds, a wider range of thermoplastics with ABS, styrene-acrylonitrile (SAN), PVC, expanded polystyrene and low-density polyethylene.

Polymers, however, accounted for only 37 percent of SIR's sales. As a petrochemical producer, the group produced in its own units or those of its subsidiaries, besides phenol and its acetone by-product, dodecylbenzene and linear alkylbenzene, formaldehyde, styrene monomer, urea, and acrylonitrile through Acril Sarda, a joint subsidiary it owned with Rumianca. In 1967, SIR launched out into the adventure of synthetic fibers producing polyester and acrylic fibers in Porto-Torres.

It was obvious that even the Italian state would not be able to sustain such a frenzy of investments for long. The prospect of painful readjustments was looming on the horizon even before the first energy crisis broke out. Italy's chemical industry, and all Europe's with it, were going to pay a heavy price for their excess capacities.

Holland's Internationally Oriented Chemical Industry

In 1959, the joint Shell and Esso exploration company Netherlands Crude Oil Company (NAM) discovered one of the world's largest gasfields in the northern part of Holland. NAM, which had the development and marketing concession for the Groningen gasfield, paired up with DSM for its exploration and development. Both NAM and DSM subsequently extended their range to the North Sea.

DSM had originally been set up to develop the Limburg coal mines in 1902. It was already converted into a chemical company at the time of the Groningen discovery, using oil as a feedstock, for there was no future in coal for the state-owned company. In 1961 naphtha had started replacing coking gases for the production of ethylene needed for the high-density polyethylene made in Beek. Natural gas with an 80 percent methane content served as a source of hydrogen and CO_2 for the synthesis of ammonia carried out in Geleen.

DSM had developed, in 1964, an original process to manufacture phenol from benzoic acid, which was itself obtained through toluene oxidation. Under a method developed in its own research laboratories, DSM produced from the phenol made in Botlek the cyclohexanone that led to caprolactam. A unit producing acrylonitrile from propylene came onstream in Geleen in 1969, the very year when DSM set up a subsidiary with EMC, Produits Chimiques de Tessenderloo, to manufacture vinyl chloride in Belgium. The monomer was polymerized in Beek. DSM was, therefore, involved in different sectors of heavy chemicals (fertilizers, olefins, aromatics), in a number of the basic intermediates (phenol, acrylonitrile, urea, melamine, caprolactam), and in a significant range of polymers (polyolefins, PVC, ABS, EPT rubber). The company also began to show an interest in fine chemicals (benzoic acid derivatives, perfumery chemicals, and lysine from caprolactam).

The proven technologies developed by its engineering subsidiary Stamicarbon in melamine, urea, caprolactam, and high-density polyethylene processes, boosted DSM's expansion abroad either through subsidiaries like Nypro (UK), Columbia Nipro (USA), Univex (Mexico) for caprolactam, and Columbia Nitrogen (USA) for fertilizers, or through licenses.

But DSM was too closely involved in heavy chemicals to remain unscathed by the general unfavorable environment, even though it had direct access to hydrocarbons and even though its good technology gave it an edge over its big competitors by cutting down its production costs.

The other Dutch chemical giant, *AKZO*, is the result of a merger that took place in 1969 between AKU, one of the world's leading chemical fiber producers and an offshoot of ENKA, and KZO, itself the result of various mergers and acquisitions made during the 1960s.

AKZO's international base came from ENKA's chemical fiber operation carried out not only in the Netherlands but also in Germany, Italy, Spain, and the United States through direct subsidiaries, and in Brazil, Colombia, Mexico, and India through association with local partners. The range of products included rayon as well as polyester and polyamide fibers. As a leading European producer of conventional artificial and synthetic fibers, AKZO set about developing in the 1970s the new aramide engineering fibers, causing a clash with Du Pont in this area.

AKZO had inherited from KZO a number of chemical processes essentially based in Holland. Thus AKZO Zout Chemie extracted its own salt to feed its Demfzijl and Rotterdam electrolysis units from Hengelo brine. Partly owned subsidiaries, Methanor and Delamine, produced methanol from natural gas and ethylene amines respectively. Shell/AKZO's association within the Rovin Company produced vinyl chloride monomer and PVC. AKZO Zout Chemie also produced monochloracetic acid which served as a precursor for carboxymethylcellulose (CMC) made in Arnhem.

AKZO had become the world's leading salt producer and owned deposits in Northwestern Europe, the United States, Canada, and the Caribbean. A further entity was AKZO Chemie, which produced a range of chemicals including sulfuric acid, sodium silicate, and the catalysts for the Ketjen refineries, organic peroxides from the former Noury Van der Land, as well as a series of additives for polymers. AKZO Chemie also became heavily involved in oleochemicals.

Through AKZO Coatings set up after the acquisition of Sikkens, Lesonal, and Astral Celluco, AKZO was internationally involved in protective coatings. It had significant paint works in the Netherlands, in Germany, in France, as well as in other European countries, and in North and South America. KZO's acquisition of Organon in 1953 had given the group a foothold in hormone, estrogen, and corticosteroid drugs. The subsequent purchase of Chefaro, Dyosynth, and Intervet added to AKZO Pharma's range of pharmaceutical and veterinary products.

With its assortment of units acquired in a variety of areas, AKZO was more in the nature of a conglomerate, where very profitable operations were conducted in addition to more marginal activities. But with its sound international base and the dominant positions it held in special areas, the group, once restructured, could face the future with confidence.

Holland's chemical industry also drew benefits from the presence in the country of the chemical division of the *Royal Dutch Shell* group dating back to 1929 when the Mekog Chemical Fertilizer Company was set up. The group's Pernis refinery developed after the war into one of Europe's main petrochemical sites.

The other Anglo-Dutch giant, *Unilever*, had developed a chemicals industry in line with its activities in fatty acids and with areas it wished to boost such as synthetic resins and adhesives. Such was the Unilever-Emery subsidiary set up in Gouda, which was the second unit in the world to supply carboxylic acids for polyamides, in this case azelaic and pelargonic acids produced under Emery USA's ozone process. Through its purchase of the Dutch company Synrès, Unilever obtained access to a whole range of synthetic resins and vinyl emulsions.

Because of its exceptional geographic situation and the supply network it provided for industrial investors anxious to settle there, postwar Holland attracted many international chemical groups such as ICI in Rozenburg, Dow in Terneuzen and Hoechst in Breda.

Diversification and Expansion in Belgium's Chemical Industry

Set up in 1863 by the brothers Ernest and Alfred Solvay in the form of a limited partnership company, *Solvay* remained under the control of the founding family until 1967, when it was opened to outside shareholders. The postwar years were marked by a major program of diversification subsequent to the 1947 decision to get into PVC through a joint subsidiary set up with ICI. The Société Solvic's first production unit came on stream in Jemeppe on the Sambre River near Charleroi. To the high-polymer range was subsequently added low-pressure polyethylene when a Phillips Petroleum license was bought in 1955.

The switch from acetylene to ethylene for the production of vinyl chloride, starting in 1967, made Solvay one of the largest ethylene consumers in Europe. It served both for PVC and the polyethylene units located in Sarralbe, France, and in Rosignano, Italy. This explains the stake Solvay took in the Feyzin cracker and the supply contract signed with Dow in Terneuzen, with Petrochim in Antwerp, and with ANIC in Priolo.

Less fortunate were the acquisition of the Celanese polyethylene unit in Deer Park, Texas, in 1974 followed by the establishment of Soltex and the association with ICI and Champlin Petroleum in the Corpus Christi steam cracker. The Solvay group was better inspired when in 1977 it invested in polypropylene in Europe and in the United States.

Diversification in peroxidized products proved a successful move. The crowning touch was the joint venture with Laporte Industries in a subsidiary, Interox, to become the world's leading hydrogen peroxide and derivatives producer. By 1975, under the leadership of Jacques Solvay, who had succeeded Baron Boël, Solvay was a group which had remained faithful to its still profitable traditional operations, but had diversified in chlorine and peroxide derivatives on the one hand, and in plastics on the other. While Solvay's main units henceforth were in France, Germany, and the United States, Belgium was still an important base as were Brazil, Italy, the Netherlands, and Spain.

The group got into pharmaceuticals quite by chance when in the 1950s it acquired the Germany company Kali Chemie, which had a sodium carbonate unit in Hanover in which Solvay was interested. Kali Chemie was conducting research on gastrointestinal drugs, hormones, and sedatives; and this led Solvay further into pharmaceuticals when it subsequently acquired Latema, Sarbach in France, Giulini Pharma in Germany, and finally in 1980 Philips-Duphar.

Unlike Solvay, a multinational company managed from Brussels, *Union Chimique Belge* (UCB), had most of its productions in Belgium. Set up in 1928, UCB had spread its chemical operations after the war over three sectors. Concentrated in Ostend and Ghent were the sulfur and phosphoric acid inorganic chemicals, to which were gradually added methylamines and their derivatives (thiurames, dimethylformamide). Organic chemicals were located in Ostend and Schoonaarde within the UCB-Ftal subsidiary and included phthalic anhydride and phthalates, adipic acid, cyclohexanol, maleic anhydride produced by the company's own process, a range of xanthates for mineral flotation, as well as special intermediates (benzene sulfochloride, butylbenzene sulfonamide, and sodium vinyl sulfonate). Chemical specialties were produced in the Drogenbos unit, one of the oldest in Belgium. They initially included acrylic polymers, polyesters, and silicones, to which were added polyurethanes as well as resins and dilutants used in radiation-curing inks.

Under the wise management of the Janssen family, which had links with Solvay, UCB managed to withdraw from crisis areas like fertilizers and Fabelta's chemical fibers, to refocus on new chemistry; on pharmaceuticals, where its research teams had proved their mettle; and on films, where the group attained international stature through the Sidac division.

Other Belgian companies developed in less spectacular fashion because of inadequate diversification. *Société Carbochimique*, set up in 1928, remained faithful to ammonia and urea produced on its Tertre site. It had set up a fertilizer joint venture with *Société de Prayon*, which also kept to its specialty, completing its range of nitrogen fertilizers with superphosphates.

Over the years, Société Carbochimique gradually withdrew from dyes, which were regrouped in 1969 within Althouse Tertre. It also withdrew from the Tensia surface agents which were taken over by BP.

Les Poudreries Réunies de Belgique, which had associated with Nobel in

France to produce dynamite, concentrated their efforts on polyurethane foams but had not tried to get into chemicals proper.

It was through *Petrofina*, an oil company founded in 1920, that Belgium took a foothold in petrochemicals. Its subsidiary *Petrochim* had set up a petrochemicals site in Antwerp to produce olefins, aromatics, cumene, ethylene oxide, and derivatives. American Petrofina bought Cosden in the United States, thereby introducing the company into a new area with polystyrene. Through its subsidiary Oleochim, Petrofina was also involved in fatty amines. It had also taken large stakes in polyethylene, polybutene, and lubricating additive units.

Belgium, like Holland, had encouraged foreign chemical companies to set up units in the country. Major international groups, such as Hercules (polypropylene, hydroperoxides, and hydrogenated rosins), Monsanto (rubber additives, plant protection products, and polyvinyl butyral sheets), and Union Carbide, built units there after the war. Union Carbide subsequently sold ethylene oxide and derivatives units to BP.

Specialization in Switzerland

Faithful to tradition, Switzerland's chemical industry maintained and even emphasized its specialization in high added-value products devised from its own research and sold on world markets. The major firms all had their headquarters in Basel, although their research centers and production units were scattered worldwide.

Hoffmann-La Roche, which had been strongly involved in vitamins between the two world wars, pursued its founder Fritz Hoffmann's policy of devoting considerable funds to research. Marked success was achieved, in particular with heparin, the izoniazide of nicotinic acid and L-dopa. The company also opened up a new way to treat anxiety through a study of benzodiazepins leading to the discovery of Librium and Valium. To diversify away from pharmaceuticals and vitamins, Hoffmann-La Roche acquired two firms that specialized in flavorings and perfumes, Givaudan S.A. in Vernier near Geneva, and Roure Bertrand Dupont in France. Through the more recent acquisition of R. Maag's company in Dielsdorf, Hoffmann-La Roche gained a foothold, albeit a modest one, in plant protection products. With its plants and research centers located in Europe, the United States, and now in Japan, Hoffmann La Roche is a multinational that has kept hold of its best assets in its traditional field of action despite the expiration of a number of basic patents, and that has successfully managed to diversify in perfume chemicals.

Ciba-Geigy also, through its emphasis on research and the patience of its leaders and main shareholders, who have been concerned more with long-range achievement than with immediate success, has gained its first-rate position among world fine-chemical companies. Founded in 1758, Geigy drew substantial profits from Paul Müller's 1939 discovery of the insecticidal properties of DDT. A major part of the revenue obtained from the product was invested in pharmaceutical research after the war with fortunate results, principally in the areas of antidepressants (Imipramine) and antirheumatoids (phenylbutazone). The company also broke new ground in herbicides with products belonging to the triazine group, which became immensely popular. Geigy had also started investigating plastic additives and managed to become a world leader in this area.

CIBA had originally been set up by the French emigrant Alexandre Clavel, who, in 1859, had associated with the Renard family in Lyon to develop the Verguin fuchsin patent. The company's first achievements in patent drugs were due to Max Hartmann. After discovering Coramin in 1918, he put on the market a particularly active sulfonamide called Cibazol. CIBA was more diversified than Geigy, producing epoxy resins (Araldites) and venturing, not very successfully it must be said, into the area of photography by purchasing the English company Ilford and Société Lumière.

Despite its higher revenues, Geigy's future seemed more at risk than CIBA's once a number of its patents such as triazines had expired. At the end of the 1960s a merger of the two companies was contemplated by Robert Käpelli, CIBA's elderly boss, and Louis von Planta, the clever lawyer who was head of Geigy. The merger in 1970 did not yield immediate benefits because of the difficulty in getting the teams of what had been rival companies to work together. Personal rivalries turned out to be particularly harmful in pharmaceutical research, and major breakthroughs became rarer. The new group was also affected during the 1970s by the disputes that arose in Japan over Entero-Vioform, which was finally withdrawn from the market.

After the merger, Ciba-Geigy became the world's leading dye producer. The sector was rounded off with the purchase of the U.S.-based American Color and Chemical Corporation and the acquisition of Sandoz' 20 percent stake in the Toms River, New Jersey, unit. From Geigy the group had inherited, besides the Huningue and Manchester production centers, the Schweizerhalle plant located a few kilometers from Basel involving semiheavy chemicals. The last stages of the synthesis of specialties were carried out in units located abroad from intermediates supplied from Switzerland. The acquisition of Airwick in 1974 was a move away from Ciba-Geigy's traditional area of activity, and that activity was subsequently sold.

Apart from Ilford, which can hardly be regarded as one of the group's achievements, Ciba-Geigy holds enviable international positions in all sectors of fine chemicals and specialties in which it is involved—in pharmaceuticals, plant protection products, plastics, pigments, polymer additives, and dyes. The variety of its bases and the intensity of its research ensure its competitive stature in an increasingly harsh environment.

Sandoz, always jealous to preserve its independence, was born in 1886 out of the association of one of CIBA's chemists, Alfred Kern, with a businessman, Edouard Sandoz, who had been with Durand & Huguenin. From dyes, Sandoz became involved in pharmaceuticals through Arthur Stoll, an assistant of the Nobel prize-winner Richard Willstätter, who isolated ergotamine from ergot. Arthur Stoll had joined Sandoz at the end of World War I. In 1929, Sandoz set up a chemicals division to round out its dye business with additives for the leather, textile, and paper industries. Ten years later the company began taking an interest in agrochemicals. After World War II, Sandoz concentrated its production on the Muttenz site, then enlarged its food product range built around Ovomaltine, by purchasing the Berne-based Wander Laboratory in 1967. Finally, it diversified in the seed business by acquiring the U.S.-based Roger Bros. and Northrupt King in 1975.

Present on all continents, Sandoz can take credit for innovative products brought out through its research in all the areas in which it was involved: Foron dyes for polyesters, ergotamine derivatives used for the prevention of thrombosis, Elcar biological insecticides, optical brighteners for textiles and paper. Its Pratteln subsidiary, Industries Suisse des Goudrons, supplied Basel chemists with local derivatives from coal-tar distillation.

Another Basel firm, *Lonza*, had been set up in 1897 to use local hydroelectric power to produce calcium carbide, acetylene, and calcium cyanamide. It had subsequently become a subsidiary of Alu Suisse. While keeping a number of heavy electrometallurgical and electrochemical productions, Lonza sold its PVC business to ICI and concentrated increasingly on more elaborate chemicals in its Visp unit. It produced, in particular, diketenes and derivatives, HCN, and a whole series of malonates, pyrazole and pyrazolone products, pyridines, nicotinates, and nitriles. Lonza also has subsidiaries in Italy. Ftalital is involved in phthalic and maleic anhydrides, and Distillerie Italiane produces plasticizers and polyesters. In 1969 it bought several medium-sized companies in the United States involved in fine chemicals and specialties: Baird Chemicals, Bio Lab, Quad Chemicals and Glyco, Inc. By refocusing on other products, Lonza has become a first-rate international company in a number of specialties, particularly in synthetic intermediates supplied to the pharmaceutical and agrochemical industries, as well as to the producers of perfume chemicals.

While less important than the four Basel groups, a number of Swiss companies still in family hands continue to play a significant role, each in its own area. Such is the case of *Rohner*, founded in Pratteln in 1906 and involved in both dyes and diazo products; of Cilag, set up in Schaffhouse in 1936 to synthesize derivatives for the pharmaceutical industry; of *Firmenich*, which has a worldwide reputation in base materials for perfumes.

One of the world's oldest chemical companies is also established in Switzerland. *Chemische Fabrik Uetikon*, dating back to 1818, is mainly involved in inorganic chemicals (mineral acids, electrochemistry, fertilizers, silica gels, molecular sieves). *Fabrique Suisse d'Explosifs*, established in Dottikon in 1913 to provide civilian and military explosives for the country, took advantage of its nitration and catalytic hydrogenation knowhow to diversify into synthetic intermediates, in particular in aromatic amines and their derivatives, and it still supplies chlorates and perchlorates produced in its electrolysis units.

The Development of the Chemical Industry in Scandinavia

Because of their individual features, the four countries of Northern Europe provide very distinct models of development as regards the chemical industry. When the war ended, it became obvious that *Sweden*, with its national resources restricted to pyrites, wood pulp, and plentiful hydroelectricity, could not nurture great ambitions in basic chemicals, for it lacked natural gas and oil and even coal. There was the Stockholms Superfosfat Fabrieks, set up in 1871, which had taken the name of Fosfatbolaget in 1964 before becoming *Kemanord* six years later. Kemanord had pioneered in chlorates and had its own hydroelectric resources. Besides electrolysis products, the company produced nitrogen fertilizers, chlorine derivatives, and also PVC. It merged in 1978 with Nitro Nobel, the company set up in 1864 by Alfred Nobel. The resulting entity, *Kema Nobel AB*, thus extended its range to cover explosives and detonators. Another company established in 1924 in Bohus near Göteborg, *EKA AB*, also supplied chlorine, alkalis, and other inorganic chemical derivatives such as hydrogen peroxide and silicates. Sweden also produced ethyl alcohol through fermentation of the sugars contained in sulfite pulp waste.

With oil becoming the raw material for major organic intermediates, a steamcracker was brought on stream by Esso in 1963 in Stenungsund near Göteborg, which became a new petrochemical site. The ethylene produced was used as feedstock in the neighboring *Unifos Kemi* polyethylene unit, a joint subsidiary of Union Carbide and Kema Nobel, as well as in the ethylene oxide unit of a company which was to be known as *Berol Kemi* in 1974 and was a joint organic chemical venture of Mo och Domsjö, SOAB, and Berol AB. The propylene was used to produce butyraldehyde and derivatives by *Beroxo*, a company which, like Berol, became part of the state-owned Swedish group Statsföretag.

Because of the inadequate oil integration and lack of critical size, some of these petrochemical units could not compete with the international giants, and they were subsequently taken over by Norwegian and Finnish state-owned oil groups, which enjoyed certain advantages and wished to develop downstream. Thus it was only in the area where its lack of natural resources did not inhibit its competitive ability that Sweden succeeded in building up sound international enterprises.

AGA was founded in 1904 by Gustaf Dalen to build lighthouses that used acetylene. It became one of the six leading world producers of industrial gases with 19 subsidiaries scattered over Europe and Latin America, and 200 plants that produced liquefied gases, dissolved acetylene, and carbon dioxide.

Bofors-Nobel is the chemicals division of the arms and munitions manufacturer AB Bofors which Alfred Nobel had bought in 1902. Its Karlskoga plant is one of the most modern in the world for the manufacture of explosives and propellants. A significant part of the division's activities in nitration technologies is geared to the synthesis of organic derivatives for the pharmaceutical industry, and to intermediates for pigments and dyes. Plastics such as polyurethanes, and reinforced polyesters, only account for a small part of the turnover but nitrocellulose is still produced in Karlskoga. In 1977, Bofors-Nobel bought up an organic intermediaries producer, the Michigan-based Lakeway Chemicals. The company encountered serious effluent problems and, as a result, became involved in the treatment of waste waters for other companies.

The policy of *Perstorp* intensively to develop research in a number of specialized sectors has gained it worldwide respect. It was originally set up in 1881 by Carl Wendt, who owned beech forests, to extract acetic acid and methanol from charcoal. In 1905 it began manufacturing formaldehyde on which it based its subsequent development.

Perstorp, one of the leading European producers of formaldehyde, for which it has developed a concentration process patented worldwide, over the years has added to its range a unique series of polyols including pentaerythritol, trimethylolpropane, and their derivatives. The company is a major producer of phenoplasts and aminoplasts as well as of decorative laminated plastics derived from these resins.

Through Pernovo, its development subsidiary, Perstorp has taken a foothold in plastics additives by acquiring Dry Color AB in the United States and Société Synthécolor in France. It is also involved in carbohydrates. The company exports the major part of its production at competitive prices made possible by its good upstream integration. Although it produces formaldehyde and laminated plastics in Brazil, it has remained cautious about setting up industrial units abroad.

Through Statsföretag AB, the Swedish state began taking over *Berol Kemi* and *Beroxo* in 1973. From the companies merged to form it in 1973, Berol Kemi had inherited surfactant specialties as well as a soluble gum, ethyl hydroxyethylcellulose. The company owned an ethylene oxide unit in Stenungsund from which it produced an interesting range of ethylene amines under an efficient process. Although it possessed good technologies as well, Beroxo's range of products was an unsophisticated one (alcohols, phthalic anhydride, plasticizers, and solvents), and this had a negative impact on profits.

Sweden is involved in other chemical and chemical-related sectors through *Svenska Rayon* for fibers, *Svenska Polystyren*, a subsidiary of A. Johnson and Company and Hüls for polystyrene, *Boliden* and its Supra AB subsidiary, supplying almost the whole of the country's sulfuric acid and superphosphate needs, *Rexolin*, bought by W. R. Grace in 1965 for fine chemicals (sequestering agents, piperazine derivatives), and *Bergvik Kemi*, a member of the Stora Kopparberg group for tall oil.

Sweden's international reputation, however, was mainly built on its pharmaceuticals sector. The most important company in the field was *Astra*, set up in 1913. Its rise began with the discovery of xylocaine in 1948, a painkiller still used in dental medicine. Between 1965 and 1970, Astra's reseachers brought out Aptine, one of the first betablocking agents used in cardiology, then the antiasthmatic Bricanyl. The quality of Astra's research led to association with Merck in the United States and with Fujisawa in Japan.

AB Fortial (Pharmacia), located in Uppsala, dates back to 1911. It specialized in the course of time in diagnostic products, in supplies to hospitals of dextrin solutions as a plasma substitute, and in biotechnologies.

AB Kabi also was involved in diagnostic agents. It was taken over by the Swedish state in 1969 and became *Kabi Vitrum* in 1978. It has production units in Sweden, France, and the United States (the Cutter-Vitrum subsidiary in Clayton, North Carolina) and has developed in three main areas: blood plasma proteins, growth hormones and parenteral nutrition.

AB Ferrosan built its reputation on two noteworthy breakthroughs, the dicoumarol anticoagulant, and para-amino salycilic acid (PAS), for the treatment of tuberculosis. Other specialties, used to fight functional sterility and senile derangement, attest to the high quality of the laboratory's research.

Two other laboratories have expanded significantly outside their own frontiers by dint of their valuable research work, *Ferring*, involved in the synthesis of polypeptides, and Léo specializing in antitumor and psychopharmacological products. While Sweden's chemical industry has always been in private hands except for Berol, sold to Nobel Industries in 1986, and Kabi Vitrum, which is still controlled by the Statsföretag state holding company, this was not the case of the industry in *Finland* and *Norway*.

In Finland, *Kemira* had been set up in 1920 by the state to provide the country's needs in sulfuric acid, ammonia, and fertilizers (nitrates and superphosphates).

The other Finnish major chemical company, *Nesté*, was also a public company. Its development has been speeded by Soviet oil supplies provided under privileged conditions. The oil is refined by Nesté in Provoo, near Helsinki, where a steamcracker supplied the olefins needed for Nesté's petrochemicals. Recently, Kemira and Nesté, which until 1979 only held a 50 percent stake in the local low-density polyethylene producer Pekema, have started to carry out a vast program of industrial acquisitions both in Europe and in the United States.

In *Norway*, the state controls two powerful entities, both anxious to make the most of their crude oil reserves to boost their petrochemicals and plastics divisions. *Statoil*'s prosperity is based on its share in the North Sea oilfields. This is also true of *Norsk Hydro* which is 51 percent state-controlled. Its oil activities account for 30 percent of its turnover, while fertilizers, petrochemicals, aluminum, and magnesium account for the balance. The two companies ended up sharing control of the Bamble petrochemical site near Oslo, where Saga, subsequently bought by Statoil, had built a polyolefins units.

Deprived of natural resources, *Denmark*, like Sweden, had developed a chemicals industry geared to a few areas where it could show some competence. Two Copenhagen firms, *Nordisk* and *Novo*, date back to the early twenties. They established their reputation on the production of insulin, which they had pioneered. Nordisk remained essentially an insulin producer, merely diversifying in the growth-hormone (HGH) field. Novo sought to widen its range with enzymes. To gain a foothold in the North American insulin market, it joined with Squibb in the United States and with Connaught in Canada. In 1982, Novo put the first semisynthetic insulin on the market.

In Denmark there is also a major fertilizer producer, *Superfos*, of international dimensions. It has an especially broad base in the United States, where it produces potash in New Mexico, potassium sulfate in Texas, and ammonium sulfate in Louisiana.

The State Chemical Industry in Socialist Countries

The Comecon Socialist countries began only in the 1960s to substitute natural gas and oil for coal as raw materials for their chemical industry, and among them only the U.S.S.R. and Rumania had such resources. This delayed development of chemical industries in the Soviet Union and its satellites. The lag was all the greater as the Comecon countries were dependent for their processes on Western enterprises, which were understandably reluctant to reveal their latest technologies to licensees on the other side of the Iron Curtain. Too much bureaucracy and recurrent shortages of strong currencies also combined to delay the startup of planned units. By repudiating market economy, the Socialist countries had also chosen to disregard actual production costs, lacking as they did any reference points to market prices.

Set up as state monopolies, the chemical enterprises of these countries had no cause to worry about competition on their home market and their goods were exported at prices decreed by political considerations. As for research geared to the strategic goals of the Administration and not to the wishes of the consumers (had they been consulted), it produced no breakthrough after the war likely to be of interest to the free world's chemical or pharmaceutical industries.

In the circumstances, the chemical industry of the Soviet Union and

Comecon countries was involved in international trade only insofar as it exported some of its basic products at a low price, as dictated by its need of foreign currencies. The contribution of this region to the progress of chemicals over the past fifty years may thus be considered negligible.

Other nations with Socialist, although not Communist leanings also reported to state chemical organizations.

In *Israel*, the greater part of the chemical industry had been assembled under the aegis of *Israel Chemicals Ltd.*, a group set up by the government to develop the Dead Sea and Negev potash, bromine, and phosphate resources, and encompassing twenty-eight different companies.

In *Austria*, too, the country's chemical industry was mainly in the hands of two state-owned companies, *OMV* with its Petrochemie Danubia subsidiary, and *Chemie Linz*. Their production was too run-of-the-mill to bring in regular profits.

The National Character of Japan's Chemical Industry

Just as was the case before the war, Japan's chemical industry after 1945 remained dependent on foreign technologies for a long time. This did not prevent it, however, from developing on strictly national lines under the watchful eye of the Ministry of International Trade and Industry (MITI). Foreign companies were not allowed until recently to develop their technologies single-handedly in Japan. They had to resort to licensing them to Japanese industrialists. The import of chemical products also encountered a whole series of obstacles besides those from tariffs, such as the solidarity among national companies, and exacting quality requirements. Investments were financed with the help of the big Japanese banks so that outside capital injection was not required.

In addition, when it came to encouraging development of a national chemical industry, the government had no qualms about fixing the price, for instance, of its naphtha feedstock at an artificially low level. Indeed, whenever some branch of activity fell victim to adverse circumstances, MITI made it its duty to form an antirecession cartel arrangement with the industries involved. While the Japanese authorities did not interfere in the capital makeup of the chemical industry, they always tried to preserve its national identity and a certain degree of financial prosperity.

This has not prevented Japan's chemical industry from suffering from a number of deficiencies due, in particular, to almost total dependence on outside supplies of raw materials, to excessive scattering of production units throughout the hilly country, to the survival of too many competing producers failing a real effort at rationalization, to research which innovated principally in biochemicals, and to a relatively modest international exposure.

For a country with its economic importance, Japan had a chemical industry that, before the first oil crisis, had no international stature. The exports of the country's six leading chemical groups accounted for only 12 percent of their sales in 1971, and none of them had set up industrial units abroad. Within Japan, however, a number of sizeable companies had established themselves.

Asahi Chemical, set up in 1931 to develop the Casale ammonia process, was involved directly or through subsidiaries in heavy chemicals (mineral acids, chlorine, soda, acrylonitrile, and Nylon intermediates), explosives, polymers (low-pressure polyethylene, polymethacrylates, Nylon-66, and synthetic rubbers), artificial textiles and synthetic fibers (rayon, polyamides, and acrylics). Associated since 1952 with Dow, within Asahi Dow, in the area of plastics, Asahi Chemical had integrated upstream in the early 1970s when it brought a petrochemical complex on line at Mizushima.

The *Mitsubishi* group had reorganized its chemical activities in 1944 by creating Mitsubishi Chemical Industries, MCI, from Japan Tar Industries founded ten years previously.

In 1951, MCI's textiles division became Mitsubishi Rayon specializing in fibers (acrylic, polyester, and polypropylene) but also in resins (methacrylates, polybutylene terephthalate, and films). The glass division became Asahi Glass Company, while chemical activities proper were concentrated within MCI Ltd. From its origins, MCI has retained its coke and tar derivative production as well as the production of fertilizers. The first ammonium sulfate unit came on stream in 1937.

Over the years, production of dyes, pesticides, organic intermediates (DMT and TPA, alcohols, acrylonitrile, acrylamide, butadiene, caprolactam, and isocyanates), polymers (polyurethanes, polybutadiene, SBR, polyacrylamide, and polycarbonates) had been added to the basic chemicals. The group set foot in petrochemicals by associating with Shell in 1956 within the Mitsubishi Petrochemical Company and by building its own steamcracker in Mizushima in 1964.

Through Mitsubishi Gas, formed with the merger in 1971 of Japan Gas Chemical and Mitsubishi Edogawa, the group became involved in other sectors of inorganic and organic chemicals: electrolysis products, peroxide derivatives, hydrazine, formaldehyde, methanol, pentaerythritol, trimethylolpropane, methylamines, phthalic anhydride, and phthalates. Through Mitsubishi Monsanto, a subsidiary established in 1952, the Mitsubishi group acquired a foothold in vinyl polymers, polystyrene, and ABS.

Benefiting from good raw materials integration, MCI has succeeded in developing original processes (1,4-butylene glycol and tetrahydrofuran from butadiene), and was well diversified. Managing the chemicals of the Mitsubishi group is no easy task because of the many subsidiaries and the multiplicity of production sites.

Sumitomo's chemicals division, encompassed within Sumitomo Chemical in Osaka, was far more coherent than Mitsubishi's despite major subsidiaries such as Sumitomo Bakelite dating back to 1911, Sumitomo Naugatuck, and Sumitomo Bayer Urethane established respectively in 1963 and 1969. The fertilizer division, which had formed the group's starting point in 1913, remained substantial, ranging from ammonia to urea to superphosphates. Sumitomo Chemical was also involved in other areas of heavy chemicals: electrolysis, mineral acids, aromatic hydrocarbons, and olefins. The company was also involved in plastics (polyolefins, PVC, polymethacrylates, polystyrene, ethylene-vinylacetate copolymers) and in elastomers (EPM rubbers and styrene-butadiene). But it was especially through the dye industry that Sumitomo Chemical had entered fine chemicals, developing a range of products for the drug and agrochemical industries as well as additives for rubber and plastics. The company was also involved in the industrial synthesis of meta- and para-cresols on the one hand, and on the other of resorcinol and hydroquinone by a process similar to that by which phenol is obtained from cumene.

Through subsidiaries established in 1959 and 1971 with Nippon Shokubai and Nippon Oxirane, Sumitomo Chemical acquired original oxidation technologies for ethylene oxide and acrylic acid, as well as the Halcon process for propylene oxide. In 1977, Sumitomo Chemical ventured outside Japan when it assumed share in setting up a petrochemical site in Singapore.

The chemicals of the *Mitsui* group took shape in 1968 with the merger of Toyo Koatsu and Mitsui Chemical Industry to produce Mitsui Toatsu Chemical. The new entity for a long time remained handicapped by the failure of the leading personnel of the two component companies to get on together.

Toyo Koatsu, established in 1933, contributed its fertilizers, while Mitsui Chemical, set up in 1941, provided its dyes and organic intermediates. The resulting group was focused on four areas of activity: fertilizers, more especially urea produced in an efficient process; basic products (methanol, formaldehyde, phenol, aniline, bisphenol A, TDI, MDI, and melamine); polymers (PVC, polypropylene, polystyrene, PVC, and polyethylene film); and fine chemicals (dyes, agrochemicals, and pharmaceuticals).

Mitsui Toatsu had its raw materials supplied by the petrochemical complex of its Osaka petrochemical subsidiary. But as early as 1955, the Mitsui group had set up one of the first petrochemical enterprises, Mitsui Petrochemicals Industries, which brought the Iwakuni unit on stream three years later. In 1960, Mitsui Polychemicals was formed in association with Du Pont to produce low-pressure polyethylene. At the same time, Mitsui Petrochemicals' own research was carrying out efficient processes in the areas of low-pressure polyethylene, petroleum resins, and synthetic *m*- and *p*-cresols. Mitsui Petrochemicals' overseas market consisted mainly of its sales of process licenses. Its project to set up a petrochemical complex in Iran became the source of much trouble for the group.

Other chemical enterprises that had no connection with the three big ''zaibatsus'' had developed rapidly after the war. Born in 1939 of the merger of Showa Fertilizer and Japan Electrical Industries, *Showa Denko* diversified in petrochemicals and polyolefins while remaining true to its electrochemical and electrometallurgical activities. In 1969, it built a major petrochemical complex in Ohita. Moreover, through its subsidiaries, Showa Denko produced phenolic resins, neoprene in association with Du Pont, acetaldehyde derivatives, and carbon black (Showa Cabot).

Known under the name of *Daicel* since 1966, the successor of Dainippon Celluloid Company, founded in 1919, had become Japan's main cellulose acetate producer. Integrating upstream, Daicel then produced acetic acid under various processes before choosing the methanol process in a common subsidiary set up with Mitsubishi Gas Chemical. Daicel also manufactured other cellulose derivatives such as CMC, and hydroxyethyl cellulose, as well as producing allyl alcohol and synthetic glycerin, alkylamines, acetic solvents and finer synthetic products (1,3-butylene glycol, α-butyrolactone, glyoxal).

From nitrocellulose, Daicel extended its range of polymers first to cellulose acetate, then to ABS and SAN. Through the medium of subsidiaries set up with Celanese and Hüls, acetal resins and Nylon-12 were also produced. The company has a films division producing cellophane and various films (polyvinylidene chloride, polyester, polypropylene), while propellants for rockets are the logical sequel to its range of explosives, originally obtained from nitrocellulose.

Another old firm, dating back to 1887, is *Nissan Chemical* which has remained strongly entrenched in its traditional fertilizer area. It has diversified toward crop protection products and silica gels. Already a urea producer, Nissan has developed an efficient process for melamine and a line of isocyanuric acid chemicals. Having come into petrochemicals through its Nissan Petrochemicals subsidiary, the group produces higher alcohols. It has also licensed worldwide its phosphoric acid and melamine processes.

Just as the traditional chemical companies had felt the need for upstream integration through petrochemicals and for diversification in polymers, the old manufacturers of artificial fibers came to chemicals through synthetic fibers. Established by Mitsui in 1926 under the name of Toyo Rayon, *Toray* began producing Nylon-6 in 1951, polyester in 1958, and acrylic fibers in 1961. Then it used its own process of cyclohexane nitrosation to produce caprolactam in 1962. It developed progressively to include para-xylene and acrylonitrile. The company, one of Japan's largest in turnover and personnel, has also become a producer of polyester and polypropylene films and of plastics (ABS, PBT, polyamide). It has pioneered in carbon fibers obtained from its own polyacrylonitrile (PAN) and has developed overseas in this area.

Set up in 1918 under the name of Teikoku Rayon, *Teijin* also went from rayon to polyesters by using ICI's technology in 1958 and by producing DMT from 1963 onwards in association with Hercules. For Nylon, Teijin has an Allied Chemical license, and its caprolactam is obtained from a subsidiary it has in common with Sumitomo and Toyobo. Since 1966, Teijin manufactures the *p*-xylene it needs through Teijin Petrochemicals. Its range of products also includes acetic derivatives as well as polyester and polycarbonate films and resins. An exception among Japanese companies, Teijin's exports account for over 20 percent of its total sales.

Other companies have also gone from fibers to chemicals. Such is the case of *Kanebo*, which has been in cosmetics and pharmaceuticals since 1972, *Kuraray*, which innovated in polyvinyl alcohol in 1957 and whose subsidiary, *Kyowa Gas*, produces acetylene by cracking natural gas; and *Toyobo*, which has launched into biochemicals and produces enzymes.

Since the end of the 1960s, the Japanese have been spared neither the difficulties arising from overcapacities in synthetic fibers nor those from soaring oil prices from 1973 onwards. Indeed, their investments in chemical fibers and petrochemicals had been unreasonable to the extent that they had control neither over technologies nor over raw materials. In the circumstances, it is not surprising that other sectors of Japan's chemicals were better able to withstand the crisis.

Through its research, Japan has acquired world renown in biochemicals both in the glutamates and lysine fields, where *Ajinomoto* and *Kyowa Hakko* have starred. More generally, Japanese firms have been successful in amino acids, enzymes, antibiotics, and antimitotics, and through them have expanded abroad through license concessions or through stakes taken in joint subsidiaries set up with local firms.

Other Japanese companies, even without the advantage of innovative technologies, have managed to stay prosperous in Japan because of the protection offered by the peculiarities of their national market, which is less than open to foreign products. This is true in the area of detergents, where old firms like *Kao Soap, Nippon Oil and Fats*, and *Sanyo* remain important, or in the area of pharmaceutical and cosmetic derivatives, where *Takeda* and *Shiseido* were able to diversify because of their strong home vantage point, and were able to extend their business abroad.

But the deliberate establishment of Japanese groups in industrial countries really began only with the yen's appreciation with regard to other currencies. In this context, *Dainippon Ink and Chemicals* (DIC), which started life in 1908 in inks and pigments and established links with Reichhold Chemicals after the war in the resins area, may be regarded as a pioneer when it purchased Polychrome in the United States, more recently Sun Chemicals and then Reichhold, USA.

Nonetheless, except in biochemicals and certain areas where Japanese technologies made breakthroughs abroad (carbon fibers, C_4 chemicals, aromatics oxidation), Japan's chemical industry on the eve of the first oil crisis was neither sufficiently innovative, nor rationalized and prosperous enough to be regarded as a serious rival to its big foreign competitors.

Chapter 6
The New Facts of World Chemicals Since 1973

Overcapacities and the Search for Remedies

The first oil shock that occurred at the end of 1973 with the Yom Kippur war served to pinpoint the crisis which world chemicals were already undergoing.

The chemical industry's soaring development after the war was due to the extraordinary burst of innovations occurring between 1935 and 1955 and coinciding with an explosion of world demand in a variety of sectors served by chemicals. Production units multiplied in Europe as well as in the United States and Japan.

Two other factors contributed to this rapid growth. The use of oil as a substitute for coal provided the chemical industry with abundant, cheap raw material that was easy to transport. With interest rates lagging behind the rate of monetary erosion over a number of years, industry leaders were tempted to carry out investments that they would not have made had currencies remained stable and interest rates higher. The fear of these leaders that competition would get the better of them if they slowed down their investments, the race for market shares advocated by a number of consultant firms like the *Boston Consulting Group*, the belief—quite widespread among world chemical leaders—that they had to keep building new units to keep up with forecast needs, all had a share in building up production overcapacities which were already becoming apparent before 1973 in certain sectors of heavy chemicals (petrochemicals, synthetic fibers, thermoplastics, and fertilizers).

The establishment of an OPEC cartel that led to a rise in the price of a barrel of crude oil from $3 to $12, then the 1979 Iranian Revolution which made it soar to $40, finally the publication of the gloomy forecasts of the Club of Rome experts which mistakenly saw oil shortages ahead when, in fact, these had been artificially engineered by the Cartel members, all these facts upset chemical leaders in industrialized countries. And yet some of them still continued to invest in new plants during the stock-building lulls that occurred in 1974 and 1979 through consumers' speculating on new price rises.

This only made the necessary adjustments much harder when they had to be carried out at the beginning of the 1980s. Companies were to suffer greatly from an error of judgment, building new plants at great expense at the same time that economic growth rates tumbled from over 10 percent to a mere 2 to 3 percent. Caught between the increasing cost of their hydrocarbon raw materials and the ever-lower prices they had to resort to selling their products on markets where offer exceeded demand, leading chemical companies in industrialized countries were forced to go through agonizing reappraisals.

This led them to act in a number of different directions. First and foremost, they had to lower their operating costs by cutting down on excess personnel and taking the measures needed to increase the productivity of each company. At the same time, they had to reduce, in a concerted way if possible, the overcapacities affecting the hardest-hit sectors. Finally, it seemed advisable to redirect production into areas that were less sensitive to economic change. This meant increasing the share of specialties in relation to commodities in overall turnover.

A new generation of leaders was called upon to carry out the socially painful and politically delicate job of rationalizing and restructuring the chemical industry through layoffs and plant closures. These same leaders were also given the more exalting, but just as difficult, task of defining the redeployment strategy that needed to be followed and of determining in a case-by-case basis the sectors that should be abandoned and those that, on the contrary, had to be invested in force.

By 1973, it was obvious that the chemical industry had reached a degree of maturity to the extent that all the companies involved in that area in industrialized countries were long established and that no discovery likely to affect its development had been made over the last two decades. While new areas of research like composite materials and biotechnologies had emerged, no immediate fallout was expected for a number of years. Thus failing any rapid internal growth brought about by major scientific breakthrough, the strategy of leaders anxious to refocus or diversify their portfolio of activities very often consisted in a kind of Monopoly game, as a range of production was shifted from one enterprise to another without anything new being created.

The Restructuring of Sectors in Distress

Priority action was required in petrochemicals, in the large thermoplastics, in fertilizers, and in synthetic fibers where the most serious investment mistakes had been made. The hardest cases were those of *petrochemicals* and *thermoplastics*. For one thing, a steamcracker cannot technically operate under 60 percent of its capacity. For another, the products that emerge are linked to one another in almost invariable proportions. Finally, a polymerization unit cannot have its pace slowed down without this affecting the upstream monomer unit to the same extent.

In addition to such rigidities, there was the need to reduce not only the quantities produced but also the number of production units. The problem then arose of sharing the sacrifices among the different producers within an economic area.

The problem was most easily solved in Japan because of the discipline which MITI managed to establish within the country's petrochemical industry. Making the most of a new law that allowed competing producers to act in concert, a cartel was set up with the object of cutting down ethylene production. Four groups of petrochemical producers were formed within which the necessary arbitrations took place. This led Sumitomo to close its Niihama units, Mitsubishi a number of its Mizushima plants, Showa Denko two of its Ohita installations.

At the same time, producers reached agreements on cutting down competing PVC and polyolefin sales networks, while MITI authorized the import of naphtha through an organization consisting of Japan's petrochemical producers. Its price served as a marker for naphtha produced in Japan.

In Europe, of course, it was difficult to show such disregard for market laws. The views of the European Economic Community Commission in Brussels had to be taken into account, and they upheld the principle of free competition as set down in article 85 of the Rome Treaty. Moreover, in Western Europe there were a number of petrochemical industries that operated according to the rules of private capitalism while there were others, as in France, Italy, Austria, Norway, and Finland, that were state-controlled and more concerned about retaining market share than ensuring profitability.

Despite such obstacles, unilateral decisions were taken and bilateral arrangements carried out among firms, leading to some measure of production rationalization. Between 1980 and 1984, twenty-five ethylene and eight polyethylene units were scrapped in Western Europe while ethylene oxide capacities were reduced by 10 percent.

The 1983 agreement between ENI and Montedison put some order in Italy's chemical industry, as ENI took over the PVC and polyethylene operations of Montedison. Previously in France, Rhône-Poulenc had sold its petrochemicals division and its thermoplastics to the Elf Aquitaine group. At the same time, steamcrackers were being shut down in Feyzin and Lavera, and a vinyl chloride unit in Jarrie. The association between BP Chimie and Atochem in polypropylene and the exchange of Atochem's Chocques unit for ICI's Rozenburg polyethylene unit were other instances of rationalization.

The Brussels Commission also gave its approval to three large-scale operations: the ICI and BP Chemicals exchange of polyethylene/PVC, the vertical integration of vinyl chloride involving AKZO and Shell Chemicals, and the recent Enichem and ICI association, which produced European Vinyls Corporation and was intended to lead to major capacity cuts in PVC.

In West Germany, rationalization measures were less spectacular because the heads of Germany's leading chemical companies had not waited for the crisis to delineate their respective fields of operation and to establish close links with international oil companies, either through long-term supply contracts or through parity associations.

A number of American companies became involved in restructuring. Union Carbide sold its Antwerp site to BP Chemicals; Monsanto, its Seal Sands acrylonitrile unit to BASF; Esso, its Stenungsund steamcracker to Statoil; while Hercules joined up with Montedison to set up the Himont company, which accounted for 20 percent of the world polypropylene market.

In the *United States* the petrochemical industry set its house in order along purely capitalistic lines. Each company involved acted alone for fear of infringing antitrust legislation and the main concern was to restore profitability. Unlike Du Pont, which acquired Conoco, other chemical companies tried to get rid of their petrochemicals. Hercules sold its DMT units to Petrofina, and subsequently its 40 percent stake in Himont to Montedison, while Monsanto was shedding its Texas City petrochemical site.

Major divestments took place, particularly in the major thermoplastics, which were taken over by individual entrepreneurs who bought up the units the chemical giants wished to get rid of. As Hoechst, Union Carbide, Du Pont, Monsanto, ICI, and USS Chemicals withdrew from a number of the major oil-based intermediates as well as from polystyrene, polyethylene, and PVC, a number of large, hitherto unknown companies emerged: Huntsman Chemical, El Paso Products, Aristech, Vista Chemical, Sterling Chemical, and Cain Chemical.

At the same time, oil companies were integrating downstream petrochemicals and polymers. Such was the case of Occidental Petroleum, which through its chemical subsidiary Hooker (later Oxychem) bought up Tenneco and Diamond Shamrock's PVC in 1986, becoming the largest American producer in this area. Likewise, BP Chemicals fully acquired its subsidiary Sohio. The long-standing petrochemical divisions of the large oil groups returned to profits in 1986 after some painful tidying up but no agonizing reappraisal, helped along by falling oil prices and dollar rates.

Most of them had cut down on operating costs and diversified to the point where they were able to face up to the economic ups and downs without too much apprehension. Productivity improvements and a better utilization of existing capacities because of higher demand put Exxon, Mobil, and Texaco on the way to prosperity in petrochemicals in 1986.

Standard Oil of California added the petrochemicals of Gulf Oil, purchased in 1984, to its subsidiary Chevron Chemical. Other United States petrochemical producers took advantage of special circumstances. Amoco was served by a strong terephthalic (TPA) base and its good performance in polypropylene; Arco, by its Lyondell subsidiary in Channelview, Texas, and by its development of the Oxirane process through which propylene oxide could be produced by direct oxidation with styrene as a coproduct. The process also led to MTBE (methyl tertiary-butyl ether), the antiknock agent used as a substitute for tetraethyl lead.

Even Phillips Petroleum, badly affected by Boone Pickens' takeover attempt, managed to make substantial profits from its petrochemicals because of drastic restructuring. New prospects were also opening up for the United States chemicals industry as needs grew for butene and hexene comonomers used to produce Linear Low Density Polyethylene (LLDPE), also as consumption of higher olefins to prepare detergent alcohols increased, and as demand for MTBE used as a gasoline additive soared.

The problem of overcapacities in *chemical fibers* in each economic region was both easier to overcome because of the small number of producers, and more complicated because of outside factors. In *Europe*, producers suffered heavy losses from 1973 onward. For one thing, the Europeans were not particularly suited to manufacture chemical fibers at satisfactory cost, a fact that was proved by growing imports from Southeast Asia. For another, the capacity increases decided upon did not tally with any comparable increase in demand in the foreseeable future.

In view of such imbalance, one might have thought that a number of producers would withdraw from the market. But this did not happen because some of them had to heed government instructions to maintain employment. Also textiles accounted for only a share of the business of the companies involved and could be kept up through the profits generated in other areas. From 1978 to 1985 two agreements were implemented with the blessing of the European Economic Community Commission. The first aimed for a linear reduction of existing capacities; the second and more important one allowed each producer to specialize in those areas where it held the best cards, giving up what amounted to marginal productions.

Thus Courtaulds withdrew from polyester and from Nylon to concentrate on its acrylics and cellulose fibers; ICI focused on Nylon and Bayer on acrylics; Rhône-Poulenc withdrew from acrylics but revamped its Nylon and polyester units well-integrated in upstream intermediates; Montedison decided in favor of polyester and acrylics; AKZO focused on polyesters and on aramide fibers while keeping up its profitable rayon sector. Such efforts, which aimed to reduce European chemical fiber capacities by 900,000 tons and to increase productivity through specialization, undoubtedly corrected the situation.

Nonetheless, European producers are still faced with two kinds of competition: first imports of synthetic fibers from Turkey, Taiwan, South Korea, and Mexico, against which it is hopeless to expect that the multifibers agreements—which contravene GATT rules—will constitute a permanent obstacle; and second, imports of natural fibers such as cotton, for which prices have fallen spectacularly in recent times.

The Japanese solution to chemical fiber overcapacities naturally involved MITI which pushed through a 17% cut in existing polyester, Nylon filament, and acrylic fiber capacities between 1978 and 1982. These were linear cuts, however, and did not restrict the range of synthetic fibers developed by each producer, contrary to the specializations that marked the second stage of Europe's approach.

The United States was faced with an additional problem because its market remained wide open to textile imports from developing countries. These imports constituted an indirect threat to American producers of chemical fibers. Their first reaction was to reduce their bases in Europe. Du Pont closed its acrylic units in Holland in 1978 and in Northern Ireland in 1980; the following year it ceased production of polyester thread in its Uentrop unit in Germany. Monsanto did likewise in 1979, shutting down its Nylon units in Luxembourg and Scotland and selling its acrylic fiber installations in Germany and Ireland to Montedison.

In the United States itself, capacity cuts were not so substantial and the 1983 upturn boosted utilization of remaining units to 80 percent of their capacities. Major American producers such as Du Pont, Celanese, and Monsanto returned to satisfactory profit margins. Other companies for which fibers were not an essential sector withdrew from this area. Chevron Chemical, for instance, shut down its Puerto Rico Nylon and polypropylene fiber units between 1980 and 1982 as well as the polypropylene fiber unit in Maryland.

The *fertilizer* market was in no better shape than the petrochemicals and chemical fibers markets, for world producers had largely allowed supply to exceed demand.

The situation in this area was further complicated by the unequal distribution worldwide of the raw materials required to produce fertilizers and the special attention which governments bestowed on agriculture. Such attention had led to a surfeit of production units and their increasing control by governments, either directly through taking a stake in the companies concerned, or indirectly through establishing ceiling prices for home sales or export subsidies. The emergence of new producers in Eastern countries and in developing areas increased the share of state-controlled companies in world production from 30 to 64 percent for ammonia, from 40 to 65 percent for potash, and from 10 to 46 percent for phosphoric acid between 1967 and 1986.

In Western Europe, nitrate fertilizer producers had deemed it expedient to set up a cartel arrangement for exporters called Nitrex. But the collapse of demand in countries outside its area had prevented it from functioning properly, sparking a fight for market shares even within the community.

As a country like Morocco switched from its long-established role as phosphate exporter to downstream ammonium phosphate and superphosphate integration, traditional fertilizer producers were forced to reappraise their strategy and take severe rationalization measures.

Japan, which had none of the required raw materials and, accordingly, had high production costs, began, as early as the 1970s, gradually to cut down capacities along the lines jointly agreed upon by the authorities and the five main Japanese producers of nitrate and phosphate fertilizers.

In *Europe*, the pressure of events disrupted the whole market as the number of producers was drastically reduced. Because of market proximity, production both from Eastern Europe of nitrates and from Africa of superphosphates were becoming dangerously competitive. Supply conditions for natural gas varied according to each country's policies. France, for instance, agreed to pay extra for Algeria's gas, while Holland's Groningen gas, which Dutch ammonia producers were getting at a very favorable price, was linked to the price of petroleum products. On the other hand, a number of Scandinavian state-controlled companies like Norsk Hydro and Kemira, were pushing ahead with ambitious fertilizer programs, taking advantage of their interests in North Sea oil or of the conditions under which they were being supplied with oil and gas from the Soviet Union.

Between mergers and acquisitions, the structure of the fertilizer industry in Western Europe was spectacularly pared down. A few giants emerged to dominate the market. In France, there was CdF Chimie, later to be known as ORKEM, which had just taken a 70 percent stake in Air Liquide's subsidiary La Grande Paroisse, and Cofaz, which was taken over by Norsk Hydro; in Western Germany, there was BASF and Ruhr Stickstoff; in Britain, ICI and Norsk Hydro, which bought up Fisons; in Italy, ANIC and Montedison's subsidiary Fertimont; in Holland, DSM's UKF and Norsk Hydro's NSM; in Finland, Kemira, which took over both Britain's Lindsay and its Kesteven facilities.

But the scene has not yet become sufficiently clear, since the competing companies do not all enjoy, within the Community, the same raw materials supply conditions, and Europe is still open to imports form other countries that do not apply the rules of market economy.

In the *United States*, the situation was in many ways different. With its large sulfur, natural gas, phosphate, and even potash resources, America's fertilizer industry rested on a sound base. It was an exporter of minerals and fertilizers, and did not have to worry to the same extent as Europe's industry about competing imports from Socialist countries. But reserves of sulfur extracted by the Frasch process are becoming depleted in Louisiana and Texas, and President Ronald Reagan's "Payment in Kind" (PIK) farm-acreage cuts has reduced the fertilizer requirement of American farmers. These farmers are also much in debt and are having trouble selling their products on saturated markets.

Consequently, very little money has been sunk into extracting phosphate rock in Florida or in increasing nitrogen fertilizer capacities; for a new ammonia and urea unit can cost as much as $250 to $500 million to build in the U.S., depending on the state of the existing infrastructure.

With such dim market prospects, it is understandable that W. R. Grace has decided to shut down its Trinidad ammonia unit, or that a company as large as International Mineral Chemicals has tried to diversify through purchase of Mallinckrodt and has put half its fertilizer assets up for sale.

The Nationalization of France's Chemical Industry

When a left-wing government came to power in 1981, France's chemical industry was in dire straits judging from the losses of the major groups: CdF Chimie was losing 1,200M F; Péchiney Ugine Kuhlmann 800M F; Rhône-Poulenc 330M F; Chloé Chimie 370M F; Atochimie 130M F; and EMC 100M F. Admittedly world chemicals were in poor shape. But while French leaders were posting losses amounting to 7 to 10 percent of their turnover, Hoechst and BASF were still making consolidated profits that year of 426M DM and 1,290M DM, respectively, even though they had noticeably slumped.

There were many reasons, some of them old, for the difficulties of France's chemical industry as illustrated by losses of 7 billion francs in seven years—4 billion francs in 1981 alone. Caught between increasingly heavy charges and price controls on the home market, France's chemical entrepreneurs never managed after the war to achieve sufficiently profitable margins. They ran up high debts to make up for their lack of funds, building up ever heavier financial costs.

A further disadvantage of France's chemical industry was its scattered production sites, originally due to the need during the two World Wars to keep plants far from the battlefields. For both social and political reasons, it was inconceivable in France to have a site like BASF's Ludwigshafen where 52,000 people are concentrated on six square kilometers with three thermal powerplants and countless production sites. The first concentrations which President Georges Pompidou sought to carry out had not changed things much, neither had they cut down increased operating costs. Indeed, the leaders of merged companies had not cared at the time to close sites down and reduce personnel, two moves that might have improved the performance of the new groups.

Although the state spent considerable sums for chemical research, particularly through CNRS and the universities, the fallout for industry was scarce because of the persistent lack of communication between industry and those doing research.

The research and development sectors of the companies themselves made few breakthroughs, so that the chemical industry had to rely for a large part on foreign technologies, a fact that left little room for maneuver,

In addition to the difficulties inherent in their environment, France's companies also suffered the effects of bad management decisions in specific areas. Rhône-Poulenc had been badly prepared for the chemical fibers slump and had sunk too much money in heavy chemicals. These did not fit in with the group's original calling, as its leaders demonstrated when they withdrew, at the height of the crisis, from petrochemicals and the base thermoplastics, concentrating on specialties. The purchase of GESA from PUK in 1978, of Sopag the following year from the Gardinier brothers, and the sale of Lautier were hardly fortunate decisions for a group that could draw no advantage from getting further into fertilizers and that could have diversified to good purpose on perfumes through Lautier.

PCUK had never managed to strengthen Francolor's international base to good purpose and had finally sold it to ICI. Also, it wasted a lot of money in belatedly trying to develop a PVC chain. In 1981, PCUK was negotiating with Occidental Petroleum the sale of its chemical division, which had long since ceased to be of interest to the group's leaders.

At no time since it was set up was CdF Chimie master of its destiny, subject as it was to political pressures rather than economic rationality. Constantly in the red despite a number of worthwhile activities, it received the final blow when the untoward decision was taken in 1978 to build, on borrowed money, a one-billion-franc petrochemical site in Dunkirk in the framework of Société Copenor set up in joint venture with the Emirate of Qatar.

Elf Aquitaine had established under Sanofi a small conglomerate with profit-making subsidiaries involved in pharmaceuticals and perfumes. But Atochem, set up on a joint basis by Total and Elf, was a loss-making concern, as was Chloé Chimie, a cast-off of Rhône-Poulenc, which retained only 19.50 percent of its capital, while Elf and Total each acquired a 40.25 percent stake in the new chemical entity.

EMC was more a mining than a chemical company. It focused on potash, having restricted its diversification to the purchase of the animal food company Sanders and to a subsidiary in Tessenderloo, Belgium.

It was in this environment that the nationalization measures decided upon by the new Socialist government took place. The state took control of 40 percent in value of production of commodity chemicals and 70 percent of petrochemicals in France, an event that had no precedent in the free world's industrial countries.

Société L'Air Liquide, which figured as one of the companies to be nationalized on the initial Socialist list, escaped this fate, no doubt because the disadvantages of taking over this star multinational had been pointed out to the President of the Republic by one of his brothers, who was advisor to the group. On the other hand, Roussel-Uclaf, which had never needed state funds, found the government partly in control of its capital in addition to the main shareholder Hoechst.

Short of the extreme solutions advocated by some Socialists in favor of a single French chemical entity, the nationalized part was cut up along the lines announced by the Ministry of Industry on November 8, 1982. The restructuring signaled the death of PCUK as an industrial enterprise. Its various sectors were shared out among the other state-controlled groups. Most favored was Rhône-Poulenc, which received the agrochemicals and pharmaceuticals sectors with Sedagri and Pharmuka as well as the Wattrelos and La Madeleine sites in the north of France, together with a plant in Rieme, Belgium. At the same time, its fluorine division was boosted. The lion's share went to Elf Aquitaine with what amounted to two-thirds of PCUK's turnover, including, in particular, the halogen and peroxide products.

Complex negotiations with Total (Compagnie Française des Pétroles) ended with the group's withdrawing from Atochimie and Chloé Chimie, after which Elf Aquitaine set up its Atochem subsidiary to encompass all its chemical activities. After a long and brilliant independent career, Rousselot was split between Atochem and Sanofi.

Already sorely tried, CdF Chimie came out the worst from the restructuring. It inherited the Oxo alcohols and organic acids of the Harnes unit and had to call upon Esso Chimie to ensure their survival; it also got an ABS unit that was too small, which it exchanged with Borg Warner for a 30 percent share in their European subsidiary company—the Villers Saint-Paul site, which could become profitable only with the help of the industries to be set up there; the polyester resins division of the Chauny unit, and the downstream activities of the Stratinor subsidiary, both open to stiff competition. Among the lot there were some profitable sectors, however, such as Norsolor's acrylics, well integrated on the Carling site, and Société Lorilleux, a small ink multinational of PCUK's. But CdF Chimie was left to manage the difficult fertilizer sector swollen by Rhône-Poulenc's and EMC's divestments (GESA and APC), as well as a petrochemical branch set off balance by the unfinished Dunkirk site. As for EMC, all it got from PCUK was the historic site of Loos, which nevertheless served to boost its chlorine and potash divisions.

This enormous restructuring job, no doubt, did produce chemical groups with sounder bases and a more promising future. But the financial cost to the country was considerable, for not only were the shareholders refunded with public money to compensate for nationalization, but the companies that were now state-controlled had to be bailed out: their losses in 1982 were even higher than those registered the previous year. Just as high was the social cost. Manpower cuts which the former company leaders had been loath to carry out had become not only absolutely necessary but also easier to implement by a left-wing administration.

Restructuring in Italy and Spain

As was to be expected, the path to overcapacities aided by state subsidies had brought *Italy's chemical industry* to the edge of the precipice. In 1981, SIR and Liquichima, on the brink of bankruptcy, had been taken over by ENI, the state-controlled oil group whose own chemical subsidiary ANIC was also losing considerable sums of money. Montedison had been able to show balanced books, only once in ten years, in 1979. Its debts had soared to $2 billion in 1984.

The rather belated restructuring measures consisted, in their first stage, in the sale of the state's 17 percent share in Montedison to private interests. Then Italy's petrochemicals and plastics companies were shared out between Montedison and ENI's chemical subsidiary Enichem.

These two groups then set out to concentrate their efforts in the fibers area on polyesters and acrylics. At the same time, Montedison gave up control of SNIA Viscosa, specialized in polyamides, to Bombrini-Parodi-Delfino (BPD). The restructuring, carried out together with manpower cuts and unit shutdowns, made it possible for Montedison in 1985 and Enichem in 1986 to post operating profits after long years in the red. Enichem received a further boost from association with ICI in PVC and with BP and Hoechst in polyethylene, for it had emerged from the restructuring in a less favorable position than Montedison because it was still saddled with commodity chemicals.

Montedison, now 45 percent owned by the Ferruzzi sugar group, reinforced its strategic sectors by purchasing Allied-Signal's fluorine polymers through its stake in Ausimont, by fully acquiring the Farmitalia and Carlo Erba pharmaceutical subsidiaries, and by buying from Hercules its 50 percent share in Himont, the joint subsidiary set up in 1983 in polypropylene.

The two Italian giants were still very much in debt, a fact that could lead to further divestments. But their leaders could nevertheless contemplate the future with some equanimity. Their heavy chemical sectors were finally merged under Enimont in 1988.

The *Spanish chemical industry* was also faced with considerable difficulties. Short of innovations, it had developed through foreign technologies and had lived a sheltered life behind customs barriers and import licenses not conducive to cost cuts. Neither Spain's petrochemicals industry, which was in the hands of the Enpetrol state group and the private company CEPSA, nor the main national companies Explosivos de Rio Tinto (ERT) and Cros, were in a position to face without transition the pressure of competition felt when Spain joined the Common Market. This was particularly true of ERT, which had missed bankruptcy by a hair, and Cros, which had remained in the red for a long time. Neither would be able to avoid severe restructuring.

Their total merger project failed through lack of financial means, and it was Kuwait in the end which, through the Kuwait Investment Office, took a 47 percent share in ERT and 24 percent in Cros in 1987 and promised to provide the necessary cash for the two groups to form a joint fertilizer subsidiary.

Arab Countries Gain a Foothold

As soon as OPEC was set up, Middle Eastern countries had sought to find ways to invest their oil revenues in downstream industries. *Kuwait's* approach was, preferably, to acquire shares in existing companies. It thus bought up Gulf Oil's interests in Europe, took a share in Germany's Hoechst, and injected considerable capital into ERT and Cros in Spain.

Qatar had chosen to associate with CdF Chimie to set up a petrochemical base in the Emirate and to build the Dunkirk site through Copenor.

Saudi Arabia's policy has been to develop a national petrochemical industry that would sell its products worldwide. More than Qatar and Kuwait, it had abundant supplies of ethane and methane extracted from gases that were being flared. The ethane separation capacities of its refineries alone accounted for a potential of 3.5 million tons a year of ethylene.

Sabic, the body in charge of the project, had cleverly involved itself with major international groups such as Mobil, Exxon, Shell, and Mitsubishi. Production would then be easier to place in Europe, North America, and Southeast Asia without wounding national feelings. The first giant methanol unit came on stream in 1983, while the other Saudi productions located in Al Jubail and Yanbu have gradually begun supplying low- and high-density polyethylene, ethylene glycol, ethanol, dichloroethane, vinyl chloride (monomer and PVC) and styrene as the relevant units came on stream.

Since 1970, Saudi Arabian Fertilizer has been producing urea and melamine in Damman, in association with Sabic; the two companies have scheduled construction of a 1,500 tons a day ammonia unit in Al Jubail.

Because of the obviously low cost of the principal local methane and ethane raw materials, and because the fixed costs of the installations are high with regard to variable costs, European petrochemical producers were afraid that Saudi Arabia with its low home consumption, would flood outside markets with its ethylene derivatives and methanol at cut prices. So far, however, Saudi exports have not shaken up the market because they have been carefully channeled through the distribution networks of Sabic's international partners.

Taking a different course than Algeria with its liquified natural gas, the Gulf States have thus upgraded their natural resources and already account for 10 percent, 5 percent, and 4 percent of world production of methanol, ethylene, and polyethylene respectively.

The American Chemical Industry Caught Off Balance

The difficulties resulting from world overcapacities were enhanced in the United States by the behavior of financial circles and the reaction to this behavior of the U.S. chemical industry leaders. America's chemical giants had reached their advanced stage of development because of the long patience of their shareholders and the acumen of their leaders based on thirty years of product and process innovation. Just like their German and Swiss counterparts, U.S. chemical industry leaders had upheld the notion of long-term interest over the more immediate concerns of the various types of shareholders.

The shock waves sent out by the two oil crises, which had not spared the United States, the growing influence of financial analysts on the behavior of shares quoted on the Stock Exchange, and the arrival at the head of the large industrial groups of graduates from glamorous business schools trained more in finance than in technology, gave the scene a new twist. Shareholders were more interested in the instant profits they could draw from breaking up a group than with the added value that could be patiently built up through its development.

Drawn along by their own convictions or under pressure from bankers and "raiders," U.S. chemical leaders were constantly redeploying their activities. The *leveraged buyout* (l.b.o.) system had already been applied by the leaders of FMC's American Viscose division when they sought to buy, with the help of the banking world in the early 1970s, the Avtex rayon and polyester producer, which thereby became a successful company. Despite the risk to buyers in borrowing from financial organizations as much as 90 percent of the amounts needed for the purchase, the system was eagerly seized upon by individuals wishing to set up their own business and taking advantage of the disenchanted mood of potential sellers. This is how *Huntsman* came to become the world's leading producer of styrene and polystyrene after buying up the relevant sectors from companies like Shell and Hoechst, which wanted to pull out of them.

Likewise, it is because Du Pont, having spent $7.4 billion to acquire Conoco, sought to reduce its debts by selling part of Conoco's chemicals and also because Monsanto, ICI, and PPG were withdrawing from petrochemicals, that firms like *Sterling Chemicals*, *Vista Chemicals*, and *Cain Chemicals* have emerged since 1984. Cain Chemicals was itself to be taken over by Oxychem (Occidental Petroleum) in 1988. Various acquisitions made at the right moment turned Vista within three years into one of the leading PVC and detergent alcohol producers in the United States. Through purchases made in its behalf by Sterling Chemicals, Cain Chemicals became a major petrochemical company with assets worth $1 billion in 1987, including ethylene, ethylene oxide, glycol, and polyethylene units, all strategically located in the Gulf of Mexico area. A further newcomer on the American scene was *Aristech*, which emerged through the takeover by its management of the heavy chemicals division of USX (U.S. Steel).

All these companies were acquired under very favorable conditions, as more often than not they were sold by the large groups at 25 percent of their replacement value. Contrary to assumed notions, individual entrepreneurs were thus able to acquire installations which until then only the most powerful groups could afford to run. These groups gave up whole sections of their traditional chemicals to redeploy in specialties for which they had no particular disposition and, at times, in areas even further removed from their original areas of competence. Thus *Diamond Shamrock* gave up its chemicals to Occidental Petroleum at the worst possible time, to devote itself exclusively to the energy sector, which in fact failed to live up to expectations.

One of the most powerful of America's chemical companies, *Allied*

Chemical, became a high-technology conglomerate under the leadership of Edward L. Hennessy, Jr., who was formerly with United Technology. After acquiring Bendix and Signal, it took on the name of *Allied-Signal* and is now focusing on electronics and space, having entrusted a large part of its chemicals to the portfolio subsidiary Henley, which will sell them to the highest bidder. As for *Monsanto*, it shed a number of fibers, plastics, and petrochemical units both in Europe and in the United States and decided to hinge its further development on biotechnologies, a new area for the group. It bought up in particular the aspartame producer *Searle* for $2.6 billion.

At the same time as these changes were being wrought by the protagonists themselves, other major changes were taking place under outside pressure. Wily businessmen acting as "raiders," with the help of financial concerns that issued high-risk and high-interest "junk bonds" to finance a large share of the targeted acquisitions, set their sights on large companies quoted on the Stock Exchange: they acted in the belief that the company's parts would be worth more sold separately than as a whole.

The raiders' takeover bids had instant attraction for shareholders, and their criticism of the way the firms they were after were being managed was often not without truth. But it stood to reason that once the raiders had bought the company, they would break it up to reduce financial charges and to refund the money borrowed for the raid. The more interesting assets were often the first to be sold off, for they found ready buyers. To counter the raiders, the managers of the targeted firms were likely to raise the ante. But this only aggravated the financial problem, and the group's dismantling was unavoidable.

The instant advantage which both shareholders and raiders drew from these operations was obvious. But their consequence was, sooner or later, to destabilize the enterprises concerned, when these did not disappear altogether. The most spectacular case was *Union Carbide*, coveted in 1985 by the real estate developer S. Hayman, who had already taken over GAF Corporation.

To fight off the raid, Union Carbide had to borrow $3 billion. To reduce such an unbearable debt, the group's management was forced to sell its best sectors (batteries, consumer products, engineering plastics, agrochemicals) and even its headquarters in Danbury, Connecticut. This was how one of the best chemical concerns in the United States, with sales amounting to $10 billion, was left with only three areas of business after divesting to the tune of $5.3 billion. Even these areas—industrial gases, petrochemicals and plastics, and graphite electrodes—were faced with stiff competition. And with debts that still remain three times as high as the industry's average, Union Carbide is in no position to invest in the short term in anything likely to push it back to its former major rank in chemicals.

Other U.S. companies involved in chemicals were also the victims of raiders in 1985. To fight off C. Icahn, *UniRoyal* was taken over by its management and was forced to sell off its chemicals to Avery, which in turn placed them on the block, before accepting a leveraged buyout by the management. *Phillips Petroleum* had to buy back its shares from C. Icahn and B. Pickens and was forced to sell $2 billion worth of assets to refund part of its debt. And what about *Gulf Oil*, which sold itself to Standard Oil of California to escape the clutches of Boone Pickens, or *Stauffer Chemical*, which changed hands three times within a single year from Cheeseborough Pond to Unilever and finally to ICI, when it was broken up among ICI, AKZO, and Rhône-Poulenc?

Attracted to the U.S. market, European investors had also joined the raider' ranks. This is how the Britain-based Hanson Trust managed to acquire *SCM*. This was a company that had just completed its restructuring; but after Sylvachem was sold off by the new owners, it retained only chemical production of titanium dioxide.

Anglo-French tycoon J. Goldsmith, unable to take control of Goodyear, nevertheless made substantial profits from his raid on the company. Goodyear was left with the sole alternative of withdrawing from all the sectors except chemicals in which it had diversified outside of tires.

In a number of cases, transactions led to an agreement between the heads of companies that had stock options and were eager to make a profit, and the potential buyers. This was how *Celanese*, an able and well-diversified company that had the means to retain its independence and competitiveness with regard to any major company, was acquired by Hoechst following a transaction that was satisfactory both to the German buyer and to the shareholders of the American group, at least for the time being.

The fear that their company might be the target of an "unfriendly" takeover bid induced the boards of directors of some of the well-managed chemical companies to guard against such attacks either through deceptively appealing offers—"poison pills"—or through purchase of their own shares. This was certainly not the best way for industrial firms to make use of their funds.

Coping with Safety and Environmental Problems

Handling chemicals has never been without danger, if only because of the unstable and harmful nature of a number of substances when they are placed in certain conditions of temperature, pressure, or concentration.

Chemists have always been haunted by the risks of explosion. The explosion which occurred on September 3, 1864, in the Heleneborg laboratory near Stockholm, where Alfred Nobel was handling nitroglycerin, caused the death of five persons, including Emile Nobel, his younger brother. The ammonia synthesis unit set up by BASF within the Oppau plant was totally destroyed in 1921 by an explosion causing the death of over 600 people. In 1946, the French cargo ship *Le Grand Camp*, carrying 2,500 tons of ammonia nitrate, exploded in Texas City, killing 512 people. Other disasters, such as that of

Flixborough in England, which took place through rupture of a Nypro caprolactam pipe within the plant in 1974, or again the one caused in a holiday camp in Los Alfraques in Spain when a tank-wagon carrying propylene exploded in July 1978, are reminders of the explosive nature of certain chemical products and of the need to handle them strictly according to the prescribed security rules.

A number of chemicals, fortunately a limited number, become dangerous either when they are used wrongly, or when they are accidentally set free. *Thalidomide*, put on the market in 1957 by the German company Chemie Gruenenthal, was indeed a powerful sedative. But it took three years to perceive that when prescribed to pregnant women, it dramatically crippled the newborn children. The synthetic intermediate for insecticides, *methyl isocyanate*, which Union Carbide has used for years without incident in its West Virginia Institute plant, caused over 2,000 deaths when it escaped in 1984 from a storage tank in Union Carbide's Bhopal plant in India.

Other products act insidiously, so that it is harder to establish their effects on human and animal health and more generally on the environment. Indeed, progress in understanding the safe dosage of minute quantities of impurities has enabled governments to fix with greater care the maximum allowed content of *vinyl chloride monomer*, *formaldehyde*, and *benzene* beyond which these products could become dangerous for workers to handle.

Lessons have been drawn from accidents caused by faulty handling of certain substances. Through the work carried on by Alfred Nobel, we know how to stabilize nitroglycerin in the form of dynamite, and since 1946 methods have been devised to avoid the spontaneous explosion of ammonium nitrate. Ammonia units with capacities of 1,500 tons a day have been operating for decades without incident.

Because of the painful thalidomide episode, long and costly tests are now carried out to study the possible secondary effects of pharmaceutically active substances. A great number of drugs that today save many lives would not have been available had they needed to go through the long periods of tests that are now required by legislation.

Likewise, in industrial countries, increasingly stringent regulations limit noxious vapor discharge from chemical plants, which are required to treat their effluents effectively. The transport of dangerous substances is also closely monitored by the authorities. Such precautions stem not only from the publicity which the media now give to any catastrophe worldwide, but also from the public's instinctive distrust of chemistry, which it still regards as a mysterious science.

But just as an air crash does not mean the end of commercial aviation, neither does the damage caused by improper use of certain substances mean the end of the chemical industry. The image of chemicals is tarnished, however. Citizens who deliberately risk their own death, when they are not actually killing others, because of speeding on the roads or because they are addicted to alcohol, tobacco, or drugs, are less and less inclined, for all that, to accept accidental security breaches when these are not caused by themselves.

Politicians in our parliamentary democracies who wish to please public opinion feel the urge to take into account demands that are more emotional than scientific, and advocate restrictions even when these go against the best interests of the citizens. The *Three Mile Island* nuclear power plant accident in the United States which resulted in no fatalities, the more recent *Chernobyl* explosion which, as of 1988 had directly caused two deaths, have, with no good reason, prevented any resumption of the U.S. nuclear program and have aroused fears in European countries in people least likely to give way to mass hysteria.

The *Seveso* leak, which occurred in Italy on July 10, 1976 in the trichlorophenol unit belonging to Hoffmann-La Roche's subsidiary Givaudan, did have an impact on the immediate environment and a number of people were temporarily affected by the dioxin vapors. But the accident caused no lasting harm. It was the publicity which the media gave to it that forced Hoffmann-La Roche to close down the unit, turning Seveso into a dead city.

The litigation, so far unsettled, over residues left in the ground by Occidental Petroleum's affiliate Hooker, in *Love Canal*, in the state of New York, has thus far led only to the evacuation of all the area's residents, beginning in 1978. But no clear explanation has yet been given of the ailments some of the inhabitants have been complaining about.

The lack of universally accepted scientific explanations for certain phenomena has often meant that the precautionary measures taken by one country do not necessarily apply in another. Where sweeteners are concerned, for instance, some governments have banned *saccharin* and other governments allow its use. The same is true of *cyclamates* and *aspartame*.

DDT was banned as an insecticide as early as 1974 by most industrial nations. But it is still widely used in many developing countries. The risks of *eutrophication* are perceived differently by governments, so that legislation applying to products for the production of detergents, like *alkylbenzene sulfonate*, *tripolyphosphate*, or *nitriloacetic acid* (NTA) differs from country to country.

The agreement which a number of nations reached in 1987 to ban the use of *chlorofluorocarbons* in aerosols is so far the only instance of harmonized legislation, even though no one has so far managed to prove scientifically that the chlorofluorocarbons really destroy the atmosphere's ozone layer.

Thus while it is understandable that authorities must be careful to soothe the fears of a public that is insufficiently informed of the dangers that threaten it, it must also be aware of the economic and social costs of refusing to accept the risks inherent in any human activity, and also conscious of the uncertainties surrounding the rules and regulations taken to satisfy its demands.

Some companies are turning the necessity of cleaning up the environment into new opportunities to improve their profitability. Thus Du Pont has found a useful application as a building material for the calcium sulfate that was piling up as a byproduct in one of its Texas plants.

Scientific and Technological Breakthroughs

Short of fundamental discoveries over the past fifteen years, the chemical industry has gone forward by systematically developing its store of knowledge in processes and products.

Process Improvement

Higher crude oil prices had revived studies in the use of coal as a chemical feedstock. But while the Fischer-Tropsch synthesis was still used in South Africa by Sasol, the only other industrial gasification unit was the one Eastman Kodak brought on stream in Kingsport, Tennessee, in 1983, to produce *coal-based acetic anhydride*. The coal came from the Appalachian mountains and was cheap enough relative to oil prices at the time to warrant such an installation, and the plant is now to be expanded.

Together with these studies on synthetic gas, some progress has been achieved in the use of a group of alumino-silicates, the *zeolites*, as selective catalysts to boost certain reactions. Half the world production of *p*-xylene and a quarter of the production of ethylbenzene, an intermediate required to prepare styrene, are carried out using the zeolite-based ZSM-5 catalysts developed by Mobil Oil, which played a pioneer role in this area.

Applications of the olefin *metathesis* reversible chemical reaction, discovered by Phillips Petroleum in the 1960s, were also developed in the subsequent years. By this reaction, Arco produces propylene from ethylene and butene-2; Hercules prepares its new plastic, Metton, from dicyclopentadiene; and Shell synthesizes its C_{12}-C_{14} SHOP (Shell Higher Olefin Process) alcohols used for detergents.

The application of *electrochemistry in organic synthesis* had already served to bring on stream in the United States in 1965 Monsanto's first industrial adiponitrile process from acrylonitrile. This was followed in 1977 by a similar installation in Seal Sands, England, which was later bought up by BASF.

The former *Reppe chemistry*, still practiced in Germany by BASF and in the United States by GAF, also led to new developments as demand for certain intermediates such as the 1,4-butanediol increased. This diol, now also obtained from maleic anhydride, is used to produce PBT polyesters through reaction with terephthalic acid and leads to other major derivates (tetrahydrofuran, butyrolactone, N-vinylpyrrolidone).

New synthetic processes for the preparation of established products were also industrially developed: in Japan the manufacture of methyl methacrylate from C_4 olefins, by Sumitomo and Nippon Shokubai; in France, the simultaneous production of hydroquinone and pyrocatechin through hydrogen peroxide oxidation of phenol by Rhône-Poulenc; in the United States the production of propylene oxide through direct oxidation of propylene operating jointly with styrene production, developed by Ralph Landau and used in the Oxirane subsidiary with Arco, which the latter fully took over in 1980; in Germany and Switzerland, the synthesis of vitamin A from terpenes, used by BASF and Hoffmann-La Roche.

Processes apparently well established were still further improved, such as the *electrolysis of sodium chloride*, dating back to the last century: diaphragm and then membrane cells were substituted for mercury cells, which were a possible source of pollution.

Important progress was also made in *chemical engineering*, such as use of rotary compressors in ammonia synthesis or ICI's fermentation reactors in Billingham to produce the Pruteen protein from methanol reactors, having no mobile parts.

Product Development

Although research was not as fruitful after 1960, new materials put on the market in the 1970s were the outcome of research in high polymers essentially conducted within industry.

It was through such research that ICI's PEEK (polyether ether ketone), one of the first high-performance aromatic polymers, was put on sale, as well as Du Pont's aramide fibers Nomex and Kevlar, more resistant than steel in like volume.

To the range of engineering plastics were added polyethylene and polybutylene terephthalates (PET and PBT), as well as General Electric's polyethers, the PPO (polyphenylene oxide) produced through polymerization of 2,6-xylenol and the Noryl plastic produced by blending PPO with polystyrene. Other special polymers, derived like the polycarbonates from bisphenol A, were added to this range: polyarylates, polysulfones, polyetherimides.

A major step forward was taken in the area of base thermoplastics with the application of Union Carbide's Unipol process. Variations of this were subsequently offered by other low-density polyethylene (LDPE) producers such as Dow and CdF Chimie (now ORKEM).

Under a process that consisted in copolymerizing in the existing high pressure installation ethylene with 5 to 10 percent of an *α-olefin* (butene-1, hexene-1), a stronger linear low-density polyethylene (LLDPE) was produced with a higher melting point than LDPE. Thinner films could thus be produced that were just as strong but required less material.

The new polymers opened up an unexpected market for producers of C_4, C_6 and C_8 α-olefins like Shell, Ethyl, and Chevron. Their higher linear α-olefins were also used either for polyalphaolefins (PAO) intended for synthetic lubricants, or to prepare detergent alcohols.

While no great new plastic has emerged over the last fifteen years, researchers in major chemical companies did their utmost to improve both the features and the performance of known polymers.

As we have just seen, they improved LLDPE by adding comonomers in the carbon chain. But also through additives they managed to render polymers more resistant to fire, to oxidation, and to alteration through ultraviolet rays.

This slowly gave rise to a new industry that consisted in supplying polymer producers and plastic processors, not only pigments and charges, but also antioxidants, light stabilizers, and fireproofing agents. Added in small doses to the polymer, they added to its value by extending its life span. Such an activity, in which the Swiss firm Ciba-Geigy plays a noteworthy role, was boosted by the spectacular development of polypropylene, a particularly sensitive polymer that has to be stabilized with appropriate additives.

Another way of improving the performance of polymers consisted in blending them either with other polymers, or with inert materials such as glass fibers, carbon fibers, or various mineral fillers. Thus were produced a series of *alloys and composite* materials. Glass fiber-reinforced polyester has long been in common use. But the possibility of introducing carbon fiber obtained through pyrolysis of polyacrylonitrile (PAN) fibers already developed in aeronautics, opened up fresh prospects, particularly in the area of sports articles. The need, in turn, to link organic polymers and mineral fillers led to coupling agents such as the silanes which Union Carbide and Dynamit Nobel have put on the market.

This is how, little by little, spurred on by the demands of the processing industries which are also under pressure from major clients like the automobile industry, a number of companies have brought a large number of improvements to plastics. While not very spectacular, these improvements have appreciably added value to existing materials.

More generally, the requirements of many downstream industrial sectors have hastened the development of derivatives that otherwise might have remained laboratory curiosities. Discoveries of new molecules have been particularly inspired by the needs of plant protection. This was because agriculture, before it became a crisis sector, offered worldwide markets for crop protection agents, and also because product approval was easier to obtain, and therefore less costly, than in the case of pharmaceuticals. The success of glyphosate, which Monsanto put on the market in 1971 under the trade name Round Up, has made it the world's leading selective herbicide, for it can be used throughout the year and becomes harmless when absorbed into the ground. A new range of synthetic pyrethroids, developed in the United Kingdom by Elliott of the National Research and Development Corporation, (NRDC), a government agency, was marketed from 1972 onward under the trademarks of Permethrin, Cypermethrin, and Decis. These wide-spectrum insecticides owe their success to the fact that they are exceptionally active in small doses and are not toxic to humans. With increasingly strict legislation and stiff competition among pesticide producers at a time of slumping agricultural markets, the golden days could well be over for crop protection products, so that the years ahead are likely to be more favorable for restructuring than for new discoveries.

Over the last fifteen years, the *pharmaceutical sector* also made great demands on the ingenuity of chemists. But from the time of the thalidomide drama, the testing times required by health authorities have increased, to the point that since 1980 ten to twelve years are needed instead of the three to four previously required to bring a drug on the market from the time of its discovery. Research and development costs, accordingly, have grown fourfold over the last ten years, dangerously reducing the number of new specialties provided for patients each year. Because of such delays, a patent protecting a new substance may be left with but a few years of validity when final approval is granted to the laboratory that made the discovery.

Such difficulties have apparently not affected the zeal of researchers. Nor have they diminished the sums devoted each year to research and development, which on the contrary have been constantly on the increase. This is because any major discovery may have worldwide portent. And in most developed countries there is a system of refunding to patients the cost of ethical drugs, so that a new active principle may provide the laboratory that has exclusive rights over it with a considerable source of profits even if such refunds are coupled with tight price controls.

And while it is also true that thirty pharmaceutical companies alone account for 60 percent of worldwide ethical drug sales, the sums of money invested in research do not always get their full return. Thus it is that a small company like Janssen's laboratory, Janssen Pharmaceuticals, in Belgium, which was acquired in 1979 by Johnson and Johnson and which has among its discoveries diphenoxylate (1963) and loperamide (1975), has proved more innovative over the last fifteen years than the Rhône-Poulenc group, which has produced no major new molecule during the same time, although it devotes far more money to its research.

Indeed, success depends at least as much on chance, the ability of researchers, and the strategy of management in that area as on the sums expended. Valium and Librium, which have been providing Hoffmann-La Roche with its largest profits since the end of the sixties, were the outcome of Leo Sternbach's acumen. Instead of merely modifying the meprobamate molecule as management had requested, he began studying the sedative properties of benzodiazepins used as dyestuff intermediates and on which he had worked for twenty years previously at Cracow University.

One of the most prolific inventors of the sixties was most certainly Sir James Black, a Nobel prizewinner in 1988. While working for ICI, he discovered the first β-blocking agent Propanolol in the early 1960s. He also discovered Cimetidine, sold under the trade name of Tagamet as an anti-ulcer agent by SmithKline & French from 1974 onward, and which has become the world's largest-selling specialty. After working successively for ICI, SmithKline & French, and for Wellcome in Britain, Sir James now has his own business, and he is convinced that small competent teams are, by nature, more innovative than the large armies of researchers which many of the big companies have set up.

Likewise, the successful ventures of Merck Sharp & Dohm cannot be dissociated from the work of its president, Roy Vagelos. This biochemist, a latecomer to research, supervised the whole process of work to bring Mevacor, the new cholesterol miracle drug, onto the market. It has just been approved by the U.S. Food and Drug Administration. Mevacor was but the crowning touch to Merck's scientific tradition with its long series of discoveries: α-methyldopa against hypertension, indomethacin and sulindac to fight arthritis, and cefoxitine, an antibiotic.

At a time when pharmaceutical research is becoming increasingly costly and the likelihood of a great discovery remains hazardous, success will come to laboratories which not only sink large sums of money into research but also rely on teams where competence does not necessarily rhyme with size, and whose management has reached a sufficient level of scientific maturity.

The Craze for Biotechnology

The catalytic action of living organisms, or rather of the proteins they contain, had received the beginnings of an explanation with the experiments of Payen and Persoz on malt amylase separation in 1833 and with J. J. Berzelius's catalyst theory in 1835. In 1897 Eduard Büchner demonstrated that a yeast extract could turn sucrose into ethyl alcohol. Fermentation took place without the presence of living organisms through enzymes. In this case zymase was the catalyst.

Ethyl alcohol, already known to alchemists, was used by industry towards the middle of the last century when continuous distillation in columns was devised by Ireland's Aeneas Coffey in 1830 and when it became exempt from excise duties on alcohol if methanol was added to it.

After alcohol, *lactic acid* was the second product obtained industrially from sugar fermentation, starting in 1880. The levo-isomer is still made this way to the tune of 20,000 tons a year.

In 1890, the Japanese chemist Jokichi Takamine had introduced a fermentation process in the United States by which an enzyme blend was produced. This takadiastase catalyzed starch and protein hydrolysis. Some years later in 1913, Boidin and Effront discovered the ''*bacillus subtilis*'' that produced an *α-amylase* stable under heat. This enzyme was used to desize cloth and later in the sugar fermentation process.

During World War I, Chaim Weizmann had succeeded in producing for the British Admiralty acetone and butanol on a large scale through anaerobic fermentation of starch. The Germans were then producing as much as 1,000 tons a month of glycerin from sugar. These war productions proved no longer competitive in peacetime. But *citric acid*, which Pfizer began producing in 1923 from sucrose, is still biochemically made today from *Aspergillus niger*, which Currie advocated in 1917.

The discovery of *penicillin* and its industrial development during World War II have led the pharmaceuticals industry increasingly to resort to *biosynthesis* for the preparation of its active principles. Through rigorous selection of the microorganisms extracted from the soil or from various molds, the cost of an antibiotic like penicillin has been brought down to $30 per kilo, compared to $25,000 per gram initially—an impossible target if the exclusively synthetic process had been used. Moreover, it became possible to extend the range of antibiotics that could be used. The antianemia *vitamin* B_{12} and most of the amino acids were prepared in the same way through culture of microorganisms in selected environments containing precursors.

In the case of *steroids*, biosynthesis permitted reactions that could not be achieved through direct synthesis. In 1952, this was how Upjohn researchers in the United States managed to introduce on carbon atom 11 of the steroid nucleus, an hydroxyl group $-OH$, using the *Rhizopus arrhizus* fungus, making the switch from the pregnancy hormone progesterone to cortisone and its derivatives.

Microorganisms are also capable of separating optical isomers. In the case of sodium glutamate, where it is necessary to start from levo-glutamic acid to obtain the desired flavor, and where synthesis produces only a racemic blend, it was a particular yeast called *Micrococcus glutamicus* that led to the required isomer through carbohydrate fermentation.

Considering that sodium glutamate, like other amino acids, is contained in soy sauce, which is a traditional Japanese food, it is not surprising that Japan should have become interested very early in this type of fermentation. Firms like Ajinomoto and Kyowa Hakko dominate the world market for amino acids and particularly for *glutamic acid* and *l-lysine*. It is also through enzymes that the resolution of *dl-methionine* into its optical isomers is achieved since its laboratory synthesis yields the racemic form.

Heat-stable amylases are frequently used in both the United States and Japan to produce *syrups with a high fructose content* from corn starch.

Single-cell proteins such as ICI's Pruteen were produced through culturing microorganisms on a bed of organic material.

Interest in biosynthesis grew still further with the discovery in 1953 of the structure of DNA, then in the 1960s of the genetic code of proteins. It then became possible to clone microbe or plant cells, through *genetic engineering*, by recombination of fragments of genetic material from different species. Thus, towards the end of the 1970s, the biotechnology firm Genentech succeeded in isolating the human insulin gene and to insert it into the DNA of the *Escherichia coli* bacteria: through reproduction, these bacteria produced the *first human insulin*, which Eli Lilly and Company has been marketing since 1982.

The *human growth hormone* (HGH), which can only be extracted in minute quantities from the pituitary glands, can now be isolated in larger quantities through genetic engineering.

Monoclonal antibodies (mabs), which replicate the antibodies in the organism with the added advantage of being "immortal," were discovered in 1975 by scientists working at the Cambridge Medical Research Council in the United Kingdom. They serve more particularly as reactive agents for medical diagnostic purposes.

Through *plant genetics*, it has also been possible to render plants resistant to chemical agents (Calgene, Monsanto) as well as to improve crop yields (Pfizer) with new seeds.

With the prospects which *biogenetics* was opening up for medicine and agriculture, a number of private laboratories sprang up in the United States between 1971 and 1978—*Genentech*, *Cetus*, *Genetic Institute*, *Biogen*, *Amgen*, and *Agrigenetics* to mention but the principal ones. These laboratories managed to finance their work with the help of venture capital, research contracts with the major chemical firms like Du Pont, Monsanto, Eastman Kodak, W. R. Grace, or shares purchased on the stock exchange.

Vast sums of money have been spent over the last ten years but with small tangible results, prompting the definition of biogenetics as a business likely to bring in a small fortune as long as a large one is invested! Thus far the only commercial fallout of biogenetic research involves human insulin (Eli Lilly), the human growth hormone HGH (Genentech, KabiVitrum), the hepatitis B vaccine (Merck, Smith, Klein-RIT), interferon (Boehringer, Ingelheim), the amylase enzyme (Novo), a number of veterinary vaccines (AKZO Pharma), and monoclonal antibodies for diagnostic reactive agents. Hopes raised by interferon and interleukin-2 as cancer cures have not materialized, but the Tissue Plasmogen Activator (TPA) as a blood clot dissolver in heart attacks has just received approval by the U.S. Food and Drug Administration (USFDA).

Plant genetic research is encountering opposition from the U.S. Department of Agriculture and the Environmental Protection Agency. Pressured by environmentalists, the U.S. administration is loath to approve developments which could affect the environment in unknown ways. In addition to these administrative obstacles, there is uncertainty over patent rights, for there are no legal precedents. Finally, the biocompanies recently set up will need to associate with large pharmaceutical groups to develop and market the products born of their research.

Generally speaking, although *biotechnology* has acquired credibility in many areas, its development is being slowed by scientific, economic and administrative obstacles. First and foremost, proteins are complex substances that cannot be handled as easily as the simple molecules involved in traditional organic syntheses.

It is true that Japan's Ajinomoto and Kyowa Hakko, in particular, have become masters of the art of producing amino acids. Likewise, enzymes have remained the specialty of Novo (now Novo Nordisk) in Denmark, Gist Brocades in Holland, and Bayer's subsidiary Miles in the United States, which together account for 60 percent of the world needs in the area.

Even when they are technologically sound, however, bioproducts may turn out to be economically uncompetitive. The profitability of l-lysine from one year to the next, for instance, depends on soy market prices. In the same way, the single cell proteins which BP produced in 1963 in Lavera from a petroleum base, using a process developed by France's Champagnat, never managed to compete with soy cakes for animal food. ICI has also just been forced to close down its 50,000-ton Pruteen unit in Billingham.

At current crude oil prices, the production of ethanol from biomass is not profitable, either. Whether produced from beets, sugar cane, or corn, it can become competitive only if it is subsidized. And these subsidies would only be forthcoming for political reasons: to please their farmer voters, the French, Brazilian, and United States governments would adopt such a policy to absorb excess agricultural products. From cereals, corn in particular, starch is produced and hydrolyzed to form glucose which ferments to ethanol.

Powerful groups like American Corn Products and France's *Roquette Frères* produce starchy matters in this way. The former is also the leading producer of *isoglucose* (a blend of glucose and fructose) in the United States, while the latter is the largest producer of *sorbitol*. Starch can, therefore, compete directly with saccharose both for foodstuffs and for industrial uses as a fermentation or enzyme-reaction base.

This gives rise to a permanent conflict in Europe between the starch manufacturers on the one hand and the sugar and beet refiners on the other, a conflict that the EEC Commission with its *Common Agricultural Policy* of quotas and subsidies has been unable to settle. The only point of agreement between the two parties is the price which they demand for their production from downstream Community industries, a price that is far higher than world rates.

Spurred on by the Italian sugar group Ferruzzi-Eridiana, Montedison's and now Enimont's main shareholder and an associate of France's Béghin-Say sugar group, there is a campaign under way to introduce ethanol into gasoline. Farmers, of course, support the move because incorporating 7 percent of ethanol in gasoline would mean for a country like France the use of two million tons of sugar or four million tons of cereals. But ethanol happens to be in competition with methanol and the new MTBE antiknock agent as a gasoline additive. More importantly, a tax rebate would be needed at current gasoline prices to induce the oil industry to incorporate ethanol in prime rate gasoline. So the "farm" lobby can receive satisfaction only at the expense of the taxpayer, whether American, Brazilian, or European.

The rules that have always governed the use of ethanol, government policy favoring one agricultural raw material over another, the new constraints that limit the marketing of genetically engineered products—all these factors serve to remind those interested in the development of biotechnology how narrow is their room for maneuver.

The Fine Chemicals Approach

In their search for products that could provide better margins than those achieved from commodity chemicals, the industry had hit upon *fine chemicals*. These typically involved derivatives from organic synthesis, obtained in multipurpose units and sold in relatively small quantities at high prices.

The German and Swiss *dye manufacturers* (Hoechst, BASF and Bayer, as well as Ciba-Geigy and Sandoz) were in the most favorable position to develop such advanced chemicals. They had a long tradition behind them of multiple-stage syntheses involving intermediate derivatives that could also serve to prepare pharmaceutically active principles or pesticides. Starting from a number of major raw materials and working according to the chemical-tree concept, these producers can work down the line to well-defined molecules which they use in their own downstream production or sell as synthetic intermediates to outside clients.

In Europe, the giant ICI group, which had retained a strong position in dyes, also became involved in this kind of chemicals.

France, with PCUK having closed down in 1980 its Société des Matières Colorantes in Mulhouse and then having sold Société Francolor to ICI, had restricted its ambitions in this area. It retained only a few products of Rhône-Poulenc and of its 51 percent subsidiary *Société Anonyme pour l'Industrie Chimique* (SAIC), located in Saint-Fons and in Mulhouse-Dornach, respectively.

As was to be expected, the U.S. chemical leaders, Du Pont, Allied Chemical, American Cyanamid, GAF, and Tenneco Chemicals, had all withdrawn between 1976 and 1979 from the dyes sector. Only three medium-sized companies were still active in this area: *Crompton & Knowles*, *American Color*, and *Atlantic Chemicals*.

Yet at the end of World War II, America's dye production had been the leading one worldwide. For over thirty years it had enjoyed high customs tariffs protection through the American Selling Price clause. But dyes were produced by giant companies used to large scale continuous productions. Their engineers were not trained to run month-long syntheses campaigns involving many stages. Moreover, American marketing executives were little attracted to the German methods for "motivating" their clients. There was also the fact that during the 1960s, U.S. dye manufacturers had come to rely on imported intermediates. With rising prices and the textile slump, they found themselves caught between rising purchase costs and falling selling prices. Finally, unlike their European counterparts, U.S. manufacturers had never given international scope to their dye business. It remained restricted to the home market.

For all these reasons and also because they were not tied down like the Germans by any prestigious tradition, they unhesitatingly gave up dyes, losing at the same time the knowhow needed to succeed in fine chemicals.

With more modest means, other firms were more successful. They either developed their own "chemical tree," or put to good use the knowhow acquired through development of certain processes.

Ethyl became a bromine and derivatives specialist and an expert in orthoalkylation (orthoalkyl phenols and anilines). Its acquisition of Dow's bromine activities has given Ethyl a leading role in this field. *DSM* developed its fine chemicals from the benzoic acid produced during manufacture of synthetic phenol by toluene oxidation. *Atochem* took advantage of the sulfur resources of its parent company Elf Aquitaine to build up successfully a thio-organic chemicals industry (thioglycol, mercaptans, DMSO). Its position will be further strengthened by the takeover of Pennwalt. *PPG* in the United States and *Société Nationale des Poudres et Explosifs (SNPE)* in France are producing a wide range of phosgene-based derivatives to be used in the most varied manner (carbonates, chloroformates). More than any other company, *Lonza* has extended its range of fine chemicals (diketenes, HCN derivatives, pyrazoles, pyrimidines). *Reilly Tar* has become a world leader in pyridine and derivatives. *Dottikon* in Switzerland and *Kema Nobel* in Sweden have put to use their nitration experience to extend their range of nitrated intermediates. Among others, *Rhône-Poulenc* and *Montedison* are involved in organic fluorine derivatives while *Hüls'* fine chemicals division has specialized in alkylation, hydrochlorination and catalytic hydrogenation.

Thus a number of firms with special knowhow in a family of products or in processes that were not among the biggest, have succeeded in taking a more than honorable place as suppliers of fine chemical derivatives alongside the organic synthesis specialists originating from the dye business.

The Attraction of Specialty Chemicals

Besides fine chemicals sold according to specifications but accounting for only a small part of the sales of major companies, *specialty chemicals* held attractions for companies wishing to diversify. These chemicals involved substances or mixtures whose composition mattered less than the function for which they were intended: the test of success lay in performance. Thus old family businesses or more recent companies born of a leader's entrepreneurial spirit had been successful in performance products, whether these were paints, inks, or glues; or in specialities, cosmetics, detergent, or electronics industries.

Indeed, not much capital is needed to manufacture specialty chemicals compared to what is required for commodity chemicals. The development of new products is both quicker and less costly than it would be to find new processes for large-volume products, or to bring to the market an original active principle for an ethical drug.

This largely explains why specialty chemicals managed to remain until the early 1970s products for medium-sized private companies. In the long run, however, the internationalization of trade, the size of advertising budgets for consumer products, and the necessary adaptation to new technologies requiring highly qualified personnel all called for funds that were not always available to family businesses. Many small owners were forced to sell out, and their need coincided with the attraction they held for large chemical groups trying to diversify away from heavy chemicals. They hoped to find in specialty chemicals the profit margins which their traditional branch of chemicals no longer supplied.

Barring a few exceptions such as *Gulf Oil* or *Diamond Shamrock*, which withdrew from downstream chemicals, all the major companies, both in Europe and in the United States, decided to make specialty chemicals a priority in their development strategy. In truth, some of them had not waited for the energy crisis for them to take a firm foothold in the specialty market.

In the United States, *Du Pont* and *PPG* had a long-established reputation in industrial and consumer paints. *W. R. Grace* since buying Dewey & Almy, and *Rohm & Haas* because of its age-old tradition in acrylics, drew substantial profits from their specialties. This was also true of *American Cyanamid* (additives for plastics, cosmetics) and of *Monsanto* (products for rubber, special polymers). Since its withdrawal from the tire business, *BF Goodrich*, aside from its PVC lines, is concentrating now on specialties.

In Europe, ICI had already acquired a large paints sector (Duco, Dulux). The three major German leaders—*Bayer*, *BASF*, and *Hoechst*—had not yet made great inroads into the specialties market, but the Swiss *Ciba-Geigy* could be said to be particularly well established in certain areas like additives for polymers, in which it was a world leader. *Rhône-Poulenc* had assembled some of its activities within a "chemical specialties" division. But on the whole, they could be said to be offshoots of fine chemicals rather than actual specialties, with the exception of the performance products brought out by subsidiaries such as Orogil, SFOS, Soprosoie, and Vulnax. Orogil is now fully owned by Chevron, however, and Vulnax has been acquired by AKZO. Failing to develop through internal growth, *AKZO* had very early developed its specialties by buying up companies involved in peroxides, paints, oleochemicals, and now rubber additives.

To increase their specialty sectors as fast as possible, the leaders of large companies found it more expedient to do so through acquisitions. The prices paid for the most interesting purchases can be considered high because, very often, they amounted to fifteen to twenty times the profits. But the financial sacrifices made by the buyers seemed worthwhile, for they gained a foothold in the market without the long preliminary work that would otherwise have been needed.

There were, of course, many companies that were sufficiently important or properous to escape being bought up. Even then their independence was often at stake. Thus *Nestlé* took a share in the cosmetics group *l'Oréal*; and in the United States, the raider Perelman managed to buy *Revlon*.

Considering that the grass always looks greener on the other side of the fence, for many leaders of the chemical giants diversification into new areas might seem more attractive than mere concentration in well-known sectors; and it was in this sense that specialty chemicals seemed a good proposition. In 1983, *Olin* began to get involved in electronic chemicals by buying up 64 percent of *Philip Hunt Chemicals*, and took a firm foothold in the sector through successive acquisitions. Other groups became interested in enhanced oil recovery and exploration, for the future of oil seemed assured at the time. In both cases, however, the electronics and oil exploration slowdown did not confirm established forecasts. The investments made in these areas have yet to prove their profitability.

Moreover, many firms were unable to contribute anything except capital to the development of sectors far removed from their traditional areas of business. They became discouraged and ended by selling out, not without suffering heavy losses. *Hercules* was seen to back out of its water treatment sector and *Rhône-Poulenc* from its very recently acquired media business.

Even when the businesses acquired are not too different, trouble can arise through disagreement between the new owners and the former boss of the purchased firm over how to manage it. The former tries to impose his own personnel and procedures, while the latter, used to making his own decisions, is unable to fit into a large unwieldy concern. As a large part of the worth of an acquisition in specialty chemicals lies in the competence of the personnel involved, some purchasers have understood that it is to their benefit to leave the day-to-day running of the business to those who have already shown their worth, and to centralize only those activities related to the financing of new investments. This was how *Witco* proceeded in the United States, most likely because the father of the current president had founded and built up the business to the point of making it one of America's leading specialty concerns.

ICI followed the same policy when it bought *Beatrice Chemicals* for $750 million in 1985. But in this case, it was important to delegate power, because Beatrice Chemicals consisted of ten distinct companies established in eighteen countries and involved in different businesses (composite materials, vinyl resins for paints, leather auxiliaries). Keeping in mind that cultural differences may produce problems that are not always easy to solve, the strong involvement of the big chemical groups in the specialties area over the past years had drastically changed the structures of the sector.

The Paint Industry

Few industries have been as affected by the restructurings of the past ten years as the paint industry. The extension of markets worldwide owing to the multiplication abroad of client factories of this industry, the technological revolutions brought about by the introduction of electrophoresis, of water-based lacquers, and of powder coatings had the twin effect of pushing the chemical leaders to expand worldwide in this area and to lead those paint companies that were still independent to sell out for want of the funds needed to develop their research base. *ICI*, which was strongly established only in Britain and in the Commonwealth, became the world's leading paint producer with 750 million liters after buying *Valentine* in France and, especially, the *Glidden* division of the U.S.-based SCM for $580 million in 1986.

PPG has been pushed back to second world position with 450 million liters. But with its 100 percent stake in France's *Corona* and its controlling share in Italy's IVI and in Germany's Wülfing, the U.S.-based PPG has maintained a comfortable technological lead in the application of cataphoresis in automobile bodies, accounting for 60 percent of the world market in this specialty.

Through its costly $1 billion purchase of America's *Inmont*, *BASF* has become the world's third-ranking paint producer, leaving behind its German Rival *Hoechst*, which was too busy bailing out its British subsidiary Berger Paints to get a foothold in the U.S. market.

AKZO, which holds an honorable place among the leaders, has not been able to penetrate the United States market, either. Most of its recent acquisitions (Blundell, Permoglaze, Sandtex, Levis) were European.

Other companies with comparable 250-million-liter paint capacities are Japan's *Nippon Paint* and *Kansai Paint*, as well as America's *Du Pont*. These three firms, however, have restricted their ambitions to filling the needs of their home markets. With a broader international base, Courtaulds' subsidiary, *International Paints*, ranks among the top ten, although it is mainly involved in the very special sector of marine paints.

Ranking fifth in the world with its 300-million-liter capacity, *Sherwin Williams* is the only large paint company that has retained its independence. It remains focused on the United States, essentially in the decoration market.

Although France is the world's third largest market for paints after the United States and Germany, none of its national manufacturers has thought of striking out beyond its frontiers. Indeed, most of the French companies involved in the sector, with the exception of *Blancomme* and *IPA*, have been taken over by foreign groups when they were not merged within state-controlled entities. *Astral Celluco* was one of the first to sell out to AKZO, *Corona* was taken over by PPG, *Celomer* by International Paints, *Bichon* and *Lefranc Bourgeois* by Sweden's Becker, *Valentine, Julien, Galliacolor* by ICI, *Ripolin Georget Freitag* became part of the *CdF Chimie* group as did *Duco* which has just been sold to *Casco Nobel*, while *La Seigneurie* was taken over by *Elf Aquitaine*.

In 1988 CdF Chimie, later known as ORKEM, took over full ownership of AVI, a profitable company specializing in decorative paints. Another subsidiary of ORKEM was *Lorilleux*, an ink manufacturer merged in early 1988 with *Coates Brothers* to become the third largest group in its field after Dainippon Ink Company (DIC) of Japan and Germany's BASF.

It must be pointed out that all the international groups involved in paints and inks on a worldwide basis produce, in addition, most of the resins and binding agents needed for their formulations. Only the solvents and pigments are likely to be partly brought in from outside sources. The restructuring of the paint industry has, accordingly, been to the advantage of the new groups. On the one hand, it reduced the number of producers and extended the range of products these producers were putting on the market, and on the other, it supplied a captive market for their resins which, until then they had mainly sold to outside customers.

Surface-Active Agents

Used for their good performance, more often than not in formulations, surface-active agents can be classified as specialties even though the quantities consumed in certain cases might connect them with commodities. The structure of the major part of the detergents industry has remained rather stable over the last few years despite some frontier adjustments. The *washing powder* sector where advertising costs are considerable, is dominated by a small number of substantial soapmakers who came into business as far back as the nineteenth century: the American companies *Procter & Gamble* and *Colgate*, the Anglo-Dutch group *Unilever*, and Germany's *Henkel*. They are all, in various degrees, involved in the major world markets.

Then there are the Japanese companies *Kao Soap* and *Lion Oil*, which remain confined to their own home territory and to a few Southeast Asian countries. Behind these giants, a number of firms catering to their home markets stand out, such as *Purex* in the United States or *Benckiser* in West Germany.

As in paints, France is curiously absent from the area. Since Germany's Henkel recently took over the detergents division of the *Lesieur-Cotelle* group and its trademarks Mir, La Croix, and Persavon, after buying up the Savon de Marseille soap flakes of *Union Générale de Savonnerie* (UGS), the French market is now 94 percent supplied by the big international soapmakers. The few remaining national firms such as *Chimiotechnic* merely sell their products through the supermarkets.

While the sector now seems to be structurally stabilized, washing powder components are fast changing to take into account the new rules and technologies laid down both by governments and consumers. For the companies which supply the soapmakers, these new rules and regulations are having major consequences throughout the world. Just as the requirement of biodegradability had doomed the use of *branched-chain alkylbenzenes* in industrialized countries in the 1960s and caused the shutdown of a large number of dodecylbenzene sulfonate units, so the new rules established by some governments against *tripolyphosphates* in Europe and elsewhere to ward off eutrophication are likely to wipe out the several-hundred-thousand-ton markets of producers like Rhône-Poulenc, Benckiser, Knapsack, or Montedison.

Replacing TPP by new formulations based on *polyacrylic acid* and *maleic anhydride* would, on the other hand, greatly boost companies like Atochem and BASF, which are very much involved in acrylic chemicals. Likewise the use in Europe of washing machines at temperatures that do not exceed 50° to 60°C, like the ones now used in the United States, should have immediate consequences for the formulation of washing powders. *Perborates*, used extensively in Europe as bleaching agents ever since Henkel invented Persil in 1907, are not very efficient at such low temperatures. Activators such as ethylenediamine tetraacetic acid (EDTA), produced by Warwick in England, are needed to hasten decomposition. *Enzymes*, which had been very popular in the 1960s in the United States and Europe, then had disappeared in 1971 because they were considered harmful to the skin, have been reintroduced in washing powder formulations because they do help remove certain stains.

The use of *liquid detergents* is more widespread in the United States, where they account for 20 percent of the market, than in Europe, where their share does not yet exceed 8 percent on an average. This has consequences on the consumption of nonionic derivatives.

Different habits as well as different regulations have therefore led to frequent changes in the chemicals supplied to soapmakers. Few industries have changed as much as the detergent industry since the end of World War II as it shifted from soap to synthetic detergents, from branched alkyl benzenes to linear alkyl benzenes, from anionic to nonionic. TPP and enzyme regulations were changed; preference was given at times to perborates, at others to chlorine-based products as bleaching agents.

To develop surface-active agents for industrial use did not require the same financing as was needed for washing powder consumer products. Therefore, producers of all sizes could become involved. Some of these producers were chemical giants who had gone into the business because they had the available raw materials or the right markets. Indeed, surface-active agents use a number of major raw materials to which suppliers attempt to add downstream value.

In Europe, for historical reasons, large chemical groups have become involved in this area. Thus the dye manufacturers had very early added to the range of products sold to the textile and leather industries, wetting agents, softeners, and dye auxiliaries. *BASF*, a pioneer in synthetic auxiliaries with its Nekal, patented in 1917, *Hoechst*, *Bayer*, and *ICI* were in fact interested at the same time in the markets which surface-active agents opened for their ethylene oxide, higher alcohols, sulfonating agents productions, and in the fact that they help provide better services for their traditional textile clients.

Hüls, the subsidiary of the German holding company VEBA, had no dye tradition. But it nevertheless acquired the Dutch surface-active unit Servo to ensure captive use for at least part of its ethylene oxide and alkyl benzene production. *BP* followed a similar line when it took over the Belgian company Tensia, selling back some of its product lines to ICI. Already involved in surface-active products through its Lissapol for many years, *ICI* has expanded in this sector by buying *Atlas Powder* and its special range of Tweens and Spans. *Shell*'s interest in surface agents went back to the development of its Teepol. It completed its range with ethoxylates, the "neodols" which used both its higher alcohols and its ethylene oxide. *Montedison* was also involved in surface-active agents through its stake in *Mira Lanza*.

In France, however, there was no vertical integration between the great national chemical industry and the surfactant sector. Producers of the latter had to find the necessary feedstock—whether ethylene oxide, alkylphenols, fatty acids or higher alcohols—from rival companies, while for instance a medium-sized company like Berol Chemie in Sweden, recently acquired by Nobel Industries, had its own source of ethylene oxide, amines and nonylphenol in Stenungsund to feed its surfactants division.

In the *United States*, vertical integration was not as thorough as in Europe. Although ethylene oxide producers like *Union Carbide*, *Dow*, or *Texaco* also had their range of ethoxylates, it was mostly specialized firms that produced the surface agents for industrial uses. The same was true in Japan, although a number of producers such as *Nippon Oil & Fats* for fatty acids, *Kao*

Corporation and *Lion Corporation* for fatty alcohols and amines had direct access to their main raw materials.

Thus in addition to the large chemical and petrochemicals companies that had chosen downstream integration, there were a number of important surfactant producers that, in varying degrees, were integrated upstream. The most striking example of this, besides the three Japanese companies just mentioned, is Germany's *Henkel*. Its natural fatty alcohol production exceeds 170,000 tons capacity, and besides fatty acids, it produces its own range of carboxymethylcellulose-based thickeners. Recently Henkel even associated in this area with Hercules within a company called *Aqualon*, now fully owned by Hercules, and acquired from Quantum Chemicals in the United States their fatty acids subsidiary *Emery Industries*.

A number of surfactant specialists have also chosen the market approach. Because they are not tied down by their own produced raw materials, they can use those that are the most suitable for the type of surfactant they wish to offer their clients.

An independent producer like the U.S.-based *Stepan* is in a position to provide a complete range of anionic, cationic, and nonionic agents because it has flexible units in four areas of the United States as well as one in southeastern France in Voreppe.

Witco is in the same position, but its own policy has been to develop through acquisitions rather than through internal growth, buying Humko Chemical and Onyx Chemical.

Right from the start *GAF* acquired, from IG Farben, experience in surfactants still of use today. This activity sector, however, was sold to Rhône-Poulenc in 1989.

With a market lacking the uniformity of the United States market, the European producers serve in greater numbers clients with standards and habits varying from country to country. The Tenneco group's *Albright & Wilson* has had to cover France, Italy, and Spain with its *Marchon* subsidiaries. Germany's *Hoechst*, *Henkel*, and *Schering* which bought up *Rewo*, also have a number of subsidiaries abroad that produce their surfactants. Hüls's subsidiary *Servo* has only the single production unit in Delden, Holland. But because of the high concentration of its products, Servo manages to carry out three-quarters of its sales abroad.

While the range of products offered by these companies is very wide, some of them, nevertheless, focus on specific sectors. Thus the cationic technology acquired in the United States from Armour by *AKZO* and from Ashland by *Schering* has given both these companies a dominant position in the market of textile softeners both in Europe and in the United States.

Companies like *Rhône-Poulenc*, *Berol* and Witco are, for their part, interested in the pesticide formulation market. Fatty amines are in the hands of such European firms as *AKZO*, *Kenobel* (Nobel Industries) and *CECA* (Atochem).

Other European companies, such as *ICI* through Atlas and Tensia, *Th. Goldschmidt*, *Rewo*, and *Servo*, have particularly targeted the lucractive area of beauty care. In the United States, *Miranol* has been very successful with the amphoterics (imidazolines, betaines) for baby shampoos, an activity acquired by Rhône-Poulenc in 1989.

America's *Du Pont* and *3M* and Japan's *Sanyo* pay particular attention to the development of fluorine-based surfactants. *Air Products* with its acetylene derivatives Surfynol and *W. R. Grace* with its sarcosinates (Hampshire Chemicals) have also focused on well-defined segments of the business. With world demand exceeding two million tons, the market of surfactants for industry is of a nature to attract a large number of operators, raw material suppliers, processors of these raw materials into anionic, nonionic and cationic derivatives, or downstream industries that use these different surfactants in various formulations.

Flavors, Fragrances, and Beauty Products

The sector of flavorings, perfumes, and beauty products has also had its share of restructuring and technological changes over the past ten years.

Although many of the raw materials needed in this area still come in the form of essential oils from natural sources like jasmine from Grasse, roses from Bulgaria, ylang-ylang from Madagascar, oak moss from Yugoslavia, an increasingly significant role is now being played by semisynthetic or fully synthetic products.

Thus *terpenes* (α-pinene, β-pinene) can be produced from natural turpentine, as is traditionally done by rosin producers such as *Hercules*, *Glidden*, or *Union Camp* in the United States, or on a smaller scale in France by *Société des Dérivés Résiniques et Terpéniques, DRT*. BASF and Hoffmann-La Roche, however, have demonstrated that starting from acetylene or isobutylene, terpene chemicals can be synthetically reproduced. Both companies are able to produce both their vitamins and perfume bases in this way.

Likewise, *vanillin* is now largely produced synthetically. The world leader in this area is *Rhône-Poulenc*, which has a unit in Saint-Fons to which was added a unit bought from Monsanto in 1986 on the West Coast of the United States. In the latter plant, vanillin is still extracted from paper pulp liquor. *Menthol* from plantations in Brazil and China is also produced by synthesis since *Haarmann & Reimer*, bought by Bayer in 1954, managed to carry out industrially the resolution of racemic menthol, thus isolating the levo-isomer. *Anethole*, synthesized by Hercules from pine oil, is two to three times cheaper than when it is extracted from star anise.

Instead of identically reproducing natural products, chemists have also succeeded in making cheaper substitutes with similar features. Thus nitrated musks and later macrocyclic musks have become substitutes for more rare natural musk. Major chemical companies became interested in the firms that specialized in perfume chemicals. But their involvement in this area was not always successful, for their business views did not necessarily apply to this new activity.

While Bayer's association with Haarmann & Reimer proved successful, it took Hercules several years, from 1973 on, to understand properly how its *Polak Frutal Works* (PFW) had to be managed. Today it is autonomous and prosperous. In contrast, Rhône-Poulenc ended by selling *Lautier* to *Florasynth* in 1981.

When *Tenneco* bought Albright & Wilson in England, it did not see the point of keeping its *Bush Boake Allen* (BBA) aroma chemicals division. BBA, itself the outcome of a merger of several family businesses, was finally sold to the U.S.-based Sylvachem in 1982. Sylvachem[1] already owned *George Lueders*, an essential oils concern Monsanto sold failing proper management. It would seem, therefore, that among the major chemical companies, only Bayer, Hercules and, more recently, BASF, which bought *Fritzsche Dodge & Alcott* in the United States in 1980, have achieved their downstream breakthrough in the flavor and perfume sector.

On the other hand, the pharmaceuticals group Hoffmann-La Roche, which purchased *Givaudan* in 1963, then *Roure Bertrand Dupont* a little later, has managed to rank third in the world in this difficult area. But the leader is undoubtedly *International Fragrances & Flavors* (IFF), an American company that accounts for 10 percent of the world market. Set up in 1929 by a Dutch immigrant, A. L. van Ameringen, IFF acquired its current form in 1958 and, pushed along by the creative invention of its perfumers and the quality of its compositions, has never ceased growing.

Close on the heels of IFF is the *Unilever* group, which developed in the field through acquisitions. After consolidating in 1983 its three perfume and flavor subsidiaries—PPL, Food Industries, and Bertrand Frères—to form *PPF International*, the group acquired a foothold on the U.S. market in 1984 with *Norda*. Three years later it merged PPF with Holland's *Naarden*, which was on the decline. Called *Quest International* (Unilever) the new company accounts for over 7 percent of the world market in its area.

Amongst the world leaders, the only privately owned company, the Swiss-based *Firmenich*, ranks fourth. It has retained its independence both because it was held together by the heirs of the founding family and because it produces quality products based on strong research. A number of smaller companies that do not belong to any multinationals are highly competitive. They include Japan's *Takasago*, which began in 1920, America's *Florasynth*,

[1] Sylvachem belongs to the Union Camp group.

which took over *Lautier*, and Britain's *Pauls Flavours & Fragrances*, which has just established a hold in the United States market by purchasing *Felton International*.

France, which had in its favor the age-old reputation of Grasse and the world image of its perfumes linked to its haute-couture prestige, is nevertheless absent from the fray of large suppliers in this area, even though it has some Grasse-based companies like *Mane* and *Robertet* and despite the efforts made by the Elf Aquitaine group which has assembled, around Sanofi, firms like *Méro et Boyveau*, *Tombarel* and *Chiris*.

Flavors account for a substantial share of the sales of these firms: 30 percent for IFF, 40 percent for Givaudan, 50 percent for Unilever and 100 percent for Sanofi-Méro. They are increasingly being used in foodstuffs since the fashion of fruity yogurts and instant desserts began between 1965 and 1970. The internationalization of food habits and the growing industrialization of the food sector have contributed to the development of demand for flavors and to the gradual substitution of natural substances by synthetic products that are less costly to produce and more active in small doses. Just as the perfume industry composes fragrances for its clients, subtle blends of flavors are now devised for the large food companies. Demanding customers, together with stringent regulations and sophisticated technologies, all combine to build up research costs. This explains the restructuring that has taken place in the sector as family businesses have been taken over by powerful international chemical, pharmaceutical, or food-industry groups, leaving only a few independents willing and able to make the necessary research efforts.

Although it still clings to a long tradition, the *world of perfumes* has also changed both in its structures and in its technologies. The highest volume comsumption derives from products of the soapmakers. *Procter & Gamble* prepares its own compositions, but its competitors mostly rely on the laboratories of their suppliers for fragrance preparation.

With a few rare exceptions, such as *Guerlain*, *Chanel*, and *Patou*, the great names as well as the small perfumemakers do likewise. One of the world's largest-selling perfumes, "Anais-Anais" by *Cacharel* (l'Oréal), is prepared by Firmenich, while *Roure-Bertrand-Dupont* has signed two other recent successful perfumes, *Dior* 's "Poison" and *Saint-Laurent*'s "Opium." Launched in 1921, *Chanel No. 5* was the first perfume to carry a synthetic aldehyde note and is still one of the ten world best-sellers. But the market has now moved to floral and oriental fragrances. Perfumes for men with stronger notes have developed spectacularly and now account for 25 percent of alcoholic perfumery. In addition, the aerosol format has boosted sales of toilet waters and deodorants.

The most varied distribution systems have been developed, ranging from door-to-door sales, which *Avon* started, to sales by mail, a specialty of *Yves Rocher*'s, to sales in large department stores, to sales in selected areas such as perfume shops and pharmacies. Few "nonessentials" have become so indispensable. If they cannot be dispensed with, it is through the efforts of the industry, which relies upstream on the suppliers of both contents and containers, who adapted to all requirements, and downstream on efficient marketing networks. It can also devote to advertising the money that it need not spend on research conducted on the industry's behalf by the chemists.

Although it originated in France, the perfume industry is now mostly in the hands of foreign firms. While *Parfums Dior* and *Givenchy* (belonging to the Moët-Hennessy-Louis Vuitton group), as well as *Guerlain, Lanvin, Nina Ricci*, and *Patou* are still under French control, *Cardin Parfums* belongs to American Cyanamid, *Orlane* to Norton Simon, *Chanel* to the Swiss Pamerco group. *Rochas* was owned by Hoechst, which has now sold it, and *Parfums Saint Laurent* is now controlled by Italy's Carlo de Benedetti, who bought it from the U.S.-based Squibb. As for the L'Oréal group, which had taken over the *Lancôme*, *Jacques Fath*, *Guy Laroche*, *Ted Lapidus*, *Cacharel*, and *Courrèges* perfume brands, it has been within the Nestlé orbit since 1973, although it was arranged that until 1993 it would be managed by those representing the interests of the founding family. Not surprisingly, the same great names recur in the area of *beauty products*, including, besides perfumes, *hair care products* and *cosmetics*. Each firm, indeed, wishes to complete its range by acquiring complementary businesses.

The cosmetics industry was born in the United States with the three great "ladies," *Elizabeth Arden*, *Harriet Hubbard-Ayer*, and *Helena Rubinstein*. In their wake are now *Estée Lauder*, the giant *Avon*, *Max Factor*, and *Revlon*, founded by Charles Revson. A number of these firms did not survive their founders. Elizabeth Arden was first bought by Eli Lilly and now belongs to the United States *Fabergé* groups; Helena Rubinstein has disappeared after being taken over by *Colgate Palmolive*; following ten years of poor management and uncontrolled diversification, particularly in pharmaceuticals, Revlon has been grabbed by Pantry Pride, a chain store group belonging to the raider Perelman; Max Factor now belongs to the *Norton Simon* group.

While all this was taking place on the American scene, two groups, one Japan's *Shiseido*, and the other France's *L'Oréal*, were climbing to the rank of leading world producers, raised there by dint of good management and competent research and marketing skills. Although Shiseido was unsuccessful in its bid in the United States to take over the famous *Giorgio* of Beverly Hills, which was acquired by Avon, and the skin care company *Charles of the Ritz*, which Yves Saint-Laurent had sold back to Revlon, it nevertheless ranks second in the world after *L'Oréal*, and has very strong positions throughout Asia.

L'Oréal's founder, Eugene Schueller, graduated as a chemist from Institut de Chimie de Paris. He resigned from his job at the Sorbonne to produce a "harmless" hair dye called l'Auréole. The trade name l'Oréal was adopted the following year. A skillful businessman and a true pioneer of ad campaigns,

Schueller bought the Monsavon soap factory in 1928 and, before the war, brought on the market the O'Cap hair lotion, then Ambre Solaire. When he died in 1957, his successors managed to develop the business both through internal growth and an efficient research base and through a series of acquisitions.

In 1961, Monsavon was sold to Procter & Gamble, and L'Oréal purchased the *Cadoricin* firm, which extended its range of hair products, to which were added *Garnier* and *Roja*. Then Lancôme was purchased, introducing high-class products. This was followed by the purchase of other perfume-makers. Tempted by the pharmaceuticals market, the company bought *Synthélabo* in 1973, consisting of four medium-sized laboratories. It is still too early to say whether the money sunk into the sector since then will bring in returns as large as those of the perfumes and skin-care business. Mixing the two has not always been successful.

In the United States the marriages between Pfizer and Coty, Colgate and U.S. Vitamin, Eli Lilly and Elizabeth Arden, Squibb and Charles of the Ritz, Avon and Mallinckrodt, Revlon and Armour Pharmaceuticals all ended in divorce. There was, of course, Bristol Myers' successful venture with Clairol, and American Cyanamid with its Shulton subsidiary. But these exceptions only confirm the general rule of failure.

In France, while Sanofi can draw satisfaction from its association with Yves Rocher, which enjoys great management freedom, the sector comprising Roger & Gallet, Stendhal, and Charles Jourdan has not yet lived up to the parent company's expectations. Only the British seem to have succeeded in combining such different businesses, possibly because from the start the skin-care activities were intimately associated with pharmaceuticals within large groups like Beecham, Glaxo, and BDH.

The Chemistry of Additives

Used in small doses to improve the products in which they are incorporated, additives are to be regarded as specialties with well-stated functions even if, in many cases, they are well-defined chemical entities sold according to specifications. Because of this ambivalence, chemical companies have approached the sector of additives sometimes through the markets they serve, sometimes through the chemicals from which they derive, even at times from both ends.

Additives for Plastics

Additives for plastics have experienced the double approach. The opening up of the markets leading to uniform production of plastics gave worldwide scope to some additive producers. Tasks were shared since polymer producers did not consider it useful to prepare the additives they needed, while additive producers were, as far as possible, careful to avoid competing with their clients in the area of base thermoplastics.

It is true that a major polyolefin producer like *Hoechst* sells its own range of antioxidants and its subsidiary *Riedel de Haen* produces ultraviolet ray absorbers. Likewise, the world ABS leader *Borg Warner*, now acquired by General Electric, has been marketing, since it took over *Weston*, a series of organic phosphites for the stabilization of high polymers. In Japan, *Sumitomo Chemical* is a supplier of large-volume plastics as well as of a rather complete range of stabilizers.

These are exceptions, however. The world's largest additives producer for plastics, *Ciba-Geigy*, remains, for its part, at the sole service of its downstream customers and tries not to appear as a competitor. This is also the position of other additive suppliers like *American Cyanamid*, *Ferro*, *Witco*, *UniRoyal Chemical* in the United States, *AKZO*, *SFOS* (Rhône-Poulenc) in Europe, *Adeka Argus* and *Dai-ichi Kogyo Seiyaku* (DKS) in Japan.

Ciba-Geigy owes its leading position to a number of factors: long perseverance in the specialty, an efficient research base through which the universally used *Irganox* antioxidants were developed, application services adapted to all the polymers requiring stabilization, worldwide production units established within large comsumption areas (Europe, America and Japan). Even where Ciba-Geigy did not invent a product but took a license on it as with *HALS* (Hindered-Amine Light Stabilizers), licensed from Japan's *Sankyo*, it developed it to the point of acquiring world supremacy in the area. Ciba-Geigy's success in this activity is all the more remarkable as it has no upstream integration on raw materials used in the synthesis of phenol antioxidants, of phosphites, of thioesters, of substituted benzophenones, of benzotriazoles, or of HALS. But this apparent weakness is fully compensated by the dominant position Ciba-Geigy has acquired in the different types of additives for plastics in its range, either through internal growth by its research, or through license acquisitions (Sankyo), or through purchase of relevant companies (Chimosa in Italy), or again through complementary activities (range of Goodrich's Goodrite antioxidants).

The other producers of plastics additives trail far behind Ciba-Geigy in variety of range or in market coverage. *UniRoyal Chemical* produces antioxidants and blowing agents and has production units in the United States, Latin America, Italy, and Taiwan; but its recent restructuring has cut short its development. *American Cyanamid* which pioneered a number of additives (substituted benzophenones, 2246) sold its European business to Ciba-Geigy in 1982 and now operates only in the American market. *AKZO*'s range is restricted to antistatic agents, PVC stabilizers, and peroxide catalysts, which it acquired through *Armak*, *Interstab*, and *Noury van der Lande*. *Borg Warner* is mainly focused on phosphites, which it produces solely in the United States; Elf Aquitaine's subsidiary *M & T* is focused on organotins; *Witco*, through its purchase of *Argus Chemical* and *Humko Products*, is involved in heat stabilizers, antistatic agents, and lubricants. *Ferro*, which also produces master batches, has developed specialities such as fireproofing agents and stabilizers for PVC, and has recently joined forces with Italy's Enichem to produce and market new lines of polymer additives in the United States.

Other companies came to additives through the chemical tree, such as *Société Française d'Organo-Synthèse* (SFOS), a subsidiary of Rhône-Poulenc, which by isobutylating phenols produces a whole range of phenolic antioxidants as well as special phosphites, or *Ethyl*, which approached the Irganox family of antioxidants and bromine fireproofing agents through its orthoalkylation technology and its access to bromine. Similarly, because the U.S.-based *Olin* was an important hydrazine manufacturer, it became interested in blowing agents like azodicarbonamide and bought up National Polychemicals, which also provided it with a range of phosphites. The blowing agent line has since been sold to UniRoyal Chemicals.

But the interest which major firms like ICI, Bayer, or even Hoechst still have in the sector is restricted by the small number of additives they supply to plastic producers. In the circumstances, it is more than likely that Ciba-Geigy's lead in the variety of products offered, in research, in customer service, or in geographic coverage will be hard for competitors to catch up with. Indeed, their narrow approach to the market would hardly warrant the heavy investment to fulfill any high ambitions they might have in the area.

But favored by their access to certain raw materials or by their specialization in a very specific range, such competitors can, at least, be assured of a degree of prosperity inasmuch as the standards required for optimum use of plastics are closely related to incorporation in the high polymer of effective additives at a reasonable cost.

Rubber Additives

The specialists in *rubber additives* are distinctly different from the specialists in additives for plastics, even though the same products are sometimes used in both industries: blowing agents (azodicarboamide), phenol antioxidants (BHT, 2246), phosphites (tris-nonylphenyl phospite). In the first place, additives for elastomers, unlike those which might come into contact with foodstuffs, do not require official approval, which makes it easier to put them on the market. In the second place, most of them are well-known products sold to specifications by a number of producers. The development of new products protected by patents is rarer than in the case of additives for plastics. In fact, while consumption of plastics has been constantly increasing, stagnant demand for both natural and synthetic rubber has not warranted any significant recent research efforts by suppliers of this industry.

If one considers that the automobile sector accounts for 75 percent of rubber consumption in developed countries, it stands to reason that the long-life radial frame and smaller diameter tires of modern vehicles should require smaller amounts of rubber for the same number of cars produced; in the United States alone, rubber consumption has fallen from 3.2 million tons in 1977 to a little over 2.6 million tons in 1989. This implies a consumption of some 150,000 tons of organic additives.

Faced with such a situation, producers of additives for rubber have either restructured, or else rationalized production. In rarer cases, others have offered new products with higher added value than the conventional additives. In the United States, *American Cyanamid* in 1982 halted production in Bound Brook of its accelerators; *Goodyear* terminated its substituted *p*-phenylenediamine production in Houston in 1984. In 1985, *Allied-Signal* took over *UOP*'s antiozonant unit, while in 1986 *UniRoyal Chemical* became part of *Avery, Inc.*, before becoming the object of a leveraged management buyout (lmbo) in 1989. In Europe, Rhône-Poulenc and ICI merged their rubber divisions within a subsidiary called *Vulnax* and then finally sold it to AKZO in 1987. Atochem, meanwhile, was taking a minority stake in *Manufacture Landaise de Produits Chimiques*, henceforward leaving France and Britain with no significant producer with international clout.

In this changed environment, three major additives manufacturers emerged: *Monsanto* with its plants in the United States, Canada, Britain and Belgium; *Bayer*, which owns two sites in Europe and produces antiozonants in Pittsburgh through Mobay; and *UniRoyal Chemical*, which has production units in Naugatuck, Connecticut, and Geismar, Louisiana, as well as in Canada, Brazil, and Italy.

UniRoyal Chemical was separated in 1966 from U.S. Rubber, which had provided it with a captive market. But two other tire manufacturers had retained their traditional activities in additives. They were *Goodrich*, which produced only in the United States, and *Goodyear*, which also operates in Europe in its antioxidant units in Le Havre, France. Both these giants sell part of their production through a rubber blend specialist, R. T. Vanderbilt. Goodrich, however, has recently withdrawn from the tire business in order to concentrate on its chemical activities, so that only Goodyear enjoys today the advantage of a captive outlet for the rubber chemicals it produces.

Although Monsanto can rely on only two of its own raw materials, tert-butylamine and *p*-nitrochlorobenzene, for its range of additives, it is regarded as an efficient producer and a pioneer in antiozonants based on *p*-phenylenediamine and prevulcanization inhibitors. It has one of the most complete ranges of additives for rubber and the most modern units to manufacture them.

Because of its long experience in organic synthesis intermediates, *Bayer* is possibly better integrated upstream than Monsanto. Its range of products is just as large, but its production units are essentially restricted to Leverkusen and Antwerp. With the exception of *AKZO*, which, through its purchase of

Vulnax, seems to want to improve its range of additives and its geographic coverage, no other major European chemical group has gone beyond a small range of special products.

Like their competitors in Europe and in America, the Japanese producers have focused their attention on accelerators (vulcanization activators and agents) and on antiaging agents (antiozonants, stabilizers). Their automobile exports provide a market for tires that their counterparts in other countries cannot claim to the same extent.

Japan's additives production, however, is too scattered among a large number of producers to be truly profitable. With the exception of two principal companies in the area, *Sumitomo Chemicals* and *Mitsubishi-Monsanto*, firms like *Ouchi, Shinko*, *Kawaguchi*, and *Seiko Chemical*, which were the first to get into the business in 1930, do not have the required size to be competitive on international markets.

Additives for Lubricants

Additives for lubricants are also greatly dependent on the automobile industry, which alone uses some 60 percent of the lubricants produced worldwide, whether lube oils for engines (gasoline and diesel oil) or for gear boxes. Since the oil-price rise in 1973, lube oil consumption has been affected by a number of factors: smaller vehicles and therefore smaller engines, falling automobile production, larger intervals between oil changes, implying a higher additives dosage to extend oil efficiency. To these various changes should be added increasing use of diesel fuel in Europe because of favorable taxation. The generalized use of multigrade oils and the introduction of unleaded gasoline, and consequently of catalytic exhaust pipes, should lead to enhanced engine oils. In the circumstances, world consumption of additives for lubricants is likely to remain at around two million tons a year over the next few years, with higher additive doses compensated by extended lube oil efficiency and smaller casing size.

With the exception of *Lubrizol*, the world leader in this area, and *Ethyl*, which came to lube oil additives by buying Edwin Cooper off Burmah Oil in 1968, the main suppliers with extensive ranges of additives are the international oil companies *Exxon*, *Chevron*, *Amoco*, and *Shell*. The business was a natural extension of their lube oil production, which serves as a captive market.

All these oil companies market their additives as a package, the efficiency of which has been extensively tested. Most of the ingredients in the package are produced by the companies themselves: *detergents* (sulfonates, phenates, naphthenates), *dispersants* (succinimides, polybutene, succinates), *antiwear agents* (zinc dithiophosphates, chlorinated paraffins, sulfur and phosphate hydrocarbons), *anticorrosion agents* (substituted amines, succinic acid derivatives, nitrites). On the other hand, the antioxidants are often supplied separately, as are pour-point depressants (polymethacrylates), and additives to improve the viscosity index of multigrade oils (polymethacrylates, olefin copolymers).

Originally called Graphite Oil Products Cy., *Lubrizol* was founded in 1928 near Cleveland with a capital of $25,000 by six associates. Their close ties with the Case Institute of Technology gave the concern a strong technical orientation. In this manner, *Lubrizol* played a pioneer role in developing lube oil additives and is still today a world leader in its area, with fourteen plants installed worldwide, including four in the United States, and testing sites in Wickliffe (Ohio), Hazlewood (Britain), and Atsugi (Japan). In its attempts at diversification, the company recently became interested in biotechnology with the purchase of *Agrigenetics* in 1985 and of a stake in *Genentech*. But it is too early to state whether this choice will bring the same long-term satisfactions as the company's traditional business.

Exxon came to chemical additives for lubricants by producing its Paraflow range of freezing-point depressants as early as 1930 in Bayonne, New Jersey. Through the agreements signed in 1937 with IG Farben, Standard Oil of New Jersey (later to become Esso and then Exxon) acquired the thickeners and additives based on polyisobutylene that improve the oil viscosity index. In the 1960s, Exxon further enlarged its range of lubricant additives and in 1979 set up the Paramins special division, which marketed a series of olefin copolymers (OCP) based on the chemistry of the group's ethylene-propylene elastomers. The object was to compete with the polymethacrylates (PMA) in improving multigrade oils (VI improvers). Based in Houston, Texas, and involved in all world markets, Paramins has become Lubrizol's most dangerous rival.

Chevron approached the oil additives market in the thirties by supplying metal naphthenates to its parent company Standard Oil of California. Some of these additives were marketed under the trademark Oronite from 1948 onwards. Chevron kept its main research center in Richmond, California, even when, in the 1950s, it spread to international markets through subsidiaries set up with local partners: *Orobis* with BP and *Orogil* with Progil in Europe; *Karonite* in Japan, *AMSA* in Mexico. In 1986, BP bought Chevron's 50 percent share in Orobis, and more recently Rhône-Poulenc sold to Chevron its 50 percent share in Orogil.

The interest shown by *Ethyl Corporation* in Edwin Cooper stems from its desire to diversify into the oil sector as unleaded gasoline begins to threaten the future of tetraethyl lead. But in a business in which it is a newcomer, Ethyl still has much to learn before attaining the efficiency and international coverage of its three main rivals. The same is true of *Amoco* and of *Texaco Chemicals*, although they are endowed with a significant captive market through their parent companies, Standard Oil of Indiana and the group made up of Texaco, Caltex, and Getty Oil respectively.

For its part, the *Royal Dutch Shell* group came to additives after the Second World War in the United States with a range of alkaline sulfonates. Subsequently it enlarged its range with new additives (detergents, dispersants, VI improvers) and fuller geographic coverage through production centers located in Berre, France; in Stanlow, England; and in Marietta, Ohio, and in Martinez, California, in the United States. More recently, a common subsidiary with Lubrizol was set up in Brazil.

Besides these large companies, which offer a range of additives as extensive as possible, if only to recoup research expenses and the high cost of tests required to obtain approval of the "packages," there are a number of chemical companies that have also established a foothold in the market of lube oil specialties. Their reason for doing so was that they had acquired knowhow in the chemical sector leading to the products marketed.

Rohm & Haas in Philadelphia developed additives to lower the freezing points of oils and to improve their viscosity index through work carried out as early as 1934 by the chemist Herman Bruson on the properties of polymethacrylates (PMA) produced from higher methacrylates. Other companies, such as *Röhm* in Darmstadt and Melle-Bezons (whose Persan unit in France was bought from Rhône-Poulenc by *Société Française d'Organo-Synthèse* [SFOS] in 1978), also supplied PMA for such applications. Through the chemistry of phenol isobutylation, *Ethyl* and *SFOS* took a foothold in the phenol antioxidant market of oil companies, while *Ciba-Geigy* is developing a significant program in this area. But the need to be thoroughly acquainted with the lube oil business and to be well introduced in the world oil circles, narrows the scope of chemical firms that have only a small range of additives to offer and precludes their taking a significant place in such a specialized market.

Food Additives

Because their nature, their uses, and their origins are extremely varied, *food additives* are supplied by a large number of different firms. In what is a fragmented industry, some producers stand out more because of the major place they occupy on the market than because of their range of additives.

In the United States, there are only two producers of *citric acid* (Pfizer and Bayer's subsidiary, Miles) and of *vitamin C* (Pfizer and Hoffmann-La Roche) and a single producer of *saccharin* (PMC, which bought Maumee from Sherwin Williams), *sorbates* (Monsanto), and *carrageenates* (FMC since it acquired Marine Colloids).

Because of the very strict rules that in industrialized nations govern additives used in human food, it has become very expensive to introduce new products. In some countries even some of the older derivatives that used to be considered nontoxic have been questioned. This is the case with saccharin, discovered by Ira Remsen in 1879 and used without drawbacks since then. Because of such limitations, few new producers have ventured into the area over the past few years except through purchase of existing companies that already had approved additives. The giants in the business are generally satisfied with being dominant in certain market sectors through their special technologies (fermentation, extraction, synthesis). The problems of excessively high sugar consumption, however, have induced a number of researchers to look for low-calorie substitutes for sucrose other than saccharin. Accordingly, new *synthetic sweeteners* have been discovered: *cyclamate* (sodium cyclohexylsulfamate), synthesized in 1937 and put on the market by Abbott in 1950; *aspartame*, isolated in 1965, produced by reaction of aspartic acid with phenylalanine methyl ester, and developed by Searle, which was susequently purchased by Monsanto); and Hoechst's *acesulfame K*. Despite lack of coordination in this area among the different national legislatures, these synthetic sweeteners, with their low calorie content and a sweetening power that is fifty to two hundred times as great as that of sugar, should sooner or later take root on international markets.

The use of *gelling* and *thickening* agents in foodstuffs goes back to earliest times. In the last few years, progress has been made in the extraction and purification of plant-based hydrocolloids used for the purpose. In addition, the polysacharide *xanthane*, produced through fermentation, has been developed over the past twenty years to take its place among the water-soluble gums supplied to the food industry. At the same time, a semisynthetic gum, *carboxymethyl cellulose* (CMC), used in a number of industrial applications, was allowed in its purified form, in human foodstuffs.

The U.S.-based *Hercules*, which started by producing precisely this CMC of which it is the world's leading producer, has gradually extended its range of products by purchasing companies. It is now involved in *pectin*, extracted in Denmark and in the United States from lemon peel, and *guar*, prepared in Italy from a bush that grows in India and Pakistan, *carob* developed in Spain, *carrageenates* extracted from algae growing along the Atlantic and Pacific coastlines. With the exception of pectin, these various gums recently became the business of *Aqualon*, a common subsidiary of Hercules and *Henkel*, its German partner, already involved in the guar and CMC market. In 1989 Hercules became the sole owner of Aqualon.

Hercules never did succeed in developing *Xanthane* through its association with the British company Tate & Lyle in 1979. This gum has remained a specialty of *Rhône-Poulenc* which produces it in Melle, France, and of *Kelco*, a San Diego, California, subsidiary of Merck that also owns *Alginates Industries*.

Other groups have likewise specialized in particular sectors. The Stein Hall subsidiary of Celanese, taken over by the British-based *RTZ Chemicals*, now part of Rhône-Poulenc, has focused on guar, while *Marine Colloids*, a subsidiary of FMC, and *Satia*, of the Sanofi Elf Bio Industries group, specialized in carrageenates. There are a great number of industrial applications for gum, and thus gum producers are usually drawn to the food industry because of their knowhow in gum. It is seldom that they have chosen to manufacture thickeners because of their experience in foodstuffs.

The same is true of *antioxidants* like *BHT* (butylhydroxytoluene). Although it is used in purified form in human and animal food, its more common use is as a stabilizer for polymers and lubricants. Only *BHA* (butylhydroxyanisole), α-*Tocopherol* (vitamin E), *TBHQ* (tertiary-butylhydroquinone) and *propylgallate*, which are marketed by *Eastman Kodak*, can be considered as purely food antioxidants for the two reasons that they are not toxic and that they are high-priced. In fact, Eastman Kodak is the only chemical leader to produce an extended range of food additives: mono- and diglycerides and vitamins.

Producers of *acidulants* came to food applications through chemistry or biochemistry. *Malic acid* is produced by Denka, now owned by Mobay, in the United States and by Croda in England. Like *fumaric acid*, it is a derivative of maleic anhydride production. The major acidulant is *citric acid*, which is also used as a stabilizer. It is a fermentation product that is produced by a few traditional specialists—in Europe by La Citrique Belge, which was bought by Hoffmann-La Roche; in Britain by Sturges, taken over by RTZ Chemicals and now owned by Rhône-Poulenc; and in the United States by Pfizer and Miles.

Phosphoric acid, used in fizzy drinks, is produced in its food quality only by a small number of firms such as FMC and Stauffer, an activity taken over by Rhône-Poulenc in the United States, and by Prayon in Belgium.

Food conservation generally requires the use of chemical additives, although the problem can be solved at times through temperature control (pasteurization or sterilization through heating, freezing, or control of water content [dehydration]). Chemical additives act by working on the metabolism of the microorganisms responsible for food deterioration. More often than not they involve organic acids and their salts, *propionic acid*, *potassium sorbate*, *sodium and calcium propionates*, and *sodium benzoate*, traditionally used to preserve cheese, jam, cakes, and fatty materials.

Here again a few large companies such as Monsanto for sorbates and Pfizer for propionates have acquired a leading place on the markets. On the whole, the food additives sector is less open to restructuring and rationalization because it is made up of enterprises that are fundamentally different in size, technologies and in objectives pursued.

Photochemicals

Since the early 1980s, the major photographic companies have made efforts to bring changes to their basic technologies, which had long remained unchanged. The U.S.-based *Eastman Kodak* became interested in reprography, setting up its own range of photocopying machines. It also became involved in electronics and video to counter competition from new Japanese equipment (such as Sony's Mavica filmless cine-camera). The other photographic giants like Bayer's subsidiary *Agfa-Gevaert* and *Fuji Photo Film* have also invested heavily in new areas, the former in magnetic tape and reprography and the latter in photo disks. *Polaroid*, whose founder Edwin Land remained to the day of his retirement an advocate of specialization, is also starting to put a range of videocassettes on the market. These changes, however, are essentially intended for the amateur and mainly concern camera manufacturers. Overall, the sensitive surfaces market should receive no shakeup from these new ventures, for there is still a high demand in a number of areas where photography remains irreplaceable (press and publishing, scientific research, industrial applications, radiographic control devices). The industry's structure reflects this stability. It is not likely to be upset in the immediate future because of the power acquired by the few large multinationals, which vie with one another on international markets and give any newcomer little chance of success.

Unable to compete with Kodak on the American market, *GAF* withdrew from the film industry in 1982. Previously, the first European merger had taken place between Belgium's Gevaert and Germany's AGFA, producing Agfa-Gevaert. Its early years were hard ones, and it is now fully owned by Bayer. Italy's *Ferrania* was taken over by America's *3M*, while Ciba-Geigy was bringing together Britain's *Ilford*, France's *Lumière* and Switzerland's *Telko* within the Ilford group based in Britain.

Following these restructurings, which, in many cases, took place sometime ago, the photographic film industry is now dominated by three giants: *Eastman Kodak*, with units in Rochester, New York, in the United States, in Châlons, France, and in Hemel Hempstead, England; *Agfa-Gevaert*, which produces its photochemicals in Antwerp, Belgium, and Vaihingen, West Germany; and *Fuji Photo Film*, which produces in Japan and has recently set up a film unit in Holland. *Ilford* and *Polaroid*, which went through difficult periods of readaptation; *3M*, which is involved in other areas besides photography; *Konishiroku* (*Konika*) in Japan, which bought Fotomat in the United States, cannot be regarded as dangerous rivals to the Big Three.

The three major companies follow different policies in matters of raw materials. Fuji Photo film, which has no links with the chemical industry, buys 80 percent of its supplies outside, while Eastman Kodak and Agfa-Gevaert supply half their needs through their own production. They all produce their most ''sensitive'' organic derivatives, which are kept secret since they form the basis of emulsion quality.

Although polyester film, introduced by Du Pont under the trade name Mylar in the sixties, has been added to the traditional supports like paper and cellulose acetate, the principle of photographic film preparation has remained unchanged since "daguerrotype" was developed. The sensitive surface always contains a silver halide crystal emulsion with a *gelatin* binder. Despite all the efforts to replace them, silver salts remain the basis of these emulsions, and film manufacturers still require gelatin, which they consume at the rate of 20,000 tons a year. The suppliers are few, and they are carefully selected. The world leader in this area is *Rousselot*, now a subsidiary of Sanofi Elf Aquitaine, with four units in Europe and one in the United States.

Reducing agents such as *hydroquinone*, *metol* (*p*-methylaminophenol) and *p-phenylenediamine* are generally purchased from outside producers. Eastman Kodak produces its own hydroquinone, however. The other producers get their supplies from Rhône-Poulenc or from Japanese firms like Sumitomo Chemicals or Mitsui Petrochemicals.

Color photography, now fully perfected, requires a *developer* like N,N′-diethylphenylenediamine which reacts with silver salts. The oxidized derivative obtained reacts with a coupling agent made up of groups ($-CH=$) or ($-CH_2-$) to produce the desired color.

Formulations for sensitive surface emulsions also include *accelerators* (alkaline carbonates, borax), *stabilizers* (sodium bromide, benzotriazoles), *conservation agents* (sodium sulfite), *hardeners*, which improve gelatin behavior (chloromucic acid, substituted 2,4-dichlorotriazines).

The great variety of products used by film manufacturers, their stringent quality standards, and their secretiveness, which prevents them from subcontracting their most advanced formulations, are all factors combing to keep photochemical producers apart from the rest of the chemical industry and keep newcomers out of their sector.

The Alliance of Chemicals and Electronics

Chemical products used today in electronics seem, at first glance, to be a very ordinary kind. They are different from those generally offered, however, by reason of the extraordinary degree of purity which their producers must achieve in order to satisfy the stringent requirements of the electronics industry. The maximum dose of impurities tolerated in monocrystalline silicon amounts to one part in 10^{13}.

Polycrystalline silicon, produced from silane (SiH_4) or trichlorosilane ($SiHCl_3$) forms the upstream part of the *semiconductor sector*. Monocrystalline silicon is extracted from polycrystalline silicon and sliced into wafers 25 microns thick and 8 to 10 centimeters in diameter.

Hoechst's subsidiary Wacker is the world's leading polycrystalline silicon producer, with a capacity exceeding 2,000 tons. The overcapacities that began affecting the electronics industry in the early 1980s forced Monsanto, one of the largest wafer producers, to slow down its silicon production units in 1984. It has since sold this business to Germany's Hüls. Rhône-Poulenc, which had ambitions in the area but lacked the right technology, has withdrawn from the business.

There are enough suppliers of this type of silicon, including, for instance, Dow-Corning, Dynamit Nobel, Shin-Etsu, Tokuyama Soda, Motorola, and Texas Instruments. A possible substitute for the silicon used to produce wafers is *gallium arsenide*, in which Rhône-Poulenc, ICI, and Shinetsu are already involved.

This situation shows how closely suppliers of electronic chemicals need to monitor the very rapid developments taking place in the area; otherwise, their productions run the risk of becoming obsolete before the full payoff.

Photosensitive products are also used for the production of wafers. These *photoresists* polymerize through X-ray treatment. They are called positive or negative according to whether or not they are soluble in solvents when exposed to light. The miniaturization of printed circuits tends to give a boost to positive resins. Germany's Hoechst has pioneered in such photosensitive resins. They are also supplied by Eastman Kodak, Olin Hunt, Ciba-Geigy, E. Merck and Tokyo Ohka Kogyo.

A great number of chemical firms have set up special divisions to manufacture products for the electronics industry, essentially through acquisitions. For example, *Du Pont* bought *Berg Electronics* in 1972, and a little later *Olin* purchased *Philip A. Hunt*. Some companies, such as Du Pont, Olin, and Ciba-Geigy, have chosen an "integrated systems" approach in this area by providing as wide a range as possible of products and services for the electronics industry. Others have elected to remain strictly within the special areas in which they excel through long experience or proper chemical integration. Thus it was the work carried out before the war with AEG that led *BASF* to make its range of magnetic tapes and gave it the supremacy in chromates which it shares with Du Pont. *Hoechst* came to silicon through Wacker and to gases through Messer Griesheim, and now provides, besides high-purity special gases, a range of photosensitive polymers. Rhône-Poulenc became involved in printed circuits through its polyimide resins and Ciba-Geigy through its epoxy resins.

Most of the companies already producing diethylene glycol terephtalate polymers have launched into the applications of polyester film to video and data processing, Hoechst through its Kalle subsidiary, ICI, Rhône-Poulenc, Du Pont, Japan's Toray, Teijin, and Toyobo, the latter in association with Rhône-Poulenc in Nippon Magphane.

Although Rhône-Poulenc has given up direct upstream development after fruitless association with Dysan in magnetic supports and Siltec in silicon, it still believes it can use its knowhow in rare earths to develop their electronics applications. Today, Rhône-Poulenc is the indisputable leader in rare earths, accounting for 40 percent of the world market. At its units in La Rochelle, France, and Freeport, Texas, it is capable of extracting from lanthanide sands the fourteen elements they contain. Over the last few years, *samarium*, for instance, has become essential for microelectronics to the same degree that *europium* and *yttrium oxides* already are for color television.

Whether they approach electronics directly, or through chemicals, or both, chemical companies involved in this business can hope to reap the fruits of their efforts in this area, providing, however, that the sector is spared the technological and economic jolts it has suffered over the past ten years.

Catalysts

Ever since England's Humphry Davy observed in the early 1800s that water was formed when hydrogen and oxygen react in the presence of a red-hot platinum wire, the phenomenon which Berzelius was to call catalysis has intrigued chemists. The uses of catalysts in industry were first consciously demonstrated by Peregrine Phillips in 1832 when he used platinum to oxidize sulfur dioxide (SO_2) to form sulfur trioxide (SO_3) and by Frédérick Kuhlmann in 1837, when he produced nitric acid from ammonia.

Early in the twentieth century, Germany's Wilhelm Ostwald, France's Paul Sabatier, and America's Irving Langmuir had advanced a step in interpreting the phenomenon of catalysis by showing that it was characterized by an acceleration of the rate of reactions and that it was conditioned by the state of the catalyst's surface. From then on, chemical technology made striking progress through use of catalysts. Between 1905 and 1920, and more particularly in Germany, there was a spurt of new industrial-scale processes, for example, Fischer-Tropsch synthesis and BASF use of vanadium oxide to produce sulfuric acid.

It is no exaggeration to say that without catalysts Germany would have been in no condition to pursue its war effort until November 1918. Likewise, if Houdry had not developed in the early days of World War II, its "catalytic cracking" process, the United States would have found it very hard to provide its bombers with light fuel. It was also through catalytic reforming that the United States managed to obtain from petroleum the toluene needed to produce TNT between 1941 and 1945.

Since then, catalysts have played an essential role, particularly in the production of ethylene oxide from ethylene (Shell, Scientific Design), in the synthesis of hydrogen cyanide and acrylonitrile through ammoxidation (oxidation in the presence of ammonia), of formaldehyde (from oxidation of methanol), and, of course, in the polymerization reactions to produce plastics, elastomers, and synthetic fibers. It is not surprising, in the circumstances, that a catalyst industry should have developed after World War II through internal growth or through acquisitions. The very diversity of catalysts and of their uses has necessarily led to a fragmented sector.

Some oil companies became involved in the production of catalysts because they needed them in their own refineries. *Mobil* has developed the ZSM 5 catalyst based on zeolite following studies which began as early as 1936 on catalytic cracking; *Shell* has used its own technology to develop the sales of its catalysts for hydrogenation cracking. Other companies became involved in catalysts because of their precious metals business. *Johnson Matthey*, *Engelhard*, and *Degussa* applied their knowhow in platinum metals to industrial catalysts. Chemical firms, for their part, approached the area in different ways. *ICI* made the most of its acquired knowhow, particularly in methanol and ammonia, by associating with *Nalco* to form *Katalco*, a catalyst supplier; *American Cyanamid* has set up a subsidiary in Holland with Ketjen; *Rhône-Poulenc* has formed *Procatalyse* in joint venture with Institut Français du Pétrole.

In other cases, the involvement in catalysts has been through acquisitions. *W. R. Grace* bought *Davison Chemical* in 1953, and in 1984 Union Carbide purchased *Katalistics International*, BV. One of the three leading United States companies in cracking catalysts, together with Engelhard and Davison, is *Harshaw-Filtrol*, which is the result of the merger of subsidiaries of Gulf Oil and Kaiser Aluminum & Chemical that have specialized in the area.

The developers of new processes have found it at times more expedient to set up their own separate entities to supply the catalysts they were advocating. Allied-Signal's subsidiary *UOP* did so for its platforming; *Houdry* for its catalytic cracking; *Ralph Landau* for the silver catalyst used for direct ethylene oxidation, which was marketed by *Halcon SD* and subsequently taken over by Denka, then by Bayer; and *Phillips Chemical* for its polyolefin catalysts, sold through its subsidiary, Catalyst Resources.

Through inert supports, a number of firms have succeeded in creating a niche in catalysts—for instance, *Crosfield*, a subsidiary of Unilever in England and a silica producer; or the German *Südchemie* group, which specializes in hydrogenation and polymerization catalysts; or again *Condea*, which produces in West Germany alumina of high purity. The sector also includes a few firms which are only involved in a very special sector. Denmark's *Haldor Topsoe* makes catalysts for the synthesis of ammonia and methanol; and *Lithium Company of America*, an FMC subsidiary, produces lithium, while *Du Pont* makes boron derivatives.

Linked to the oil industry, to petrochemicals, and to the large commodity chemicals, the catalyst industry can hardly escape the economic ups and downs affecting these three large sectors. Its clients are understandably both demanding and prudent, for the catalytic system is basic to the good running of production units. This explains why it is an area of business that is so difficult to penetrate and run profitably. Its structure should therefore remain rather stable even with the development of catalytic exhaust systems. Introduction in the United States and in Europe of unleaded gasoline and the use of bimetallic systems for catalytic reforming should open up new markets for platinum and rhodium.

Retrospect and Prospect

The economic slump that started in 1973 when OPEC pushed up crude oil prices challenged what were until then regarded as indisputable truths.

First came the realization that just as no tree can climb as high as the sky, so *no growth can be guaranteed to be continuous*. Suddenly investments made at a time of high inflation and low interest rates turned out to be disastrous as demand slowed down simultaneously with monetary erosion.

The scale effect, which until then was assumed to be cost-saving, showed its weaknesses as the giant steamcrackers proved more expensive to run at low capacity than smaller units already written off and working at full capacity.

The notion that production costs could be improved by grabbing a greater share of the market turned out to be fatal as the gain in sales was wiped out by severe price erosion.

Likewise the assumption that the fruits of research would be proportionate to the funds devoted to the sector was totally invalidated, for never had the world's chemical industry spent so much money in *research and development* to so little avail. At the same time, the venture capital poured into biotechnology companies has yet to bring in the returns expected. The managers of chemical plants, wary of world petrochemical and heavy chemical overcapacities, believed they would find in a switch to *specialties* at least partial compensation for the losses incurred through traditional productions. Although they were not all disappointed in their hopes, some of them found that results obtained fell short of expectations, for until then specialties had been the special field of firms that had acquired experience in what were specific and as yet uncrowded sectors.

Manufactured by too many producers, some specialties were becoming commonplace. For a manufacturer, there are only two kinds of products, *those that make money* and *those that do not*. The profits that can be made on a sale are closely related to the number of producers on the market and to the day-to-day relationship between supply and demand. Whether a product is deemed a "commodity" or a "specialty," it is all the more profitable for its being offered by a smaller number of producers for a demand that remains unchanged. In this context, there are some pharmaceutically active materials protected by patents and some secret formulations that are genuine profit centers for their producers. Likewise, should a base product become scarce on the market because of an accident on a petrochemical site or because of sudden high demand, prices soar and the fortunate producer can turn out the product to maximum capacity and profit.

Over the last few years, the high cost of installations and of the money needed to finance them was not conducive to the building of new plants on any large scale in industrialized countries. But as demand trends have been moving upwards lately, *petrochemicals* have at long last *returned to profitability*.

The specialties rush of chemical leaders is, on the contrary, more likely than not to produce a surfeit of products, at least insofar as some specialties are concerned. These will shed their "added value," and consequently lose their attraction for the too numerous industry leaders that had decided to follow that path.

Other disappointments are likely to come from the organizational and managerial differences between a purchaser and the specialties firm acquired. The many divestments that have often followed upon hasty acquisitions show how difficult it is to force on an entrepreneurial company the management methods of a large multinational.

One of the paradoxes of the last few years has precisely been that specialties suitable for medium-sized firms capable of being flexible in their approach to daily matters should have fallen into the hands of chemical giants with necessarily heavier structures, while in the United States, for instance, through the "leveraged buyout" procedure, a few strong-minded individuals have succeeded in taking over large petrochemical and thermoplastic production units considered until then as the rightful field of the industry's greats.

It is not certain that the errors of the past will not be repeated in the future. The thirst for power could indeed lead some company heads to overinvest, especially if they have public funds at their disposal. They would then recreate the overcapacities that have been so harmful to fertilizer, petrochemicals, synthetic fibers, and plastic producers over the last few years. It is also likely that specialties will continue to attract industry leaders anxious to develop fresh prospects.

Let us hope that all the decision-makers will bear in mind that capital funds, whether provided involuntarily by the taxpayer or willingly by the shareholder, are a rare resource that must be judiciously allocated, and that success in all things comes from mastery acquired through long patience. In this respect, Germany's chemical industry, which has shown continuity from the time it was established in the last century to the present under the guidance of professionals, is a tried and tested model, showing profits even in the most adverse circumstances.

Drawing inspiration from this example for long years, the United States chemical industry, under the pressure of financial analysts and raiders, has in recent years undergone many upheavals. While they provided new opportunities for the fortunate few, they changed the environment and made people forget that to operate efficiently any industry must set its sights on the long term.

For reasons that were more political than financial, France's and Italy's chemical industries have also undergone too frequent changes over the past twenty years—in their structures, their strategies, and their management teams—to have had a chance of getting through the economic slump unscathed. It is only very recently that they have returned to profits by recovering a measure of stability.

Worldwide, in 1987 through 1989 the industry, whether in specialties or in basic chemicals, has certainly had its most prosperous years ever. The chemical industry, on the whole, *does not, however, enjoy a very favorable image* in the eyes of the public. The harmful spillovers caused by untoward accidents are given wider publicity by the media than the benefits the industry provides. In consequence, administrations that were anxious to soothe the more or less justified fears of their citizens, have brought out a *spate of regulations* often more restraining and therefore more costly than is really necessary.

Since one cannot work simultaneously toward a thing and its opposite, no great spate of discoveries useful to humanity should be expected at a time when everything is being done to make it difficult to bring new products onto the market. For a long time the chemical industry was left free to apply its own safety standards and could devote most of its time to the development of new products. In the last few years, it has had to submit to increasingly costly and prolific rules and regulations that require its attention and delay the development of innovations that could save human lives or at least improve our living conditions. Some balance will have to be found between safety requirements and the wider interest of the public.

As in all history, the story of chemicals recalls past events and makes an attempt to explain them. But it can neither create them nor prevent them from recurring. While such history, therefore, teaches us the essential facts that have taken place within two richly endowed centuries, it does not tell us which major facts will form the threads of the next years. It is this unknown factor which makes up the spice of our professional life. We can at least hope that if we conform to reason, to ethics, and to scientific and economic laws for all that is within our scope, each of us will have served this wonderful science that is chemistry to the best of our capacities and in the interests of the greatest number of people.

Appendix

The 250 Leaders of the World Chemical Industry
(Millions of 1988 U.S. dollars)

Rank/Group[a]	*Nationality*[b]	*Turnover chemicals*	*Turnover global*	*Net profit*	*Investments*	*R & D expenditures*	*Total payroll*
1 Bayer	WG	22,694	22,824	1,076	1,774	1,387	165,700
2 Hoechst	WG	21,948	23,105	1,136	1,568	1,363	164,527
3 BASF	WG	21,543	24,743	795	1,971	1,010	134,834
4 ICI	GB	21,125	21,125	1,591	1,465	1,020	130,400
5 Du Pont de Nemours	USA	19,608	32,917	2,190	4,207	1,319	140,949
6 Dow Chemical	USA	16,659	16,682	2,398	1,267	772	55,500
7 Unilever	GB-NL	12,338	30,950	1,509	1,504	600	291,000
8 Royal Dutch Shell	GB-NL	11,848	79,643	5,311	7,665	773	134,000
9 Ciba-Geigy	SWI	11,018	11,753	883	1,076	1,197	88,757
10 Procter and Gamble	USA	(11,000)	21,400	1,210	1,030	–	72,000
11 Rhône-Poulenc	F	10,802	10,802	571	1,584	632	79,670
12 Exxon	USA	9,892	87,542	5,260	7,842	551	101,000
13 Union Carbide	USA	8,324	8,324	662	671	59	43,992

Source: *Chimie actualités* (October 1989).

Rank/Group[a]	*Nationality*[b]	*Turnover chemicals*	*Turnover global*	*Net profit*	*Investments*	*R & D expenditures*	*Total payroll*
14 Elf Aquitaine	F	8,216	20,848	1,191	2,219	467	72,183
15 Mitsubishi Kasei Corp.	J	(8,095)	5,401(*)	103	–	424	19,000
16 Akzo	NL	7,846	8,283	421	634	405	71,100
17 Montedison	I	7,725	10,763	481	817	395	47,115
18 Monsanto	USA	7,453	8,293	591	436	590	45,635
19 Solvay	B	6,836	6,836	408	639	291	44,301
20 Eastman Kodak	USA	6,724	17,034	1,397	1,914	1,147	145,300
21 Sumitomo Chemical	J	6,532	7,178	313	362	347	–
22 Sandoz	SWI	5,960	6,761	507	557	607	48,079
23 Merck & Co.	USA	5,939	5,939	1,210	–	750	–
24 British Petroleum	GB	5,876	59,773	2,185	5,922	481	128,450
25 Asahi Chemical Industry	J	5,799	7,696	263	–	–	22,908
26 Henkel	WG	5,782	5,782	198	386	174	35,943
27 EniChem	I	5,592	5,592	378	560	–	27,690
28 Bristol-Myers	USA	(5,500)	5,972	829	–	–	–
29 Hoffmann-La Roche	SWI	5,416	5,788	427	434	806	49,671
30 Dainippon Ink and Chemicals	J	5,412	5,412	56	–	–	–
31 Takeda Chemical Ind.	J	5,219	5,219	295	–	–	–
32 Toray Industries	J	5,208	5,985	255	481	–	26,211
33 Showa Denko	J	5,076	5,076	148	–	–	–
34 3M	USA	(5,000)	10,581	1,154	–	690	82,800
35 Lyondell Petrochemical	USA	4,700	4,700	543	–	–	–
36 DSM	NL	4,689	5,056	311	449	179	28,625
37 Hüls	WG	4,645	4,645	213	831	171	27,167
— Atochem	F	4,640	4,640	365	385	100	14,218
38 Occidental Petroleum	USA	4,617	19,417	302	1,073	–	52,500
39 Cyanamid	USA	4,592	4,592	306	374	365	35,501
40 Norsk Hydro	NO	4,573	9,151	523	1,347	113	39,000

Rank/Group[a]	*Nationality*[b]	*Turnover chemicals*	*Turnover global*	*Net profit*	*Investments*	*R & D expenditures*	*Total payroll*
41 Beecham	GB	4,532	4,532	544	–	243	29,000
42 Abbott Laboratories	USA	4,500	4,937	752	–	455	39,000
43 American Home Products	USA	(4,500)	5,500	932	–	–	–
44 Colgate-Palmolive	USA	(4,500)	4,734	318	–	–	–
45 Johnson and Johnson	USA	(4,500)	9,000	974	–	–	–
46 Pfizer	USA	4,485	5,385	791	344	473	40,900
47 Kao	J	4,335	4,335	133	569	186	–
48 Amoco	USA	(4,300)	23,930	2,060	–	–	–
49 Teijin	J	4,210	4,210	155	–	–	–
50 Eli Lilly	USA	4,070	4,070	761	–	–	–
51 Mobil	USA	3,922	54,361	2,087	3,194	230	69,600
52 L'Oréal	F	3,906	4,042	217	199	181	27,570
53 Sekisui Chemical	J	3,832	3,832	138	–	–	–
54 The BOC Group	GB	3,792	4,627	328	531	84	38,813
55 Glaxo Holdings	GB	3,705	1,912	1,049	497	415	26,423
56 Kanebo	J	3,660	3,660	22	–	–	–
57 Grace	USA	3,579	5,786	334	418	119	45,700
58 Orkem	F	3,561	3,561	412	172	54	13,492
59 L'Air Liquide	F	3,539	4,238	261	380	120	27,000
60 Formosa Plastics	TAIW	(3,500)	5,400	–	–	–	–
61 SmithKline-Beckman	USA	3,497	4,749	229	–	570	41,000
62 General Electric	USA	3,345	49,681	3,386	–	–	–
63 Shiseido	J	3,210	3,210	76	–	–	–
64 Asahi Glass	J	3,160	7,896	369	–	224	–
65 W. R. Grace	USA	3,150	5,786	234	–	–	47,500
66 Avon Products	USA	3,063	3,063	–404	–	–	–
67 Ube Industries	J	3,062	3,062	48	–	–	–
68 PPG Industries	USA	3,053	5,617	468	410	239	36,300
69 Allied-Signal	USA	3,033	11,909	463	602	647	109,550

Rank/Group[a]	Nationality[b]	Turnover chemicals	Turnover global	Net profit	Investments	R & D expenditures	Total payroll
70 Baxter Travenol Lab.	USA	(3,000)	6,861	388	499	238	64,000
71 Mitsui Toatsu Chemicals	J	2,991	2,991	81	176	150	–
72 Chevron	USA	2,976	27,722	1,768	2,459	–	53,675
73 Schering	WG	2,974	2,974	89	232	355	24,685
74 Schering-Plough	USA	2,969	2,969	390	–	–	–
75 Courtaulds	GB	2,907	4,713	463	313	–	63,200
76 Hercules	USA	2,802	2,802	120	251	145	22,718
77 Upjohn	USA	2,746	2,746	353	–	382	–
78 Arco Chemical	USA	2,700	2,700	494	199	39	–
79 Degussa	WG	2,663	7,673	82	253	211	–
80 Quantum Chemical Corp.	USA	2,571	3,274	383	334	38	–
81 Petrofina	B	2,544	13,127	544	1,740	–	23,000
82 Rohm & Haas	USA	2,535	2,535	230	338	156	12,444
83 Phillips Petroleum	USA	(2,500)	11,304	650	950	–	–
84 S. C. Johnson	USA	2,500	2,500	–	–	–	–
85 Warner-Lambert	USA	(2,500)	3,908	340	–	260	–
86 Yokohama Rubber	J	2,488	2,488	44	–	–	–
87 Nobel Industries S.	S	2,458	3,501	107	177	156	22,101
— Sanofi	F	2,408	2,408	127	158	222	19,851
88 Toyobo	J	2,398	2,398	55	–	–	–
89 Sabic	S ARAB	(2,100)	2,368	902	90	–	8,219
90 Sankyo	J	2,345	2,345	–	175	–	–
91 Boehringer Ingelheim	WG	2,342	2,342	41	141	381	22,206
92 Wellcome	GB	2,258	2,258	230	239	295	20,236
93 RWE	WG	2,256	15,150	432	440	–	72,800
94 Air Products and Chem.	USA	2,237	2,432	214	542	72	13,300
95 Mitsubishi Petrochem. Ind.	J	2,221	2,221	126	215	159	5,144
96 Shionogi	J	2,216	2,216	90	–	–	–

Rank/Group[a]	*Nationality*[b]	*Turnover chemicals*	*Turnover global*	*Net profit*	*Investments*	*R & D expenditures*	*Total payroll*
97 Squibb	USA	2,216	2,586	426	–	294	16,000
98 Kemira	FI	2,162	2,229	60	349	–	14,857
99 Otsuka Pharmaceutical	J	2,148	2,148	75	–	–	4,106
100 Borg Warner	USA	2,145	2,145	419	–	–	–
101 Revlon	USA	2,100	3,000	–	–	–	–
102 Ashland Oil	USA	(2,000)	7,753	224	–	–	–
103 Mitsui Petrochemical Ind.	J	1,991	1,991	122	–	–	–
104 Rütgerswerke	WG	1,964	2,257	25	–	–	12,700
105 B. F. Goodrich	USA	1,930	2,417	196	–	–	–
106 Unitika	J	1,898	1,898	21	–	–	–
107 Shin-Etsu Chemical	J	1,879	1,879	83	–	–	–
108 Borden	USA	1,862	7,244	312	–	–	46,000
109 E. Merck	WG	1,839	1,839	91	143	143	21,017
110 Tanabe Seiyaku	J	1,761	1,761	72	96	127	–
— Roussel Uclaf	F-WG	1,751	1,751	84	73	206	14,759
111 Goodyear	USA	1,735	10,810	350	492	–	–
112 Sequa	USA	1,713	1,713	69	–	–	–
113 Fujisawa Pharmaceutical	J	1,686	1,709	64	58	166	–
114 EMC	F	1,684	2,645	40	118	27	13,136
115 Snia-PBD	I	1,649	1,798	64	–	–	13,094
116 AECI	SO AF	1,640	1,640	–	–	–	–
117 Petrobras	BR	1,623	16,486	265	2,172	54	59,210
118 Tosho Corporation	J	1,611	1,946	47	403	–	–
119 Kuraray	J	1,604	1,604	21	–	–	–
120 Nova	CAN	1,575	3,310	356	–	–	–
121 Reckitt and Colman	GB	1,562	2,510	217	–	–	–
122 Beiersdorf	WG	1,541	1,943	55	196	–	17,998
123 Daiichi Pharmaceutical	J	1,526	1,526	148	149	98	4,646

Rank/Group[a]	*Nationality*[b]	*Turnover chemicals*	*Turnover global*	*Net profit*	*Investments*	*R & D expenditures*	*Total payroll*
124 Mitsubishi Rayon	J	1,523	1,523	33	–	–	–
125 Wacker-Chemie	WG	1,518	1,518	34	157	120	12,074
126 Mitsubishi Gas Chemical	J	1,515	1,515	65	–	–	–
127 FMC	USA	(1,500)	3,287	129	–	–	–
128 Olin	USA	(1,500)	2,310	98	–	–	–
129 Texaco	USA	(1,500)	35,138	1,304	–	–	–
130 Witco	USA	1,500	1,586	72	–	–	–
131 Yamanouchi Pharmaceutical	J	1,481	1,481	204	–	–	–
132 International Minerals & Chem.	USA	1,471	1,471	113	–	–	–
133 Cookson Group	GB	1,445	2,813	208	116	–	14,351
134 Japan Synthetic Rubber	J	1,444	1,444	–	–	–	–
135 Eisai	J	1,417	1,417	84	–	–	–
136 Neste	FI	1,409	6,434	102	863	–	10,855
137 Huntsman Chemical	USA	1,400	1,400	–	–	–	–
138 Morton International	USA	1,357	1,407	97	132	38	8,400
139 Kyowa Hakko Kogyo	J	1,352	2,202	107	583	110	–
140 Ethyl	USA	(1,300)	2,011	231	–	–	–
141 Syntex	USA	1,271	1,271	297	111	218	10,000
142 Wella	WG	1,241	1,241	40	–	–	–
143 Ecolab	USA	1,212	1,212	44	–	–	–
144 Unocal	USA	1,204	10,085	480	1,100	52	182,035
145 Tenneco (A & W)	USA	1,183	13,234	822	–	–	–
146 Thiokol Corp	USA	1,168	1,168	18	–	–	–
147 Lonza	SWI	1,164	1,196	26	142	55	7,656
148 Lubrizol	USA	1,126	1,126	140	55	65	–
149 Taisho Pharmaceutical	J	1,119	1,119	–	–	–	–
150 Pharmacia	S	1,115	1,115	161	–	–	–
151 AGA	S	1,106	1,610	91	253	–	13,543
152 IMC Fertilizer Group	USA	1,086	1,086	112	–	–	–

Rank/Group[a]	Nationality[b]	*Turnover chemicals*	*Turnover global*	*Net profit*	*Investments*	*R & D expenditures*	*Total payroll*
153 Israël Chemicals	ISR	1,080	1,080	47	100	–	–
154 Aristech Chemical	USA	1,065	1,065	188	–	–	–
155 Georgia Gulf	USA	1,061	1,061	194	–	–	–
156 Repsol	SP	1,059	8,251	498	–	–	18,000
157 Röhm	WG	1,052	1,052	61	67	35	–
158 Rorer Group	USA	1,042	1,042	62	–	100	–
159 Henley	USA	1,036	1,036	−47	–	–	–
160 Pennwalt	USA	1,024	1,024	179	–	–	5,000
161 Gechem	B	1,021	1,196	−137	–	–	9,943
162 Ferro	USA	1,009	1,009	106	–	–	–
163 NL Chemicals	USA	1,007	1,007	163	–	–	–
164 Boehringer Mannheim	WG	1,003	1,003	34	75	186	8,651
165 Cabot	USA	(1,000)	1,677	61	–	–	–
166 Astra	S	998	1,031	111	84	198	6,977
167 Nalco Chemical	USA	994	994	106	62	–	–
168 Chugai Pharmaceutical	J	985	998	58	54	147	3,731
169 GAF	USA	961	961	100	–	–	–
170 Total CFP	F	959	13,771	245	2,773	100	41,862
171 Boots	GB	947	4,883	378	–	–	–
172 Gist-Brocades	NL	942	942	48	115	47	6,459
173 UCB	B	942	942	38	84	45	7,602
174 A. H. Robins	USA	934	934	58	–	–	–
175 Laporte Industries	GB	932	932	96	40	–	4,600
176 Marion Laboratories	USA	930	930	–	–	–	–
177 Harrisons & Crosfield	GB	903	3,070	–	–	30	–
178 Chemie Holding	A	895	895	–	69	53	5,509
179 CEPSA	SP	879	3,123	107	–	–	–
180 ERT	SP	879	1,605	46	–	34	–

Rank/Group[a]	*Nationality*[b]	*Turnover chemicals*	*Turnover global*	*Net profit*	*Investments*	*R & D expenditures*	*Total payroll*
181 Sherwin-Williams	USA	850	1,950	101	–	–	–
182 Perstorp	S	844	844	45	89	29	5,991
183 Benckiser	WG	841	841	40	–	–	–
184 IFF	USA	840	840	129	–	–	–
185 RTZ	GB	835	9,015	772	1,347	33	82,078
186 Dexter	USA	827	827	35	28	32	–
187 Vista Chemical	USA	781	781	113	–	–	–
188 Dainippon Pharmaceutical	J	780	780	–	–	–	–
189 Nippon Zeon	J	780	780	–	–	–	–
190 Novo	DK	766	766	93	87	98	6,002
191 VIAG	WG	761	5,329	123	620	–	33,427
192 PCD	A	734	734	–	34	–	1,758
193 UniRoyal Chemical	USA	734	734	–	–	–	–
194 Fisons	GB	729	1,487	–	–	–	–
195 Banyu Pharmaceutical	J	728	728	52	–	–	–
196 Metallgesellschaft	WG	705	8,593	88	190	–	25,132
197 Sterling Chemicals	USA	699	699	214	–	–	–
198 The Green Cross Corp.	J	696	696	19	17	70	–
199 Indian Petrochemicals	IND	686	686	–	–	–	–
200 H. B. Fuller	USA	685	685	21	40	15	–
201 Engelhard	USA	675	2,351	64	–	–	–
202 Statoil	NO	670	7,894	52	1,503	–	11,167
203 Linde	WG	664	2,632	36	215	62	21,222
204 LVMH	F	653	2,718	331	–	–	–
205 Coates Brothers	GB	623	623	–	16	–	–
206 Feldmuehle Nobel	WG	620	4,456	89	181	–	33,500
207 Th. Goldschmidt	WG	620	620	20	21	–	–
208 Williams Holdings	GB	616	1,492	154	–	–	–

Rank/Group[a]	Nationality[b]	Turnover chemicals	Turnover global	Net profit	Investments	R & D expenditures	Total payroll
209 Great Lakes Chemicals	USA	616	616	103	47	28	–
210 Altana	WG	612	1,092	27	47	–	8,695
211 Servier	F	580	580	–	–	133	4,000
212 Ono Pharmaceutical	J	561	561	93	–	–	–
213 Croda International	GB	551	630	45	33	–	5,217
214 Lafarge-Coppée	F	550	3,750	311	253	49	22,491
215 Kerr-McGee	USA	550	2,689	110	–	–	–
216 Smith & Nephew	GB	542	1,080	159	76	–	13,832
217 Toho Rayon	J	540	540	5	–	–	–
218 Ajinomoto	J	531	3,876	120	274	40	9,532
219 Kabi	S	523	530	104	40	74	3,195
220 Noxell	USA	522	522	51	–	–	–
221 Carter-Wallace	USA	515	515	45	–	–	–
222 Schwarzkopf	WG	508	508	–	40	–	–
223 Yoshitoni Pharmaceutical	J	503	503	12	–	–	–
224 SNPE	F	489	557	8	30	33	7,120
225 Helen Curtis	USA	489	489	11	–	–	–
226 Süd Chemie	WG	468	468	18	49	–	3,587
227 Fermenta	S	455	455	18	–	–	–
228 Betz Laboratories	USA	448	448	48	44	16	–
229 Yves Saint-Laurent	F	441	441	–	–	–	–
230 EMS Chemie	SWI	434	434	43	42	–	2,400
231 Kumiai Chemical Industry	J	431	431	7	7	19	–
232 Turner and Newall	GB	422	1,894	127	111	–	40,859
233 Ares-Serono	SWI-USA	421	421	48	–	–	3,100
234 Loctite	USA	417	417	42	–	–	–
235 Orion	FI	410	741	–	–	–	6,050
236 Quimigal	P	410	682	–	–	–	–

Rank/Group[a]	*Nationality*[b]	*Turnover chemicals*	*Turnover global*	*Net profit*	*Investments*	*R & D expenditures*	*Total payroll*
237 Union Camp	USA	404	2,661	295	346	44	18,508
238 Octel	GB	400	400	30	–	–	3,000
239 Hexcel	USA	399	399	16	–	–	–
240 Grow Group	USA	383	383	0.04	–	–	–
241 Teva	ISR	381	381	22	–	–	–
242 Groupe Roullier	F	380	440	15	16	–	1,964
243 Mochida Pharmaceutical	J	378	378	17	–	–	–
244 Pierre Fabre	F	372	372	–	–	35	3,300
245 The Burmah Oil	GB	368	2,467	165	117	21	10,919
246 Cultor	FI	353	979	29	91	–	4,226
247 Gujarat State Fert.	IND	349	349	–	–	–	–
248 Fresenius	WG	347	347	–	23	–	–
249 EPSI	P	346	346	14	–	–	496
250 Genentech	USA	335	335	21	–	133	–

[a] *Notes on individual groups*

AIR PRODUCTS, ASHLAND, STERLING CHEMICALS, VISTA CHEMICAL: turnover on Sept. 30 1988
DAINIPPON PHARMACEUTICAL: estimated turnover for 12 months
EASTMAN KODAK: including Sterling Drug
GENERAL ELECTRIC: including Borg Warner (turnover $2,145 millions, net profit $419 millions)
IMC: at the closing on June 30 1988
MARION LABORATORIES: at the closing on June 30 1988
MORTON & THIOKOL: split of Morton-Thiokol in 1989
MITSUBISHI KASEI: turnover for chemicals estimated over a 12-month period
NOVA: including Polysar as of July 1988
PROCTER & GAMBLE: closing on June 30 1989. On June 30 the turnover reached $19.34 billions
QUANTUM CHEMICAL: formerly NDCC
SEQUA CORPORATION: formerly Sun Chemical
SMITHKLINE-BECKMAN: before the merger with Beecham
SQUIBB: merger with Bristol-Myers in 1989

SYNTEX: closing on July 31 1988

KUMIAI: closing on Oct. 31 1988—**SHISEIDO**: closing on Nov. 30 1988

ASAHI GLASS, CHUGAI PHARMACEUTICAL, GREEN CROSS, KYOWA HAKKO, SHOWA DENKO, SUMITOMO CHEMICAL and **YAMANOUCHI**: closing on Dec. 31 1988

Closing on March 31, 1989, for the other Japanese groups.

[b] *Abbreviations*

A	Austria	NL	Netherlands
B	Belgium	NO	Norway
BR	Brazil	P	Portugal
CAN	Canada	S	Sweden
DK	Denmark	S ARAB	Saudi Arabia
F	France	SO AF	Union of South Africa
FI	Finland	SP	Spain
GB	Great Britain	SWI	Switzerland
I	Italy	TAIW	Taiwan
IND	India	USA	United States of America
ISR	Israel	WG	West Germany
J	Japan		

Indexes

Names of Individuals

Names of Companies

Company names beginning with *Compagnie, Société*, etc. are often listed under key words.

P9-CMP-641

CIP

Library of Congress Cataloging in Publication Data

DeMont, Billie E. 1941 -
Accountability: an action model for the public schools.

Includes bibliographies and index.
1. Educational accountability. 2. Public schools — United States. I. DeMont, Roger, joint author.

II. Title

LB2806.D417 379'.15 75-2485
ISBN 0-88280-023-X

18512 Pierce Terrace
Homewood, Illinois 60430

Printed in the United States of America.

table of contents

Chapter **Page**

LIST OF TABLES

Table **Page**

LIST OF FIGURES

To Orin Graff

A dedicated scholar
and a true friend.

preface

Accountability is a political subsystem of the organization. It can serve diverse and often conflicting political purposes. Accountability can be used as a vehicle for increasing organizational efficiency; an instrument for meeting requirements of external pressure groups; or a smokescreen for control by a small group of decision makers. When accountability has served these purposes, it has been roundly attacked and rejected by substantial numbers of educators. The alternative accountability system presented in this book is value laden; that is, it also serves definable political purposes and processes. It represents an attempt to define organizational policies and practices which will be more acceptable to the profession — vehicles which bring greater professional control over the identification and realization of educational values.

The accountability system advanced does not attempt to define or classify appropriate values for individuals in the organizational setting. The system does specify organizational policies and practices which give formal recognition to the values of individuals — the professionals, community clientele and student clientele. In developing the accountability theory, the authors have assumed that organizational goals and individual goals have distinct but mutually reinforcing functions. To the contrary, several theories of administration assume that organizational and individual goals must be balanced and compromised and the resultant is a force which serves both the individual and the "greater good." It is the belief of the authors that this assumption lies at the base of numerous organizational evils. The "watering down" of individual purpose in the name of the "greater good of the organization" may result in perpetrating the will of a power elite over the working or learning conditions of the professional and consuming committees. Under these circumstances, compromise of individual and organizational needs is carried so far that no one professional recognizes (or is committed to) its conclusions.

To avoid this circumstance, an alternative theory is based on the assumption that there are appropriate organizational goals and appropriate individual goals. In this framework, individual goals pertain to those behaviors related to the delivery of an educational

service. On the other hand, it is assumed that the organization does not have substantive goals; that is, its goals should be secondary and facilitate the goals of individual professionals and clients. The organization provides a decision-making and delivery system for realizing the educational values of individuals. In this sense organizational and individual goals *cannot be meshed or negotiated*; they have different referrents and should therefore be complementary.

The accountability theory which follows deals with appropriate organizational goals — policies and procedures which do not favor or compete with individual values but provide a neutral framework for the identification, negotiation, and realization of those values by the individuals in the educational setting. In effect, this theory disclaims a paternalistic role for organizational accountability. It neither advances the interests of specialized groups nor ensures that particular values of individuals will be accepted or introduced into the educational enterprise. The accountability framework which follows will deal with a set of rules for ensuring that all points of view have their "day in court" in decision-making processes which are public and subject to the review of professional and lay persons. Within this framework, it is the individual who is ultimately responsible for selecting and implementing educational values. Whereas this system does not favor any values regardless of the status or political leanings of their proponents, it does not ensure that these factors will be eliminated from decision-making. It simply eliminates them from *formal* institutional policy and practice. This is the political framework of the accountability model which has been developed. When it is chosen and implemented in this spirit, it represents a conscious effort to eliminate indoctrination and favoritism.

Ideally, this system of accountability will be implemented when professionals are committed to the identification of novel and diverse individual goals and the application of human and material resources in achieving those ends. In this situation, the commitment is to a problem solving activity in which the educational goals of professionals and clients are given first priority. It is not a comfortable or clear cut system; rather, it is designed to disrupt and re-examine traditional goals; to diversify purposes; to draw new configurations of values and schemes; and to examine the consequences of our value systems. In other, less ideal settings in which the commitment to individual values is less pronounced, the system may be partially effective; and, in highly repressive political settings, the accountability model may be totally

ineffective. The accountability system does not replace or ensure a political community committed to a professional freedom and responsibility. Although it is not a system designed to impose the will of any individuals upon another group of individuals, it does not prohibit that possibility. It is a structure for enhancing already existing commitments. When those positive commitments exist, the accountability approach is a self-chosen structure through which the values, commitments and goals of individuals are enhanced and reach fruition in educational communities.

Billie DeMont
Roger DeMont

chapter one
a theoretical framework for accountability[1]

Introduction

Every educator has heard of the term accountability. Pick up a professional journal, attend a professional meeting or even talk to another educator and accountability will inevitably come up. Teachers, administrators, professors and even students talk about accountability, some with fear for the concept conjurs up visions of reprisals and reprimands; others with optimism because the age of accountability brings with it the hope that the profession will turn the corner from its long history of mixed success.

Most professionals and lay persons would agree that the public system of education is under attack as never before for its inability to respond to pressing demands for change. The educational institution, however, is not the sole recipient of public criticism. The demand for accountability in education matches the demands for reform in welfare systems, the "revolts" on the part of middle income taxpayers, and the boycotts on goods by consumer groups. It would be difficult to note an institution, public or private, that is not under more severe scrutiny today than it was a decade ago. Depending on one's point of view, it is the price we must pay or the harvest we reap for having a more knowledgeable, concerned population.

[1]This chapter is adapted from an article by the authors entitled, "A Practical Approach to Accountability," *Educational Technology*, 13:40-45 (December, 1973).

Any significant social movement has its rallying point, its popular outcry. The movement spurred by those who are unhappy with education's progress has rallied around the concept of accountability. The outcry for "accountability" enables a number of diverse groups with various complaints to join together under one umbrella. Groups opposed to sex education, others concerned with high property taxes, and those disgusted with poor performance in inner-city schools are able to find a common bond.[2] To the educator the number of people singing the same tune seems overwhelming. The task of satisfying demands for some semblance of accountability becomes impossible, however, when one considers the various meanings attached to it.

Diverse Rationales for Accountability

Accountability has a hundred definitions depending upon who is talking at a particular time. The profession has as many definitions as the public. Before anyone can agree to be accountable, there must be some agreement about the meaning of accountability.

A most popular definition of accountability is based on the notion that within any organization, each person is responsible for a particular function and that function must be carried out adequately. Failure to achieve specified goals and objectives should result in the transfer or release of individuals or the termination of the program. Accountability, in this sense, often means that there is a critical point which separates radical change (termination of people or programs) from minor adjustments to the existing system. Inherent in this philosophy is a belief that the function of any one individual or program is essentially divisible from other components of the system. In its extreme form it can become a case of maintaining "what is" or radically changing, a "go" or "no go" situation. The popular adaptation of this theory is often stated in terms of either "he does his job" or "he should be fired."[3] In more subtle forms, such as performance contracting, the objectives may be ranked in a set of priorities. If an individual or organization achieves one level, there exists a predefined reward; at each level

[2]Henry M. Levin, "Summary of Conference Discussion," *Community Control of Schools*, Henry M. Levin, editor, (Washington, D.C.: The Brookings Institution, 1971), pp. 276-279.

[3]C. A. Bowers, "Accountability from a Humanist Point of View," *The Educational Forum*, 35:479-486 (May, 1971).

the reward increases by a specified degree. The reward for achievement is clearly defined and by inference, the penalty for failure.[4]

A second notion of accountability is that it represents a specific alternative to our present system of education, usually through the adaptation of business methods. Thus performance contracting — the contracting for educational systems with outside agencies — and the voucher system are levers for circumventing the traditional educational bureaucracy. The employment of an outside agency with profit motives appeals to those who are disenchanted with public education. Performance contracting and the voucher system are easier to understand than our present, process-oriented system. The recipients of the system, parents and students, are afforded some choice, or at least a clear-cut definition of what will be accomplished for a given level of expenditure. These techniques are often described as a type of accountability since they assume that educators must perform at a level acceptable to the public or risk loss of credibility with clientele.[5]

A third related concept of accountability is one that streamlines educational planning through the use of specific management techniques also borrowed from industry and government. Planning programming budgeting systems (PPBS or PPBES), program evaluation and review technique (PERT) and other management tools are important applications of this accountability concept.[6] The lack of effective educational planning and the spending of vast sums of public money without any discernable measures of tangible results have led to demands for more streamlined management strategies. The use of management tools calls not so much for a redefinition of the goals and objectives of public education but a clearer definition as well as more efficient use of the resources necessary to achieve those goals. The ability to the National Aeronautics and Space Agency (NASA) to forecast future space developments on the basis of planned technological advancement is a good example of the use of management tools for planned change.

---

[4]Robert E. Roush, Dale L. Bratton and Caroline Gillin, "Accountability in Education: A Priority for the 70's," *Education*, 92:116 (September, 1971).

[5]Peter A. Janssen, "Education Vouchers," *American Education*, 6:9-11 (December, 1970).

[6]For a description of these management techniques, see Robert F. Alioto and J. A. Jungherr, *Operational PPBS for Education* (New York: Harper and Row Publishers, 1971).

Still another view of accountability which is widely appealing is the stating of measurable objectives at all organizational levels. Although the need for measurable or performance objectives is implicit in all of the other treatments of accountability, to some educators they have become an end in themselves. The teacher in many school systems has been asked to state performance objectives for students, while the administrator specifies the organizational goals under a system of management by objectives. With the growth in popularity of this approach, roles and functions become more clearly defined, allowing the practitioner to more easily perform "by the book" than in the past.

A final concept of accountability to be reviewed here is one that insures that the public has information on the goals, processes, and products of the school. This notion considers accountability to be a "public accounting" of that which is. Inherent in this belief is either that the public will understand the problems and limitations faced by the schools or that public pressure and demands will cause the school to change focus or improve results. Perhaps the expectation is a combination of the two. Nevertheless the fact that the public knows what is going on in schools would be the criteria for some individuals to maintain that the school system has been accountable.

These popular conceptions of accountability contain at least four elements: A focus on educational output rather than process; the development of alternative methods of achieving ends; the production of more effective evaluative and research models; and the inclusion of noneducators in the decision-making process. These elements are basic to any discussion of accountability, and any system that is to be effective needs to contain each one. But none of the previous conceptions of accountability suffices in the development of a practical system of accountability for the immediate future.

It is not reasonable to expect teachers and administrators to perform adequately simply because they have direct competition from a private school system. Nor is it reasonable to expect that measurable objectives alone will provide a viable system of accountability. Teachers and administrators may not perform because the school system does not provide the proper supportive services. An educational plan including public and private schools may merely provide *two* inadequate schools from which parents

may choose. Defining measurable objectives can be valueless unless the objectives are linked to the available resources, properly evaluated and used to initiate improvements. Each of these concepts is too narrow to serve as an adequate model of accountability.

A Reconsideration of the Concept of Accountability

To be more effective, the preceding notions of accountability must be incorporated into an organizational model which permits systematic scrutiny and revision of educational programs. The model must address not only the question of an improved educational product, but deal with problem-solving strategies appropriate to the priorities and development needs of a professional staff — those who must implement accountability.

The professional staff has responsibility for managing programs at the classroom, building and system-wide levels of the school district. The leadership of those programs involves four interrelated functions called program management: goal setting, programming, evaluation, and decision-making. When all of these functions serve a defined clientele, a program is institutionalized. The program has goals or objectives for the clientele; specific strategies and services which are employed to meet the objectives; an evaluation of the extent to which the objectives are met; and finally, decisions about shaping and refining the services of the program. Program management is the problem-solving framework of the professional; whereas accountability is an organizational support system which increases the problem-solving capability of individual professionals. Accountability brings program management into relief and allows improvements to be instituted in classrooms, buildings, central offices, and school systems. These criteria are consistent with the following definition of accountability:

Accountability consists of a set of organizational policies and procedures which:

1. Designate the agents accountable for particular educational authority for particular educational programs (Primary Accountability Agents);
2. Support documented reviews of educational programs by those agents (Internal Program Reviews);
3. Support documented reviews of educational programs by agents external to program operation (External Program Reviews); and

4. Utilize the reviews to diagnose needs and introduce innovation (Program Development).

An accountability effort undergirded by these provisions promises improved organizational results by better utilization of human and material resources currently available to the profession. Most school systems can redesign policies and practices to deal effectively with the tasks of designating responsibility, conducting internal and external reviews, and developing programs. In order to highlight the practicality of the accountability approach, the following discussion suggests several examples which are first approximations to full implementation of an accountability system. Succeeding chapters will include details of the four components needed in a viable system of accountability.

Components of an Accountability System

The formulation of a definition of accountability was guided by two overriding assumptions: Accountability must enable the school district to (1) reach acceptable levels of performance and (2) continuously improve its services. In achieving these ends the organization can frame policies and procedures to enable it to review and refine a plethora of widely differing programs and services. The era of accountability is one in which programs are no longer expanded and tacked on with the emergence of a need, but are planned, reviewed and related to one another in a systematic effort to improve the output of the system. The steps in achieving such an effort are as follows.

Designating the Primary Accountability Agent. Effective shaping of the great number of programs sponsored by a given district mandates the identification of employees whose job assignments give them the most knowledge about the programs under consideration. The first component of any accountability system is to designate an individual as the *primary accountability agent* (PAA) for a given program. The agent must be the chief decision maker for the program and must have major *authority* and *responsibility* for goal setting, programming, evaluating and refining the services delivered. In other words, the PAA is accountable for those programs for which both decision-making authority and operational responsibility are exercised. This

designation of authority is an asset to the agent, allowing him to assess the current output of the program and introduce revisions that will have likelihood of greater success.

To illustrate, if the accountability target were the first grade reading program in Building X, Room 100, the teacher of students assigned to that class would probably be responsible. This teacher has major decision-making authority not only about the classroom goals, the program, and the evaluation procedures, but also about the use of this information in making daily improvements and revisions in classroom strategy. On the other hand, if the accountability target were a system-wide inservice program for reading teachers, then the central office person with authority and responsibility for planning and implementation would lead the accountability effort for that program. Using this criterion, some accountable parties may be identified for a typical school system. Results of this process are summarized in Table I, which contains agents accountable for classroom, building, and system-wide programs. Although responsibility may vary according to differing patterns of decision-making authority, teachers are typically responsible for classroom programs; building level administrators must answer for building programs; and central office personnel are accountable for system programs. These officers are identified as the *primary accountability agents* for their respective programs. Dramatic deviations from the typical pattern illustrated in Table I sometimes result from withholding decision-making authority from those with operating responsibility thereby creating two situations in which accountability is weakened. In one case, there are professionals with responsibility but no authority to alter programs. In the corollary instance, there are professionals with authority but lack of knowledge and experience in program operation. Further discussions of these instances are presented in Chapter Two. A typical matching between programs and accountability agents appears in Tables I and II, although each school system can establish responsibility according to its own patterns of professional authority.[7]

While an individual is charged with the final decision-making authority and responsibility, by no means does the employee work

[7]Many school systems have catalogued decision-making authority within their systems. The list of decisions often run into the hundreds with each one assigned to a specific individual or group.

alone in terms of the goal setting, programming and evaluation components of the program. Involvement of others in these areas is a must and as such needs no defense. Consistent with the idea of involvement of others, however, is the corollary that others are responsible for providing input to the primary accountability agent. If the classroom teacher, for example, seeks aid in goal setting, a supervisor must accept responsibility for seeing that the teacher receives that service. Although the teacher has decision-making responsibility within the classroom, he can and should expect help from others in the decision-making process.

TABLE I

Agents Accountable for Major School System Programs

Primary Accountability Agent	Classroom Program	Building Program	System Program
Central Office: 1. 1. Superintendent Assistant Supt. Program Director			X
2. Performance Contractors	X	X	X
Building Level: 3. Administrators		X	
4. Teachers	X		
5. Instructional Teams	X		

This concept, as indicated in Table II, provides a slightly different way of looking at the traditional line and staff definition. Anytime the individual has control over the program operation and the decision-making process (i.e., he is the primary accountability

TABLE II

Agents Commonly Performing Functions Necessary for Program Accountability

Agent	Classroom Program	Building Program	System Program
Central Office:			
1. Superintendent Assistant Supt. Program Director	GS, P, EV	GS, P, EV	GS, P, EV, DM*
2. Instructional Supervisor	GS, P, EV	GS, P, EV	GS, P, EV
Building Level:			
3. Administrator	GS, P, EV	GS, P, EV, DM*	GS, EV
4. Teacher	GS, P, EV, DM*	GS, P, EV	GS, EV
5. Instructional Team	GS, P, EV, DM*	GS, P, EV	GS, EV
Other:			
6. Board of Education	GS, EV	GS, EV	GS, EV
7. Advisory Committee	GS, P, EV	GS, P, EV	GS, P, EV
8. Students	GS, P, EV	GS, P, EV	EV
9. Performance Contractor	GS, P, EV, DM*	GS, P, EV, DM*	GS, P, EV, DM*
10. Federal Project Officer			GS, P, EV
11. Consultant	GS, P, EV	GS, P, EV	GS, P, EV
12. Researcher	EV	EV	EV

Explanation:
GS – Goal Setting
P – Programming
EV – Evaluation
DM – Decision-Making
* – Indicates primary accountability agent

agent for that program), he is performing line functions. The same person can serve staff functions in other programs since he does not have decision-making responsibility but provides assistance in one or more of the program components. Each person within the system has line and staff functions (with the exception of a few people who may have no decision-making responsibilities with respect to any program). It is essential, under this definition, to detail not only line functions for individuals (decision-making) but staff functions as well. Table II indicates the responsibilities that might be served by various categories of people within the program. The building administrator, for example, performs line functions in a building program related to extra curricular activities and staff functions in a classroom program of an individual science teacher. In this tabular summary, the primary accountability agent for a given program is the individual responsible for all four functions of program management.

Executing the Internal Program Review. In education the review process rarely emerges from a systematic effort. Reviews are usually provided in response to state and federal agencies or when a program in the system becomes the object of widespread complaints.[8] Then the reviews are usually conducted solely by the profession, an in-house audit that is bound to lead to some suspicion and mistrust. The opposite approach, having all reviews conducted by external agents, citizen committees or other such groups, is equally fallacious. Any group composed solely of external agents will have difficulty gaining access to all the appropriate information (witness efforts by accrediting agencies).

Any appropriate review includes, as a first requirement, the systematic documentation, evaluation, and prescription by the internal agent — the primary accountability agent. This person provides documentation for the various facts and data that led to decision-making. He reviews the program continuously in terms of the input of various individuals to the program and the subsequent effects on students or other recipients of services. In other words, the primary accountability agent analyzes the input-output relationship.

---

[8]Luvern L. Cunningham, "Our Accountability Problems," *Accountability in American Education*, Frank J. Sciara and Richard K. Jantz, editors (Boston: Allyn and Bacon, Inc., 1972), p. 79.

Information provided in Table II can be used to illustrate the process of internal review. Consider a project which is a federally funded, non-graded program for fifth and sixth year students in an elementary school. Approximately 115 students are taught by a teaching team which includes two certified teachers and three half-time paraprofessionals. According to the table, the responsibility for program accountability has been assigned to the instructional team since it has been vested with authority for the goals, programming, evaluation and decisions about the nongraded instruction. Since performance contracting is not involved, and the program is not directed by a single teacher in a self-contained setting, the instructional team is the primary accountability agent. The team must conduct reviews of the four program functions, using not only its own resources but input derived from others who are both internal and external to the program operation. Column I of Table II indicates that at least ten additional agents can provide input to the review procedure. At the central office, personnel (such as an assistant superintendent for instruction or resource supervisors) may have input for goals, programs, and evaluation. At the building level the influence of the principal and individual teachers must be considered. Other sources of input may be the Board of Education, advisory committees, students, federal project officers, consultants, and researchers. These persons may perform differing services as indicated by their program functions. In the review process, the teaching team must identify the services provided, document the effects upon classroom programs, and make recommendations about program improvement. The review requires the primary accountability agent to become conscious not only of personal input but of the total range of influence on the classroom program. The agent must paint a comprehensive "picture" of the program action and suggest changes which would bring promising results.

Providing for External Program Reviews. The continuous in-house review is not sufficient, however, to provide for a viable accountability system. To supplement and check the in-house review, external agents should periodically examine the available documentation, gather other data if necessary, and provide their own evaluation of the program. This group corresponds to the independent audit frequently mentioned in the literature of

accountability.[9] There are, however, some notable differences. First, these agents need be external only to the work group directly responsible for the program operation. In the case of the classroom program previously mentioned, the officer charged with supervising the work of the instructional team could be part of the external team. In addition, parents in the community could also serve as members of a review team, as could outside consultants. This group must be free to conduct its own review of the program and make public the findings., While the review process may differ greatly from system to system and program to program, three principles seem essential: (1) that adequate documentation be required in order to speed the review process; (2) that the review be systematic and periodic (not just when trouble looms); and (3) that both external and internal agents be used in the review process.

Refining and Developing Programs. Program reviews cannot be ends in themselves. To be justified they must make some difference in the "way of life" structured by a particular program. The danger with any review is that it becomes merely an audit, an attempt to verify that which is. The review must examine critically the goals, the programming, the evaluations, and the resultant decisions; diagnose the problems within the sub-systems; as well as prescribe alternative courses of action to correct or improve the situation. Reviews must include alternatives to what exists. For example, the alternatives can be program modifications, program termination, or personnel changes. Of course the review team, internal or external, which is responsible for making these recommendations needs to be sure that any alternative falls within existing legal and contractual obligations.

The viability of the review is dependent upon its influence over the decision-making which brings the recommendations to fruition. The final step in the accountability process is the consideration of the recommendations and the implementation of the needed changes. While elements of this refinement process may have been suggested by the review team, the decision maker must be accountable for any action taken. This final step is the critical link, which establishes accountability not only as "organizational trouble

[9]For a discussion of the independent audit see Leon M. Lessinger, "The Independent Educational Accomplishment Audit," *Every Kid a Winner,* (Palo Alto, California: Science Research Associates, 1970), pp. 75-88.

shooting" but as the practical business of pioneering institutional improvement.

Summary

Accountability is an organizational commitment, manifest in policies and procedures that lead to four types of actions. These are the interdependent practices of designating agents who are accountable for specific programs, conducting internal program reviews, ensuring external reviews, and instituting program innovations. Accountability provides a problem-solving framework in which professionals negotiate their roles and corresponding institutional practices for each of the four components. In the first component, the organization establishes who is accountable by selecting, as a primary accountability agent, the program officer who has major decision-making authority and responsibility for particular programs at either the classroom, building, central office or system-wide levels. Secondly, the accountable person documents and reviews the goals, programming, evaluation and decisions associated with the program. Periodic independent reviews are conducted by professionals and lay persons not directly responsible for the program. These reviews diagnose and prescribe improvements in the form of recommendations which serve as catalysts for the final step in which the decision makers refine program operations.

In subsequent chapters, each component of the accountability model will be described in detail, and the implications for both professional roles and organizational practices will be drawn. Finally, the last chapter will deal with procedures for implementing accountability systems in the local school setting.

chapter two designing the primary accountability agents

Introduction

During the first phase of accountability, the organization designates accountable parties — those persons with decision-making authority and responsibility for the goals, programming, and evaluation of particular educational programs. This designation of accountable parties defines professional responsibilities in the accountability system — responsibilities that include conducting internal reviews; facilitating external reviews; and using the resulting recommendations in program development. By this definition, the primary accountability agent is accountable for institutional arrangements over which he exercises these key decision-making responsibilities.

In some cases the designation of the primary accountability agent is a fairly obvious, but highly abstract guideline to responsibility. As an example, the classroom teacher is the agent responsible for managing the unique aspects of the classroom program (neither building nor system programs). This global designation, however, must further be refined so that more specific program areas can be matched with responsibility and decision-making authority. These designations of authority require an analysis of formal authority and its relationship to particular program decisions. As a guide to this analysis, several questions will be explored in the context of the scope and interrelationships of authority: How is accountability

affected by traditional and emerging concepts of authority? How are classroom, building and system programs defined in terms of decision-making authority? How is responsibility defined for programs which are interdependent? What are the responsibilities of line and staff officers who make decisions about the same program? Are there appropriate organizational rewards for the primary accountability agent? Each of these questions poses a central issue for the identification of those who are accountable and the development of a clear definition of the objectives, resources and activities which are their appropriate focus.

Assignment of Responsibility According to Organizational Layers

In the preceding chapter the thesis was developed that the primary accountability agents for the instructional programs are the teacher, principal, and either an assistant superintendent for instruction or the superintendent. The thesis is based on the assumption that there are three definable and distinctive layers of instructional programs. First, there is the classroom program directed by the teacher. Presumably, the teacher has a personal set of instructional goals and objectives which partially define the classroom program. In addition, the principal has responsibility for coordinating the efforts of all teachers into a building program with goals and objectives of its own. The expansion of these efforts leads to a system-wide program encompassing the efforts of all individual buildings and built upon common goals and objectives. Thus the classroom program consists of those activities that are designed to meet goals and objectives not common to similar classrooms in the building. The building program maintains the same relationship vis-a-vis the system program.

At the classroom level, it is therefore the unique goals and objectives that establish the teacher's decision-making responsibility. At the building and system levels it is the common goals and objectives that establish a principal or central office administrator as the primary accountability agent. To illustrate this decision-making pattern, the typical "accountability" pattern for a school system can be contrasted with an emerging approach.

The Typical Approach to Program Responsibility

The Hawthorne Elementary School has three classrooms staffed

by teachers A, B, and C. The classroom activities of these teachers are designed to meet goals and objectives developed by the teachers or cooperatively with others. Some of the goals and objectives are dictated by school board policies; some by the principal's request; and still others by the teacher's values. Some goals are common; some unique. Schematically, the building program looks like:

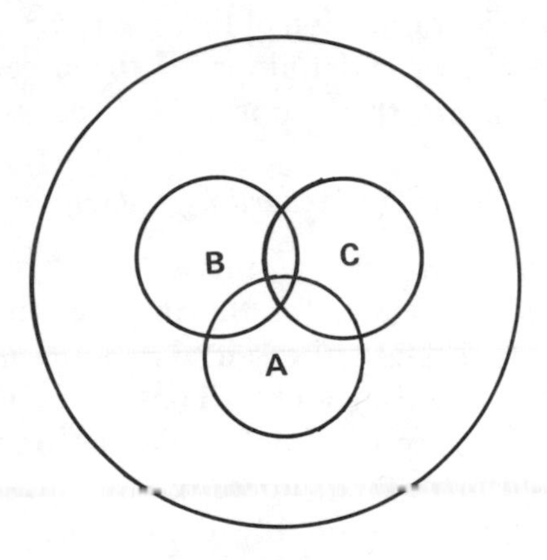

FIGURE I

Typical Building Program

In Figure I, Classroom teacher A maintains a program that has some objectives shared by all the other staff members. (Presumably these givens are dictated by the board, the administration, past practice, etc.) "A" also has some goals shared with "B", and others shared with "C". Those classroom goals and objectives that are shared but not universal are probably related to grade level or community of interest. Likewise, teacher A has unique goals and objectives devised to fit personal style and values. The building principal in cooperation with others has also developed activities that are separate from classroom programs but relate to achievement of goals and objectives. Among these activities are assemblies, remedial programs, work-study arrangements, etc. Thus, the building program has goals and objectives that are separate from the classroom goals or may be related to common goals that are evident in all the classroom programs. The traditional approach to building authority recognizes the

principal's responsibility for *all* activities since they take place within "his" building. He is responsible for all non-classroom programs and all classroom programs as well. The decision-making authority would thus look like:

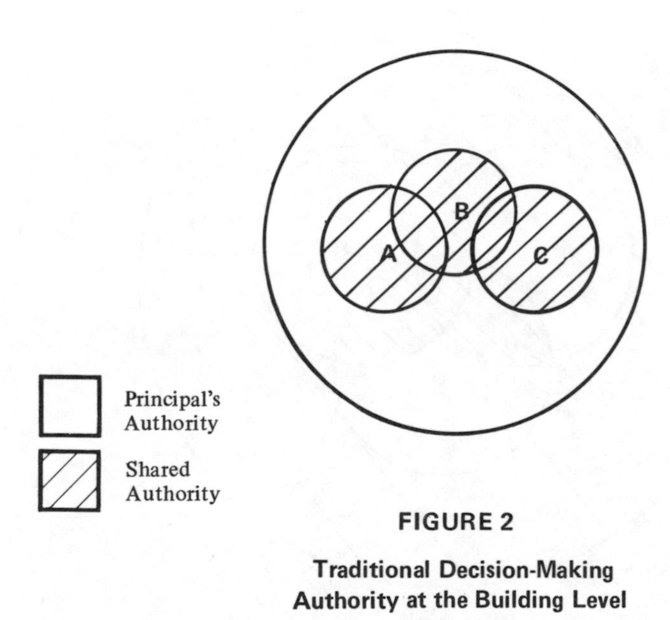

FIGURE 2

Traditional Decision-Making Authority at the Building Level

In reality the principal delegates much of this shared authority.

An Accountability Approach to Program Responsibility

The notion that a principal is accountable for all that goes on within the building is incompatible with the definition of accountability proposed in Chapter One. On a daily basis, the teacher is the decision maker for classroom goals and objectives while the principal's decision-making capacity lies with the goals and objectives that are common to a group of teachers in the building.

Using a decentralized concept of authority, teacher A in Figure 2 becomes the primary accountability agent for the classroom program. The principal becomes the primary accountability agent for all non-classroom building components as well as the activities

common to all classroom programs.[1] This differentiation of responsibility is illustrated in Figure 3.

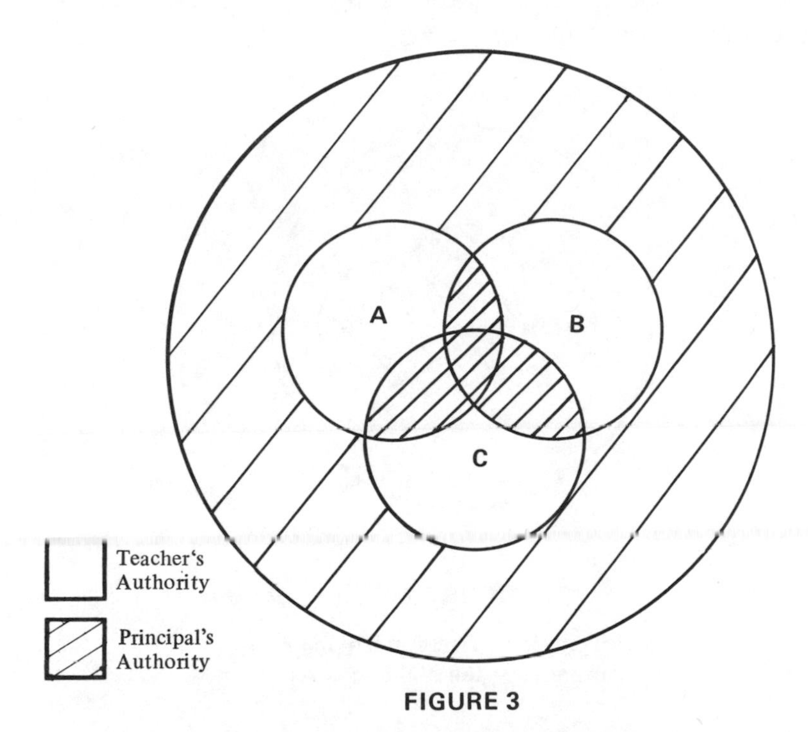

FIGURE 3

Teacher-Principal Accountability

The principal is therefore the primary accountability agent for those aspects of the instructional program that are not contained in classroom activities as well as those activities and practices dictated by institutional policies common to a group of classrooms. (Activities arising from informal arrangements among teachers are considered part of the classroom program. The guideline of institutional commitment as opposed to informal arrangements separates the teacher's accountability from the principal's accountability in the activities common to more than one classroom.)

[1]A reminder — the principal may not be the primary accountability agent for those aspects of the instructional program that are a unique part of classrooms A, B, or C, but he does have a staff responsibility to those teachers for aspects of the program for which they are the primary accountability agents. The process of staff accountability is developed in another section.

The relationship between the building program and the classroom program is comparable to program responsibility at the district level, illustrated in Figure 4.

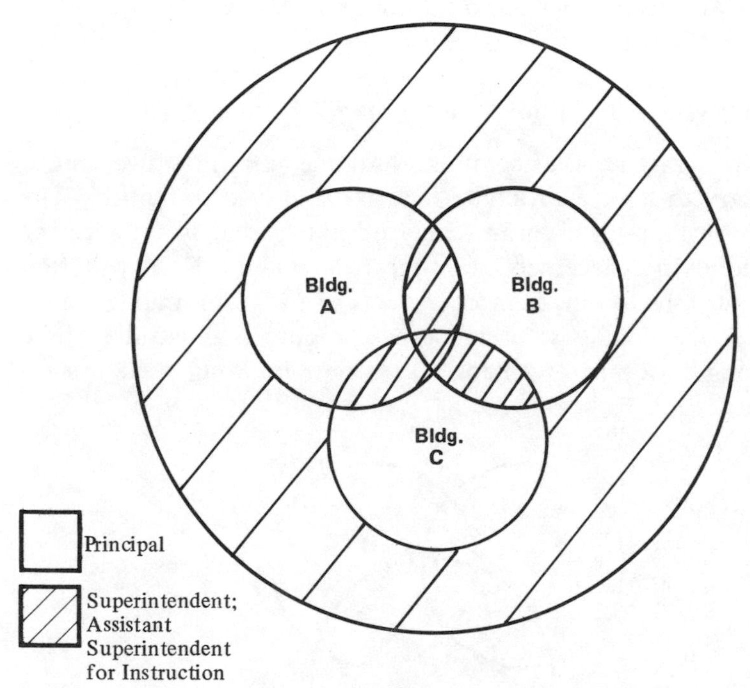

FIGURE 4
Principal-Superintendent Accountability

The unique aspects of building programs are the responsibility of the principal. He is the PAA. All non-building instructional activities, such as the supervision of curriculum directors, become the responsibility of the assistant superintendent of instruction or the superintendent, depending on role definitions. All programs common to two or more buildings likewise become the responsibility of the assistant superintendent as long as those programs arise out of an institutional commitment of resources.

Thus, three primary programs — classroom program, building program and system program — all have unique and common program elements. The primary accountability at each organizational level rests with the individual who has operational responsibility and formal authority for decision-making in the program: the teacher in the classroom; the principal in the

building; and the assistant superintendent or superintendent in the system. In other words, accountability implies that authority is dispersed across all organizational levels.[2] A school district structured in this manner has a minimum of layers — it becomes a flat rather than a tall organization.

Defining Responsibility for Overlapping Program Areas

When program areas affect more than one administrative unit, it is necessary to have a rationale for assigning accountability. The areas of overlapping programs may be illustrated by intersecting circles signifying three sets of program elements. Each program element pertains to one or more aspects of program management, such as goals, activities, or resources. Figure 5 illustrates these overlapping program areas for classroom, building, and system programs.

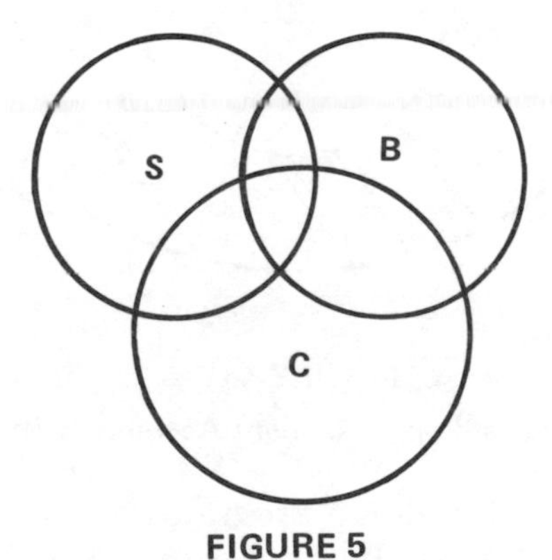

FIGURE 5

Overlapping Program Areas at Three Organizational Levels

The set "S" signifies all program aspects for a given system service such as elementary reading supervision. "B" is all building programs for a given school, such as George Washington Elementary School, and "C" is the program for a first grade classroom in that building. Each of the sets shares some elements in common with each of the other sets. In other words, there are

[2]For a discussion of decentralized authority see Rensis Likert, *New Patterns in Management* (New York: McGraw-Hill, 1961); and Rensis Likert, *The Human Organization* (New York: McGraw-Hill, 1967).

program aspects common to system and building; building and classroom; system and classroom; as well as all three organizational layers. These common program aspects are contained in the intersections of the three sets. Figure 6 illustrates the intersection of the system and building program.

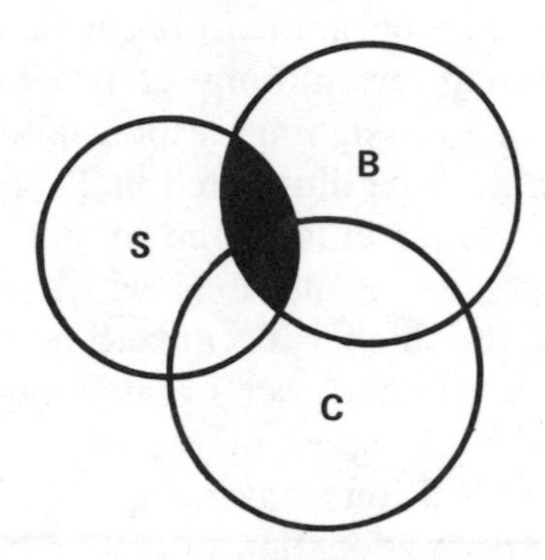

FIGURE 6
Intersection of System and Building Program

As an example of the relationship in Figure 6, the shaded area may signify the program of elementary reading supervision as it is practiced in George Washington Elementary School. This program

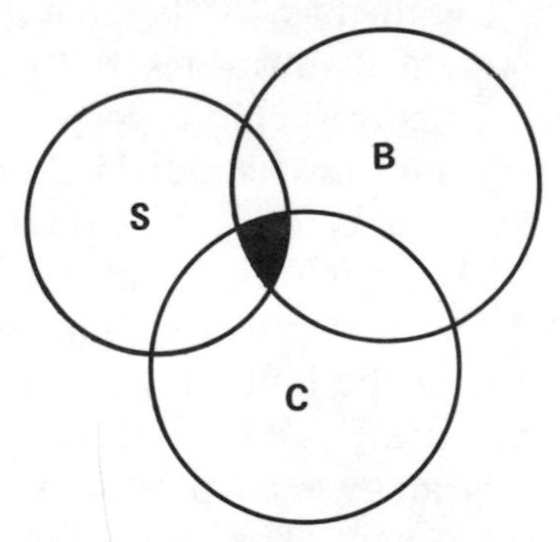

FIGURE 7
Intersection of Programs Common to Three Organizational Layers

area can also be symbolized as a set of program elements, $S \cap B$, which is read "the intersection of S and B." $B \cap C$ is the set of program elements common to building and classroom, and $C \cap S$ contains program areas for both classroom and system levels. The shaded area in Figure 7 illustrates the intersection of program features common to all three organizational layers. This set is symbolized by $S \cap B \cap C$.

Examples of program features common to two or three organizational layers are illustrated in the summary contained in Table III. There are four basic combinations of overlapping programs; and common program features are exemplified in goals-objectives, programming, evaluations or program refinement.

As a general guide to assigning responsibility for overlapping program areas, such as those illustrated in Table III, the following principle is advanced: Elements in the program intersections are those goals, strategies, or evaluations which are subject to the decision-making of the higher organizational levels but affect program management at the lower organizational level. In these cases, the officer who is higher in the organizational hierarchy typically has major decision making perogatives for program purpose, execution and reallocation of resources. Since this is the criteria for designation of the PAA advanced in Chapter One, it follows that in those cases of overlapping programs, the accountability rests with the line officer at the higher organizational level. This scheme is illustrated by the diagrams in Figure 8. In this illustration diagram 1 shows overlapping programs at the three organizational layers, signified as "S" (system), "B" (building) and "C" (classroom). Diagram 2 illustrates the unique programs at those levels (signified by corresponding lower case letters). In the third diagram, programs are subdivided according to the organizational level with the primary accountability responsibility. In this diagram, all program officers at the system level have primary responsibility for the system's unique programs and system programs implemented in the buildings and classrooms. Set "b′" illustrates the building principal's responsibility, which includes unique building programs and programs shared with the classrooms but not the system. The classroom responsibilities are "c" — its unique set of program features.

The scheme in Figure 8 also shows relationships between *micro-accountability* and *macroaccountability* for a given program. Microaccountability focuses on the accountability process which can be implemented within one administrative unit only; whereas, macroaccountability deals with programs that affect several units. Microaccountability is based on the notion that certain program changes in a classroom, system, or building can be brought about

TABLE III

Examples of Program Features Common To Two or More Organizational Levels

Set	Program Interpretation	Common Goals Objectives	Common Programming	Common Evaluations	Common Programs Refinements
S∩B	Program elements common to system and building	*State curriculum standards *Reduction of building vandalism *100% participation in free-lunch program	*Procedures for hiring new teachers *Procedures for filing budget requests	*Study to determine attendance zones *Study of school enrollment projections	*Improvement of building grounds *Alter criteria for selection of building principals
B∩C	Program elements common to building and classroom	*Increase numbers of teacher-parent conferences *Increase numbers of college scholarships to graduates	*Procedures for student scheduling *Community development activities	*Principal's evaluation of teaching performance *Building study of curriculum innovation	*Reassignment of classroom spaces *Offer new courses for adults in community
S∩C	Program elements common to system and classroom	*Criteria for selection of textbook committee *Increase supervisory services to special education teachers	*Implement fringe benefit program for teachers *Implement federal grant in select classrooms	*Evaluation of federal grant *Determination of teacher's contract status	*Install classrooms as demonstration sites *Begin grants for classroom innovations
S∩B∩C	Program elements common to system, building and classroom	*Achieve average reading scores at national norms *Institute management by objectives *Achieve purpose of statement on student rights and responsibilities	*Inservice training for teachers *Transportation program for students	*Evaluation of teacher grievance *Preparation of contract negotiation positions	*Reallocation of teacher positions by subject areas *Develop new board policies

FIGURE 8

Designation of Accountability for Unique and Overlapping Program Features

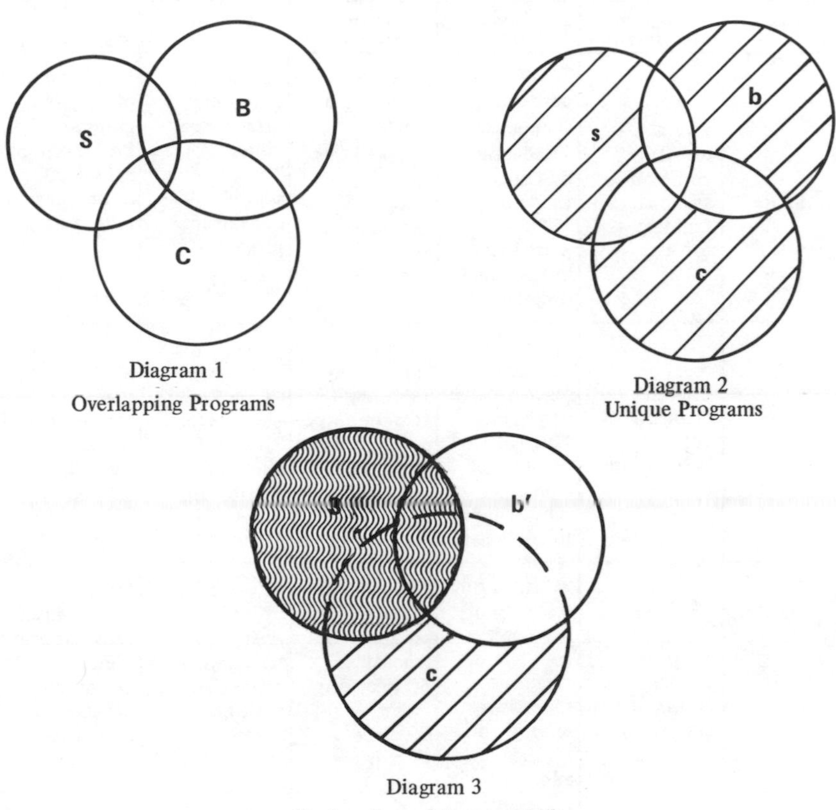

Diagram 1
Overlapping Programs

Diagram 2
Unique Programs

Diagram 3
Designation of Accountability

exclusively by the program officer at that level while officers at other levels stand in a staff relationship to those program developments. On the other hand, program changes affecting one level may require decisions at another organizational level. Reallocations of transportation funds, utilization of supervisory assistance, and the initiation of joint programs among classrooms, are examples of changes which affect classroom programs but require program decisions by the officers of building or system programs. These developments are examples of macroaccountability.

The distinction between the two types of accountability illustrates the importance of *staff* assistance in the accountability process.

Both the classroom and building programs are affected by system services for which they do not have the *primary accountability responsibility.* The classroom is likewise affected by programs under the jurisdiction of the building principal, who makes decisions about those services. In these cases, the different organizational levels have line and staff relationships. The primary accountability agent has a line responsibility which includes seeking the counsel, advice and recommendations of personnel whose programs are affected. When the system and classroom programs overlap ($S \cap C$), the system officer is accountable for internal and external reviews, as well as program development decisions; however, the classroom teacher has a staff role which entails the development of evaluations and recommendations which are submitted to the primary accountability agent. The exchange of information emerges from a mutual responsibility of the two units. In other words, the primary accountability agent at the system-level is responsible for seeking counsel in the form of the external program review while the classroom teacher has a recommending role. The teacher must analyze the influence of external resources on the classroom and make recommendations to those parties with decision-making prerogatives. The assignment of line and staff roles for two types of accountability within system, classroom or building programs is contained in Table IV. The information in Table IV may be summarized: (1) for all program elements in intersecting sets, the program officer at the higher organizational level is the primary accountability agent; whereas, the program officer in the other unit(s) of the intersection serves in a staff capacity; (2) the PAA is accountable for *seeking* and *utilizing* staff counsel through the administrative mechanism of the external program review; (3) in turn the staff officer, in the external program review, may develop recommendations which can be acted upon by the primary accountability agent; (4) within microaccountability, staff relationships may be developed with other organizational levels even though definite program links are not established.

Conflict Between Instructional, Non-Instructional Programs

The previous examples in this chapter have focused upon resolving accountability questions among various levels of

TABLE IV

Line and Staff Responsibilities for Macroaccountability and Microaccountability

Set of Program Elements*	Interpretation	Type of Accountability	Primary Accountability Agent	Agent(s) Responsible for Staff Assistance
$C \cap S$	Programs common to all classrooms in system	macro	system officer	classroom officer
$S \cap B$	Programs common to all buildings in the system	macro	system officer	building officer
$C \cap B$	Programs common to all classrooms in building	macro	building officer	classroom officer
$S \cap C \cap B$	Programs common to all classrooms and buildings in the system	macro	system officer	classroom and building officers
s+	Programs unique to the system	micro	system officer	classroom and building officers
b+	Programs unique to the building	micro	building officer	system and classroom officers
c+	Programs unique to the classroom	micro	classroom officer	system and building officers

+Each of s, b, and c may be defined as all program aspects not included in programs of the other two organizational levels. Using set theory, the following equations signify this relationship:

$$s = \overline{B \cup C} \qquad c = \overline{B \cup S} \qquad b = \overline{C \cup S}$$

In set terminology, microaccountability is the complement of the union of the other units.

*All symbols refer to sets defined in Figure 8.

instructional programs. The principles forwaıded seem to be applicable to non-instructional programs as well. These theories, however, can not be used to resolve the following situation:

FIGURE 9
Overlapping System-wide Programs

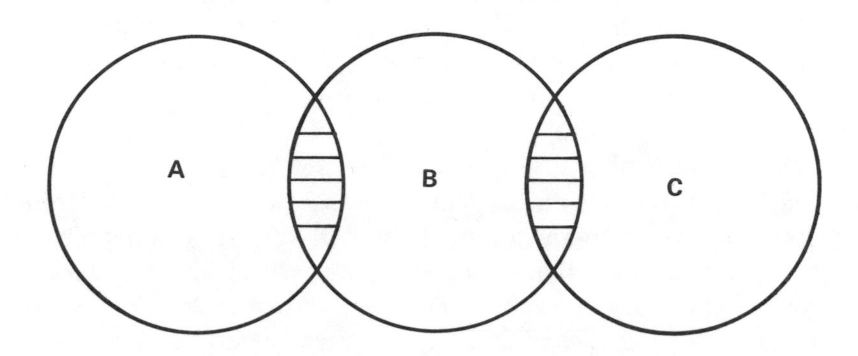

In the above example the program "A" is the personnel program under the direction of an assistant superintendent; "B" is the instructional program; and "C" is the business management program. The latter two areas are likewise under the direction of an assistant superintendent. Who is the primary accountability agent for the overlapping areas? The values of the authors dictate that the instructional program must be superordinate to any of the supportive programs. The primary purpose for schools is the instruction of the young; therefore the overlapping areas should be under the jurisdiction of the chief instructional officer.

The implications of such a policy are clear. The second in command in a school system should be the instructional leader. One could conclude that when the budget officer is accountable for the overlapping areas, the budget program is more important in that district than the instructional program. Any district that values the instructional program more than other programs should provide the organizational arrangements that support those values.

Implications of Line and Staff Roles in Accountability

The officers of the school district are accountable for programs under their jurisdiction, i.e., the programs for which they exercise decision-making authority. Table I in the first chapter refers to the various types of programs and the primary accountability agents for each as well as the various people who should have input but do not

have line authority over the decision-making. As previously mentioned, anyone who has these ancillary responsibilities serves in a staff relationship to that program. Further, an individual may have line authority or staff responsibility depending upon the particular program or task involved. The fixed notion of an individual being either line or staff on any organizational chart is an outmoded concept.

While line responsibilities for decision-making have been defined, staff roles are not as clear. Obviously a number of different people at different organizational levels will have various degrees of input to the accountability agents. To have a workable system of accountability the responsibility of each staff person to the program must be specified. To illustrate, a classroom science teacher is developing a new unit in biology. Previous evaluations indicate that the concept of osmosis has not been well understood by most of the students in the class. The teacher, in searching for a better way to teach the topic, asks the science coordinator for the district to aid in developing such a unit. But such aid could require anything from locating materials to completely developing the unit. The *nature* of the staff responsibility must be spelled out, in a manner similar to a good job description.

Staff accountability carries with it an authority-responsibility requirement similar to the obligations of the primary accountability agent. The staff role entails a responsibility to provide help to the PAA within the limits prescribed. Likewise the staff person may hold others accountable for seeking his aid and counsel about a program. The concept of staff accountability is much like community involvement in that the absence of such involvement becomes obvious when things go wrong. The science teacher who develops a new biology unit without consulting the science coordinator, and who finds failure in the new program, can lose some autonomy over the program in question. It is incumbent upon the organization, therefore, to specify the various roles assumed by individuals, including the necessity for the line officer to consult with others before making decisions that alter programs.

Accountability for Support Services

The three layers of line responsibility in the instructional program are assigned to system, building and classroom officers —

the only primary accountability agents for those programs. All the other personnel are staff officers in their relationship to instruction and in this relationship are responsible for services they provide or do not provide to the primary accountability agents. In education, the primary objectives are related to instruction; however, each of the instructional officers needs human and material resources in an effort to achieve instructional goals. As officers for instructional programs, they should be able to call upon available staff resources such as department chairmen, curriculum consultants or budget experts. These resource people within their designated roles and responsibilities aid the primary accountability agent in attempts to execute tasks. The list of personnel that are available to teachers includes a variety of people with expertise in curriculum, guidance, purchasing, personnel, research, administration, etc.

The need for supportive services to the instructional process has stimulated the growth of many ancillary but independent administrative units. At the central office level of large districts, there are typically budgetary units, research operations, pupil personnel offices, and employee relations units. These units are established to serve the instructional divisions but are typically under the direction of assistant superintendents who are funded independently of the instructional arm of the organization. In this organizational framework, the assistant superintendent for employee relations typically has line responsibility for hiring, evaluating and terminating teachers in the district. The assistant superintendent for business affairs is responsible for the budgeting criteria for each administrative unit in the school system. Each of these officers is the primary accountability agent for a particular supporting service to the instructional arena. For example, the assistant superintendent for instruction is responsible for the purposes and quality of a program of classroom supervision which is available to all classrooms on a system-wide basis. In general, the assistant superintendents become primary accountability agents for the supporting service programs and have line responsibilities with respect to the services which are commonly available to all classroom units. With respect to these classroom services, the classroom teacher or the building principal serves in a staff or advisory capacity to the primary accountability agent — the assistant superintendent. Thus, although there are specific primary accountability agents for each organizational level, supportive

services and other cooperative ventures create programs that affect one or more of these instructional levels. For each overlapping program area, the line and staff responsibilities must be spelled out at three levels in the school system.

Additional Criteria for Delimiting Accountability Responsibilities

In executing their responsibilities, the primary accountability agents first need to indentify the limitations placed on the program by forces which are not under the agents' control. These restraints and resources are myriad and the PAA cannot define his task without being aware of them. As an example, the building principal should consider some of the limitations illustrated in Figure 10.

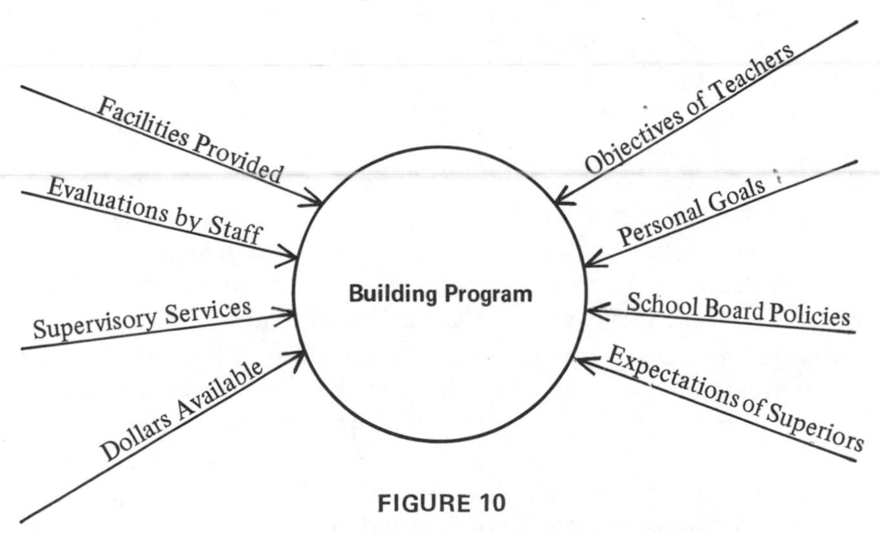

FIGURE 10

Some Limitations on the Accountability Responsibilities of the Building Principal

There are more restraints and resources than those listed. The principal must have a working definition of these limitations before he defines his program responsibilities and authority as primary accountability agent. Thus he provides a framework (he can expect his superior, the primary accountability agent for the district, to define some of these restraints for him) for his building program. Those forces that impinge upon the classroom program should be communicated to the teacher who will also identify forces which are unique to the classroom and influence its goals and objectives.

Under such a system, therefore, the accountability agent has three responsibilities for assessing input before defining personal program accountability:

1. He negotiates with his superiors those purposes, restraints, and resources impinging upon the program;
2. He establishes other restraints and resources that are unique to the program;
3. He defines with subordinates the purposes, restraints and resources impinging upon their programs.

For the development of items 1 and 2 above, the primary accountability agent has an *active* role in designating areas of program responsibility. This role defines responsibilities that are not automatically determined by organizational levels or relationships between ongoing programs. In negotiating the superior's expectations and the PAA's expectations for subordinates, the primary accountability agent brings elements of uniqueness to program responsibility. A mutual understanding of the superior-subordinate expectations emerges as a type of contract between the two individuals. The contract is best exemplified in management by objectives, which requires the two parties to reach mutually acceptable goals subject to evaluation at some point in the future.[3] In drawing this contract the superior and subordinate have a peer relationship, and the contract contains *mutually* acceptable terms. The same relationship extends downward to the PAA's subordinates, who must negotiate their contracts. Finally, the primary accountability agent determines program areas which will be unique to the services provided by the agent's work group. The three tasks outlined above afford the greatest outlet for the creativity of the primary accountability agent whose personal goals may be translated into practice.

Under such a system a necessary framework is established to make accountability possible. For each of the three steps, the program management model can be a guide for identifying limitations at each organizational level. In this connection, superiors and subordinates can state their goals and objectives, programming resources and activities, evaluations and program

[3]For an appraisal of the philosophical basis and prognosis for management by objectives see George Strauss, "Management by Objectives: A Critical View," *Training and Development Journal*, (April, 1972), pp. 10-15.

refinement decisions. When these limitations have been revealed, the accountability agent may then establish additional goals and objectives for the program and methods for implementing them. In both cases the PAA must have the authority to utilize resources provided (both human and material) in any way he sees fit. *After the framework is set*, the PAA has complete autonomy to make the decisions necessary to achieve those goals. This is the necessary professional autonomy that makes accountability work.

Organizational Rewards for the Primary Accountability Agent

The primary accountability agent for a set of program tasks has the responsibility for setting goals, developing programs, evaluating the goals and making the decisions about the program's refinements. The buck stops with the PAA about those particular program aspects. The agent may delegate much of this authority to others, but he is still the individual who is responsible for the progress of the program.

But how is the PAA responsible? Does the agent, as noted earlier in the differing definitions of accountability, get fired or transferred if the program does not achieve the stated goals and objectives? Or is the PAA merely responsible for a public "accounting" for the program? In other words, does he make public the strengths and weaknesses of the program and in doing so fulfill his function as the accountable person? Does the agent spell out the differing options that are available for improving the program and does the fulfillment of that responsibility satisfy the accountability mandate?

The definition of accountability forwarded here is not solely related to public accounting, or enumerating options about programs. It is related to the output of a program in which the primary accountability agent increases his authority (autonomy) as a final decision maker to the degree that previous decisions about the program decreased the gap between actual and desired output. The converse is also true — that the accountability agent decreases his autonomy in decision-making to the degree that his decisions did not improve the educational output. *The key to developing an adequate system of accountability is predicated on this assumption. The organization ensures that those individuals who make wise decisions will increase their decision-making authority.* Those whose decisions have failed to improve performance will gradually

lose their authority, i.e., some of their decision-making will be lost or curtailed.

A fundamental complaint in the literature on accountability is that educators do not know what they are accountable for. Response: They are accountable for decreasing the gap between the actual and the desired output of the programs in which they are the chief decision makers. Another complaint is that accountability seems to be a one way street; the educator loses but cannot win. Response: Educational accountability must allow for specified rewards and the greatest reward to most professional educators is the knowledge that improved results will mean increased autonomy.

A third complaint is that the individual teacher or principal is often "saddled" with rules, regulations, budget, etc., determined by other persons who fundamentally affect programs. There is no guarantee that under an accountability system such might not still be the case. The restraints, however, must be spelled out and the primary accountability agent is accountable only for those aspects of program over which he wields decision-making power. In addition, part of the accountability obligation is to examine restraints external to his authority and recommend possible changes. Finally, the educator can be relieved of the restraints on programs through the school district's methods for rewarding individuals who are improving programs. The autonomy of those educators can be increased by such options as decentralized budgets, exemptions to rules and regulations, etc. Accountability offers a most important asset for dealing with unreasonable program restraints external to the decision maker. These restraints become subject to continuous review and refinement as do all other program areas.

Summary

One of the basic assumptions underlying the accountability framework, proposed herein, is the notion that within the school system, authority is widely dispersed. Personnel at the top of the organizational hierarchy are not the only agents who are accountable for educational programs. Quite the contrary, each work group in the system is vested with decision-making authority for setting goals and objectives, designing and implementing services, evaluating program results and refining programs

according to these results. This program responsibility and decision-making authority is the basic criterion for designating the primary accountability agent for a given set of programs. The designated individual is responsible for conducting internal reviews of the programs, facilitating external reviews and using the recommendations in program development. The officers for the system, building and classroom programs share equally in fulfilling these responsibilities for the respective programs shaped by their decisions.

In the process of designating the primary accountability agent, each school system must decide (within its decision-making structure) which employees have authority for managing particular programs. Although the designation must ultimately be guided by local needs, there are some important guidelines for designating the PAA. Except for program aspects which overlap organizational levels, the classroom teacher is accountable for the classroom program; the building principal is the primary accountability agent for the building program; and the assistant superintendent or the superintendent is accountable for system operations. For those services that affect programs at more than one organizational level, the primary accountability agent is the officer who is higher in the organizational hierarchy.

Programs common to various organizational levels set the stage for a cooperative framework in the accountability process. This framework is organized around the line and staff relationship. Although the primary accountability agent has line responsibility for particular programs, other employees are affected by those services and serve in a staff capacity as they make recommendations for developing the services. The staff role and the line role are equally important in making accountability work. The line-staff concept implies that each employee has both line and staff responsibilities. A classroom teacher has line responsibilities for evaluating a student's daily progress in reading comprehension. And the building principal serves in a staff capacity that provides resources for the evaluation. On the other hand, the roles are reversed with respect to certain building programs such as the development of student schedules. Line and staff roles carry a mutual responsibility. The line officer is responsible for seeking staff recommendations in the external program review while the staff officer is required to make recommendations to the PAA in

the process of conducting internal program reviews.

The program responsibilities of the primary accountability agent are shaped by three sources:

1. organizational level of the program;
2. functional relationships with work units at other organizational levels;
3. unique programs personally adopted by the PAA.

Program accountability emerges from contracts negotiated between the PAA and superiors or subordinates as well as programs based on the PAA's personal goals and objectives. A necessary condition for accountability is that the professional has decision-making authority as well as operational responsibility for educational programs. Both criteria are essential to ensure that the primary accountability agent is in a position to control and adjust programs for which he exercises stewardship. This delegation of authority and responsibility can be controlled by both the individual and the organization. In other words, it is a negotiable item. Accountability provides a framework for the negotiation; it increases the district's capability of reaching agreement on specific program assignments. For example, it provides a mechanism to ensure that teachers are not held accountable for programs mandated by the policies, data, or preferences of principals or central office personnel. The definitions of both line and staff accountability are negotiable and may be continuously revised through such procedures as management by objectives.

The principle reward to the primary accountability agent is an increase in program autonomy. The scope of decision-making influence may be increased by developing workable programs which are either unique to the work unit or recommendations which spur the development of supportive services at other organizational levels. The school district which is committed to accountability must insure that planning decisions, budgetary allocations, salary and fringe benefit decisions are shaped by recommendations of accountable parties and reward achievement by increasing the autonomy of responsible individuals.

chapter three
the internal program review: a link between individual judgment and collective action

Introduction

Demands for educational accountability have caused increased numbers of professional personnel to question the nature of their roles in the accountability process. Role confusion, which is evident both in theory and practice, is related to two types of new expectations for professional behavior. In the first case, role confusion is based on the uncertainty about appropriate responses to external pressures for change — requests to meet new expectations of community groups or educational clients. In the other case, confusion is related to uncertainty about appropriate measures for coping with the professional's personal dissatisfaction within the educational system. In many local settings, workable solutions have not been forthcoming, and professional confusion has been followed by the adoption of adversary tactics or disenfranchisement from decision-making. A case study of both conditions is the recent controversy surrounding accountability practices in Michigan.[1] Teacher groups were targets of pressure

[1]For an evaluation of Michigan's accountability program, see Ernest House, Wendell Rivers and Daniel Stufflebaum, "An Assessment of the Michigan Accountability System," *Phi Delta Kappan*, 55: 663-669 (June, 1974).

from local and state administrators as well as community forces to accept teacher evaluation programs with an accountability theme. In the Detroit School System the teachers assumed adversary roles in response to both sets of pressures, and a teachers' strike ensued. Teachers' groups strongly argued that the Detroit School System had not met their needs for appropriate decision-making provisions or resource allocation. Although the contract the following year contained a provision for teacher accountability, professional roles within such an arrangement have yet to be specified or tested.

Increased clarity and consensus about professional roles will be required for the success of future accountability efforts. To some extent, confusion about professional rights and responsibilities is related to the absence of clarity and consensus within accountability theory. Theory must deal with formal avenues for (1) professional initiation of new policies and procedures for the internal audit of educational practice and (2) professional response to external demands for accountability. Systems for dealing with the first consideration are contained in this chapter; whereas, a discussion of the second issue is reserved for Chapter Four.

The diversity in accountability practice is matched by the variety of accountability approaches appearing in professional literature.[2] Browder estimated that 4,000 books or articles on accountability appeared during the period between 1969 and 1974.[3] Any attempt to bring clarity to the accountability movement must attend to the problem of defining roles which are reasonably acceptable and workable in educational practice. In this chapter, the discussion of professional roles is based on an inquiry into the generic problem of accountability theory and its relationship to a familiar professional tool — the internal program review. Although accountability theory is still embryonic in terms of detailing and specifying professional roles, several guidelines bring clarity to their initiation and practice. Some of these guidelines focus on the task of developing the internal program review (IPR) and using it as a vehicle for introducing program innovations.

---

[2]Gene V. Glass, "The Many Faces of 'Educational Accountability' " *Phi Delta Kappan*, 53: 636-639 (June, 1972).

[3]Lesley H. Browder, "A Developing Concept of Accountability," Paper Read at 1974 Conference of Professors of Educational Administration (NCPEA), Marquette, Michigan.

Purposes of the Internal Program Review

Developing Programs. An adequate system of accountability can be recommended for its rationality as well as its political feasibility.[4] Whereas many definitions of accountability emphasize the latter, the internal program review lends an emphasis on the former since it provides rational means for making program decisions.

As the critical second component of the accountability process, the internal program review presents the professional's point of view about increasing the efficiency and effectiveness of school programs. Accountability policies and procedures *formally* incorporate these points of view into decision processes that introduce educational innovations with a good prognosis for success. In this decision-making context, the IPR serves functions which deviate from some of the traditional uses of program reviews. Traditional reviews are often motivated by political priorities or the need to reduce conflict. Reviews associated with public relations efforts, the need to satisfy federal funding regulations, evaluations of teaching merit, or school-community conflicts are seldom attempts to systematically improve programs. On the other hand, the accountability mission emphasizes the need for the professional review with program development as its primary purpose. The primary accountability agent is required to describe the program and make recommendations about future goals and strategies within that program.

The internal program review is a necessary link between the individual's governance of his own professional behavior and the collective governance of the organization. These links have often been informal. In the case of the relationship between a teacher and a building principal, the links may be maintained by discussions at faculty meetings or informal conversations in a hallway. Although effective administrators have long taken pride in efforts to secure and utilize such input, there is clearly a need to formalize and organize the recommending role of the individual. In an accountability effort, the IPR provides formality and structure for the recommendations of professional personnel. The primary

[4]Lesley H. Browder, Jr., William A. Atkins, Jr., and Esin Kaya, *Developing an Educationally Accountable Program* (Berkeley, California: McCutchan Publishing Corporation, 1973), pp. 10-20.

accountability agent's role in producing the IPR is an opportunity to formally speculate about causes of program weakness and options for improving future programs. The document is a medium through which the PAA identifies exemplary and substandard program aspects, communicates program needs, and recommends alternatives to ineffective strategies. The PAA's position on these matters is partially determined by recalling and evaluating the components of the program and their interactions. In turn, this analysis forms a basis for judgments about introducing new program elements which serve needs of professionals, lay persons and clients.

Coordinating Programs. The IPR is the individual's lever for influencing two types of collective action. Through the internal program review, the PAA's recommendations influence program decisions within one unit or across several program units. In either case, the program officer utilizes experiences in an immediate work group to influence its program development or its relationship to other work units.

One type of influence may be restricted to the accountability practices of the PAA. For example, the teacher, who is responsible for day-to-day decision-making in a classroom program, conducts internal program reviews and makes recommendations for decisions about scheduling daily activities, space utilization, or criteria for evaluating students' reading skills. When these recommendations are made in the internal review framework, the PAA links personal judgments with evaluations of co-workers who assist in the program. In such cases the internal program review is an important factor in building professional consensus within one administrative unit. The review influences a program's internal consistency, efficiency and effectiveness — those aspects that pertain to the uniqueness and idiosyncratic dimensions of program management.

The internal program review also influences program development which requires cooperation across two or more administrative units. In these cases, a decision group considers several internal program reviews and installs new services that affect those programs or the school system as a whole. For example, the IPR produced by a given instructional team may suggest that a particular reading program is beneficial to highly imaginative

students who enjoy creating stories and learning to read from these materials. This information can be used at the system level to institute a district-wide program for identifying groups of these students and enrolling them in the exemplary programs. In addition to broadening the scope of successful programs, internal reviews indicate needs for improved supportive services such as inservice training programs or altered budgetary priorities. By revealing relationships between programs at many levels of the organization as well as pinpointing needed relationships, internal reviews are a point of departure for developing programs in the larger school system.

The two foci of program coordination — within and across administrative units — suggest a definition for the two types of accountability. When the reviews are used to coordinate program development within one administrative unit, the process is defined as *microaccountability*. This definition indicates that one PAA can institute accountability practices without a large scale organizational commitment. On the other hand, *macroaccountability* involves program improvements that affect two or more administrative units.

Personnel Implementing the Internal Program Review

The internal program review is produced by the primary accountability agent, the officer responsible for program management in a given administrative unit. The superintendent, assistant superintendents and program directors are typically responsible for reviewing system-wide programs. At the building level, the principal is accountable for the building program, and the teacher or instructional team is responsible for a classroom program. In addition, teams of professionals may be accountable within a performance contract or federal project. These professionals or management teams provide direction for the program goals and objectives, programming, evaluation and decision-making — the four basic elements of program management. In their review of defined programs, primary accountability agents focus on two fundamental tasks. The first task is a descriptive process detailing and reviewing the features of program management. Secondly, the PAA recommends needed program changes or innovations based upon those reviews.

A Methodology for Program Description

The first step in the internal program review is to describe the program under the direction of the primary accountability agent. The four dimensions of program management are an analytical scheme for the descriptive task of the IPR — recounting the program plan of classrooms, buildings, and central office units. Programs at each of these levels consist of goals and objectives, programming or strategies, evaluations, and decision-making.

During the first phase of describing programs, the PAA re-captures the major features of goals and objectives. The documentation involves stating both the long-range and short-range program aims. Primary goals and objectives are concerned with the ultimate aim of the educational enterprise; namely, the behavior of students. Secondary goals and objectives are those aims believed to be instrumental in producing desired behaviors. For classroom purposes, primary goals and objectives pertain to behavioral changes in cognitive or affective areas and secondary goals concern instructional or administrative services which are instrumental to achieving the primary goals.[5]

The second area of documentation focuses on the strategies or means of reaching the program goals. At the classroom level, these means include teaching methodologies, group activities, individual assignments, physical facilities and media, instructional materials, and supporting services such as counseling or parent and community assistance. The thrust of this step is to describe all physical and human resources which are utilized in goal achievement.[6]

The PAA makes judgments about the extent to which outcomes compare with program goals and objectives. In making these judgments, some type of program evaluation is essential. The evaluation involves measuring the results of services rendered.[7] For

[5]For references on the structure of instructional objectives, see Benjamin S. Boom, editor, *Taxonomy of Educational Objectives, Handbook I: Cognitive Domain* (New York: David McKay Company, Inc., 1956); David R. Krathwohl and others, *Taxonomy of Educational Objectives, Handbook II: Affective Domain* (New York: David McKay Company, Inc., 1964).

[6]For further discussion on program structure see Paul L. Brown, "Establishing a Program Structure," *Planning, Programming, Budgeting — A Systems Approach to Management*, Fremont Lyden and Ernest G. Miller, editors, second edition (Chicago: Markham Publishing Co., 1972), pp. 183-195.

[7]See Daniel L. Stufflebaum, "Evaluation as Enlightenment for Decision-Making," *Improving Educational Assessment and an Inventory of Measures of Affective Behavior* (Washington D.C.: American Society for Curriculum Development, 1969), pp. 41-73.

example, the classroom evaluation can focus on behavioral changes of students in the areas of attendance, time on task, or numbers of correct math problems. For the building principal, it may involve monitoring parent participation, student vandalism or annual gains in school-wide reading achievement. At the central office level, evaluation can take the form of measuring teacher attitudes about inservice training or principals' receptivity to innovation. System-wide evaluations are frequently included in Planning-Programming - Budgeting - Evaluation - Systems, cost studies, or needs assessment. Regardless of its focus, the evaluation is a basic tool for monitoring program effectiveness.

Evaluations allow the primary accountability agent to continuously adjust tactics according to their workability. These adjustments comprise the final dimension of program management; they are the decision-making component.[8] In the classroom, the decisions affect small groups of students. At the central office level, program decisions may have an impact on most students in the system. In the IPR, both the content and procedures of decision-making should be described and reviewed by the PAA.

The scope of program reviews should be limited according to priorities of local districts and work groups within those districts. The task of reviewing programs in their entirety is a complex and time consuming procedure. Prior to beginning the reviews, professionals should engage in planning which delimits reviews according to high priority goals, procedures or problem areas. By establishing these priorities, review procedures become manageable and are more likely to result in success within a timeframe which is acceptable to the professional. Attempts to review *all* aspects of programs may be a task of such proportions that the intensity of effort necessary for success is sacrificed by reviews that provide superficial coverage of the total program.

Sources of Data for Program Reviews

Securing Perceptions of Resource Persons and Clients. In recalling the basic features of program management, the PAA can use several sources of information. Other persons directly or indirectly associated with the program are one source of thc

[8]The process of program refinement is discussed by Ronald C. Doll, *Curriculum Improvement: Decision-Making and Process*, second edition (Boston: Allyn and Bacon, Inc., 1970).

required program descriptions. Within the administrative unit, employees and clients may give their perceptions of goals and objectives, programming, evaluation and decision-making. External resource persons include central office personnel, Board of Education members, consultants, researchers or advisory committee members. The process of securing perceptions of the work group and resource persons can be framed by brainstorming and planning implemented by the primary accountability agent.[9]

Executing Content Analyses. A second source of descriptions is based on content analyses of the work team, outside consultants and other support personnel.[10] The PAA can perform content analyses with summaries of program evaluations, logs, program materials, taped meetings, lesson plans, or management control instruments (e.g., PERT). Details of each document or record can provide data about the objectives, activities, evaluations and decisions which influenced the nature of the program.

There are, of course, other sources of information for describing the program of the IPR. Each district must develop a format acceptable for its purposes. Whatever the format and data base, however, the purpose of the description is to succinctly note the program features within a particular timeframe.

Methodologies for Making Program Recommendations

Changes in Program Management. The professional's review of program descriptions is designed to reveal successful program aspects as well as problem areas. Professional interpretation and judgment are critical elements in making recommendations based on the successes and failures which are identified. Several guidelines will be suggested for the exercise of such judgment.

The review process is used to recommend program adjustments within the work group or in the working relationship of two or more work groups. It should culminate in a formal statement of the PAA's recommendations for program development. Based on available evidence, these changes may pertain to substantive alterations in program management — goals and objectives,

---

[9]Richard Schmuck, "Developing Collaborative Decision-Making: The Importance of Trusting, Strong, and Skillful Leaders," *Educational Technology* 12:43-47 (October, 1972).

[10]For further information on content analysis see Otto N. Larsen, editor, *Violence and the Mass Media* (New York: Harper and Row, Publishers, 1968), pp. 97-100.

programming, evaluation or decision-making structure. Secondly, rather than substantive changes, recommendations may suggest alterations in the processes by which planning is accomplished, the program or evaluation is executed, or decisions are made. Recommendations about program management may require adjustments in the support system such as the need for new resources or the reallocation of resources. Resource needs should be linked to suggested changes in program substance and working relationships. Personnel additions, increased supportive services, and physical facility improvements are resource allocations which may be translated into economic terms and finally into budgetary decisions and educational practice.[11]

Continuity in Program Management. Another strategy which may be used by the professional is to analyze the logical continuity across the four components of program management. Are the program events, materials and activities designed to measure program goals and objectives? Are program decisions formulated on the basis of information secured in the evaluations? The sequential nature of program management is a logical basis for decisions. For example, programming must be shaped by desired goals; in turn, the evaluation should indicate the effectiveness of program services and provide the impetus for decisions about future goals and services. If decisions were made to discontinue system-wide IQ testing, an analysis of the goals of the testing program and its effects could support the decision which was made or it could indicate possible fallacies or limitations of that decision. If there was no logical continuity in program management, recommendations may suggest substantive changes in the processes ofdecision-making.

Capability for Program Change. A third guideline for making recommendations is based on the workability of program management as a control for educational change. For example, is it possible to implement the changes which are suggested by program evaluation? Can these changes be implemented within the immediate work group? Can they be implemented when cooperation of one or more work groups is required? Negative

[11]Harry J. Hartley, *Educational Planning-Programming-Budgeting: A Systems Approach* (Englewood Cliffs, N.J.,: Prentice-Hall, Inc., 1968), pp. 6-7; 97-98.

answers require scrutiny of each step in program management as well as the full range of decision-making processes which can be influenced by the work unit. Recommendations may suggest not only immediate changes in institutional practice but continuing investigation of problems which the PAA discovers.

Conflict Management. A final route for the program officer is to analyze the extent to which the roles of individuals and the work group facilitate cooperative working relationships. Cooperative relationships as contrasted with competitive relationships are characterized by

1. Open as opposed to closed communication;
2. Emphasis on common goals and interests rather than differences;
3. Trust and helpfulness as opposed to hostility; and
4. Problem solving based on needs of all parties rather than vested interests.[12]

In the management of any conflict that may exist, the program officer must recognize the roles of many other agents who provide input to the goal setting, programming and evaluation stages of program management. Each of these agents holds certain expectations for the nature of program management in the administrative unit. These expectations influence the demands which the agent places on the work of the unit or the nature of the service which the agent delivers. For the individual classroom program, there may be agents at six organizational levels who could provide input which should be considered in the IPR. The summary of Table V, illustrates roles through which various agents may influence the classroom program. Conflicts in values or roles may exist among members of the work group and any of the supporting personnel.

An analysis of the supporting roles, such as those in the above example, may lead the primary accountability agent to make recommendations for conflict management. Unproductive procedural or goal conficts can exist across administrative units, between the work group and its clients; among work group members; between the work group and community persons, or

[12]Martin Deutsch, "Conflict and Its Resolution," *Conflict Resolution: Contributions of the Behavioral Sciences*, Claggett G. Smith, editor (Notre Dame: University of Notre Dame Press, 1971), pp. 39-40.

TABLE V

ROLES OF VARIOUS AGENTS WHO INFLUENCE THE CLASSROOM PROGRAM

Agent	Roles Affecting Classroom Programs
System Level:	
Superintendent Asst. Superintendent Program Director	GS, P, EV
Instructional Supervisor	GS, P, EV
Building Level:	
Administrator	GS, P, EV
Teacher	GS, P, EV, DM
Instructional Team	GS, P, EV, DM
Other:	
Board of Education	GS, EV
Advisory Committee	GS, P, EV
Students	GS, P, EV

Key: GS––goal setting; P––programming; EV––evaluation; DM––decision-making

between the work group and other community institutions. A final area of discrepancy may exist between work group productivity and that group's expected level of achievement. With respect to conflicts across these groups, the PAA can recommend program

changes or group procedures through which conflicts may be resolved by the concerned parties.

Professional roles in conducting internal program reviews are determined according to the review's purposes, tasks, and methodology, which may be summarized as follows:

Purpose of the IPR	Methodology of the PAA	Tasks
Describe and Review Program Management	Utilize existing program descriptions Secure perceptions of professionals and clients: brainstorming interviewing administering questionnaires Perform content analyses of: logs evaluation-research summaries program materials observational records planning documents video/audio reproductions	Give Description of: Goals-objectives Programming Evaluation Decision-making
Make Recommendations for Program Improvement	Analyze logical continuity across four steps of program management Analyze workability of each step in program management Analyze work group's ability to introduce program innovations Analyze extent to which cooperative relationships support program management	Make Recommendations for: Improving each step in program management Building functional relationships among four aspects of program management Improving decision-making relationships with other units Changing resource allocations Continuing study of specific instructional practices

Development of an Administrative Structure for the IPR

The development of a system to support and facilitate the internal program review requires some additions and/or modifications to the reward system for participants. Employee benefits derived from producing internal reviews are related to broader institutional rewards for accountability; however, there are several

provisions which apply specifically to the review process. Five categories of employee support include the provision of (1) time for planning and decision-making, (2) staff educational benefits, (3) increased administrative and clerical services, (4) budgetary rewards, and (5) promotional criteria based upon the use of the IPR. Many services in these categories may be suggested by staff involved in the initial planning of school system accountability. The substance of services described below is intended to illustrate two approaches to school system changes — increasing staff ownership of planned changes and improving working conditions of those who change their work roles in order to participate in reviews.

To ensure the functional value of the IPR, the work group or the larger institution can provide time for the planning, decision-making, and writing that are required for the review. The manner in which recommendations are organized and documented by the PAA has substantial effect on decision makers who must respond to those requests. (Guidelines for the recommending process are detailed in Chapter Four).

In addition to the time provided to the PAA and the work group, the other officers in the organization must commit some of their time to formally respond to the recommendations of the IPR. The formal response is necessary for every IPR or the accountability report will go the way of many other reports in public institutions — to the dead letter file.

The institutional commitment to provide time for individuals and groups to execute reviews does not necessarily imply that the other duties must be left unattended. Existing structures, such as inservice training days, planning periods, university classes, and workshops provide at least some of the needed time for the descriptive and prescriptive tasks of the IPR.

School districts also support review efforts through educational approaches to training staff, such as the creation of specialized supervisory assistance[13] or provisions for released time for visitations to exemplary programs. Each service may be offered through general programs established on a calendar year basis or according to demands of individual administrative units.

A third example of positive support is administrative services which are delivered during various stages of the internal program

[13]For an extensive discussion of this topic see Browder, Atkins, and Kaya, "Staff Development and Program Implementation," *op. cit.*, pp. 224-254.

review. Assistance may be provided for developing materials such as program monitoring devices, which are useful in the descriptive phase of the reviews. In connection with its services for conceptualizing and building monitoring devices, the district can reproduce and disseminate materials which have been successfully used by various work groups. Finally, administrative and clerical services may be needed to reproduce the IPR recommendations and disseminate them to appropriate work groups.

The viability of internal program reviews is also dependent upon their usefulness in program development, not only in the PAA's immediate work group but also in the development of support programs. For example, if classroom teachers find that positive action is taken on the recommendations which are directed to other work groups, interest in conducting future reviews will be heightened. In this manner, a given unit may influence the total array of school programs including budgeting, supervision, instructional services and community development. To increase the likelihood of this result, a district can reward units which develop programs according to recommendations of employees serving staff as well as line functions. For example, budgetary requests can be favored according to evidence of staff support from other work units as well as the immediate work group. Secondly, the organization can reinforce review utilization by including those considerations in decisions about promotion and employee evaluations. Such personnel actions can be based on a history of employee participation in reviews and employee responsiveness to recommendations of other work groups.

Each of the preceding rewards can be important in establishing a viable system of accountability. All of them are related to *the central reward for participation in accountability: An increase in autonomy and sphere of influence.* A reward system built upon increased autonomy would undoubtedly provide the greatest incentive for the continuous review of programs.

Summary

In the context of the accountability movement, the internal program review achieves a specific level of distinction: It is a vehicle for developing professional judgments about program innovation and channeling those judgments into decision-making processes.

The IPR becomes a basic link between the judgment of professionals who are accountable for programs and the collective decisions of either (1) the immediate work group or (2) the larger organization. The IPR is a medium through which the primary accountability agent in a particular unit influences educational practice in the larger organization. As it serves these purposes, the internal review is an important option for dealing with professional disenfranchisement from the governance of the educational institution. Professional roles in the review process require skills in describing program management, developing sound judgments about program effectiveness and communicating these judgments as recommendations for program improvement.

The school district can ensure broad participation in reviews by establishing institutional rewards in five areas: educational benefits, increased administrative and clerical services, organized group and individual planning times, and budgetary-promotional rewards. Ultimately, the final test of internal program reviews is whether they influence decisions about program improvement — effective alterations of policies, practices and resource allocation.

chapter four
the external program review: broadening the decision base

Introduction

The internal program review is a continuous professional activity. The professionals who are accountable for the program develop the means and methods for a systematic and on-going review and interpretation of findings. Their professional knowledge and expertise are essential in establishing sound review procedures. On the other hand, professional involvement becomes a limiting factor if only those individuals with vested interests in the program are responsible for the reviews. A unique but complementary review procedure is needed.

The external program review (EPR) is a formal procedure for ensuring an *open* educational system — one that is responsive to the purposes and activities of individuals and interest groups in the broader social order. Although this purpose has been a longlasting tenet of educational theory, current practice has been found wanting by parents and lay groups who express dissatisfaction and desire greater involvement in the affairs of schooling. Public representation is provided through the board of education which has neither sufficed as an articulator of clients' concerns nor as a vehicle for building innovative and workable policy. New structures are needed for accommodating markedly diverse interests and a high degree of professional/community interaction. In this chapter

the external program review is described as a viable response to these needs. The approach must be detailed according to its personnel, purposes and methodologies.

The formal EPR has neither been widely practiced in education nor subjected to evaluations that lead to standardization of operating procedures. The broadest experience with independent audits has been in connection with federally funded programs. Drawing upon his experience with the independent educational accomplishment audit, Leon Lessinger outlined a six-stage audit process including production of: (1) a pre-audit of program objectives, (2) translated objectives, (3) instrumentation, (4) review calendar, (5) assessment process and (6) public report.[1] In the first stage the auditor meets with those who are responsible for a program or affected by it and reaches consensus on the program's objectives, methodology and priorities. In the second phase of the process, the auditor and the local team outline standards for determining when objectives are met. The instrumentation phase involves joint decisions between auditor and local clients — in this case, on the topic of data collection devices and procedures. The review calendar specifies a description of the events, responsible agent, timing and location. The fifth stage is execution of the audit, and the final step is a public reporting of the review and auditor's recommendations. In this chapter stages 1, 2, 5, and 6 are related to the accountability framework. Stage 3, instrumentation, is discussed in Chapter Three, and in the final chapter, calendar procedures are presented.

Purposes of the External Program Review

The need for an independent audit is not peculiar to program accountability. There are many other situations in education dictating independent judgments. Accreditation studies are done primarily by outsiders, who are chosen for their different perspective not because they alone have the necessary expertise. Within all academic institutions, different points of view are essential in evaluating a program. Universities often have written (or unwritten) policies that restrict the hiring of their own graduates. Instead, they try to hire outstanding graduates from a

[1]Leon Lessinger, *Every Kid a Winner* (Palo Alto, California: Science Research Associates, 1970), pp. 84-5.

number of different institutions and build a staff which has differing viewpoints.

Those who are accountable for programs have structured those programs according to personal and professional interests and therefore must necessarily view them with a specialized and limited focus. The logic of the external program review rests on more than the prevailing opinion that someone needs to curb professional error or that there is a need for a public inspection. It rests on the assumption that a periodic review by outsiders is essential to provide the additional information and viewpoints needed to evaluate the program in a larger context of educational and social values.

To illustrate, the authors were involved several years ago with a project including kindergarten teachers and the parents of kindergarten children. The purpose of the joint sessions was to define the expectations of both groups in terms of the kindergarten experiences of the children. In 95 percent of the cases, the professionals were able to demonstrate that some expectations of parents were unrealistic. And they were able to do so in a spirit of cooperation. There were significant areas, however, in which the professionals were enlightened by the lay public. One case involved teachers' expectations that the children would know each other's name within a particular timeframe. A parent mentioned that her son had erroneously been called James throughout his kindergarten year. Each of the teachers could explain how they guarded against such incidents, but in most cases their methods were unworkable unless the child voluntarily stated that he was addressed by a different name. The outcome of the parent-teacher session was that the school introduced a new method of registration, which ensured that each child would be called by the name designated by the parent. This action accommodated important considerations that the professionals had overlooked.

The primary purpose of the external program review, therefore, is to secure diverse perspectives and use them to improve the quality and scope of the programs. If the EPR team includes parents and students, there can be additional by-products such as wider acceptance of program changes by non-professionals.[2] Many lay

[2]For a discussion of the relationship between participation and the success of change see Lesley H. Browder, Jr., William A. Atkins, Jr., and Esin Kaya, *Developing An Educationally Accountable Program* (Berkeley, California: McCutchan Publishing Company, 1973), pp. 106-127.

persons are determined to have a greater voice in the public institution that so vitally affects them. In this regard, administrators note the dramatic change in expectations of school boards in recent years. So while the essential purpose of the EPR is to gain a different perspective on program development, the credibility of the program is enhanced by the involvement of others.

External program reviews provide different views of individual programs and program relationships within the school system and the larger community. These diverse viewpoints improve the review by highlighting the program's relationship to other programs in the district. Sometimes the best of programs is disjunctive with other efforts of the system. The implication is not that programs should be uniform; indeed the concept of accountability as developed dictates uniqueness and autonomy. Nevertheless, the PAA must be aware of complementary or conflicting effects on other programs. This comparison enables the agent to change (or not change) program efforts in light of these considerations. If, for example, the seventh grade math program as taught in school X, is completely out of phase with all other seventh grade math courses except in the fundamentals dictated by system-wide policy, at least two possibilities exist. First, the students may be having no problem with high school math in which case the difference might be encouraged. Second, problems may arise at the high school as indicated by an EPR and either the primary accountability agent of the seventh grade math program must change the program or a system officer must direct some changes so that seventh grade math increasingly becomes a common system program.

While the internal program reviews deal with the long-range success of a given program, the external program review must utilize such data to make recommendations for coordination across many programs. In achieving this goal, the reviewers have the advantage of considering program priorities at several organizational levels and recommending changes in their relationships. In the above case, the reviewers could recommend changes in all the other seventh grade programs as well as the high school program because the Building X program is the "best" format on which to build.

Often the EPR can serve yet another function which is based on the assumption that the school's goals and programs should not duplicate but complement programs of other community agencies.

In an era when social and professional expectations have led the schools to take on many other roles beside teaching the three R's, it is essential to eliminate duplication between the school and other institutions. If the EPR team has representatives from other community agencies, they may recommend that the overlapping areas be eliminated or replaced by cooperative programs. Ideally, duplicate programs can be housed in those community agencies to which they most naturally belong. Such coordination can only exist as the school opens its doors to representatives of outside agencies and vice versa.

Personnel of the EPR

The internal program review is conducted by the primary accountability agent and designated staff. The function of that review is to provide the staff that is most deeply involved in the program with a systematic way of reviewing and initiating changes in the program under its jurisdiction. It serves in effect as an internal audit. On the other hand, the external program review is conducted by individuals not accountable for the program. Unlike the internal team, their sole function is to evaluate and make recommendations, not to implement program changes.

The term "external program review" indicates that the review team is not involved in the operation of the program. This does not mean that the team members should not identify with the program by virtue of giving support or receiving its services. It simply means that the EPR members are not accountable for program operation. Nor does the term imply that the review team is necessarily external to the school system or disinterested in the success of the program. The argument can and has been forwarded that lay participation on such a team is desirable. As conceptualized here, however, the EPR team is composed of individuals not involved in the direction of the program under consideration. In effect, the team's composition could be based on the exclusion of school personnel, a combination of internal and external representatives or total involvement of noneducators.

Although the basic considerations have been outlined, some suggestions about the selection of personnel in the EPR team seem warranted. The key position is held by the leader of the EPR team. Generally, the convener of the EPR team should be the immediate superior of the primary accountability agent. If for example, the

review is of a building program for which the principal is the designated PAA, then the logical head of the EPR would be an assistant superintendent for instruction (assuming that the assistant superintendent is the immediate superior). If the superior, however, were held responsible for the actions of his subordinates, he could not be the head of the team. He would have a vested interest in the results of the review. It should be noted, however that the concept of accountability which has been presented is predicated on the assumption that people are not responsible for the actions of others. Therefore, the line officer can direct the review since he is not even by inference responsible for the quality of decisions made by the subordinate. The superior is responsible for (1) providing the staff support for the program discussed in Chapter Two and (2) ensuring that the EPR takes place. Within the model which has been described, the superior is therefore responsible for *a systematic process of review.* The primary accountability agent is responsible for decisions about program operation and refinement based on the review process.

Who are the other participants in the EPR? The make-up of any such committee will vary greatly from district to district and situation to situation. There are, however, some guidelines which aid in determining the membership of the EPR team.

Who is affected by the programs under review? Perhaps some group or groups such as the PTA have a special interest in the program. Any EPR team will be more effective if the members, especially the lay members, are vitally concerned about the topic at hand.

Who has expertise about the content represented by the program? It may be that the appropriate expertise is held only by those involved in the program. Perhaps, under these circumstances, the district will choose to hire an outside consultant to aid the EPR team in its review.

How do you avoid being arbitrary in selecting members for the EPR? If, for example, the PTA has a vital interest in the program under review and accordingly requires representation on the review team, then the PTA should select its own member. As much as possible, groups should be chosen for representation in the review team and not individuals. In turn, the group assumes responsibility for the individual who will represent its interests.

What restrictions should be placed on the size of the EPR team?

The authors are not committed to a specific number of individuals for the team. However, blue ribbon committees of more than ten members may not be able to ensure significant roles for all participants. The size of the committee should probably be dictated by the considerations of diversity listed below. The team should, however, be small enough so that each member has the opportunity to participate *fully* in the review processes.

How do you ensure diversity? Unless these are well-articulated, diverse opinions among groups who are affected by the program, it is difficult to ensure diversity. When special interest groups cannot be identified, school personnel or community representatives may be selected according to the diversity of their individual points of view. However, since the need for diversity is essential if the review is to be maximally successful, several additional criteria can be used in the selection process. The following sources of diversity can be used to identify groups which should be represented.

Diversity can be based upon:

(1) Program relationships among various administrative levels such as special interests of classroom, building or central office;
(2) Program overlaps based on cooperative relationships among different work groups;
(3) Needs to be served among different subpopulations of students or parents;
(4) Program relationships with different community agencies;
(5) Program knowledge of resource persons including
    (a) first-hand experience with similar, highly successful programs and
    (b) recognized expertise with programmatic considerations;
(6) Membership in racial or ethnic groups who have special interests in the program.

Since the EPR team is a recommending body, public acknowledgement and action on their recommendations are essential if membership on the team is to be an honor rather than an unwanted, meaningless task. In addition, however, the system must further enhance the image of the team. This can be done, in part, by choosing outstanding staff members as participants in the EPR. The work of the team should be supported either by released time or extra compensation for participants from the school system as well as external consultants.

School system support for individuals working on EPR teams is not sufficient for the proper functioning of the review process. Unless the EPR team is composed of individuals who are committed to the spirit of the EPR, the professional staff may view it as a witch hunt. Members of the EPR team must actively support (1) the concept of the autonomous professional; (2) a review process based on stated goals and objectives; and (3) an advisory relationship with the primary accountability agents. Professionals and lay persons with these commitments can help the review gain viability and influence among the professional staff.

Methodology

A team composed of individuals external to the program under review has been formed. How does it proceed? The first step is to develop a clearly defined charge *before* the group is constituted. In other words, each member of the committee, upon accepting the invitation for membership, in turn accepts the ground rules for the operation of the review team. For example, the major function of the EPR team is to make recommendations about the program under review. The team does not function as a super school board for that program, i.e., it does not pass program policy that binds the decision makers. Acceptance of an invitation to serve on the team implies an acceptance of the recommending role. Beyond this consideration, however, is the need for any committee to be properly charged regarding its role and functions.

To be successful in fulfilling its role, the EPR team must have the recognition and support of the school staff and thereby gain access to any and all information concerning the program. One means of gaining immediate information is by first-hand observations of the program and other related programs. Indeed it would be difficult for the EPR team to perform reviews based only on a "paper" familiarity with the program.

An additional source of information is the content of the internal program review (IPR). Since the IPR is the professional staff's account of the development and evaluation of the program, it is essential that the EPR team consider it in reviewing the program. It may provide a framework for the EPR by revealing areas that need further study or aspects of the program that should be expanded or eliminated.

Observation of the program and the review of the IPR can provide an orientation to the program. A most important by-product of this process is that lay persons and professionals as well become familiar with the *staff's* goals and objectives for the program. A review team member will have some (probably unspecified) goals and objectives for the program and, in the absence of an orientation, these goals can cloud the review process. While it is well within the province of the review team to suggest alternative goals and objectives, the review should not be geared to a program that does not exist. The framework for the review is the evaluation and development of the program in light of its *stated* objectives.

After gaining a general agreement about the program's goals, a major task for the external review team is to audit the internal program review. The audit can have several dimensions, including an investigation of the following questions.

(1) Was the internal review logical? That is, on the basis of stated goals and objectives, did the recommendations follow from evidence presented in the review? Was the evidence valid and complete?
(2) Were other, more promising recommendations ignored? Were the recommendations promising in terms of projected consequences? Were they realistic in terms of required resources, both human and monetary?
(3) Within the framework of the program's objectives, did the IPR ignore important program aspects or problems?

Answers to the above questions will shape the reaction of the EPR team to the internal review and may require the team to adopt strategies such as gathering additional data, exploring alternative recommendations, and investigating uncharted program areas. If, for example, the IPR relied heavily on hard data such as cost analyses, achievement scores and the like, the EPR team might extend that work by including attitudinal data from the professionals and clients of the program. In any case the EPR should be separate from the IPR and should provide another view of the program, not the same view revisited.

Following the audit of the internal review, the EPR team can address issues which were beyond the scope of internal studies. The unique goals and objectives of the members may enter the review at

this point. Their criteria may concern (a) program coordination across various organizational levels, (b) comparisons with similar exemplary programs, (c) coordination with community agencies, or (d) responses to priorities held by program clientele including subpopulations of students and community members. Prior to beginning its review, the EPR team should reach some consensus about unique goals and objectives which will be investigated. In order to keep all concerned parties informed, these should be shared with the work group, which can participate in designing strategies for studying those issues and gathering data. The nature of the data gathered by the EPR team will vary from program to program and system to system and will depend on types of materials produced in the IPR as well as unique problems perceived by team members. Chapter Three contains explanations of methods available for investigations designed by the EPR team. Following the gathering and interpretation of evidence, the EPR team must develop recommendations for the decision makers.

Recommendations

As previously stated, the EPR team has the responsibility for making recommendations about the program. The recommendations are the end product of the external review mission, and several guidelines can be used in their formulation.

The purpose of the EPR is to provide an independent review of the program under consideration and forward recommendations for improvement. The impact of the recommendations upon the program depends upon the team's influence on decision makers as well as the quality of the recommendations. It is evident, however, that if accountability is to work, recommendations in the EPR must influence more than committee discussions. They must shape institutional policy and practice. To ensure that the work of the EPR becomes a viable part of the accountability process, three standards are suggested:

(1) The recommendations should be communicated in a manner that provides a clear description of what is intended;
(2) The recommendations should be made public; and
(3) The EPR process should require a definitive response from the decision makers.

Organization of Recommendations. Many recommendations contained in reports are lists of things that should be done but seldom include discussions of the interrelationships or feasibility of the recommendations. They rarely project probable consequences of carrying out the recommendations. Without further criticism of present methods of recommending, several suggestions can be applied to EPR recommendations. The EPR, as a minimum, should make recommendations that have been:

(1) Grouped according to interrelationships of recommendations;
(2) Prioritized and listed in order of importance to the team;
(3) Clarified by listing available and needed resources;
(4) Analyzed so that several alternatives are suggested; and
(5) Justified according to specific consequences which are predictable.

The grouping of recommendations becomes necessary whenever the achievement of one determines whether another ought to be implemented. To use an obvious example, the purchase of specialized individual testing materials for a guidance program could depend on the hiring of a school psychologist. Other recommendations may not be as closely linked, yet the EPR team could conclude that one depends upon the other. Hence, grouping is necessary.

In addition to grouping techniques, the team should prioritize the recommendations. The priorities are determined by the EPR team's prediction of the effect the recommendation will have in meeting the program's stated goals and objectives. The prioritized recommendations may parallel the goals and objectives; i.e., the highest priority indicated in the goals may correspond to the highest priority in the recommendations. Another means of prioritizing may be accomplished by ordering recommendations according to the relative gap between the actual achievement of goals and the intended level of achievement. On the other hand, the priorities may be based on nothing more than subjective feelings of the EPR team.

Decision makers will consider recommendations in light of the available and needed resources. By noting the resources necessary to carry out the recommendations, the EPR team further clarifies its position and the practicality of its recommendations. These

notations add an *action* dimension to the recommendations and assist the decision makers in conceptualizing their impact on the school system. In fact, the recommendations can parallel the program management process by including specifications of prioritized goals and objectives, material resources and activities and appropriate evaluative measures.

For most recommendations, an alternative or alternatives should be listed. Alternative recommendations may include some ideas that are more imaginative and expansive as well as other recommendations that cut back on the original plans and recommendations. A third possibility is an option that achieves the same goal with similar resources but represents an alternative method altogether. In rare cases there may be no feasible alternative available in which case the recommendation is either accepted or rejected.

The most difficult part of the recommending function is the prediction of possible consequences. Many times the direction is easily predicted while the magnitude is not. Yet some notion of the outcome is necessary when the recommendation is made, and this prediction must be forwarded to the decision maker. In addition, any suggested evaluations of changes based on the recommendation should consider the recommended *degree* of change as well as the direction of change. The ability to predict consequences is improved by implementing recommendations and revising judgments since most program recommendations only approach the desired goals rather than reach them.

Publication of Recommendations. Recently a survey performed at the request of a state highway department "leaked" to the public. The report remained under cover solely because the final recommendation of the report conflicted with the departmental position. The results of such cover-ups are obvious: loss of confidence in the department and an overwhelming belief that the conclusions of the survey were accurate. So many reports and recommendations are overtly, covertly and accidentally filed in the dead letter drawer that one wonders why this type of reporting exists at all. In any accountability system the reports generated by the EPR must be public documents. If valuable time and effort are spent in reviewing programs, then the reviewers have a right to expect their recommendations to at least see the light of day.

The necessity for good communication must progress beyond the "good intentions" stage. Structures must be established so that decision makers receive and respond to the recommendations. In addition to dissemination of written reports, public hearings and discussions of the EPR may be held.

Response to Recommendations. The recommendations in the EPR are not intended to bind the decision maker. Accountability must respect the autonomy of professionals to make progress toward goals in a manner they deem necessary. As previously stated, when their judgment does not lead to the expected progress, they lose some of that autonomy to mandated curricula and, eventually, if the lack of progress continues, mandated methods. But they must have complete autonomy as a beginning point.

The PAA is free to accept or reject any of the recommendations made by the EPR team. The decision maker must, however, respond to each of them either by accepting or rejecting those that could be implemented by existing resources, or by requesting the reallocation of resources necessitated by some of the recommendations. The decision maker thus has at his disposal the recommendations of the EPR and the IPR as well as the data generated by the two reviews and can make decisions accordingly.

To ensure that a basic requirement of accountability is not eliminated by the external review system, it should be noted that the decision maker is *solely* responsible for the results occurring from the adoption of an EPR recommendation. It is not a case of retaining responsibility if the PAA does not adopt it and losing it when the recommendation is adopted. The purpose of the EPR is not to shift accountability within the system, but rather to offer a different point of view and different sets of data to the decision maker. The PAA must be free to respond to recommendations solely on the basis of what is judged to be appropriate for the program involved.

Some of the recommendations will require a reallocation of resources which are unavailable to the primary accountability agent. If the recommendation seems feasible to the PAA, then the agent in turn must ensure that the recommendation is considered by decision makers who have the authority to reallocate resources. While such a system may seem like an attempt to protect oneself by

shifting responsibility whenever possible, the truth is that the system ensures that those persons charged with making decisions do indeed make those decisions. Too often in a bureaucratic structure decisions "happen" because of a lack of communication among the various levels in the organization. In an accountability system, requests are directed to the appropriate decision maker who must make a formal response. Definitive answers and reasons for those decisions must be forthcoming.

Summary

The external program review is conducted by a team of individuals employed by the school district or representing non-educational institutions and the lay public. The team is charged with reviewing specific programs and making recommendations about continuing programs, expanding services, and introducing innovations. These recommendations are based on evidence of the degree to which stated goals and objectives were met as well as the appropriateness of those goals and objectives.

External reviews bring fresh perspectives to an accountability thrust — making possible more informed and acceptable decisions as well as better coordination among subunits and community agencies. To reap each of these benefits, the groups which should be represented must be carefully chosen, according to political, organizational, technical, and programmatic requirements.

The methodology of the external review is two-fold and consists of an audit of the internal program review as well as studies of different objectives germaine to interests of the groups represented by team members. As a team external to the primary accountability agent, the reviewers or auditors publicly recommend the changes regarded as most beneficial. The recommendations should be prioritized and include any available alternatives as well as the predicted consequences. The PAA is not bound by the recommendations since the agent remains accountable whether or not the judgments of the EPR team are acceptable. However, the primary accountability agent is accountable for the quality of a formal response to each recommendation, whether they are implemented completely, in part or not at all.

chapter five
program development: the action phase of accountability

Introduction

Program development is the introduction of innovative educational practices and support systems. In this sense, it is a major component of the schools' efforts to improve educational services. In recent years, professional issues have included several topics related to program development such as theories of organizational change, the management of change, and alternative educational models. In this chapter program development is viewed in a particular perspective — as the aim or end result of accountability. The accountability context relates the concept of program innovation to methodologies of program change, responses to requests of noneducators, definitions of professional roles and responsibilities, and decision-making theory. In short, accountability provides the framework to marshall institutional and extra-institutional resources in the service of program growth.

Accountability may be described as the professional's response to demands for program change — demands which may be diverse, conflicting, ill-defined, unrealistic, and removed from the politics of decision-making. Accountability acts as a professional mechanism for testing the workability of demands for change; sorting out inconsistencies; compromising or refining ideas with input from all interest groups, and basing organizational decisions on complete

evidence. When demands for change are raised, accountability provides answers to several questions: Who is accountable for those changes? To what extent are the proposals workable and effective according to professional criteria and independent opinion? And finally, what organizational decisions and program innovations respect the full array of evidence? In this chapter program development is described both as an independent professional activity and as a component of educational accountability.

Program Development and Institutional Change

Current Demands for Change. The press for change in educational institutions is one of the most noteworthy social forces of the last decade. Sources of these demands are both internal and external to the profession since clients and professionals are demanding improvement in the quality and quantity of the educational product. External controls on the educational program have been sought through such measures as:

1. court action requiring cross-district busing for the integration of the classroom;
2. community demands to decide on the hiring or firing of school principals;
3. lawsuits demanding redress for poor instruction to high school graduates;
4. national publicity condemning racist and sexist literature in the elementary and secondary grades.

The drama of these actions is equalled by pressure for change within the education profession:

1. collective bargaining over the accountability issue;
2. organizational thrusts for higher productivity such as management by objectives, performance contracts, voucher systems;
3. implementation of performance objectives systems within classrooms;
4. use of cost-benefit analyses in program decision-making;
5. development of free school and alternative school approaches.

These demands for change are interwoven with the same social fabric that shapes the accountability movement, and they share one element in common with that movement. Each of the particular

attempts for change deals with program improvement — the introduction of new standards for policies and procedures. This is the action phase of the accountability model.

Characteristics of the Change Process: Alvin Toffler, among others, has documented change as a contemporary social force dramatically affecting the American public. Change is not a phenomenon to be turned on or off at will. It is omnipresent within our culture, and its impact on the educational institution is only one outcropping of its pervasive effects.

According to Toffler, change is manifest in several social conditions:

1. widespread rejection of traditional social institutions;[1]
2. increasing transience of new social and material relationships;[2]
3. "overchoice" or the fact that individuals are required to make more decisions than they can comfortably execute;[3] and
4. the emergence of a multiplicity of alternate life styles and institutional arrangements.[4]

Individual withdrawal from reality or apathy and institutional chaos are frequent results of these four social conditions. As in any institutional situation, in education the disruptive effects are multiplied. Not only must the profession identify new learner goals, it must design corresponding professional rules and behaviors. Rules from the past no longer suffice for many decisions which are required or desirable.

Educational change can be both reactive and proactive; that is, it can be designed to reduce adverse effects of social change or to build more satisfactory institutional and personal arrangements. In either case, the success of the change strategy can run the gamut of workability and produce negative or positive effects on worker and client satisfaction and on organizational productivity. Educational change can be orderly or chaotic — healthy or cancerous — coordinated — or disjointed. It can facilitate individual purposes or subvert them. To increase the likelihood of highly positive results, some controls must be exercised on the change process itself.

[1] Alvin Toffler, *Future Shock* (New York: Bantam Books, Inc., 1971), pp. 185-186.

[2] *Ibid.*, 51-181.

[3] *Ibid.*, 263-283.

[4] *Ibid.*, 284-322.

Accountability provides a framework of control for the planning and execution of change.

Accountability as a Control for Educational Change. Within an accountability system, educational change occurs during the fourth phase, which is program development. This is the adoption and implementation phase; and it is, relatively speaking, more future-oriented than the first three phases of accountability. At this point, program development creates institutional changes suggested by the intensive study and reflection of the internal and external review processes. These changes are not restricted to increases in the size of the operation but encompass qualitative changes in the nature of the program, roles of employees and work groups, and cooperative arrangements among individuals and work groups.

As it culminates in the program development phase, accountability makes a fourfold contribution to the control of educational change. The contribution is based on:

1. the use of rational means for making decisions about educational change;
2. the ownership of change by those who are affected;
3. the use of empirical workability for assessing the success of change; and
4. the understanding and coordination of changes across all organizational units.

Each of the above conditions has been documented as an important condition in producing satisfactory and productive change.[5] In the first place, the likelihood of rational decision-making is increased by assigning accountability to those who, by virtue of their job responsibility, have access to the requisite information. Secondly, accountability ensures that formal reviews of program management and prescriptive processes are systematically linked to the politics of district decision-making. Both provisions are steps toward basing program decisions on educational issues and evidence as opposed to personal biases or expediency.

Another major strength of program development, as defined in the accountability model, is that it links educational change to the reflection and judgment of those persons engaged in program

[5]Lesley H. Browder, Jr., William A. Atkins, Jr., and Esin Kaya, *Developing an Educationally Accountable Program* (Berkeley, California: McCutchan Publishing Company, 1973), pp. 106-127.

leadership. In turn the accountability system requires program managers to consider recommendations of (1) professional resource persons who serve in a staff relationship to the program and (2) lay persons and students who participate in external program reviews. Accountability procedures respect the principle that those who are affected by educational change should have ownership of those changes. This provision permits the authors of change to introduce new practices that will be accompanied by rewards for themselves, work groups and students.

As a third condition, program development must demonstrate empirical workability. In the accountability model, changes are first suggested by practice and secondly, when institutionalized, pass the scrutiny of internal and external reviews. These reviews build a case for approval or disapproval of change or continuous refining and reshaping of those changes.

A final advantage to educational change within the accountability framework is that it avoids isolated actions which advantage one group to the detriment of others. This is accomplished by a system of basing accountability on both line and staff functions within specific programs. Changes in those programs are based not only on recommendations of those who direct the program but emerge from recommendations of those *affected by* the program. The consideration of recommendations in the internal and external reviews respects consensus and cooperation and reduces the likelihood of unilateral decision-making. Decisions can be appropriately linked and coordinated; thereby, reducing the risk of change that satisfies one work group but produces chaotic results at another organizational level.

Personnel Responsible for Program Development

Who is responsible for program development? In the accountability model which has been presented, program development requires both line and staff participation within every work unit. Recommendations for altering goals-objectives, programming and evaluation emerge naturally from the internal and external reviews. For this reason, the lines of responsibility for program development correspond with the professional roles in accountability. The primary accountability agent is responsible for using the recommendations to make educational changes for those programs

under that agent's jurisdiction. The assignment of line and staff responsibilities for program development is illustrated in Table VI. The responsibility is defined according to line and staff roles in accountability, i.e., accountability roles for particular program aspects are carried forward to the program development task. In Table VI these responsibilities are illustrated for programs in several administrative units. Within a particular administrative unit the primary accountability agent has line responsibilities for program development; whereas, those with staff responsibilities may include subordinates in the work unit, superiors not in the work unit, members of an external review team or other advisory groups. The role of the PAA in program development includes the clarification, selection and introduction of educational values.

Program Development and Educational Values

Program development is defined as the full range of activities that change educational policies and practices. Such educational changes are based on decisions that structure "ways of life" for individual employees, work groups and clients. The range of influence may include not only the decision makers, but the school district as a whole as well as the broader social system. Program change involves value choices which may be motivated by a desire to alter unacceptable practices or the need to introduce totally new practices into educational programming. On the other hand, program development does not refer to the daily, routine decisions which are components of program management. A convenient way to distinguish between the two types of decisions is based on definitions of programmed and non-programmed choices.[6]

Programmed decisions concern the daily routine of educational practice — those actions that are derived from conscious or unconscious values and thereby built into daily practice. Compared to non-programmed decisions, these actions require less professional energy because a body of rules has been developed for reaching those decisions. Examples of decisions which are typically programmed are (1) determination of students' attendance records; (2) assignment of classes to rooms in the building; and (3) scheduling of school system bus routes. On the other hand, non-programmed decisions require *new* behaviors and institutional

[6]For a discussion of the two types of decisions, see Toffler, *op. cit.*, pp. 355-358.

TABLE VI

PERSONNEL RESPONSIBLE FOR VARIOUS TYPES OF PROGRAM DEVELOPMENT

Organizational Level of Programs	Examples of Program Development Decisions	Responsible Personnel	
		Line	Staff
Classroom Programs	*Introduce Student Evaluation of Teaching *Develop Study Unit for Environmental Protection *Establish Peer-Group Tutoring	Teacher	*Principal *Paraprofessionals *Student Teachers *Parents
Building Programs	*Reschedule Daily Activities *Develop Parent Involvement Project *Reutilize Physical Spaces	Principal	*Teacher *Other Principals *Superintendent *Parents
System Programs	*Create Secretarial Pool *Initiate Systemwide Reading Tutorial *Establish Accountability	Superintendent (or Central Office Designee)	*Principals *Teachers *Assistant Superintendent *Lay Persons

arrangements. Typical examples of these decisions include (1) introduction of occupational studies into a formerly "traditional" curriculum; (2) involvement of community members in decision-making about classroom activities or (3) adoption of PPBES at the central office level. The above examples present distinctions which appear logical; however, the final determination of programmed or non-programmed decisions is a perceptual matter and this may vary from one professional or work group to another. Decisions affecting program development concern non-programmed practices and the clarification, selection and introduction of educational values.

Clarification of Values

Classification and Comparison of Values. Many times program values are held in an implicit manner. That is, the standard itself may be an unconscious one. The reconsideration of daily practices in educational programs may require explication of several sets of standards. A traditional approach to raising program values to the level of consciousness has been to study philosophies and their particular applications in the concrete educational setting. The five basic world views, including Idealism, Realism, Pragmatism, Logical Positivism, and Existentialism, offer principles about the nature of reality, knowledge, man, and ethics; and each philosophy has specific applications to education. In addition to these basic philosophical categories, values may be organized into aesthetics, economics, politics, religion, law, etiquette, or intelligence.[7] By examining educational practice through the study of traditional value groupings, educators have studied the dominant standards pervading the educational setting and made choices among them.

An alternative and increasingly popular approach to identifying and choosing program values requires the simultaneous study of behavior and the practices which reward it. In this case the choice among program values is based on the outcomes of different practices. Current interest in classroom applications of behavior modification is an example of the trend to identify values as those practices that reinforce particular behaviors. B.F. Skinner argues that it is impossible to change motives or values per se. He further

[7]For a discussion of other ways of classifying values, see Paul W. Taylor, *Normative Discourse* (Englewood Cliffs, N.J.: Prentice Hall, Inc., 1961). pp. 299-333.

maintains that abstract concepts such as "patriotism," "loyalty," and "freedom" can be destructive because they do not deal with behavior and its consequences. To reach goals or objectives, it is necessary to change the reinforcers of particular behaviors.[8] In this framework, practices, as opposed to philosophical principles, are the major focus because they serve as rewards of desired behavior. This premise implies that the development of programs is directed by a study of the extent to which educational practices are rewarding and can become increasingly rewarding to the program's clients. In this process, educational values are clarified by focusing on specific practices and their influence over student and professional behavior.

The Scope of Program Values. A second aspect of value clarification is an analysis of the scope of educational values. In the process of conducting internal and external values, some practices are questioned according to their reward value for a specific behavior. For example, a teacher's practice of giving weekly math quizzes may be unsatisfactory in terms of its value for rewarding student efforts to correct weekly homework assignments. The practice may be discontinued in favor of a more effective "help" session in which students receive tutoring in homework corrections from the teacher and other students. In this situation, the scope of program development is relatively narrow as compared to an effort to revamp teaching methodologies for the entire eighth grade mathematics program. In the later case, it is necessary to justify educational practice by its correspondence with a set of multi-dimensional and complex outcomes. The appropriateness of a practice is not judged solely by relating it to a single outcome but according to the importance of that outcome within a larger set of educational benefits. For example, the teacher's lectures in theorem derivation must be rewarding for students' abilities to prove theorems but also beneficial or consistent with a full range of expected student behaviors including (1) participation in group problem solving, (2) applications of theorems to solving mechanical problems, (3) ability to use theorems in the identification of mechanical errors, and (4) ability to employ theorems in developing higher-order mathematical concepts. In this example, the

[8]B.F. Skinner, *Beyond Freedom and Dignity* (New York: Bantam Books, Inc., 1971), p. 140.

particular educational practice is judged according to its importance for behaviors comprising a broad learning style rather than its relationship to a specialized result.

Priorities Among Program Values. Program development is not only affected by the scope of values, which has been described, but also by the priorities among values which are acceptable to professionals and educational clientele. The clarification of program values depends, in part, on whether the practices have intrinsic or instrumental value. Instrumental values have merit because of their function in achieving other values; whereas intrinsic values have independent merit and higher priority for the individual who holds these values.

The practices which characterize a lifestyle or working style are verified through evidence of meeting a full range of educational-human needs. For example, Abraham Maslow organized human needs in a hierarchical fashion, and presented them in a sequence in which the latter needs are not met without a certain degree of achievement of the preceding needs. The stages of need satisfaction are: (1) bodily; (2) physical safety; (3) love; (4) self esteem; and (5) self-actualization.[9] An educational practice can thus be verified according to its consequences within a lifestyle (or workstyle) which meets various levels of these human needs; however, their priorities to the individual must be taken into account. When practices with instrumental value are subjected to an evaluation, they may be justified by showing their relationship to intrinsic values acceptable to those who are auditing or questioning the program. This methodology requires a validation of the relationship between the daily practice and a given set of desired behavioral standards. If acceptable standards do not "back up" practices, then new instrumental practices may be substituted so that they will conform to acceptable standards. Any aspect of a program may be validated by demonstrating that it reinforces the desired objectives. The evaluation process itself is a validation of the logical relationship between goals-objectives and programming activities. If the objectives are met, then corresponding program activities are judged to be valuable. If the intrinsic values which "back up" the practices are not acceptable, it becomes necessary to

[9]Abraham H. Maslow, *Motivation and Personality,* second edition (New York: Harper and Row, 1970), pp. 35-58.

explore alternative, intrinsic values which, necessarily, are non-programmed.

Educators are becoming increasingly involved with the second type of program development — the identification and introduction of practices based on emerging, intrinsic values. The literature abounds with case studies of innovative practices as well as alternative educational models. These developments encompass a broad spectrum of cultural and interpersonal values which must be identified and prioritized as a basis for programmatic change.

Acceptability of Program Values. The decision to introduce new values as educational practices is affected by the social acceptability of those values. A culture is defined by interrelated values which serve various agencies and institutions and reinforce particular lifestyles of its individual members. Skinner characterized the healthy culture — one which has survival value — as a social system which makes the individual increasingly sensitive to the long-range consequences of behavior. A stable and healthy culture embodies a set of reinforcers that further the survival of the members — that expand the range of satisfying practices beyond the opportunities available to the individual who acts in isolation from such cooperative ventures.[10] These reinforcers reward (1) the individual, (2) the individual's relationship to peers, and (3) the individual's interaction within the culture or institution. The second and third areas concern the social feasibility of the particular practice. Educational innovations must be feasible in the sense that a minimal level of consensus or acceptability exists within the individual's reference group — the group in which that individual aspires to be a significant participant. In the case of the educational institution it may be the faculty of a given building. This criterion also concerns the relationship of the immediate group to other groups. The degree of acceptability or consensus about values is an important consideration in efforts to introduce new institutional patterns. Since there is dramatic change in the number and diversity of intrinsic rewards to individuals, it is important to involve those individuals in evaluating, redesigning, and testing new cooperative ventures. This test for the acceptability of new values is one aim which is served by internal and external program reviews.

---

[10]Skinner, *op. cit.*, pp. 36-37.

In summary, decisions about program development are based on judgments about the reward value of educational practices — rewards for professionals, students and lay persons. Desirable behaviors may be achieved by altering the reward system (program practice) which is built into program management. The reward value of various practices is clarified by the individuals' priorities among instrumental and intrinsic values, the scope of the values and the acceptability of values among professional and social groups. The required expenditure of effort to change educational practices must approximate perceived benefits or rewards which participants will receive. Primary accountability agents clarify program values through the internal and external reviews and can influence the introduction of new values by engaging staff in negotiating, executing, and revising program development decisions. After clarifying educational values, the primary accountability agent is responsible for the selection and introduction of educational values.

Negotiating Program Development Decisions — the Selection of Educational Values

At each organizational level primary accountability agents are responsible for using recommendations in internal and external reviews to select those practices which will be introduced into the educational program. One of the first tasks is to compare reviews to determine areas of agreement or disagreement. Descriptions of program management may differ, i.e., there may be differing perceptions of the program itself. Other differences may exist in perceptions of logical relationships across various steps in program management. For example, a group of parents may differ from the teacher in their belief that program decision-making is not based on information secured in evaluations. Further discrepancies may exist in perceptions of the workability of each step in program management; or the extent to which program management satisfies the needs of clients, parents, work group members, etc. In other words, conflict may arise due to different values of the several interest groups. These differences can lead to recommendations which may be conflicting, mutually exclusive or complementary. Such contrasts highlight critical issues affecting the workability of program decisions and the extent to which those decisions are satisfying for professionals and program clientele.

Conflicts in perceived needs and suggested recommendations must be negotiated in the program development process. The management of the negotiation is the responsibility of the officer accountable for the program aspects under scrutiny. When an administrative unit recommends program changes which differ from those recommended by the external review team, there are several options for seeking consensus or cooperation on the issues identified in the two reviews. Many approaches may be used for conflict resolution. Each of the five which will be discussed has different advantages depending upon the nature of the conflict and the timeframe in which decisions are required. The five approaches are: (1) the Charette; (2) Delphi; (3) ombudsmen; (4) organizational development; and (5) implementation of model programs.

Charettes may be structured by holding group discussions with members of both review teams and attempting to clarify issues or seek mutually acceptable strategies. During these meetings, each group may present new information for conflict resolution, or propose compromise positions. If the decision does not require immediate resolution, research methodologies such as the Delphi Technique may be employed. In the Delphi Technique, a representative cross section of employees and clients indicate their program expectations and have the opportunity to sequentially alter individual opinion on the basis of choices made by others who are surveyed. On a periodic basis, the respondents receive information about others' opinions and reasons for holding those opinions as they move toward consensus. In the face of continuing conflict, other employees and lay persons may be included as mediators in the discussions or written forms of communication.[11]

The use of a neutral party or ombudsman may be accomplished for each problem as it arises, or an ombudsman may be a full-time employee of the school system. Deutsh maintained that a third party is most successful in achieving compromise when that person has institutional status and resources which can be used in conflict resolution.[12] This notion of a full-time, specialized employee for conflict resolution is magnified in the use of the organizational

[11]A detailed discussion of the Delphi Technique is presented in Richard W. Hostrop, *Managing Education for Results* (Homewood, Illinois: ETC Publications, 1973), pp. 67-87.

[12]Martin Deutsch, "Conflict and Its Resolution," *Conflict Resolution: Contributions of the Behavioral Sciences*, Claggett G. Smith, editor (Notre Dame: University of Notre Dame Press, 1971), p. 55.

development model. Organizational development is a year-round program designed to improve decision-making relationships within and between the work groups of the school system. Rather than constantly reacting to impasse situations, OD creates learning situations in which employees (1) explore and test different decision-making strategies; (2) use group problem solving; and (3) sharpen interpersonal and group communication skills. The adoption of the OD approach indicates a high system priority for continuous improvement in making decisions by consensus or within a cooperative framework.[13]

A final area for the mediation of conflict is empirical testing of alternate approaches. In some cases the disparate points of view may be compared in research settings within the school system. The advantages of this research capability are many, notably the rigorous intellectual and empirical comparisons of differing program features. On the other hand, several research features may prohibit its use for many program development decisions. Cost of researcher services and the length of time for project completion may limit research usefulness to those problems whose long-range effects must be studied. An alternative to the research model is the evaluation of model programs. Implementation of a program proposal on a small scale can give rapid and continuous indications of the success of its instructional goals and objectives as well as procedural advantages and disadvantages. This information may then be used by decision makers for expanding or dismantling the alternative or for refining and continuing the approach. The advantage of this model for change is that innovation is "contained" on a small scale (with limited resources) until enough information is gathered to allow decision makers to proceed comfortably with expanding the new program approaches.

Planning Program Development — the Introduction of Educational Values

As with any new or refined programs, program development must be proposed, approved and planned before the changes are implemented in the administrative units. Using the framework of PPBES, the primary accountability agent provides leadership for

[13]The conceptual framework for OD and organizational case studies are presented by Chris Argyris, *Management and Organizational Development* (New York: McGraw-Hill Book Company, 1971).

establishing goals and objectives for the changes, corresponding program activities and resources, budget, and evaluations of the new programs. A planning document is useful as (1) a proposal to key decision makers including those who provide budgetary support; (2) an information device; and (3) a training document for personnel who implement the changes.[14]

The primary accountability agent must analyze a given program change to determine specific decisions which are required for its implementation. These decisions may be either inside or outside the decision-making authority and responsibility of the PAA. For example, if a high school building principal proposes to hire consultants for curriculum development in career education, it would probably be necessary to secure budgetary support from the central office; gain approvals for the allocation of teacher inservice days; and determine responsibilities of teaching personnel. Although preliminary approvals may have been gained as responses to recommendations in the internal and external reviews, specific details must be negotiated with other professionals. Prior to such negotiation, the PAA must describe the nature of program changes with sufficient clarity for those whose services are required. These details along with their justification can be presented in the form of a program prospectus. The first step in developing such a planning proposal is to delineate aims and objectives of the changes.

Two types of objectives may be served by program change: refinement of currently accepted programs or the introduction of totally new approaches. In the first case, the program changes are primarily deductive since they take place in a framework of acceptance and familiarity. Examples of such changes would be refinements in (1) personnel training, (2) utilization of physical facilities, (3) development of materials, equipment and media and (4) utilization of resource persons. The second level of change is one which introduces new aims and policies. Within an instructional unit, such changes might include (1) the introduction of new subject matter such as a foreign language into an elementary school or (2) alterations in learning theory such as the change from traditional to modern mathematics.

---

[14]Suggestions for implementing curriculum changes are contained in Ronald C. Doll, *Curriculum Improvement: Decision Making and Process*, second edition (Boston: Allyn and Bacon, Inc., 1970), pp. 159-191.

The PAA can use recommendations in the internal and external reviews as a beginning point for delineating goals and objectives as well as ways of implementing them. On the other hand, the planners of change must go beyond the recommendations to specifically determine methods, budgeting, and evaluation strategies. The budgeting requirements may be specified as normal budget requests or may be treated as a special case, especially when seeking funding outside the school district.

Evaluation strategies should also be included in the program plan. These can be specified for the regular operation of program management (i.e., built into the daily evaluation of the proposed change) or they can be constructed for the independent auditor whose recommendations would constitute an external program review.

The PAA can consider several guidelines for developing a planning document that includes goals and objectives, methods, activities, resources, budget, and evaluation strategies:

(1) Detail the program values which will be served by the changes. Whenever possible, state the objectives in behavioral terms. In other words, indicate the probable benefits or rewards of change practices.
(2) Indicate target groups who will be recipients of the rewards including personnel and clients in other units.
(3) If possible, state long-range as well as immediate consequences of proposed practices.
(4) Indicate the degree to which proposed changes deviate from currently acceptable practices. State reasons for these deviations.
(5) Indicate the degree to which proposed changes are consistent with currently accepted values. Specify the scope of proposed changes.
(6) Give evidence which was presented in internal and external reviews to support the objectives.
(7) Cite professionals' formal responses to recommendations contained in internal and external reviews. Specify priorities when possible.
(8) Use evidence from internal and external reviews to support the resources, activities or methods which are proposed.
(9) Give a timetable for implementing the changes and indicate personnel responsible for each event in the sequence.

(10) Present a line item budget for the proposed changes. State the economic contribution of various units.
(11) Detail evaluations according to information to be secured, evaluators, and plans for dissemination of results.

Developing Reward Systems for Program Development

Program accountability culminates in program development — a process requiring a substantial professional commitment of both time and energy. Although program development may be practiced in accordance with the interest and commitment of the individual professional, institutional rewards can promote wider and more intense participation in this aspect of accountability. Among the institutional arrangements giving positive support to program development are team approaches for the planning of program development and the formulation of guidelines. Under the leadership of line officers at the classroom, building and system level, program development gains social approval from advisory teams who create guidelines for the necessary administrative arrangements. These arrangements reflect the needs of various work groups who participate in the accountability effort. There are several options which the planning groups may consider. These include:

1. Released time;
2. Arrangements for granting academic credit;
3. Recognition and promotion;
4. Reimbursement;
5. Provision of additional support personnel; and
6. Other arrangements to ensure the workability of program development.

Personnel may be granted released time through provisions that earmark time slots for program development within the work day or work week. For personnel with year-round appointments, larger blocks of time such as a month or two may be devoted to program development. Consortium arrangements between public schools and colleges or universities may allow the granting of college credit for participation in program development projects. In addition to rewarding the employee, such consortia can provide additional support personnel such as university faculty members or consultants.

Another means of reward is the granting of professional recognition for achievements in program development. Such

recognition may extend anywhere from public announcements and demonstrations to promotion decisions, reimbursements, and merit pay.

A final area of reward emerges from the professional's ability to successfully execute program development. These are the rewards based on making program management a more satisfying educational experience for professionals and clients. The larger organization has partial responsibility for these successes and can exercise its responsibility by committing institutional resources to supporting services. Three of the supporting services pertain to: (a) executing model projects; (b) institutionalizing program changes; and (c) refining innovations. Types of administrative arrangements may include supervision of work group, supervisory assistance, and special centers for change. Supporting services may be developed through supervisory programs or administrative units established for program experimentation and demonstration. These units may be an integral part of the building program or located within a central office. Centers for program development or curriculum building[15] may provide opportunities for individual work groups to:

1. experiment with various program changes in a controlled setting;
2. receive specialized counsel in designing and evaluating program changes;
3. observe the effects of various program options;
4. cooperate with other work groups on areas of mutual concern;
5. work with other agents to negotiate program refinement decisions; and
6. assist in refining innovations.

In addition to serving as an "action" center for testing the feasibility of various approaches to program development, the center can also coordinate program development information emerging on a local and national level. Information specialists in these settings receive, organize and disseminate information about program development throughout the school system. In this manner, the accountability experience of many programs is shared

[15]Suggestions for administering curriculum-making units in public schools are summarized by Harold Rugg and George S. Counts, "A Critical Appraisal of Current Methods of Curriculum-Making," *Curriculum-Making: Past and Present*, Harold Rugg & others, editors (New York: Arno Press & the New York Times, 1969), pp. 425-447.

with those in similar situations. Such information indicates not only promising program options but facilitates the negotiation of any conflicting recommendations which were contained in the internal and external reviews.

Summary

The program development task is one of the basic duties of the professional who is accountable for an instructional or supportive service program. At the classroom level, program development involves the full range of decisions about introducing and refining classroom goals and objectives, teaching strategies and utilization of physical and human resources in the teaching-learning process. Regardless of the program's organizational level, program development is a promising tool for the control of change within the education profession.

Change is a way of life for our contemporary institutions; and in education, it is manifest in changing expectations of clientele, a reduction of professional consensus about goals and objectives, and rapid increases in the numbers of diverse and independent programs. In many instances change produces chaos and confusion within the educational enterprise. Few professionals — whether teacher or administrator — have not felt the stress associated with conflicting demands of students, peers and community interests. Program development is a structure for controlling change by a deliberate process of clarifying, selecting, and introducing educational values.

The program development emphasis within accountability is a systematic, cooperative approach to the refinement of services and the introduction of new programs. Program development requires a study of the relationship between educational practices and the results of implementing those practices. In this process, each reviewer asks the question: "How reasonable are my daily practices?" Answers to this question depend upon the extent to which these practices serve the values of the leader, administrator, and clients. In its simplest form the connection may be drawn by logical deduction of the practice from the value; however, greater rigor is more acceptable and is achieved by an evaluation of the extent to which results of the practice satisfy the needs of program participants. This process of determining the reasonableness of practice is an empirical problem which is addressed in the internal

and external reviews. Program values must be explicit and connected with results of educational practice.

Educational change may also come about by adopting comprehensive, interrelated sets of values. When a current practice is brought into question it may not be possible to justify it by appeal to a particular outcome. In these cases, the practice can be examined by determining its priority within a lifestyle or set of practices which satisfy a full range of human-educational needs; which consist of compatible and mutually reinforcing practices; and which are realistic in terms of acceptability within the social groups in which participants hold membership. Changes in educational practices must be accompanied by rewards which are reasonable to those called upon to effect the changes.

Internal and external reviews contain recommendations about introducing new values into the educational setting. Various decision makers must respond to these recommendations which represent conflicting, alternative or complementary values. The PAA is responsible for negotiations leading to consensus about program development decisions.

Implementation of decisions about program innovations requires a period of program planning and institutionalization of supporting services. The planner-manager is the primary accountability agent for the program under consideration. The responsibilities of the accountability agent involve planning the programming, budgeting and evaluation strategies for the program changes and negotiating these plans with other key personnel.

An important aspect of the institutional commitment to program development is the establishment of rewards which will encourage the various professionals to appropriate and focus their time and energy. In addition to the general approaches of released time, recognition, promotion and reimbursement, there are several innovative options including the granting of academic credit and the development of centers for program development. Such centers can assist in executing model projects, institutionalizing program changes and refining innovations. This approach provides supportive services to the innovator and permits experimentation with program options. When experimentation is conducted prior to making long-range commitments to a program option, the probability is increased for "good" program decisions — those which are both workable and satisfying to professionals and clients.

chapter six implementing accountability practices in the school district

Introduction

Like all other new educational endeavors, building a system of accountabiity demands careful preparation and practice before and during the period of implementation. A detailed plan should be developed and enunciated to all who are involved. In this chapter, a step-by-step process of implementation will be forwarded. The process stresses the activities that develop staff readiness for participation; concrete plans for execution; and the capability to revise the accountability system.

Step I — Assessing Present Accountability Needs

Before launching an accountability effort, there are several types of needs which should be assessed and used to provide direction to the implementation of accountability practices in the school system. An accountability needs assessment is a determination of the degree to which current practice conforms to agreed-upon standards of accountability. The needs assessment should be designed for the standards or model of accountability to be utilized. The four-phase model presented in the preceding chapters is used as the standard for the needs assessment described below.

Identifying the Primary Accountability Agents. The first principle in an accountability system is the delegation of authority

and responsibility to those professionals who are accountable for specific program tasks. It has been argued that the current practice and experience of all professionals must be fully utilized in shaping future program directions. In the preceding chapters, examples have been presented to show that each organizational level has unique program responsibilities and that professionals at each level must have authority to execute those responsibilities. This line of reasoning supports the need to involve teachers, principals and central office personnel in those activities which meet accountability obligations. In turn, their involvement indicates that authority for accountability is dispersed throughout the organization. Professionals are accountable for those programs for which they exercise operational *responsibility* and decision-making *authority*. Personnel in the central office, building level and classroom units should share alike in meeting the accountability challenge. Accountability cannot be imposed or dominated by one particular organizational level; on the other hand, there are appropriate programs for which each level has primary accountability.

Professional reciprocity across organizational levels can be matched with reciprocity in the exercise of line and staff roles of the personnel. For a given program task, professional X assumes line responsibilities and professional Y assumes staff or advisory functions. However, there could be other program tasks for which these roles are reversed: Professional Y is the line officer and professional X performs staff functions. In the accountability effort, line and staff functions are equally important. Furthermore, each professional performs some line and some staff functions depending upon the particular program focus. In order to assess the extent to which these professional roles are clearly defined and accepted in the school district, a checklist of items has been provided in Table VII. These items secure information about teachers' perceptions of the first accountability need: Designation of professionals with authority and responsibility for programs at the classroom, building and system levels. Information about the degree of consensus across teachers', principals', and Superintendent's responses can be secured by administering comparable checklists to principals and the Superintendent (See Appendix).

Analyzing the Internal Review Process. Within an accountability system, the internal program review is a critical procedure for

TABLE VII

TEACHER'S CHECKLIST FOR PROGRAM AUTHORITY & RESPONSIBILITY

ITEMS+		
Classroom	**Building**	**System**
1. My responsibilities for directing the classroom program are: (a) clearly defined to me (b) clearly defined to the principal; (c) clearly defined to the Superintendent; (d) acceptable to me; (e) acceptable to the principal; (f) acceptable to the Superintendent.	4. My responsibilities for providing staff assistance to building-wide programs are: (a) clearly defined to me; (b) clearly defined to the principal; (c) clearly defined to the Superintendent; (d) acceptable to me; (e) acceptable to the principal; (f) acceptable to the Superintendent.	7. My responsibilities for providing staff assistance to system-wide programs are: (a) clearly defined to me; (b) clearly defined to the principal; (c) clearly defined to the Superintendent; (d) acceptable to me; (e) acceptable to the principal; (f) acceptable to the Superintendent.
2. The building principal's responsibilities for providing staff assistance to the classroom programs in this building are: (a) clearly defined to me; (b) clearly defined to the principal; (c) clearly defined to the Superintendent; (d) acceptable to me; (e) acceptable to the principal; (f) acceptable to the Superintendent.	5. The building principal's responsibilities for directing building-wide programs are: (a) clearly defined to me; (b) clearly defined to the principal; (c) clearly defined to the Superintendent; (d) acceptable to me; (e) acceptable to the principal; (f) acceptable to the Superintendent.	8. The building principal's responsibilities for providing staff assistance to system-wide programs are: (a) clearly defined to me; (b) clearly defined to the principal; (c) clearly defined to the Superintendent; (d) acceptable to me; (e) acceptable to the principal; (f) acceptable to the Superintendent.
3. The Superintendent's responsibilities for providing staff assistance to the classroom programs in the building are: (a) clearly defined to me; (b) clearly defined to the principal; (c) clearly defined to the Superintendent; (d) acceptable to me; (e) acceptable to the principal (f) acceptable to the Superintendent.	6. The Superintendent's responsibilities for providing staff assistance to building programs in my school are: (a) clearly defined to me; (b) clearly defined to the principal; (c) clearly defined to the Superintendent; (d) acceptable to me; (e) acceptable to the principal (f) acceptable to the Superintendent.	9. The Superintendent's responsibilities for directing system-wide programs are: (a) clearly defined to me; (b) clearly defined to the principal; (c) clearly defined to the Superintendent; (d) acceptable to me; (e) acceptable to the principal (f) acceptable to the Superintendent.

+Each of the above statements is rated according to the respondent's experience in a current teaching assignment. The following scale is used for each item:

1-Always, 2-Almost Always, 3-Sometimes,
4-Almost Never, 5-Never, 6-Uncertain.

guaranteeing the input of the individual professional. The review not only recaptures basic program features but includes recommendations for growth and innovation within that program. The internal program review must be regarded as an effective tool by school system personnel. That is, IPR recommendations must be an integral part of introducing innovation at every level of the organization. As they serve this function, internal reviews are the "official" position of the primary accountability agent for a particular program. Accountability practices give status to the IPR. In the first place, they become official documents of the institution. Secondly, they require a *response* on the part of appropriate program officers in the district.

In any school program, with or without an accountability system, some form of internal review takes place. The second step in the needs assessment is to determine the roles of various professionals in the internal review process. Survey items appropriate for teaching personnel are exhibited in Table VIII. Corresponding checklists for principals, and the Superintendent are in the Appendix.

Analyzing the External Review Process. A third phase of accountability involves the formal recognition of the school district's relationship to its many publics. In the past, formal recognition has centered on the Board of Education as the major source of input from the larger society. Whereas the first accountability assumption requires the dispersal of professional authority throughout the organization, this commitment involves a dispersal of formal avenues for input from lay persons, community institutions, and professionals serving staff functions for particular programs. The external program review is the recognized channel for building independent recommendations from groups of persons not involved in the daily responsibilities of program operation. When external reviews are taken seriously by the school district, it is a declaration of building an *open* system — an institution whose growth is influenced not only by its diverse internal programs but with input from representatives of a network of private and public interest groups. Together with the internal review, this principle recognizes the recommending process as a critical dimension of decision-making.

To assess the district's status in providing external reviews, illustrative checklist items are contained in Table IX. These items give a profile of the principal's ratings of external reviews. Similar checklists for other professionals in the school district are in the Appendix.

TABLE VIII

TEACHER'S CHECKLIST FOR THE INTERNAL PROGRAM REVIEW

ITEMS+		
Classroom	**Building**	**System**
1. I keep records of the following aspects of classroom programs for which I am responsible: (a) goals and objectives; (b) programming (activities, resources, methods, etc.); (c) evaluations; (d) decision-making based on evaluations.	7. The principal in my building keeps records of the following aspects of his/her program responsibilities: (a) goals and objectives; (b) programming (activities, resources, methods, etc.); (c) evaluations; (d) decision-making based on evaluations.	13. The Superintendent keeps records of the following aspects of programs for which he/she is responsible: (a) goals and objectives; (b) programming (activities, resources, methods, etc.); (c) evaluations; (d) decision-making based on evaluations.
2. I review my goals & objectives according to: (a) their workability; (b) their compatability with programs directed by other professional personnel.	8. The principal in my building reviews his/her goals and objectives according to: (a) their workability; (b) their compatibility with programs directed by other professionals.	14. The Superintendent reviews goals & objectives according to: (a) their workability; (b) their compatibility with programs directed by other professional personnel.
3. I review my programming (methods, activities, resources, planning procedures, etc.) according to: (a) its workability; (b) its compatibility with programs directed by other professional personnel.	9. The principal in my building reviews his/her programming (methods, activities, resources, planning procedures, etc.) according to: (a) its workability; (b) its compatibility with programs directed by other professional personnel	15. The Superintendent reviews programming (methods, activities, resources, planning procedures, etc.) according to: (a) its workability; (b) its compatability with programs directed by by other professional personnel.
4. I review my evaluation activities according to: (a) their workability; (b) their compatibility with programs directed by other professional personnel.	10. The principal in my building reviews his/her evaluation activities according to: (a) their workability; (b) their compatibility with programs directed by other professional personnel.	16. The Superintendent reviews evaluation activities according to: (a) their workability; (b) their compatibility with programs directed by other professional personnel.

TABLE VIII (Continued)

TEACHER'S CHECKLIST FOR THE INTERNAL PROGRAM REVIEW

ITEMS+		
Classroom	**Building**	**System**
5. For programs under my jurisdiction I use reviews to propose changes in:	11. For programs under his/her jurisdiction, the principal uses reviews to propose changes in:	17. For programs under his/her jurisdiction, the Superintendent uses reviews to propose changes in:
(a) goals and objectives; (b) programming (activities, resources, methods, etc.) (c) evaluations; (d) decision-making based on evaluations; (e) study of program areas.	(a) goals and objectives; (b) programming (activities, resources, methods, etc.); (c) evaluations; (d) decision-making based on evaluations; (e) study of program areas.	(a) goals and objectives; (b) programming (activities, resources, methods, etc.); (c) evaluations; (d) decision-making based on evaluation; (e) study of program areas.
6. For program under the direction of other professionals. I use reviews to propose changes in:	12. For programs under the direction of other professionals, the principal uses reviews to propose changes in:	18. For programs under the direction of other professionals, the Superintendent uses reviews to propose changes in:
(a) goal setting: (b) programming (activities, resources, methods, etc.); (c) evaluations; (d) decision-making based on evaluations; (e) study of program areas.	(a) goal setting; (b) programming (activities, resources, methods, etc.); (c) evaluations; (d) decision-making based on evaluations; (e) study of program areas.	(a) goal setting; (b) programming (activities, resources, methods, etc.); (c) evaluations; (d) decisions based on evaluations; (e) study of program areas.

+Each of the above statements is rated according to the respondent's experience in current administrative assignments. The following scale is used for each item:

1-Always	3-Sometimes	5-Never
2-Almost Always	4-Almost Never	6-Uncertain

Describing Program Development. The fourth aspect of a needs assessment is a survey of the current capability of the school district to engage in productive program change. The final step in an accountability process is the introduction of program refinements or innovation within the educational mission — the education of youth or the continuing education of an adult population. This step includes an examination of school district policies and practices — a rethinking of current values and the projection of optional values for the future. Some of the "sacred cows" of educational practice are questioned and replaced. The commitment to program development thrusts the school system into the business of developing and controlling change strategies. Changing roles and

TABLE IX

PRINCIPAL'S CHECKLIST FOR THE EXTERNAL PROGRAM REVIEW

ITEMS+		
Building	**Classroom**	**System**
1. I receive recommendations about the building program from other professionals and groups: (a) teachers; (b) other principals; (c) the Superintendent; (d) parents; (e) other lay persons; (f) consultants.	6. The teachers receive recommendations about the classroom programs from other professional groups: (a) other teachers; (b) principals; (c) the Superintendent; (d) parents; (e) other lay persons; (f) consultants.	11. The Superintendent receives recommendations about the system-wide programs from other professionals and groups: (a) teachers; (b) principals; (c) central office personnel; (d) consultants; (e) parents; (f) other lay persons.
2. I consider making program changes based on the recommendations of: (a) teachers; (b) other principals; (c) the Superintendent; (d) parents; (e) other lay persons; (f) consultants	7. The teachers consider making program changes based on the recommendations of: (a) other teachers; (b) principals; (c) the Superintendent; (d) parents; (e) other lay persons; (f) consultants.	12. The Superintendent considers making program changes based on the recommendations of: (a) teachers; (b) principals; (c) central office personnel; (d) consultants; (e) parents; (f) other lay persons.
3. I communicate my reactions to the recommendations initiated by other professionals and groups: (a) teachers; (b) other principals; (c) the Superintendent; (d) parents; (e) other lay persons; (f) consultants.	8. The teachers communicate their reactions to the recommendations initiated by other professionals and groups; (a) other teachers; (b) principals; (c) the Superintendent; (d) parents; (e) other lay persons; (f) consultants.	13. The Superintendent communicates his/her reactions to the recommendations initiated by other professionals and groups; (a) teachers; (b) principals; (c) central office personnel; (d) consultants (e) parents; (f) other lay persons.
4. When an external group is reviewing the building program, I share program information with those groups: (a) teachers; (b) other principals; (c) the Superintendent; (d) parents; (e) other lay persons; (f) consultants.	9. When an external group is reviewing the classroom programs in this building, the teachers share program information with those groups: (a) other teachers; (b) principals; (c) the Superintendent; (d) parents; (e) other lay persons; (f) consultants.	14. When an external group is reviewing the system-wide program, the Supintendent shares program information with those groups: (a) teachers; (b) principals; (c) central office personnel; (d) other lay persons; (e) consultants; (f) parents.

TABLE IX (Continued)

PRINCIPAL'S CHECKLIST FOR THE EXTERNAL PROGRAM REVIEW

ITEMS+		
Building	**Classroom**	**System**
5. When I receive recommendations which involve programs directed by others, I refer those recommendations to the appropriate professional:	10. When the teachers receive recommendations which involve programs directed by others, the teachers refer those recommendations to the appropriate professional:	15. When the Superintendent receives recommendations which involve programs directed by others, the Superintendent refers the recommendations to the appropriate professionals:
(a) teachers; (b) other principals; (c) supervisors; (d) assistant superintendents; (e) Superintendent; (f) other Central office personnel.	(a) myself; (b) other teachers; (c) supervisors; (d) other principals; (e) assistant superintendents (f) Superintendent.	(a) myself; (b) other principals; (c) teachers; (d) assistant superintendents; (e) supervisors; (f) other central office personnel.

+Each of the above statements is rated according to the respondent's experience in current administrative assignments. The following scale is used for each item:
1-Always, 2-Almost Always, 3-Sometimes,
4-Almost Never, 5-Never, 6-Uncertain.

expectations can be counterproductive for the school system as well as the larger society. To encourage productive change, the school system must take steps to ensure that:

1. ownership of change is held by those who are affected;
2. change is carefully planned and controlled by those who must implement it; and
3. desired changes in professional behavior are reinforced by appropriate institutional rewards.

Another dimension of program development is the use of accountability as a tool for increasing the autonomy of professionals whose "good" decisions shape programs under their jurisdiction. In this sense, accountability is a tool which rewards the individual professional and collective relationships formed by professionals. On the other hand, it should not be a punitive system which builds negative cases against individuals or groups. It simply realigns institutional rewards (budget allocations, institutional goals, promotions, etc.) in favor of those individuals and groups whose decisions have been workable in meeting the educational needs of the students and other clients of the school system. Since accountability is a vehicle for giving institutional support to those

efforts which serve the professional staff and the clients, the school district needs to assess and appropriate resources for achievements in program development. These achievements are identified by the items in Table X which illustrates a profile of program development

TABLE X

SUPERINTENDENT'S CHECKLIST FOR PROGRAM DEVELOPMENT

ITEMS+		
System	**Classroom**	**Building**
1. I make changes in the system-wide program based on: (a) my own review of system-wide goals, objectives and policies; (b) my own review of system-wide activities, methods, and materials; (c) my own review of system-wide evaluations; (d) my own review of system-wide decision-making.	6. The teachers make changes in the classroom program based on: (a) their review of classroom goals and objectives; (b) their review of classroom activities, methods and materials; (c) their review of classroom evaluations; (d) their review of classroom decision-making.	11. The principal makes changes in the building program based on: (a) the principal's review of building goals, objectives and policies; (b) the principal's review of building activities, methods and resources; (c) the principal's review of building evaluations; (d) the principal's review of building decision-making.
2. I make changes in the system-wide program according to the recommendations of: (a) teachers; (b) principals; (c) assistant superintendents; (d) parents & other lay persons.	7. The teachers make changes in classroom programs in accordance with recommendations of: (a) other teachers; (b) principals; (c) the Superintendent; (d) parents and other lay persons.	12. The principal makes changes in the building program in accordance with recommendations of: (a) teachers; (b) parents & other lay persons; (c) other principals; (d) the Superintendent.
3. I use reviews to make program changes in existing: (a) system goals, objectives and policies; (b) system activities, methods and materials; (c) system evaluations; (d) system decision-making based on evaluations.	8. In order to introduce changes into the classroom program, the teachers request new services from other personnel: (a) other teachers; (b) principals; (c) the Superintendent; (d) other central office personnel.	13. The principal uses reviews to make program changes in existing: (a) building goals, objectives and policies; (b) building activities, methods and materials; (c) building evaluations; (d) decision-making based on evaluations.

TABLE X (Continued)

SUPERINTENDENT'S CHECKLIST FOR PROGRAM DEVELOPMENT

ITEMS+		
System	**Classroom**	**Building**
4. In order to introduce changes in the system-wide program, I request new services from other personnel: (a) teachers; (b) principals; (c) assistant superintendents; (d) other central office personnel.	9. The teachers use reviews to make program changes in existing: (a) classroom goals & objectives; (b) classroom activities, methods, and materials; (c) classroom evaluations; (d) classroom decision-making based on evaluations.	14. In order to introduce changes in the building program, the principal requests new services from other personnel: (a) teachers; (b) other principals; (c) the Superintendent; (d) other central office personnel.
5. I plan program changes by specifying: (a) their goals & objectives; (b) ways to evaluate them; (c) human and material resources needed to implement them; (d) their structure, methods, activities, etc.	10. The teachers plan program changes by specifying: (a) their goals & objectives; (b) ways to evaluate them; (c) human and material resources needed to implement them; (d) their structure, methods, activities, etc.	15. The principal plans program changes by specifying: (a) their goals & objectives; (b) ways to evaluate them; (c) human and material resources needed to implement them; (d) their structure, methods, activities, etc.

+Each of the above statements is rated according to the respondent's experience in a current administrative assignment. The following scale is used for each item;

1-Always,	2-Almost Always,	3-Sometimes,
4-Almost Never,	5-Never,	6-Uncertain

rated by the Superintendent. Checklists for other professional personnel are in the Appendix.

Educational leaders who take action to improve accountability practices can assess the status of their district in meeting the four areas which are outlined above. The assessment will suggest a level of readiness for embarking on accountability ventures. Undoubtedly, individual and group ratings will vary along a continuum between the two extremes of highly receptive or highly negative toward the accountability model. However, staff education in the four need areas can occur simultaneously with staff participation in the accountability process.

Step II — Establishing Appropriate Advisory Groups

Planning for accountability can take place at two levels in the district. It can proceed at the level of line officers (probably the Superintendent, assistant superintendents, principals or directors of federal projects) or it can occur through the work of advisory groups. The pattern of leadership necessary for accountability, however, requires both access to institutional rewards and broad institutional participation in the changes required by accountability. The need to meet both goals suggests that planning should proceed with a combination of support from administrative officers as well as advisory groups.

To be successful, accountability should be initiated, or at least fully supported, by the Superintendent and the Board of Education. Commitment at the level of the Superintendent and Board may entail general agreement with the concept of accountability as well as the basic procedures which support that concept. This initial support, however, must be translated into practice through the establishment of an advisory system. The advisory system creates detailed policies and procedures which govern the accountability system and provide rewards for professional groups who implement accountability. The advisory system which is chosen could include representation from every level in the school system as well as other important sectors that will participate in external program reviews. These latter groups include central office personnel, principals, classroom teachers, students, parents and other lay persons, professional consultants, and representatives from community interest groups including businesses and agencies.

Numerous models may be used for developing an advisory system. In Figure 11, several options are illustrated. Option A is a unicameral advisory group with representation from the central office, buildings, classrooms and community groups. A second option consists of a steering committee with the same composition as in Option A but supplemented by similar groups operating at the building level. Option B formally decentralizes the system-wide thrust to include concerns of individual buildings. In the third model, Option C, advisory groups are organized around particular accountability goals, which are described in the next section. A fourth option (D) would be multi-cameral, system-wide representation.

The members and constituency of advisory groups will vary from one system to another. If only one system-wide group — a steering

FIGURE 11

Optional Advisory Patterns for Accountability

Option A
Unicameral System-wide Representation

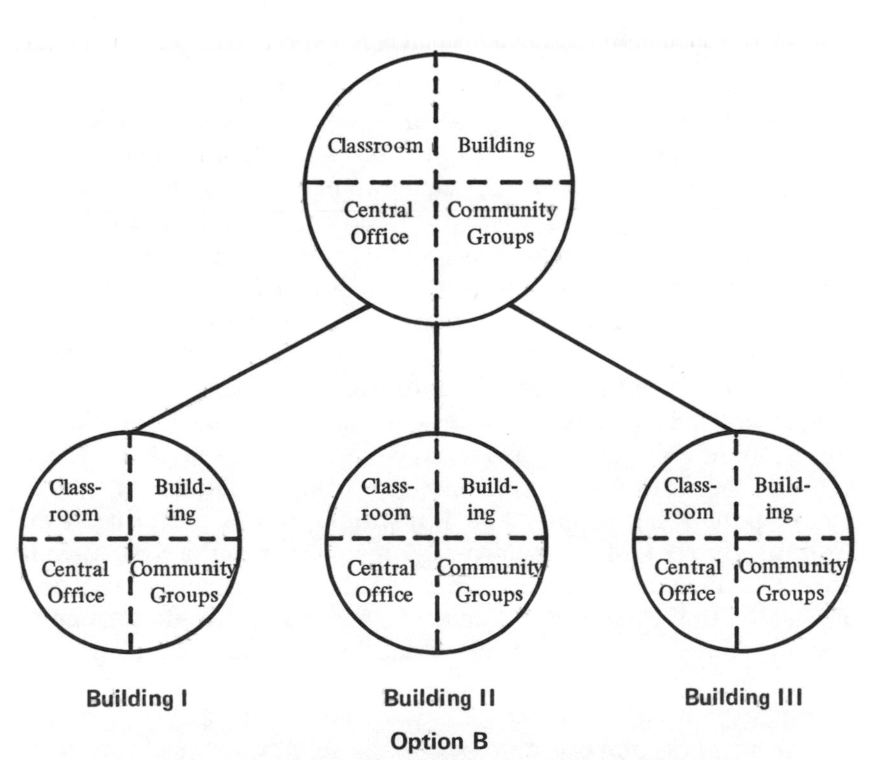

Option B
Unicameral Representation Decentralized by Building

FIGURE 11 (Cont.)
Optional Advisory Patterns for Accountability

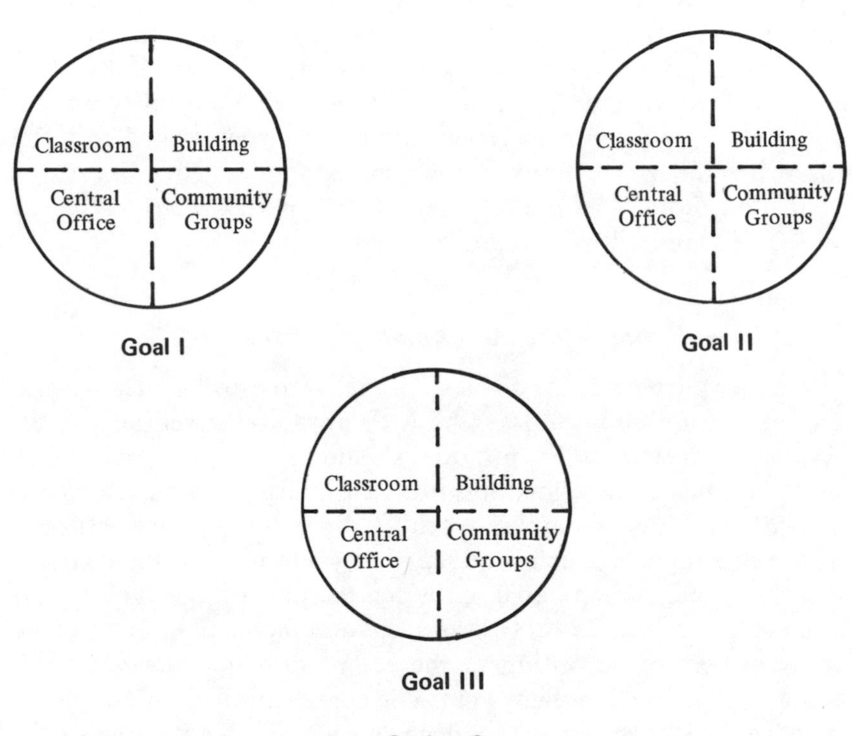

Option C
Unicameral Representation Organized by Goals

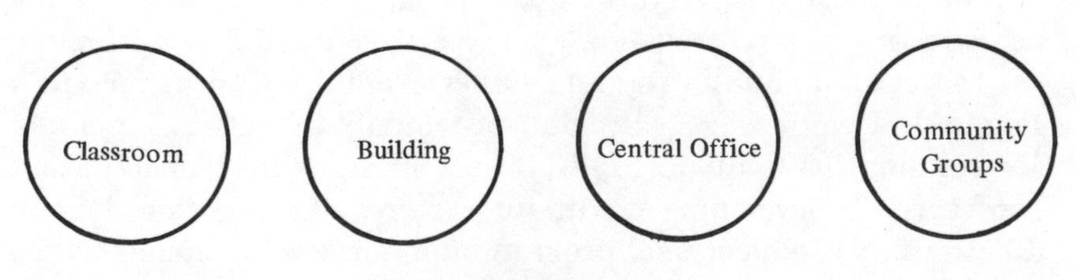

Option D
Multicameral, System-wide Representation

committee — is selected, the decision-making authority is clear (Option A). If some form of Option B is utilized, it likewise implies a form of centralized decision-making authority. Option C and D provide a diffuse system, which may be seen by some as a more cumbersome but effective, decision-making model. Regardless of the type of advisory plan chosen, however, two principles must undergird the system. First, the decision-making authority among the various groups must be clear. Second, the group vested with the authority to make final recommendations to the Board of Education should include representation from all groups affected by the accountability system.

Step III — Defining the Accountability Framework

A major purpose of the advisory groups is to provide a framework for the accountability effort. As with any other program in the system, the accountability program should have its own goals which serve as a benchmark for those who are charged with making the system work. For example, a school district may have a list of fifteen broad goals which were previously adopted by the Board of Education and disseminated to the community at large. The Superintendent and Board adopted the accountability plan in an effort to adjust programs according to the extent to which the goals are being implemented in the system. The degree to which instructional practice achieves diverse, broad goals such as "Knowledge of the Basic Skills" and "Citizenship" is difficult to assess. Yet one of the functions of the advisory group will be to provide guidelines so such assessment is possible. It may be, for example, that the first guideline would indicate the desirability of beginning the accountability effort with those objectives that support the "Knowledge of Basic Skills" goal. Thus, one major duty of the advisory group is to provide a goal orientation and basic direction for the accountability program. Other examples of practices that might be emphasized in the accountability effort include: (1) building instructional units based on student/teacher/parent contracts, (2) developing credit for career and work experiences, (3) introducing educational programming for new clientele such as working adults or senior citizens, (4) developing cross-age instructional programming, or (5) integrating the counseling function into the instructional program.

A second area in which the advisory group provides a framework is the development of procedures to be followed. These procedures should come forth as both the essential mandated procedures and the suggested practices. The former, for example, might insist upon the inclusion of parents as members of the external review team while the latter could include suggestions about other groups that *might* be represented on the team. It is essential that the number of mandates (such as community input) be kept to a minimum and that they reflect a point of view that has substantial justification.

The primary function of the advisory group, therefore, is to establish the direction of the accountability program by defining the major goals and objectives of the program as well as some mandated and suggested procedures which are considered vital to the effort. In establishing this framework, it is necessary from the outset to involve various interest groups whose support is essential for the success of the effort.

Step IV — Establishing an Accountability Timetable

From the beginning of the accountability program, a basic timetable should be adopted as a guide to the effort. Since any worthwhile system involves some planning and coordination, each member involved in the program ought to be working under a common timeframes. As with any new system, the timeframe will have to be altered as events progress. These alterations, however, should result from events within the accountability effort itself rather than the priority placed on accountability by individuals in the system. Any timetable adopted should therefore be based upon the school system's commitment to allow individuals to fully develop their roles in the accountability effort.

All things considered, the planning phase of the accountability effort will probably consume a minimum of one year.[1] At the end of the year, the planning will have progressed, in at least some of the programs, to the point where implementation is possible. When plans in some work settings have been completed, it is probably preferable to begin the accountability process with those programs serving as models rather than delaying until the entire system

[1]An interesting case study of the Louisville Public Schools describes the relationship between dramatic program change and staff planning and preparation time. The Louisville story is described and interpreted in: Richard W. Hostrop, *Managing Education for Results* (Homewood, Illinois: ETC Publications, 1973), pp. 30-46; 50-61.

"catches up." This strategy provides a psychological edge for aggressive work groups and ensures that others have sufficient time to develop a readiness for change. Whatever the circumstances, planning efforts should run the full course before accountability is implemented, and the timeframe should ensure sufficient flexibility so that plans can be developed without placing excessive demands on program staff. The following schedule is an example of accountability events and timing for a two-year span.

FIRST YEAR

May 1........Secure accountability commitment from Board of Education
........Develop plan for involving professional and lay groups in an advising system for accountability

May 15........Select accountability advisory group and begin planning

June 1........Select all planning groups

July 1........Develop an accountability framework
........Begin needs assessment

August 1......Begin staff development and staff planning

August 15.....Conduct system-wide orientation to accountability

September 1 .. Identify all primary accountability agents
.. Specify line program responsibilities of PAA
.. Specify program responsibilities for which PAA serves staff role

November 1 .. Identify classroom accountability goals
.. Identify building accountability goals
.. Identify system accountability goals

April 1....... Develop system of institutional rewards (e.g., released time, clerical services, academic credit, salary and promotion, budget increases)

April 1....... Plan for staff development (consultation, in-service training, etc.)

April 1....... Begin planning of internal reviews
Program targets
Processes
Timing

April 1....... Begin planning external reviews including program targets, team membership, internal coordination, and timing

May 1....... Establish time schedules for internal and external reviews
.... Identify members of external review teams

June 1 .... Coordinate recommendations for institutional rewards across classroom, building and system levels

June 1 .... Coordinate time schedules for internal and external reviews

June 1 .... Establish methods and times for responding to recommendations in reviews

June 1........ Identify key decision points and times for program development

SECOND YEAR

September 5... Begin accountability projects

January 15.... Complete internal and external reviews
.... Disseminate recommendations

February 1... Begin program development based on recommendations in reviews

February 15 .. Begin discussions of program development across overlapping program areas

March 1 ..... Complete responses to recommendations in internal and external reviews

May 1 .... Complete plans for program development for following school year

May 15 ...... Complete evaluations and research studies on accountability (received by planning groups and steering committee)

June 1 .... Refine accountability strategies for continued implementation in following school year

Accountability schedules such as the above example can be revised according to local needs and studies of the effects of the accountabilty events.

Step V — Training and Development of Staff

Support services for accountability include the area of staff development and training. Since the success of the accountability model depends upon its acceptability and usefulness to professionals at *each* organizational level, planning and training should include inductive approaches as well as the more typical deductive approaches. Deductive planning and training involve centralized or system-wide decisions about implementing accountability and staff training approaches. These decisions are then implemented in various work units. On the other hand, inductive tactics decentralize these decisions and permit individual

groups to address their training needs according to specialized expectations. A proper balance between the two approaches can be insured by planning that involves checkpoints for coordination across system-wide and decentralized planning groups. The accountability steering committee or an administrator can be responsible for this type of coordination.

The planning that takes place in preparation for identifying accountability goals, events, and schedules can be a significant part of staff training. For example, each work unit or its representative can be involved in creating a prospectus which includes: (1) identification of primary accountability agents, (2) specification of program tasks and decisions for which the PAA is responsible, (3) specification of related program areas for which the PAA has staff responsibilities, (4) identification of accountability goals, (5) specification of time schedules for internal and external reviews, (6) specification of time schedules for responding to recommendations contained in reviews, (7) identification of staff development needs and (8) selection of appropriate institutional rewards for the participants.

The school district can provide special assistance to each planning group in the form of demonstrations, special courses and workshops, and services of consultants who participate in the planning process. As the training and planning continue, the steering committee can review system, building and classroom prospectuses and conduct sessions for the coordination of various plans.

Time allocated to the training and planning function should be sufficient for the full development of each work unit's plan and the coordination of plans across all organizational levels. The coordination should result in plans which are complementary, efficient and cooperatively based. The timeframe should also ensure sufficient flexibility so that comprehensive plans can be developed without placing excessive demands on program staff. Intensive training and planning may be scheduled for the summer months or proceed more gradually during the school year.

Step VI — Initiating the Accountability Program

As stated earlier, the accountability program will probably begin with a few programs which have produced well-developed plans at an early stage. These programs will initiate accountability as

professionals in other areas complete their accountability planning. It is crucial that these early efforts meet with some success in terms of the accountability process itself. To ensure such success, the administration must:

1. provide the decentralized decision-making which is essential to accountability;
2. provide the necessary resource persons for the proper review, evaluation, and development of the programs in question;
3. provide clear and concise results of the accountability effort to others in the school system; and
4. reward the efforts of the primary accountability agents in the model programs.

These pilot efforts will probably reveal imperfections in the system. Some of the problems will need to be resolved immediately. One of the critical questions that often occurs during the first stages of innovation is the degree to which plans should be changed in order to meet some initial success. Within an accountability system, this consideration is crucial since initial success is a must.

Dramatic changes in the basic accountability framework should be avoided if at all possible, and capricious decisions to change tactics should also be guarded against. To prevent such occurrences, the group may decide that changes should have the agreement of a substantial number of individuals affected by the change. (One school district, for example, maintains its objectives until 70 percent of the teachers affected by the objective vote to alter it). In the execution of such decision-making, care must be taken to ensure that changes do not violate the basic principles of the accountability system. If, for example, the external program review were scheduled every other year for each program and this proved to be too time consuming, the review could be rescheduled for three or four-year intervals. But external program reviews are *necessarily* time consuming and professionals need to avoid the expedient but erroneous assumption that internal program reviews are sufficient. Rather than violate the fundamental concept of independent auditing, the steering committee must devise conditions which are acceptable and rewarding to staff.

Step VII — Evaluating and Refining Accountability Practices

While numerous means of evaluation of the accountability program are possible, three methods of review seem pertinent. One

approach for evaluating an accountability program is to assess the extent to which all steps in the schedule of events (either the above example or a similar one tailored to local needs) are implemented. The second strategy assesses the extent of participation in the implementation of the schedule by the full range of professional staff and lay persons. Finally, there is the question of the quality of participation and/or the success of the particular event.

Data for the accountability review can include a continuous log which records the occurrence of events, degree of participation, outcomes and evaluations of participants — evidence needed to assess the status of accountability practices. A second source of evidence for evaluating accountability goals is documentation of the extent to which accountability goals have been achieved. In this sense, the degree and nature of program development, participation of new sub-populations in decision-making, and alterations in institutional reward systems are changes which are systematically sought through accountability efforts. If the school system begins to achieve new goals and consistently maintains or accelerates this level of achievement, there is substantial evidence that accountability is having an impact on educational programs. This evidence may be supplied by primary accountability agents or planning groups.

Records of accountability processes and achievements can be maintained internally by planning groups and primary accountability agents and other checks may be conducted by independent auditors. Internal monitoring can be conducted according to monthly or even weekly schedules. Independent audits, which are less concerned with frequent refinements, should probably focus on annual or semi-annual changes in accountability procedures. In this case, reviews can be scheduled to allow feedback at two or three checkpoints extending from the point of initiating accountability efforts to the end of the first year of their operation.

The evaluations described above are formative in nature and probably have greater immediate utility for the school district than research studies designed to compare different accountability models or to test predictive relationships. The aim of basic research on accountability is primarily to increase generalizable knowledge about accountability. In contrast, formative evaluations proceed within program operations in which the primary goal is to improve service to clients continuously. In this developmental context there

is less long-range control over the relationship between a well defined program strategy and its effects. On the other hand, in basic research, program strategy must be defined in advance and adhered to throughout the duration of the project. This is essential for comparing the program with other programs or testing a cause-effect relationship with other variables. In a formative evaluation setting, unanticipated program change may be instituted according to new knowledge of program effects on clients. Thus, formative evaluations are compatible with a fluid program, which is highly desirable in the operational setting. Provided below is a summary of the different questions which may be addressed by formative evaluations and research studies of accountability practices.

Formative Evaluations	Research Studies
*To what extent were events in the accountability schedule implemented?	*What institutional practices may be isolated as predictors of educational accountability?
*To what extent were accountability goals achieved? program changes participation of staff, clients & lay persons implementation of institutional rewards	*Which accountability practices may be employed to create specific changes in educational output?
*To what extent were accountability events successful as perceived by staff, clients, lay persons and independent auditors?	*What are the comparative advantages of differing accountability strategies for producing specific educational outcomes?

In both types of studies, planning groups, primary accountability agents and external auditors may be involved. However, the involvement of program officers in basic research will obviously be limited unless released time can be provided or unless their services are merely supportive to the researchers with full-time involvement in the project.

Recommendations from both types of accountability studies can be used to introduce refinements or new approaches to accountability. The different sets of questions which distinguish the two types of studies indicate that research implications usually apply to the broader, long-range program features; whereas, formative evaluations suggest changes which can be made on a daily or weekly basis.

Summary

Accountability is a commitment which does not overlay the "rest" of the school program. It cannot be an external requirement which is imposed on the professional or the school district. Testing programs required by "outsiders" or standards for professional "survival" are far cries from the accountability model which has been described. To be successful, accountability must *on its own merits* become an increasing part of professional ethics and practice. Such integration requires broad participation in planning and executing accountability practices. It requires a recognition of distinctive program responsibility and authority at every organizational level; a respect for staff as well as line responsibilities; a system for rewarding individuals whose decisions create or expand services to clients; a willingness of each professional to review programs for which responsibility is exercised the development of formal input from lay persons and professionals not directly responsible for programs; and commitments to introducing program changes with beneficial effects.

Using the above criteria as indexes of accountability, it is clear that a degree of accountability commitment currently exists (to greater or lesser extents) in the school districts of this nation. A basic question becomes: To what extent is the district accountable? And how can the district increase its capacity for accountability? The model proposed in this book specifies that there are four types of interrelated practices which determine the extent to which accountability is realized. These are (1) the identification of primary accountability agents and their respective program responsibilities, (2) the execution of internal program reviews by those program officers, (3) the completion of external program reviews by independent auditors including professionals and lay persons, and (4) the use of reviews to develop programs by introducing refinements and innovations.

Professionals in a given school district can control the extent to which the above practices are a part of daily practice and long-range planning. A receptive leadership and endorsement of the Board of Education provide important access to institutional rewards, but they are only part of the required resources for building accountability systems. Professional involvement at every organizational level and representation for each work unit should

initiate the planning stages of each accountability project. Planning includes the development of (1) an advising system for guiding accountability practices, (2) a set of programmatic goals including instructional objectives, (3) goals for a broadened base of decision-making, (4) a system for staff training and development, (5) appropriate institutional rewards, and (6) a schedule of required accountability events and responsible personnel. As with every instructional strategy, accountability practices should be selectively shaped according to their degree of workability and the existence of promising alternatives. Continuous evaluations and accountability research provide evidence for these adjustments of accountability practices.

selected bibliography

Browder, Lesley H., Jr. *An Administrator's Handbook on Educational Accountability.* Washington, D.C.: American Association of School Administrators, 1973.

Browder, Lesley H., Jr. *Emerging Patterns of Administrative Accountability.* Berkeley: McCutchan Publishing Company, 1971.

Browder, Lesley H., Jr., William A. Atkins, Jr., and Esin Kaya. *Developing an Educationally Accountable Program.* Berkeley: McCutchan Publishing Company, 1973.

Combs, Arthur W. *Educational Accountability: Beyond Behavioral Objectives.* Washington, D.C.: Association for Supervision and Curriculum Development, 1972.

Hostrop, Richard W., J.A. Mecklenburger and J.A. Wilson. *Accountability for Educational Results.* Hamden, Connecticut: Linnet Books, 1973.

Hostrop, Richard W. *Managing Education for Results.* Homewood, Illinois: ETC Publications, 1975.

Knezevich, Steven J., editor. *Creating Appraisal and Accountability Systems.* San Francisco: Jossey Bass, 1973.

Leight, Robert L., editor. *Philosophers Speak on Accountability in Education.* Danville, Illinois: The Interstate Printers and Publishers, 1973.

Lessinger, Leon M. *Every Kid A Winner.* Palo Alto, California: Science Research Associates, 1970.

Lessinger, Leon M. and Associates. *Accountability: Systems Planning in Education*, Creta D. Sabine, editor. Homewood, Illinois: ETC Publications, 1973.

Lessinger, Leon M. and Ralph W. Tyler. *Accountability in Education.* Worthington, Ohio: Charles A. Jones Publishing Company, 1971.

Olson, Arthur V. and Joe Richardson, editors. *Accountability: Curricular Applications.* San Francisco: Intext Educational Publishers, 1972.

Wynne, Edward. *The Politics of School Accountability.* Berkeley: McCutchan Publishing Company, 1972.

Sciara, Frank J. and Richard K. Jantz, editors. *Accountability in American Education.* Boston: Allyn and Bacon, Inc., 1972.

appendix

Accountability Checklists

Accountability checklists were designed to provide information about the four components of the accountability model which have been developed and presented in this book. In the assessment of accountability needs, these surveys can be administered to teachers, principals and the Superintendent whose perceptions are indicators of the nature and extent of accountability practices. Similar checklists can be constructed for other groups of professionals such as school counselors or central office supervisors. For purposes of illustration, the following checklists are included:

1. Teacher's, Principal's and Superintendent's Checklists on Program Responsibility;
2. Teacher's, Principal's and Superintendent's Checklists on Internal Program Reviews;
3. Teacher's, Principal's and Superintendent's Checklists on External Program Reviews; and
4. Teacher's, Principal's and Superintendent's Checklists on Program Development.

Each of the above checklists is printed in its entirety following a discussion of checklist scoring and interpretation.

Checklist Scoring and Interpretation

Each checklist surveys perceptions of accountability practices as they pertain to the three levels of programs (classroom, building and system) that exist in a district. Thus, each checklist has three subscores corresponding to the three program levels, as well as an overall score. For example, in Checklist I, on page 115, a teacher's responses indicate perceptions of professional responsibility at three levels: classroom program, building program and system-wide program. The composite score for a teacher on this checklist indicates a perception of overall program responsibility in the district.

To obtain a subscore of classroom program responsibility in Checklist I "Teacher's Ratings of Program Responsibility," it is necessary to average the scores or responses to subsets of the checklist items. Specifically, items 1, 2, and 3 on page 115 provide eighteen ratings of professional responsibility for the classroom program. Each rating ranges from a score of "1," which indicates a

relatively high degree of accountability, to a score of "6," which designates the absence of an accountability system. The average of the eighteen responses provides a subscore reflecting the teacher's, principal's and Superintendent's responsibility to the *classroom* program. The second subscore on program responsibility pertains to *building* programs and it is calculated by averaging responses to items 4, 5, and 6 on page 116. A third subscore, program responsibility in the *system*, is determined by averaging the eighteen responses to items 7, 8, and 9 on page 116. Finally, an overall score on this checklist is determined by averaging all fifty-four ratings. In a similar manner, the other checklists (internal reviews, external reviews, and program development) provide three subscores — one for each of the three program levels — and an overall score. Table A details the item numbers used to obtain subscores for each of the checklists. The average of *all* responses on a given checklist yields the total score for the individual rater.

TABLE A

Item Numbers for Subscores on Accountability Checklists

Checklist	Subscore	Item Numbers		
		Form A (teachers)	Form B (principals)	Form C (Superintendent)
Program Responsibility	Classroom PR	1 - 3	4 - 6	4 - 6
	Building PR	4 - 6	1 - 3	7 - 9
	System PR	7 - 9	7 - 9	1 - 3
Internal Reviews	Classroom IPR	1 - 6	7 -12	7 -12
	Building IPR	7 -12	1 - 6	13-18
	System IPR	13-18	13-18	1 - 6
External Reviews	Classroom EPR	1 - 5	6 -10	6 -10
	Building EPR	6 -10	1 - 5	11-15
	System EPR	11-15	11-15	1 - 5
Program Development	Classroom PD	1 - 5	6 -10	6 -10
	Building PD	6 -10	1 - 5	11-15
	System PD	11-15	11-15	1 - 5

The subscore and total score may be interpreted* as follows:

Score	Interpretation
1.0 - 2.0	Always or Almost Always Accountable
2.1 - 3.9	Sometimes Accountable
4.0 - 5.0	Almost Never or Never Accountable
5.1 - 6.0	No Accountability System

Table B summarizes the scores that are provided by the checklists on program responsibility. Similar sets of scores are available with each of the other checklists.

TABLE B

Accountability Scores for Program Responsibility

Target Groups	Classroom PR Subscore	Building PR Subscore	System PR Subscore	Total PR Score
Teacher	–	–	–	–
Principal	–	–	–	–
Superintendent	–	–	–	–

Several methods of analyzing the data are possible. Three methods, however, are suggested for the systematic assessment of accountability practices. The first method is to ignore — in the beginning — all results except those subscores directly related to the individual in his role as primary accountability agent, i.e., the teacher in the classroom program, the principal with the building program and the Superintendent with the system program. When the individual examines and evaluates his own role in program responsibility, internal and external reviews, and program development, some personal strategies for accountability can be devised. For example, this approach requires teachers to interpret their responses to items applying only to their own roles in the *classroom* program; the principals to use items that pertain to their roles in *building* programs; and the Superintendent to examine items that assesses his personal impact on the *system* programs.

*Score interpretations are based on the definition of the response scale of the survey instrument rather than scaling results, which are not currently available.

A second method of assessing the status of accountability is to analyze the individual's relationship to programs for which he does not have primary accountability. In this case teachers would use their self-ratings about their roles in the building and system programs; principals would analyze item scores about their roles in classroom and system programs; and the Superintendent would focus on his scores for supporting classroom and building programs.

A third method enables professionals to make judgments about differing perceptions of subgroups of educators. In this case the scores of one group of professionals are compared with scores of another group by computing their difference. The difference scores provide a measure of agreement-disagreement between two groups of professionals. For example, in a given building the average teacher's score on classroom program responsibility may be 1.5 (always or almost always accountable). The principal's subscore of 3.0 yields a difference score of 1.5, which reflects substantial disagreement. Information provided by this difference score may lead to productive discussions of its causes and selection of alternative approaches to classroom accountability. For each subscore or total score, the following system is used to compute difference scores:

Difference Score	Formula
Teacher/Principal	Teacher Score - Principal Score
Principal/Superintendent	Principal Score - Superintendent Score
Teacher/Superintendent	Teacher Score - Superintendent Score

Although national forms do not exist for score interpretation, it may be useful to compute difference scores and compare them within a district. For example, buildings which have relatively small teacher/principal difference scores as well as favorable accountability subscores for both teachers and principal, may be singled out for initial accountability efforts.

Accountability Checklist I: Teacher's Ratings of Program Responsibility

This checklist is designed to survey your perceptions of professional responsibility in your classroom, your building program and the system-wide program for your school district. A professional has primary accountability for those programs for which he exercises both decision-making authority and responsibility. In addition to these direct responsibilities, a professional has staff or advisory relationships to other programs. This survey provides information about the professional responsibility exercised by you; the building principal and the Superintendent of the school district.

Please rate each of the following statements according to your experience in your current teaching assignment. Use the following scale for each item: 1-Always, 2-Almost Always, 3-Sometimes, 4-Almost Never, 5-Never, 6-Uncertain. Place the number of the appropriate rating in the blank next to the item.

PR (Form A)

1. My responsibilities for directing the classroom program are:
   _____ clearly defined to me; _____ acceptable to me;
   _____ clearly defined to the principal; _____ acceptable to the principal;
   _____ clearly defined to the Superintendent; _____ acceptable to the Superintendent.

2. The building principal's responsibilities for providing staff assistance to the classroom programs in this building are:
   _____ clearly defined to me; _____ acceptable to me;
   _____ clearly defined to the principal; _____ acceptable to the principal;
   _____ clearly defined to the Superintendent; _____ acceptable to the Superintendent.

3. The Superintendent's responsibilities for providing staff assistance to the classroom programs in the building are:
   _____ clearly defined to me; _____ acceptable to me;
   _____ clearly defined to the principal; _____ acceptable to the principal
   _____ clearly defined to the Superintendent; _____ acceptable to the Superintendent.

4. My responsibilities for providing staff assistance to building-wide programs are:

_____ clearly defined to me; _____ acceptable to me;
_____ clearly defined to the principal; _____ acceptable to the principal;
_____ clearly defined to the Superintendent; _____ acceptable to the Superintendent.

5. The building principal's responsibilities for directing building-wide programs are:

_____ clearly defined to me; _____ acceptable to me;
_____ clearly defined to the principal; _____ acceptable to the principal;
_____ clearly defined to the Superintendent; _____ acceptable to the Superintendent.

6. The Superintendent's responsibilities for providing staff assistance to building programs in my school are:

_____ clearly defined to me; _____ acceptable to me;
_____ clearly defined to the principal; _____ acceptable to the principal;
_____ clearly defined to the Superintendent; _____ acceptable to the Superintendent.

7. My responsibilities for providing staff assistance to system-wide programs are:

_____ clearly defined to me; _____ acceptable to me;
_____ clearly defined to the principal; _____ acceptable to the principal;
_____ clearly defined to the Superintendent; _____ acceptable to the Superintendent.

8. The building principal's responsibilities for providing staff assistance to system-wide programs are:

_____ clearly defined to me; _____ acceptable to me;
_____ clearly defined to the principal; _____ acceptable to the principal;
_____ clearly defined to the Superintendent; _____ acceptable to the Superintendent.

9. The Superintendent's responsibilities for directing system-wide programs are:

_____ clearly defined to me; _____ acceptable to me;
_____ clearly defined to the principal; _____ acceptable to the principal;
_____ clearly defined to the Superintendent; _____ acceptable to the Superintendent.

Accountability Checklist II: Principal's Ratings of Program Responsibility

This checklist is designed to survey your perceptions of professional responsibility in your building program, classroom programs, and system-wide programs of the school district. A professional has primary accountability for those programs for which he exercises both decision-making authority and responsibility. In addition to these direct responsibilities, a professional has staff or advisory relationships to other programs. This survey provides information about the professional responsibility exercised by you, the teachers in the building, and the Superintendent of the school district.

Please rate each of the following statements according to your experience in your current administrative assignment. Use the following scale for each item: 1-Always, 2-Almost Always, 3-Sometimes, 4-Almost Never, 5-Never, 6-Uncertain. Place the number of the appropriate rating in the blank next to the item.

PR (Form B)

1. My responsibilities for directing building programs are:

_____ clearly defined to me; _____ acceptable to me;
_____ clearly defined to teachers; _____ acceptable to the teaching staff;
_____ clearly defined to the Superintendent; _____ acceptable to the Superintendent.

2. The teaching staff's responsibilities for providing staff assistance to building programs are:

_____ clearly defined to me; _____ acceptable to me;
_____ clearly defined to teachers; _____ acceptable to the teaching staff;
_____ clearly defined to the Superintendent; _____ acceptable to the Superintendent.

3. The Superintendent's responsibilities for providing staff assistance to building programs are:

_____ clearly defined to me; _____ acceptable to me;
_____ clearly defined to teachers; _____ acceptable to the teaching staff;
_____ clearly defined to the Superintendent; _____ acceptable to the Superintendent.

4. My responsibilities for providing staff assistance to classroom programs are:
   _____ clearly defined to me; _____ acceptable to me;
   _____ clearly defined to teachers; _____ acceptable to the teaching staff;
   _____ clearly defined to the Superintendent; _____ acceptable to the Superintendent.

5. The teaching staff's responsibilities for directing classroom programs are:
   _____ clearly defined to me; _____ acceptable to me;
   _____ clearly defined to teachers; _____ acceptable to the teaching staff;
   _____ clearly defined to the Superintendent; _____ acceptable to the Superintendent.

6. The Superintendent's responsibilities for providing staff assistance to classroom programs in the building are:
   _____ clearly defined to me; _____ acceptable to me;
   _____ clearly defined to teachers; _____ acceptable to the teaching staff;
   _____ clearly defined to the Superintendent; _____ acceptable to the Superintendent.

7. My responsibilities for providing staff assistance to system-wide programs are:
   _____ clearly defined to me; _____ acceptable to me;
   _____ clearly defined to teachers; _____ acceptable to the teaching staff;
   _____ clearly defined to the Superintendent; _____ acceptable to the Superintendent.

8. The teaching staff's responsibility for providing staff assistance to system-wide programs are:
   _____ clearly defined to me; _____ acceptable to me;
   _____ clearly defined to teachers; _____ acceptable to the teaching staff;
   _____ clearly defined to the Superintendent; _____ acceptable to the Superintendent.

9. The Superintendent's responsibilities for directing system-wide programs are:
   _____ clearly defined to me; _____ acceptable to me;
   _____ clearly defined to teachers; _____ acceptable to the teaching staff;
   _____ clearly defined to the Superintendent; _____ acceptable to the Superintendent;

Accountability Checklist III: Superintendent's Ratings of Program Responsibility

This checklist is designed to survey your perceptions of professional responsibility in your system-wide programs, building programs and classroom programs. A professional has primary accountability for those programs for which he exercises both decision-making authority and responsibility. In addition to these direct responsibilities, a professional has staff or advisory relationships to other programs. This survey provides information about the professional responsibility exercised by you, the teachers, and principals in the school district.

Please rate each of the following statements according to your experience in your current administrative assignment. Use the following scale for each item: 1-Always, 2-Almost Always, 3-Sometimes, 4-Almost Never, 5-Never, 6-Uncertain. Place the number of the appropriate rating in the blank next to the item.

PR (Form C)

1. My responsibilities for directing system-wide programs are:
   _____ clearly defined to me; _____ acceptable to me;
   _____ clearly defined to teachers; _____ acceptable to the teaching staff;
   _____ clearly defined to principals. _____ acceptable to principals.

2. The teaching staff's responsibilities for providing staff assistance to system-wide programs are:
   _____ clearly defined to me; _____ acceptable to me;
   _____ clearly defined to teachers; _____ acceptable to the teaching staff;
   _____ clearly defined to principals; _____ acceptable to principals.

3. The principals' responsibilities for providing staff assistance to system-wide programs are:
   _____ clearly defined to me; _____ acceptable to me;
   _____ clearly defined to teachers; _____ acceptable to the teaching staff;
   _____ clearly defined to principals; _____ acceptable to principals.

4. My responsibilities for providing staff assistance to classroom programs are:

_____ clearly defined to me;
_____ clearly defined to teachers;
_____ clearly defined to principals;
_____ acceptable to me;
_____ acceptable to the teaching staff;
_____ acceptable to principals.

5. The teaching staff's responsibilities for directing classroom programs are:

_____ clearly defined to me;
_____ clearly defined to teachers;
_____ clearly defined to principals;
_____ acceptable to me;
_____ acceptable to the teaching staff;
_____ acceptable to principals.

6. The principals' responsibilities for providing staff assistance to classroom programs are:

_____ clearly defined to me;
_____ clearly defined to teachers;
_____ clearly defined to principals;
_____ acceptable to me;
_____ acceptable to the teaching staff;
_____ acceptable to principals.

7. My responsibilities for providing staff assistance to building programs are:

_____ clearly defined to me;
_____ clearly defined to teachers;
_____ clearly defined to principals;
_____ acceptable to me;
_____ acceptable to the teaching staff;
_____ acceptable to principals.

8. The teaching staff's responsibilities for providing staff assistance to building programs are:

_____ clearly defined to me;
_____ clearly defined to teachers;
_____ clearly defined to principals;
_____ acceptable to me:
_____ acceptable to the teaching staff;
_____ acceptable to principals.

9. The principals' responsibilities for directing building programs are:

_____ clearly defined to me;
_____ clearly defined to teachers;
_____ clearly defined to principals;
_____ acceptable to me;
_____ acceptable to the teaching staff;
_____ acceptable to principals.

Accountability Checklist IV: Teacher's Ratings of Internal Program Reviews

This checklist was designed to survey your perceptions of internal review processes in your classroom, your building and the system-wide program. A professional conducts internal reviews of programs under his direction when three steps are followed: (1) records are used to provide descriptions of those programs, (2) programs are reviewed according to workability and other criteria and (3) recommendations are made for improving those programs. This survey provides information about the internal reviews conducted by you, the building principal and the Superintendent in the school district.

Please rate each of the following statements according to your experience in your current teaching assignment. Use the following scale for each item: 1-Always, 2-Almost Always, 3-Sometimes, 4-Almost Never, 5-Never, 6-Uncertain. Place the number of the appropriate rating in the blank next to the item.

IPR (Form A)

1. I keep records of the following aspects of classroom programs for which I am responsible:
   _____ goals and objectives;
   _____ programming (activities, resources, methods, etc.)
   _____ evaluations;
   _____ decision-making based on evaluations.

2. I review my goals and objectives according to:
   _____ their workability;
   _____ their compatibility with programs directed by other professional personnel.

3. I review my programming (methods, activities, resources, planning procedures, etc.) according to:
   _____ its workability;
   _____ its compatibility with programs directed by other professional personnel.

4. I review my evaluation activities according to:
   _____ their workability;
   _____ their compatibility with programs directed by other professional personnel.

5. For programs under my jurisdiction I use reviews to propose changes in:
   _____ goals and objectives;
   _____ programming (activities, resources, methods, etc.);
   _____ evaluations;
   _____ decision-making based on evaluations;
   _____ the study of program areas.

6. For programs under the direction of other professionals, I use reviews to propose changes in:
   _____ goal setting;
   _____ programming (activities, resources, methods, etc.);
   _____ evaluations;
   _____ decision-making based on evaluations;
   _____ the study of program areas.

7. The principal in my building keeps records of the following aspects of his/her program responsibilities:
   _____ goals and objectives;
   _____ programming (activities, resources, methods, etc.);
   _____ evaluations;
   _____ decision-making based on evaluations;

8. The principal in my building reviews his/her goals and objectives according to:
   _____ their workability;
   _____ their compatibility with programs directed by other professionals.

9. The principal in my building reviews his/her programming (methods, activities, resources, planning procedures, etc.) according to:
   _____ its workability;
   _____ its compatibility with programs directed by other professional personnel.

10. The principal in my building reviews his/her evaluation activities according to:
   _____ their workability
   _____ their compatibility with programs directed by other professional personnel.

11. For programs under his/her jurisdiction, the principal uses reviews to propose changes in:
   _____ goals and objectives;
   _____ programming (activities, resources, methods, etc.);
   _____ evaluations;
   _____ decision-making based on evaluations;
   _____ study of program areas.

12. For programs under the direction of other professionals, the principal uses reviews to propose changes in:
    _____ goal setting;
    _____ programming (activities, resources, methods, etc.);
    _____ evaluations;
    _____ decision-making based on evaluations;
    _____ study of program areas.

13. The Superintendent keeps records of the following aspects of programs for which he/she is responsible:
    _____ goals and objectives;
    _____ programming (activities, resources, methods, etc.);
    _____ evaluations;
    _____ decision-making based on evaluations.

14. The Superintendent reviews goals and objectives according to:
    _____ their workability;
    _____ their compatibility with programs directed by other professional personnel.

15. The Superintendent reviews programming (methods, activities, resources, planning procedures, etc.) according to:
    _____ its workability;
    _____ its compatibility with programs directed by other personnel.

16. The Superintendent reviews evaluation activities according to:
    _____ their workability;
    _____ their compatibility with programs directed by other professional personnel.

17. For programs under his/her jurisdiction, the Superintendent uses reviews to propose changes in:
    _____ goals and objectives;
    _____ programming (activities, resources, methods, etc.);
    _____ evaluations;
    _____ decision-making based on evaluations;
    _____ study of program areas.

18. For programs under the direction of other professionals, the Superintendent uses reviews to propose changes in:
    _____ goal setting;
    _____ programming (activities, resources, methods, etc.);
    _____ evaluations;
    _____ decision-making based on evaluations;
    _____ study of program areas.

Accountability Checklist V: Principal's Ratings of Internal Program Reviews

This checklist was designed to survey your perceptions of internal review processes in your building program, classroom programs and system-wide programs. A professional conducts internal reviews of programs under his direction when three steps are followed: (1) records are used to provide descriptions of those programs, (2) programs are reviewed according to workability and other criteria, and (3) recommendations are made for improving those programs. This survey provides information about the internal reviews conducted by you, classroom teachers in your building, and the Superintendent of the school district.

Please rate each of the following statements according to your experience in your current administrative assignment. Use the following scale for each item: 1-Always, 2-Almost Always, 3-Sometimes, 4-Almost Never, 5-Never, 6-Uncertain. Place the number of the appropriate rating in the blank next to the item.

IPR (Form B)

1. I keep records of the following aspects of building programs for which I am responsible.
   _____ goals and objectives;
   _____ programming (activities, resources, methods, etc.);
   _____ evaluations;
   _____ decision-making based on evaluations.

2. I review my goals and objectives according to:
   _____ their workability;
   _____ their compatibility with programs directed by other professionals.

3. I review my programming (methods, activities, resources, planning procedures, etc.) according to:
   _____ its workability;
   _____ its compatibility with programs directed by other professionals.

4. I review my evaluation activities according to:
   _____ their workability;
   _____ their compatibility with programs directed by other professional personnel.

5. For programs under my jurisdiction I use reviews to propose changes in:
   _____ goals and objectives;
   _____ programming (activities, resources, methods, etc.);
   _____ the study of program areas;
   _____ evaluations;
   _____ decision-making based on evaluations.

6. For programs under the direction of other professionals, I use reviews to propose changes in:
   _____ goal setting;
   _____ programming (activities, resources, methods, etc.);
   _____ evaluations;
   _____ decision-making based on evaluations;
   _____ the study of program areas.

7. The teachers in my building keep records of the following aspects of their program responsibilities:
   _____ goals and objectives;
   _____ programming (activities, resources, methods. etc.);
   _____ evaluations;
   _____ decision-making based on evaluations.

8. The teachers in my building review their goals and objectives according to:
   _____ their workability;
   _____ their compatibility with programs directed by other professional personnel.

9. The teachers in my building review their programming (methods, activities, resources, planning procedures, etc.) according to:
   _____ its workability;
   _____ its compatibility with programs directed by other professional personnel.

10. The teachers in my building review their evaluation activities according to:
   _____ their consistency with programming (activities, resources, methods, etc.);
   _____ their compatibility with programs directed by other professional personnel.

11. For programs under their jurisdiction, the teachers use reviews to propose changes in:
   _____ goals and objectives;
   _____ programming (activites, resources, methods, etc.);
   _____ evaluations;
   _____ decision-making based on evaluations;
   _____ the study of program areas.

12. For programs under the direction of other professionals, the teachers use reviews to propose changes in:
    _____ goal setting;
    _____ programming (activities, resources, methods, etc.);
    _____ evaluations;
    _____ decision-making based on evaluations;
    _____ the study of program areas.

13. The Superintendent keeps records of the following aspects of programs for which he/she is responsible:
    _____ goals and objectives;
    _____ programming (activities, resources, methods, etc.);
    _____ evaluations;
    _____ decision-making based on evaluations.

14. The Superintendent reviews goals and objectives according to:
    _____ their workability;
    _____ their compatibility with programs directed by other professional personnel.

15. The Superintendent reviews programming (methods, activities, resources, planning procedures, etc.) according to:
    _____ its workability:
    _____ its compatibility with programs directed by other professional personnel.

16. The Superintendent reviews evaluation activities according to:
    _____ their workability;
    _____ their compatibility with programs directed by other professional personnel.

17. For programs under his/her jurisdiction, the Superintendent uses reviews to propose changes in:
    _____ goals and objectives;
    _____ programming (activities, resources, methods, etc.);
    _____ evaluations;
    _____ decision-making based on evaluations;
    _____ the study of program areas.

18. For programs under the direction of other professionals, the Superintendent uses reviews to propose changes in:
    _____ goal setting;
    _____ programming (activities, resources, methods, etc.);
    _____ evaluations;
    _____ decision-making based on evaluations;
    _____ the study of program areas.

Accountability Checklist VI: Superintendent's Ratings of Internal Program Reviews

This checklist was designed to survey your perceptions of internal review processes in your system-wide programs, classroom programs and building programs. A professional conducts internal reviews of programs under his direction when three steps are followed: (1) records are used to provide descriptions of those programs, (2)programs are reviewed according to workability and other criteria, (3)recommendations are made for improving those programs. This survey provides information about the internal reviews conducted by you, principals, and teachers in the school district.

Please rate each of the following statements according to your experience in your current administrative assignment. Use the following scale for each item: 1-Always, 2-Almost Always, 3-Sometimes, 4-Almost Never, 5-Never, 6-Uncertain. Place the number of the appropriate rating in the blank next to the item.

IPR (Form C)

1. I keep records of the following aspects of system programs for which I am responsible:
   _____ goals and objectives;
   _____ programming (activities, resources, methods, etc.);
   _____ evaluations;
   _____ decision-making based on evaluations.

2. I review my goals and objectives according to:
   _____ their workability;
   _____ their compatibility with programs directed by other professional personnel.

3. I review my programming (methods, activities, resources, planning procedures, etc.) according to:
   _____ its workability;
   _____ its compatibility with programs directed by other professional personnel.

4. I review my evaluation activities according to:
   _____ their workability;
   _____ their compatibility with programs directed by other professional personnel.

5. For programs under my jurisdiction I use reviews to propose changes in:
   - ____ goals and objectives;
   - ____ programming (activities, resources, methods, etc.);
   - ____ evaluations;
   - ____ decision-making based on evaluations;
   - ____ the study of program areas.

6. For programs under the direction of other professionals, I use reviews to propose changes in:
   - ____ goal setting;
   - ____ programming (activities, resources, methods, etc.);
   - ____ evaluations;
   - ____ decision-making based on evaluations;
   - ____ the study of program areas.

7. The teachers in the system keep records of the following aspects of his/her program responsibilities:
   - ____ goals and objectives;
   - ____ programming (activities, resources, methods, etc.);
   - ____ evaluations;
   - ____ decision-making based on evaluations.

8. The teachers in the system review their goals and objectives according to:
   - ____ their workability;
   - ____ their compatibility with programs directed by other professional personnel.

9. The teachers in the system review their programming (methods, activities, resources, planning procedures, etc.) according to:
   - ____ its workability;
   - ____ its compatibility with programs directed by other professional personnel.

10. The teachers in the system review their evaluation activities according to:
    - ____ their workability;
    - ____ their compatibility with programs directed by other professional personnel.

11. For programs under their jurisdiction, the teachers use reviews to propose changes in:
    - ____ goals and objectives;
    - ____ programming (activities, resources, methods, etc.);
    - ____ evaluations;
    - ____ decision-making based on evaluations;
    - ____ the study of program areas.

12. For programs under the direction of other professionals, the teachers use reviews to propose changes in:

_____ goal setting;
_____ programming (activities, resources, methods, etc.);
_____ evaluations;
_____ decision-making based on evaluations;
_____ the study of program areas.

13. As a group, principals keep records of the following aspects of programs for which they are responsible;

_____ goals and objectives;
_____ programming (activities, resources, methods, etc.);
_____ evaluations;
_____ decision-making based on evaluations.

14. As a group, principals review their goals and objectives according to:

_____ their workability;
_____ their compatibility with programs directed by other professional personnel.

15. As a group, principals review their programming (methods, activities, resources, planning procedures, etc.) according to:

_____ its workability;
_____ its compatibility with programs directed by other professional personnel.

16. As a group, principals review their evaluation activities according to:

_____ their workability;
_____ their compatibility with programs directed by other professional personnel.

17. For programs under their jurisdiction, principals use reviews to propose changes in:

_____ goals and objectives;
_____ programming (activities, resources, methods, etc.);
_____ evaluations;
_____ decision-making based on evaluations;
_____ the study of program areas.

18. For programs under the direction of other professionals, the principals use reviews to propose changes in:

_____ goal setting;
_____ programming (activities, resources, methods, etc.);
_____ evaluations;
_____ decision-making based on evaluations;
_____ the study of program areas.

Accountability Checklist VII:
Teacher's Ratings of External Program Reviews

This checklist is designed to survey your perceptions of external reviews in your classroom, your building program and the system-wide program. External reviews are conducted when professionals and lay persons *not responsible for the operation of a program* review that program and make recommendations for its improvement. This survey provides information about the external reviews conducted in programs under your direction, the direction of the building principal, and the direction of the Superintendent in the school district.

Please rate each of the following statements according to your experience in your current teaching assignment. Use the following scale for each item: 1-Always, 2-Almost Always, 3-Sometimes, 4-Almost Never, 5-Never, 6-Uncertain. Place the number of the appropriate rating in the blank next to the item.

EPR (Form A)

1. I receive recommendations about the classroom program from other professionals and groups:

_____	the building principal;	_____	parents;
_____	other teachers;	_____	other lay persons;
_____	Superintendent;	_____	consultants.

2. I consider making program changes based on the recommendations of:

_____	the building principal;	_____	parents;
_____	other teachers;	_____	other lay persons;
_____	Superintendent;	_____	consultants.

3. I communicate my reactions to the recommendations initiated by other professionals and groups:

_____	the building principal;	_____	parents;
_____	other teachers;	_____	other lay persons;
_____	Superintendent;	_____	consultants.

4. When an external group is reviewing the classroom program, I share program information with those groups:

_____	the building principal;	_____	parents;
_____	other teachers;	_____	other lay persons;
_____	Superintendent;	_____	consultants.

5. When I receive recommendations which involve programs directed by others, I refer those recommendations to the appropriate professional:
   ____ other teachers; ____ assistant superintendents;
   ____ the building principal; ____ Superintendent;
   ____ other principals; ____ supervisors.

6. The principal receives recommendations about the building program from other professionals and groups:
   ____ teachers; ____ parents;
   ____ other principals; ____ other lay persons;
   ____ the Superintendent; ____ consultants.

7. The principal considers making program changes based on the recommendations of:
   ____ teachers; ____ parents;
   ____ other principals; ____ other lay persons;
   ____ the Superintendent; ____ consultants.

8. The principal communicates his/her reactions to the recommendations initiated by other professionals and groups:
   ____ teachers; ____ parents;
   ____ other principals; ____ other lay persons;
   ____ the Superintendent; ____ consultants.

9. When an external group is reviewing the building program, the principal shares program information with those groups:
   ____ teachers; ____ parents;
   ____ other principals; ____ other lay persons;
   ____ the Superintendent; ____ consultants.

10. When the principal receives recommendations which involve programs directed by others, the principal refers those recommendations to the appropriate professional;
    ____ myself; ____ other principals;
    ____ other teachers; ____ assistant superintendents;
    ____ supervisors; ____ the Superintendent.

11. The Superintendent receives recommendations about the system program from other professionals and groups:
    ____ teachers; ____ parents;
    ____ principals; ____ consultants;
    ____ central office personnel; ____ other lay persons.

12. The Superintendent considers making program changes based on the recommendations of:

_____	teachers;	_____	parents;
_____	principals;	_____	consultants;
_____	central office personnel;	_____	other lay persons.

13. The Superintendent communicates his/her reactions to the recommendations initiated by other professionals and groups:

_____	teachers;	_____	parents;
_____	principals;	_____	consultants;
_____	central office personnel;	_____	other lay persons.

14. When an external group is reviewing the central office program, the Superintendent shares program information with those groups:

_____	teachers;	_____	parents;
_____	principals;	_____	consultants;
_____	central office personnel;	_____	other lay persons.

15. When the Superintendent receives recommendations which involve the job responsibilities of others, he/she refers the recommendations to the appropriate professionals:

_____	myself;	_____	other teachers;
_____	principals;	_____	assistant superintendents;
_____	supervisors;	_____	other central office personnel.

Accountability Checklist VIII:
Principal's Ratings of External Program Reviews

This checklist is designed to survey your perceptions of external reviews in your building program, classroom programs and the system-wide programs. External reviews are conducted when professionals and lay persons *not responsible for the operation of a program* review that program and make recommendations for its improvement. This survey provides information about the external review conducted in programs under your directon, the direction of the classroom teachers in your building and the direction of the Superintendent in the school district.

Please rate each of the following statements according to your experience in your current administrative assignment. Use the following scale for each item: 1-Always, 2-Almost Always, 3-Sometimes, 4-Almost Never, 5-Never, 6-Uncertain. Place the number of the appropriate rating in the blank next to the item.

EPR (Form C)

1. I receive recommendations about the building program from other professionals and groups:
   _____ teachers; _____ parents;
   _____ other principals; _____ other lay persons;
   _____ the Superintendent; _____ consultants.
2. I consider making program changes based on the recommendations of:
   _____ teachers; _____ parents;
   _____ other principals; _____ other lay persons;
   _____ the Superintendent; _____ consultants.
3. I communicate my reactions to the recommendations initiated by other professionals and groups:
   _____ teachers; _____ parents;
   _____ other principals; _____ other lay persons;
   _____ the Superintendent; _____ consultants.
4. When an external group is reviewing the building program, I share program information with that group:
   _____ teachers; _____ parents;
   _____ other principals; _____ other lay persons;
   _____ the Superintendent; _____ consultants.

5. When I receive recommendations which involve programs directed by others, I refer those recommendations to the appropriate professional:

_____ teachers;
_____ other principals;
_____ supervisors;
_____ assistant superintendents;
_____ Superintendent;
_____ other central office personnel.

6. The teachers receive recommendations about the classroom programs from other professionals and groups:

_____ other teachers;
_____ principals;
_____ the Superintendent;
_____ parents;
_____ other lay persons;
_____ consultants.

7. The teachers consider making program changes based on the recommendations of:

_____ other teachers;
_____ principals;
_____ the Superintendent;
_____ parents;
_____ other lay persons;
_____ consultants.

8. The teachers communicate their reactions to the recommendations initiated by other professionals and groups:

_____ other teachers;
_____ principals;
_____ the Superintendent;
_____ parents;
_____ other lay persons;
_____ consultants.

9. When an external group is reviewing the classroom programs in this building, the teachers share program information with that group:

_____ other teachers;
_____ principals;
_____ the Superintendent;
_____ parents;
_____ other lay persons;
_____ consultants.

10. When the teachers receive recommendations which involve programs directed by others, the teachers refer those recommendations to the appropriate professional:

_____ myself;
_____ other teachers;
_____ supervisors;
_____ other principals;
_____ assistant superintendents;
_____ Superintendent.

11. The Superintendent receives recommendations about the system-wide programs from other professionals and groups:

_____ teachers;
_____ principals;
_____ central office personnel;
_____ parents;
_____ consultants;
_____ other lay persons.

12. The Superintendent considers making program changes based on the recommendations of:

_____	teachers;	_____	parents;
_____	principals;	_____	consultants;
_____	central office personnel;	_____	other lay persons.

13. The Superintendent communicates his/her reactions to the recommendations initiated by other professionals and groups:

_____	teachers;	_____	parents;
_____	principals;	_____	consultants;
_____	central office personnel;	_____	other lay persons.

14. When an external group is reviewing the system-wide program, th Superintendent shares program information with those groups:

_____	teachers;	_____	parents;
_____	principals;	_____	consultants;
_____	central office personnel;	_____	other lay persons.

15. When the Superintendent receives recommendations which involve programs directed by others, the Superintendent refers the recommendations to the appropriate professionals:

_____	myself;	_____	assistant superintendents;
_____	other principals;	_____	supervisors;
_____	teachers;	_____	other central office personnel.

Accountability Checklist IX:
Superintendent's Ratings of External Program Reviews

This checklist is designed to survey your perceptions of external reviews in your system-wide program, building programs, and classroom programs. External reviews are conducted when professionals and lay persons *not responsible for the operation of a program* review that program and make recommendations for its improvement. This survey provides information about the external reviews conducted in programs under your direction, the direction of classroom teachers, and the direction of building principals.

Please rate each of the following statements according to your experience in your current administrative assignment. Use the following scale for each item: 1-Always, 2-Almost Always, 3-Sometimes, 4-Almost Never, 5-Never, 6-Uncertain. Place the number of the appropriate rating in the blank next to the item.

EPC (Form C)

1. I receive recommendations about the system-wide programs from other professionals and groups:

_____ teachers; _____ parents;
_____ building principals; _____ other lay persons;
_____ central office personnel; _____ consultants.

2. I consider making program changes based on the recommendations of:

_____ teachers; _____ parents;
_____ building principals; _____ other lay persons;
_____ central office personnel; _____ consultants.

3. I communicate my reactions to the recommendations initiated by other professionals and groups:

_____ teachers; _____ parents;
_____ building principals; _____ other lay persons;
_____ central office personnel; _____ consultants.

4. When an external group is reviewing the system-wide programs, I share program information with those groups:

_____ teachers; _____ parents;
_____ building principals; _____ other lay persons;
_____ central office personnel; _____ consultants.

5. When I receive recommendations which involve programs directed by others, I refer those recommendations to the appropriate professional:

   _____ teachers; _____ assistant superintendents;
   _____ principals; _____ supervisors;
   _____ other superintendents; _____ other central office personnel.

6. The teachers receive information about the classroom programs from other professionals and groups:

   _____ other teachers; _____ parents;
   _____ principals; _____ other lay persons;
   _____ myself; _____ consultants.

7. The teachers consider making program changes based on the recommendations of:

   _____ other teachers; _____ parents;
   _____ principals; _____ other lay persons;
   _____ myself; _____ consultants.

8. The teachers communicate their reactions to the recommendations initiated by other professionals and groups:

   _____ other teachers; _____ parents;
   _____ principals; _____ other lay persons;
   _____ myself; _____ consultants.

9. When an external group is reviewing the classroom programs in their buildings, the teachers share program information with those groups:

   _____ other teachers; _____ parents;
   _____ principals; _____ other lay persons;
   _____ myself; _____ consultants.

10. When the teachers receive recommendations which involve programs directed by others, the teachers refer those recommendations to the appropriate professional:

    _____ myself; _____ principals;
    _____ other teachers; _____ assistant superintendents;
    _____ supervisors; _____ other central office personnel.

11. The principal receives recommendations about the building program from other professionals and groups:

    _____ teachers; _____ parents;
    _____ other principals; _____ other lay persons;
    _____ myself; _____ consultants.

12. The principal considers making program changes based on the recommendations of:

    _____ teachers; _____ parents;
    _____ other principals; _____ other lay persons;
    _____ myself; _____ consultants.

13. The principal communicates his/her reactions to the recommendations initiated by other professionals and groups:

_____	teachers;	_____	parents;
_____	other principals;	_____	other lay persons;
_____	myself;	_____	consultants.

14. When an external group is reviewing the building program, the principal shares program information with those groups:

_____	teachers;	_____	parents;
_____	other principals;	_____	other lay persons;
_____	myself;	_____	consultants.

15. When the principal receives recommendations which involve programs directed by others, the principal refers those recommendations to the appropriate professional:

_____	myself;	_____	other principals;
_____	teachers;	_____	assistant superintendents;
_____	supervisors;	_____	other central office personnel.

Accountability Checklist X:
Teacher's Ratings of Program Development

This checklist is designed to survey your perceptions of program development in your classroom, your building program, and the system-wide program. Program development concerns two types of innovation: 1) refinements in existing policies and practices; and 2) introduction of new aims and procedures into the programs. This survey provides information about program development in programs directed by you, programs directed by the building principal, and programs directed by the Superintendent in the school district.

Please rate each of the following statements according to your experience in your current teaching assignment. Use the following scale for each item: 1-Always, 2-Almost Always, 3-Sometimes, 4-Almost Never, 5-Never, 6-Uncertain. Place the number of the appropriate rating in the blank next to the item.

PD (Form A)

1. I make changes in the classroom program based on:
   _____ my own review of classroom goals and objectives;
   _____ my own review of classroom activities, methods, and materials;
   _____ my own review of classroom evaluations;
   _____ my own review of classroom decision-making.

2. I make changes in the classroom program according to the recommendations of:
   _____ other teachers; _____ the Superintendent;
   _____ the principal; _____ parents and other lay persons.

3. I use reviews to make program changes in existing:
   _____ classroom goals and objectives;
   _____ classroom activities, methods, and materials;
   _____ classroom evaluations;
   _____ classroom decision-making.

4. In order to introduce changes in the classroom program, I request new services from other personnel:
   _____ other teachers; _____ the principal;
   _____ the Superintendent; _____ other central office personnel.

5. I plan program changes by specifying:
   - _____ their goals and objectives;
   - _____ ways to evaluate them;
   - _____ human and material resources needed to implement them;
   - _____ their structure, methods, activities, etc.

6. The principal makes changes in the building program based on:
   - _____ the principal's own review of the building goals and objectives
   - _____ the principal's own review of building activities, procedures and resources;
   - _____ the principal's own review of building evaluation;
   - _____ the principal's own review of building decision-making.

7. The principal makes changes in the building program in accordance with recommendations of:
   - _____ teachers;
   - _____ parents and other lay persons;
   - _____ other principals;
   - _____ the Superintendent.

8. The principal uses reviews to make program changes in existing:
   - _____ building goals, objectives, and policies;
   - _____ building activities, methods and materials;
   - _____ building evaluations;
   - _____ building decision-making.

9. In order to introduce changes in the building program, the principal requests new services from other personnel:
   - _____ teachers;
   - _____ other principals;
   - _____ the Superintendent;
   - _____ other central office personnel.

10. The principal plans program changes by specifying:
    - _____ their goals and objectives;
    - _____ ways to evaluate them;
    - _____ human and material resources needed to implement them;
    - _____ their structure, methods, activities, etc.

11. The Superintendent makes changes in the system-wide program based on:
    - _____ The Superintendent's review of system-wide goals and objectives;
    - _____ the Superintendent's review of system-wide methods, activities, and resources;
    - _____ the Superintendent's review of system-wide evaluations;
    - _____ the Superintendent's review of system-wide decision-making.

12. The Superintendent makes changes in the system-wide program according to the recommendations of:
    ____ teachers;
    ____ principals;
    ____ assistant superintendents;
    ____ parents and other lay persons.

13. The Superintendent uses reviews to make program changes in existing:
    ____ system goals, objectives, and policies;
    ____ system activities, methods and materials;
    ____ system evaluations;
    ____ system decision-making based on evaluations.

14. In order to introduce changes in the system-wide program, the Superintendent requests new services from other personnel:
    ____ teachers;
    ____ principals;
    ____ assistant superintendents;
    ____ other central office personnel.

15. The Superintendent plans program changes by specifying:
    ____ their goals and objectives;
    ____ their structure, methods, activities, etc.;
    ____ human and material resources needed to implement them;
    ____ ways to evaluate them.

Accountability Checklist XI:
Principal's Ratings for Program Development

This checklist is designed to survey your perceptions of program development in your building program, classroom programs in your building, and system-wide programs. Program development concerns two types of innovation: 1) refinements in existing policies and practices; and 2) introduction of new aims and procedures into programs. This survey provides information about program development in programs directed by you, programs directed by classroom teachers in your building, and programs directed by the Superintendent of the school district.

Please rate each of the following statements according to your experience in your current administrative assignment. Use the following scale for each item: 1-Always, 2-Almost Always, 3-Sometimes, 4-Almost Never, 5-Never, 6-Uncertain. Place the number of the appropriate rating in the blank next to the item.

PD (Form C)

1. I make changes in the building program based on:
   ____ my own review of building goals and objectives;
   ____ my own review of building activities, procedures and resources;
   ____ my own review of building evaluations;
   ____ my own review of building decision-making.

2. I use reviews to make program changes in existing:
   ____ building goals, objectives and policies;
   ____ building activities, methods and materials;
   ____ building evaluations;
   ____ building decision-making.

3. In order to introduce changes in the building program, I request new services from other personnel:
   ____ teachers; ____ the Superintendent;
   ____ other principals; ____ other central office personnel.

4. I make changes in the building program according to the recommendations of:
   ____ teachers; ____ the Superintendent;
   ____ other principals; ____ parents and other lay persons.

5. I plan program changes by specifying:
   _____ their goals and objectives;
   _____ ways to evaluate them;
   _____ human and material resources needed to implement them;
   _____ their structure, methods, activities, etc.

6. The teachers make changes in the classroom program based on:
   _____ their review of classroom goals and objectives;
   _____ their review of classroom activities, methods and materials;
   _____ their review of classroom evaluations;
   _____ their review of classroom decision-making.

7. The teachers use reviews to make program changes in existing:
   _____ classroom goals and objectives;
   _____ classroom activities, methods, and materials;
   _____ classroom evaluations;
   _____ classroom decision-making.

8. In order to introduce changes into the classroom program, the teachers request new services from other personnel:
   _____ other teachers;
   _____ principals;
   _____ Superintendent;
   _____ other central office personnel.

9. The teachers make changes in classroom programs in accordance with recommendations of:
   _____ other teachers;
   _____ principals;
   _____ the Superintendent;
   _____ parents and other lay persons.

10. The teachers plan program changes by specifying:
   _____ their goals and objectives;
   _____ ways to evaluate them;
   _____ human and material resources needed to implement them;
   _____ their structure, methods, activities, etc.

11. The Superintendent makes changes in the system-wide program based on:
   _____ the Superintendent's review of the system-wide program, goals, objectives, and policies;
   _____ the Superintendent's review of system-wide activities, methods, and services;
   _____ the Superintendent's review of system-wide evaluations;
   _____ the Superintendent's review of system-wide decision-making.

12. The Superintendent makes changes in the system-wide program according to the recommendations of:
    ____ teachers;
    ____ principals;
    ____ assistant superintendents;
    ____ parents and other lay persons.
13. The Superintendent uses reviews to make program changes in existing:
    ____ system objectives and policies;
    ____ system activities, methods and materials;
    ____ system evaluations;
    ____ system decision-making.
14. In order to introduce changes in the system-wide program, the Superintendent requests new services from other personnel:
    ____ teachers;
    ____ principals;
    ____ assistant superintendents;
    ____ other central office personnel.
15. The Superintendent plans program changes by specifying:
    ____ their goals and objectives;
    ____ ways to evaluate them;
    ____ human and material resources needed to implement them;
    ____ their structure, methods, activities, etc.

Accountability Checklist XII: Superintendent's Ratings for Program Development

This checklist is designed to survey your perceptions of program development in your system-wide programs, classroom programs and building programs. Program development concerns two types of innovation: 1) refinements in existing policies and practices; and 2) introduction of new aims and procedures into programs. This survey provides information about program development in programs directed by you, programs directed by classroom teachers in your district, and programs directed by principals in your district.

Please rate each of the following statements according to your experience in your current administrative assignment. Use the following scale for each item: 1-Always, 2-Almost Always, 3-Sometimes, 4-Almost Never, 5-Never, 6-Uncertain. Place the number of the appropriate rating in the blank next to the item.

PD (Form C)

1. I make changes in the system-wide program based on:
   ____ my own review of system-wide goals, objectives and policies;
   ____ my own review of system-wide activities, methods and materials;
   ____ my own review of system-wide evaluations;
   ____ my own review of system-wide decision-making.
2. I make changes in the system-wide program according to the recommendations of:
   ____ teachers; ____ assistant superintendents;
   ____ principals; ____ parents and other lay persons.
3. I use reviews to make program changes in existing:
   ____ system goals, objectives and policies;
   ____ system activities, methods and materials;
   ____ system evaluations;
   ____ system decision-making.
4. In order to introduce changes in the system-wide program, I request new services from other personnel:
   ____ teachers; ____ assistant superintendents;
   ____ principals; ____ other central office personnel.

5. I plan program changes by specifying:
   _____ their goals and objectives;
   _____ human and material resources needed to implement them;
   _____ ways to evaluate them;
   _____ their structure, methods, activities, etc.

6. The teachers make changes in the classroom program based on:
   _____ their review of classroom goals and objectives;
   _____ their review of classroom activities, methods and materials;
   _____ their review of classroom evaluations;
   _____ their review of classroom decision-making.

7. The teachers make changes in classroom programs in accordance with recommendations of:
   _____ other teachers;
   _____ myself;
   _____ principals;
   _____ parents and other lay persons.

8. In order to introduce changes into the classroom program, the teachers request new services from other personnel:
   _____ other teachers;
   _____ myself;
   _____ principals;
   _____ other central office personnel.

9. The teachers use reviews to make program changes in existing:
   _____ classroom goals and objectives;
   _____ classroom evaluations;
   _____ classroom decision-making.
   _____ classroom activities, methods, and materials;

10. The teachers plan program changes by specifying:
   _____ their goals and objectives;
   _____ human and material resources needed to implement them;
   _____ ways to evaluate them;
   _____ their structure, methods, activities, etc.

11. The principal makes changes in the building program based on:
   _____ the principal's review of building goals, objectives and policies;
   _____ the principal's review of building activities, methods and resources;
   _____ the principal's review of building evaluations;
   _____ the principal's review of building decision-making.

12. The principal makes changes in the building program in accordance with recommendations of:
   _____ teachers;
   _____ other principals;
   _____ parents and other lay persons;
   _____ myself.

13. The principal uses reviews to make program changes in existing:
    _____ building goals, objectives and policies;
    _____ building activities, methods and materials;
    _____ building evaluations;
    _____ building decision-making.

14. In order to introduce changes in the building program, the principal requests new services from other personnel:
    _____ teachers;
    _____ other principals;
    _____ myself;
    _____ other central office personnel.

15. The principal plans program changes by specifying:
    _____ their goals and objectives;
    _____ ways to evaluate them;
    _____ human and material resources needed to implement them;
    _____ their structure, methods, activities, etc.

index

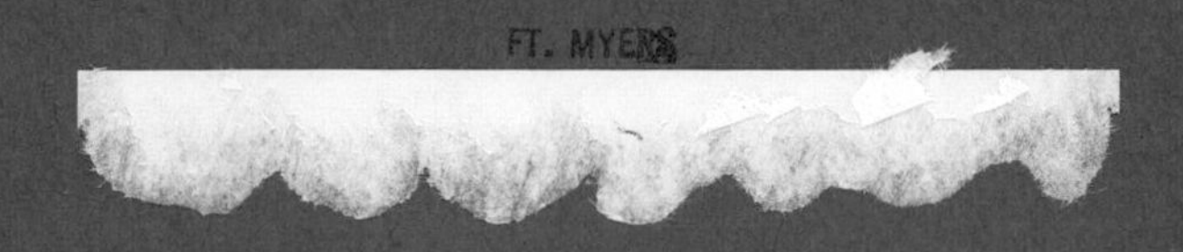

This book was set in Linotron Times. Times is a modified book version of the original Times New Roman, a newspaper typeface, commissioned in 1931 to be designed under the supervision of Stanley Morison for the London *Times*.

Printed on acid-free paper.